Neurodegenerative Diseases

Daniela Galimberti • Elio Scarpini
Editors

Neurodegenerative Diseases

Clinical Aspects, Molecular Genetics and Biomarkers

Second Edition

Editors
Daniela Galimberti
University of Milan
Fondazione Ca' Granda
IRCCS Ospedale Policlinico
Milan
Italy

Elio Scarpini
University of Milan
Fondazione Ca' Granda
IRCCS Ospedale Policlinico
Milan
Italy

ISBN 978-3-319-72937-4   ISBN 978-3-319-72938-1   (eBook)
https://doi.org/10.1007/978-3-319-72938-1

Library of Congress Control Number: 2018934851

© Springer International Publishing AG 2018
This work is subject to copyright. All rights are reserved by the Publisher, whether the whole or part of the material is concerned, specifically the rights of translation, reprinting, reuse of illustrations, recitation, broadcasting, reproduction on microfilms or in any other physical way, and transmission or information storage and retrieval, electronic adaptation, computer software, or by similar or dissimilar methodology now known or hereafter developed.
The use of general descriptive names, registered names, trademarks, service marks, etc. in this publication does not imply, even in the absence of a specific statement, that such names are exempt from the relevant protective laws and regulations and therefore free for general use.
The publisher, the authors and the editors are safe to assume that the advice and information in this book are believed to be true and accurate at the date of publication. Neither the publisher nor the authors or the editors give a warranty, express or implied, with respect to the material contained herein or for any errors or omissions that may have been made. The publisher remains neutral with regard to jurisdictional claims in published maps and institutional affiliations.

Printed on acid-free paper

This Springer imprint is published by Springer Nature
The registered company is Springer International Publishing AG
The registered company address is: Gewerbestrasse 11, 6330 Cham, Switzerland

Contents

1. **Genetics and Epigenetics in the Neurodegenerative Disorders of the Central Nervous System** 1
   Chiara Fenoglio

2. **Diagnosis of Alzheimer's Disease Typical and Atypical Forms** 21
   Laura Ghezzi

3. **Genetic Complexity of Early-Onset Alzheimer's Disease** 29
   Mahdi Ghani, Christiane Reitz, Peter St George-Hyslop,
   and Ekaterina Rogaeva

4. **Genetic Risk Factors for Complex Forms of Alzheimer's Disease** 51
   Céline Bellenguez and Jean-Charles Lambert

5. **Role of Cerebrovascular Disease in Cognition** 77
   Ana Verdelho

6. **Risk Factors and Prevention in Alzheimer's Disease and Dementia** 93
   Giulia Grande, Davide L. Vetrano, and Francesca Mangialasche

7. **Diagnosis of Frontotemporal Dementia** 113
   Giorgio Giulio Fumagalli

8. **Autosomal Dominant Frontotemporal Lobar Degeneration: From Genotype to Phenotype** 123
   Maria Serpente and Daniela Galimberti

9. **Genetic Risk Factors for Sporadic Frontotemporal Dementia** 147
   Raffaele Ferrari, Claudia Manzoni, and Parastoo Momeni

10. **Alzheimer's Disease and Frontotemporal Lobar Degeneration: Mouse Models** 187
    Lars M. Ittner, Wei S. Lee, Kristie Stefanoska, Prita R. Asih, and
    Yazi D. Ke

11. **Fluid Biomarkers in Alzheimer's Disease and Frontotemporal Dementia** 221
    Niklas Mattsson, Sotirios Grigoriou, and Henrik Zetterberg

12	Biomarkers for Alzheimer's Disease and Frontotemporal Lobar Degeneration: Imaging	253
	Marco Bozzali and Laura Serra	
13	Genotypic and Phenotypic Heterogeneity in Amyotrophic Lateral Sclerosis	279
	Nicola Ticozzi and Vincenzo Silani	
14	Lewy Body Dementia	297
	L. Bonanni, R. Franciotti, S. Delli Pizzi, A. Thomas, and M. Onofrj	
15	Rare Dementias	313
	Camilla Ferrari, Benedetta Nacmias, and Sandro Sorbi	
16	Neurodevelopmental and Neurodegenerative Alterations in the Pathophysiology of Schizophrenia: Focus on Neuro-Immuno-Inflammation	337
	Bernardo Dell'Osso, M. Carlotta Palazzo, and A. Carlo Altamura	
17	Parkinson's Disease: Contemporary Concepts and Clinical Management	349
	Vanessa Carvalho, Carlota Vicente Cunha, and João Massano	
18	Neurodegeneration and Multiple Sclerosis	379
	Axel Petzold	

Index 401

Contributors

A. Carlo Altamura Department of Psychiatry, University of Milan, Fondazione IRCCS Ca' Granda, Ospedale Maggiore Policlinico, Milano, Italy

Prita R. Asih Dementia Research Unit, School of Medical Sciences, The University of New South Wales, Sydney, NSW, Australia

Céline Bellenguez Inserm, U1167, RID-AGE—Risk Factors and Molecular Determinants of Aging-Related Diseases, Lille, France

Institut Pasteur de Lille, Lille, France

Univ. Lille, U1167—Excellence Laboratory LabEx DISTALZ, Lille, France

L. Bonanni Department of Neuroscience, Imaging and Clinical Sciences, University G. d'Annunzio of Chieti-Pescara, Pescara, Italy

Marco Bozzali Neuroimaging Laboratory, Santa Lucia Foundation IRCCS, Rome, Italy

Vanessa Carvalho Department of Neurology, Hospital Pedro Hispano, Matosinhos Local Health Unit, Matosinhos, Portugal

Carlota Vicente Cunha Department of Neurology, Hospital de Santo António/Porto Hospital Center, Porto, Portugal

Bernardo Dell'Osso Department of Psychiatry, University of Milan, Fondazione IRCCS Ca' Granda, Ospedale Maggiore Policlinico, Milano, Italy

S. Delli Pizzi Department of Neuroscience, Imaging and Clinical Sciences, University G. d'Annunzio of Chieti-Pescara, Pescara, Italy

Chiara Fenoglio University of Milan, Centro Dino Ferrari, Fondazione Ca' Granda, IRCCS Fondazione Policlinico, Milan, Italy

Camilla Ferrari IRCCS Don Gnocchi, Florence, Italy

Raffaele Ferrari, Ph.D. Department of Molecular Neuroscience, UCL Institute of Neurology, London, UK

R. Franciotti Department of Neuroscience, Imaging and Clinical Sciences, University G. d'Annunzio of Chieti-Pescara, Pescara, Italy

Giorgio Giulio Fumagalli Department of Neurosciences, Psychology, Drug Research and Child Health (NEUROFARBA), University of Florence, Fondazione Cà Granda, IRCCS Ospedale Maggiore Policlinico, Milan, Italy

Daniela Galimberti Neurology Unit, Department of Pathophysiology and Transplantation, University of Milan, Fondazione Cà Granda, IRCCS Ospedale Maggiore Policlinico, Milan, Italy

Mahdi Ghani Tanz Centre for Research in Neurodegenerative Diseases, University of Toronto, Toronto, ON, Canada

Laura Ghezzi University of Milan, Centro Dino Ferrari, Fondazione Ca' Granda, IRCCS Fondazione Policlinico, Milan, Italy

Giulia Grande, M.D. Aging Research Center, Department of Neurobiology, Care Sciences, and Society (NVS), Karolinska Institutet and Stockholm University, Stockholm, Sweden

Center for Research and Treatment on Cognitive Dysfunctions, Biomedical and Clinical Sciences Department, "Luigi Sacco" Hospital, University of Milan, Milano, Italy

Sotirios Grigoriou, M.D. Department of Neurology, Skåne University Hospital, Lund, Sweden

Lars M. Ittner Dementia Research Unit, School of Medical Sciences, The University of New South Wales, Sydney, NSW, Australia

Transgenic Animal Unit, School of Medical Sciences, The University of New South Wales, Sydney, NSW, Australia

Neuroscience Research Australia, Sydney, NSW, Australia

Yazi D. Ke Motor Neuron Disease Unit, School of Medical Sciences, The University of New South Wales, Sydney, NSW, Australia

Jean-Charles Lambert Inserm, U1167, RID-AGE—Risk Factors and Molecular Determinants of Aging-Related Diseases, Lille, France

Institut Pasteur de Lille, Lille, France

Univ. Lille, U1167—Excellence Laboratory LabEx DISTALZ, Lille, France

Wei S. Lee Dementia Research Unit, School of Medical Sciences, The University of New South Wales, Sydney, NSW, Australia

Motor Neuron Disease Unit, School of Medical Sciences, The University of New South Wales, Sydney, NSW, Australia

Francesca Mangialasche, M.D., Ph.D. Aging Research Center, Department of Neurobiology, Care Sciences, and Society (NVS), Karolinska Institutet and Stockholm University, Stockholm, Sweden

Section of Gerontology and Geriatrics, University of Perugia, Perugia, Italy

Claudia Manzoni, Ph.D. School of Pharmacy, University of Reading, Reading, UK

João Massano Department of Neurology, Centro Hospitalar São João, Porto, Portugal

Department of Clinical Neurosciences and Mental Health, Faculty of Medicine, University of Porto, Porto, Portugal

Niklas Mattsson, M.D., Ph.D. Clinical Memory Research Unit, Faculty of Medicine, Lund University, Lund, Sweden

Memory Clinic, Skåne University Hospital, Malmö, Sweden

Department of Neurology, Skåne University Hospital, Lund, Sweden

Parastoo Momeni, Ph.D. Rona Holdings, Silicon Valley, CA, USA

Benedetta Nacmias Department of Neuroscience, Psychology, Drug Research and Child Health (NEUROFARBA), University of Florence, Florence, Italy

M. Onofrj Department of Neuroscience, Imaging and Clinical Sciences, University G. d'Annunzio of Chieti-Pescara, Pescara, Italy

M. Carlotta Palazzo Department of Psychiatry, University of Milan, Fondazione IRCCS Ca' Granda, Ospedale Maggiore Policlinico, Milano, Italy

Axel Petzold Moorfields Eye Hospital, London, UK

The Neuroimmunology and CSF Laboratory, London, UK

MS Centre and Dutch Expertise Centre for Neuro-ophthalmology, VUmc, Amsterdam, The Netherlands

Christiane Reitz Departments of Neurology, and Epidemiology, Taub Institute for Research on Alzheimer's Disease and the Aging Brain, Gertrude H. Sergievsky Center, Columbia University, New York, NY, USA

Ekaterina Rogaeva Tanz Centre for Research in Neurodegenerative Diseases, University of Toronto, Toronto, ON, Canada

Department of Medicine, University of Toronto, Toronto, ON, Canada

Maria Serpente Neurology Unit, Department of Pathophysiology and Transplantation, University of Milan, Fondazione Cà Granda, IRCCS Ospedale Maggiore Policlinico, Milan, Italy

Laura Serra Neuroimaging Laboratory, Santa Lucia Foundation IRCCS, Rome, Italy

Vincenzo Silani Department of Neurology and Laboratory of Neuroscience, IRCCS Istituto Auxologico Italiano, Milan, Italy

Department of Pathophysiology and Transplantation, 'Dino Ferrari' Center, University of Milan, Milan, Italy

Sandro Sorbi IRCCS Don Gnocchi, Florence, Italy

Department of Neuroscience, Psychology, Drug Research and Child Health (NEUROFARBA), University of Florence, Florence, Italy

Peter St George-Hyslop Tanz Centre for Research in Neurodegenerative Diseases, University of Toronto, Toronto, ON, Canada

Department of Medicine, University of Toronto, Toronto, ON, Canada

Department of Clinical Neurosciences, Cambridge Institute for Medical Research, University of Cambridge, Cambridge, UK

Kristie Stefanoska Dementia Research Unit, School of Medical Sciences, The University of New South Wales, Sydney, NSW, Australia

A. Thomas Department of Neuroscience, Imaging and Clinical Sciences, University G. d'Annunzio of Chieti-Pescara, Pescara, Italy

Nicola Ticozzi Department of Neurology and Laboratory of Neuroscience, IRCCS Istituto Auxologico Italiano, Milan, Italy

Department of Pathophysiology and Transplantation, 'Dino Ferrari' Center, University of Milan, Milan, Italy

Ana Verdelho, M.D., Ph.D. Department of Neurosciences and Mental Health, Centro Hospitalar Lisboa Norte-Hospital de Santa Maria, Instituto de Medicina Molecular—IMM e Instituto de Saúde Ambiental-ISAMB Medical School, University of Lisbon, Lisbon, Portugal

Davide L. Vetrano, M.D. Aging Research Center, Department of Neurobiology, Care Sciences, and Society (NVS), Karolinska Institutet and Stockholm University, Stockholm, Sweden

Department of Geriatrics, Catholic University of Rome, Rome, Italy

Henrik Zetterberg, M.D., Ph.D. Department of Psychiatry and Neurochemistry, Institute of Neuroscience and Physiology, the Sahlgrenska Academy at the University of Gothenburg, Mölndal, Sweden

Clinical Neurochemistry Laboratory, Sahlgrenska University Hospital, Mölndal, Sweden

Department of Molecular Neuroscience, UCL Institute of Neurology, London, UK

UK Dementia Research Institute, London, UK

Genetics and Epigenetics in the Neurodegenerative Disorders of the Central Nervous System

Chiara Fenoglio

Abstract

Most of the neurodegenerative diseases share several clinical, pathologic, and molecular aspects. Clinically, these diseases are often characterized by an insidious onset during adulthood, after which they progress at different rates, ultimately leading to severe physical disability or death. The symptoms are often common among the different disorders: dementia is not only peculiar of Alzheimer's disease (AD) or frontotemporal dementia (FTD), but could occur also in Parkinson's disease (PD) or amyotrophic lateral sclerosis (ALS).

Furthermore, under a genetic point of view, many neurodegenerative diseases manifest an important family history, highlighting a relevant contribution of genetic factors to disease causation and progression.

Genetics and epigenetics, together with their new designed technologies capable of analyzing genetic variability, have disclosed an appealing scenario that will offer the biomedical sciences new insight for the study of neurodegenerative diseases, multifactorial complex, and rare diseases. In this chapter, an overview of the current genetic and epigenetic progresses in AD, FTD, PD, and ALS, reached by the application of the new genetic technologies, will be provided.

Keywords

Genetics · Epigenetics · Alzheimer's disease · Frontotemporal dementia · Parkinson's disease · Amyotrophic lateral sclerosis

---

C. Fenoglio
Centro Dino Ferrari, Fondazione Ca' Granda, IRCCS Fondazione Policlinico,
University of Milan, Milan, Italy
e-mail: chiara.fenoglio@unimi.it

© Springer International Publishing AG 2018
D. Galimberti, E. Scarpini (eds.), *Neurodegenerative Diseases*,
https://doi.org/10.1007/978-3-319-72938-1_1

Introduction

Most of neurological disorders including Alzheimer's disease (AD), Parkinson's disease (PD), amyotrophic lateral sclerosis (ALS), and frontotemporal lobar degeneration (FTLD) could be considered multifactorial diseases. A small percentage of these diseases however occur in families with a Mendelian inheritance pattern of transmission. The majority of cases results from complex interactions between a number of genetic and environmental factors, and therefore they are said to follow a multifactorial (or complex) inheritance pattern. The familial clustering can be explained by recognizing that family members share a greater proportion of their genetic information and environmental exposures than do individuals chosen at random in the population. Thus, the relatives of an affected individual are more likely to experience the same gene-gene and gene-environment interactions that led to disease in the first place than are individuals who are unrelated to the proband. The multifactorial inheritance pattern that results represents an interaction between the collective effects of the genotype at one or, more commonly, multiple loci (polygenic or multigenic effects) either to raise or to lower susceptibility to disease, combined with a variety of environmental exposures that may trigger, accelerate, or protect against the disease process. The gene-gene interactions in polygenic inheritance may be simply additive or much more complicated. Gene-environment interactions, including systematic exposures or chance encounters with environmental factors in one's surroundings, add even more complexity to individual disease risk and the pattern of disease inheritance.

Herein, main genetic variations will be described, and progress in the genetic and epigenetic knowledge of the most common neurodegenerative diseases will be reviewed considering the achievements of new technologies.

Genetics: Basic Concept

The genetic background of the individuals differs from thousands to millions of genetic variants that are the differences in DNA sequences within the genome of individuals in the populations. These variations can take many forms, including single nucleotide polymorphisms (SNPs), tandem repeats (short tandem repeats and variable number of tandem repeats), small indels (insertions and deletions of a short DNA sequence), duplications or deletions that change the copy number of a larger segment of a DNA sequence (≥ 1 kb), i.e., copy number variations (CNVs), and other chromosomal rearrangements such as inversions and translocations (also known as copy-neutral variations) [1–3].

These genetic variations are typically referred to as either common or rare to denote the frequency of the minor allele in the human population. Common variants are synonymous with polymorphisms, defined as genetic variants with a minor allele frequency (MAF) of at least 1% in the population, whereas rare variants have a MAF of less than 1% [2].

The large majority of genetic variants are hypothesized to be neutral [4] (that is they do not contribute to phenotypic variation), achieving significant frequencies in the human population simply by chance.

Types of Genetic Studies

Four strategies have governed the field of neurodegenerative diseases' genetics in the last decades: genetic linkage analysis, candidate genes studies, genome-wide association studies (GWAS), and next-generation sequencing (NGS) technology-based studies, whole-genome sequencing studies (WGS), and whole-exome sequencing studies (WES).

Genetic Studies Based on Linkage Analysis

Linkage analyses were the first kind of strategy used to unravel the genetic basis of Mendelian traits, involving families presenting autosomal dominant inheritance. Genetic linkage studies led to identify chromosomal regions associated with diseases but do not identify the causal gene associated. Linkage mapping is a powerful tool in identifying monogenic traits in Mendelian inherited neurodegenerative diseases but is less powerful to identify variants acting as risk factors for complex traits as are most neurodegenerative diseases [5].

Genetic Study Based on Candidate Genes

This kind of study aims to determine if frequencies of genetic variants of people with a specific disease differ significantly from a control population of subjects. According to that, susceptibility genes are defined when cases and controls show significant differences in occurring genetic variant frequencies. Candidate gene approach led to identify the apolipoprotein gene (*APOE*) risk alleles implicated in late-onset Alzheimer's disease (LOAD). Thousands of genes were tested in this way for neurodegenerative disease susceptibility, very often giving inconsistent results. This approach however has been very helpful and powerful since it is based on the existing knowledge on disease pathogenesis. Anyway, most of the candidate gene association studies could not be replicated mainly due to the small sample size that does not let the studies to reach the adequate statistical power [6].

Genome-Wide Association Studies (GWAS)

The advent of microarray technology era revolutionized genetics research, allowing for the contemporaneous determination of millions of SNPs in thousands of samples. GWAS is based on the testing common genetic variants in a hypothesis-free manner. Thus, it provides information on how common genetic variability confers risk for the common diseases [7]. Several susceptibility genes for common neurodegenerative disorders have been revealed by GWAS, although the odds ratios associated with these risk alleles are relatively low and account for just a small part of the estimated heritability, suggesting that there are rare variants, not causative, which cannot be captured by GWAS employing common SNPs [8].

NGS Technologies

Recent advances, collectively referred to as NGS, have allowed for high-throughput sequencing giving massive data results that need to be analyzed by specific bioinformatics software. Moreover, in opposition to the first-generation sequencing, also known as Sanger sequencing, which can take several years with relevant costs, NGS can produce the same genome sequence within a few weeks and with reduced costs. This allows for simultaneous investigation of multiple genes in one single reaction and has been demonstrated to be able to be an effective alternative for establishing the genetic base for Mendelian diseases in the research setting [9, 10] and recently also in the clinical setting [11, 12].

NGS relies upon multiple, short, overlapping reads of fragmented DNA that can be aligned against a reference genome or assembled "de novo" if no information on the reference genome is available. The regions that are amplified could include either a subset of genes (targeted approach) or all the genes in the genome. If just the protein-coding regions are amplified when sequencing all the genes, the method is referred to as WES. When the target is the whole genome, it is known as WGS.

Genetic linkage family studies have led to the determination of dominantly inherited, rare mutations in genes as presenilin 1 (*PSEN1*), presenilin 2 (*PSEN2*), amyloid precursor protein (*APP*) for early-onset Alzheimer's disease (EOAD), leucine-rich repeat kinase 2 (*LRRK2*), α-synuclein (*SNCA*), leucine-rich repeat kinase 1 (*LRRK1*) for PD, microtubule-associated protein tau (*MAPT*), progranulin (*GRN*) for FTD, and superoxide dismutase 1 (*SOD*) for ALS.

The development of whole-genome genotyping by different GWASs has allowed for the study of the involvement of common variants with low risk of disease.

Most of the neurodegenerative diseases, including AD, PD, ALS, and FTD, show rare but significant familial inherence and lower penetrance variants associated with the more common sporadic forms of diseases [13]. Thus they are suitable for both approaches: sporadic cases are amenable to GWAS, whereas patients with a definite positive family history, supporting a Mendelian transmission, are suitable candidates for NGS-based studies.

Genetic Features of Common Neurodegenerative Diseases

Most of the neurodegenerative diseases share several clinical, pathologic, and molecular aspects [13]. Clinically, these diseases are often represented by an insidious onset during adulthood, after which they progress at different rates, ultimately leading to severe physical disability or death. The symptoms are often common among the different disorders: dementia is not only peculiar of Alzheimer's disease (AD) or frontotemporal dementia (FTD), but could occur also in Parkinson's disease (PD) or amyotrophic lateral sclerosis (ALS).

Furthermore, under a genetic point of view, many neurodegenerative diseases manifest an important family history highlighting a relevant contribution of genetic factors to disease causation and progression.

The following sections will provide overviews of the current genetic progresses in AD, FTD, PD, and ALS reached by the application of the new genetic technologies.

Progress in AD Genetics

Alzheimer's disease (AD) is a multifactorial and complex neurodegenerative disorder and the leading cause of dementia among elderly people. Genetically, AD can be subdivided into a rare familial form, accounting for 5–10% of all patients and presents with autosomal dominant inheritance, and a multifactorial sporadic form in which specific environmental exposures in combination with genetic susceptibility contribute to the exacerbation of the disease. The first type of disease generally develops before age 65 years and is referred as EOAD, whereas the sporadic type of disease often occurs later in life in individuals older than 65 years and is referred to as LOAD [14]. Three genes are responsible for familial AD: *APP*, *PSEN1*, and *PSEN2*. Together, they account for 5–10% of cases of the early-onset form of the disease. They map on three different chromosomes but share a common biological pathway [13].

The proportion of cases of autosomal dominant AD explained by mutations in these genes is high but varies widely from 12 to 77% [15, 16]. This aspect suggests that there are additional genetic factors involved in the pathogenesis of EOAD. Recently, thanks to NGS approach, some new genetic mutations were found in small families with unexplained EOAD. Guerreiro et al. (2012) identified a missense mutation in *NOTCH3* (R1231C), that is, a gene previously linked to cerebral autosomal dominant arteriopathy with subcortical infarcts and leukoencephalopathy [17].

Another study [18] identified mutations in the sortilin-related receptor 1 (*SORL1*) gene in EOAD. This gene encodes a neuronal sorting protein able to bind APP, driving it towards the endosomal recycling pathways. Other studies involving EOAD found association between triggering receptor expressed on myeloid cells 2 (*TREM2*) gene and the risk of developing the disease. TREM2 is an immune phagocytic receptor expressed in brain microglia able to modulate microglial phagocytosis and inflammatory pathways [19].

Pottier et al. (2013), by using a NGS WES-based approach, identified an association between *TREM2* variants in exon 2 and EOAD in Caucasian subjects of French origin. They found an association between rs75932628T allele (R47H) with the risk of developing AD [20].

The same variant was further confirmed to be a risk factor for EOAD in a recent study conducted by Slattery et al. (2014) that determined that individuals with R47H variant-associated EOAD had significantly earlier symptom onset than individuals with no *TREM2* variants [21].

A recent GWAS identified a novel missense mutation in phospholipase D family member 3 (*PLD3*) gene in an EOAD autopsy-confirmed patient. However, further confirmatory analysis carried out in larger sample size population of European EOAD didn't let to prove significant evidence for an enrichment of rare *PLD3* variants in EOAD patients. Therefore, a genetic role of *PLD3* in AD still remains to be demonstrated.

Rare variants in PLD3 do not affect the risk for EOAD in a European Consortium cohort [22]. Recently, the use of NGS with a target panel able to analyze ten genes involved in dementia led to the identification of novel coding variants in *PSEN1* predicted to be pathogenic [23].

LOAD represents the large majority of all cases (>95%), typically presenting with an onset age higher than 65 years and involving multiple susceptibility genes. Although the advent and the application of NGS screening technologies led to identify several risk variants, their association appears to be associated with very low risk, except the ε4 allele of *APOE*. The risk associated with *APOE* ε4 allele was consistently replicated in a number of independent studies considering many ethnic groups. the presence of the ε4 heterozygous genotype confers fourfold the risk to develop AD. The risk reaches 15-fold for subjects homozygous for the ε4 allele [24].

Recent completion of several GWAS studies led to confirm that *APOE* remains the single most important genetic risk factor for AD, although other risk factors, as binding integrator (*BIN1*) and clusterin (*CLU*), emerged recently as strongly associated with LOAD [25, 26].

Recently advances in NGS add robust evidences that rare variants explain some of genetic heritability in AD. A rare variant, nicastrin gene, has recently been identified by NGS as risk factor for LOAD in Greek population [27]. TREM2 genetic variability has been investigated also with regard to LOAD susceptibility by different groups. Jonsson et al. found that rs7593628T allele variant in *TREM2* confers a significant threefold increased risk for AD in a cohort of Icelanders. This result was further replicated in other independent populations [28].

The same variant was further tested by Guerreiro et al. (2013) by WES technology and was found to cause a fivefold increase risk to develop AD. Furthermore, six additional *TREM2* variants were found in AD cases and not in controls, highlighting their possible contribution to increase AD risk [29]. *TREM2* has also emerged as associated with AD in a very recent study that took advantage of NGS technology in a wide population of the International Genomics of Alzheimer's Project (IGAP). In addition to *TREM2* rs143332484 rare variant association, other two genomic variants were found associated with AD: the protective variant rs728224905 in *PLCG2* and the risk variant rs616338 in *ABI3* gene [30].

Lastly, Kohli et al. (2016), using WES on 11 affected individuals in a large kindred with apparent autosomal dominant LOAD, found damaging missense mutations in the tetratricopeptide repeat domain 3 (*TTC*) gene in all affected individuals [31].

Progress in PD Genetics

Parkinson's disease (PD) is one of the several neurodegenerative diseases that affects aging individuals, in particular, is the second most common neurodegenerative disorder of adult onset and shows an increased prevalence with age. At histopathological level it is characterized by severe loss of dopaminergic neurons in the substantia nigra and cytoplasmatic inclusions in the remaining neurons consisting of insoluble protein aggregates of alpha-synuclein protein named Lewy bodies.

Monogenic forms of PD show both autosomal dominant and recessive pattern of inheritance. Up to date, at least five genes are known to be causative for Mendelian forms of the disease [32–36].

Alpha-synuclein gene (*SNCA*) is the first gene found to be causative of genetic PD. It encodes for a presynaptic protein that modulates neurotransmitter release and vesicle turnover.

In addition to *SNCA*, autosomal dominant causing PD mutations were found also in leucine-rich repeat kinase 2 gene (*LRRK2*), in the retromer complex component (*VPS35*), and in the eukaryotic translation initiation factor 4G gene (*EIF4G*). The two last genes were identified through NGS techniques by exome sequencing in PD-affected families [37, 38].

VPS35 is a component of the retromer complex involved in retrograde transport from endosomes to trans-Golgi network. A recent NGS-based study carried out in 213 patients with PD found three novel non-synonymous variants which could contribute to PD pathogenesis [39].

Conversely, genes involved in recessive transmitted PD forms probably result in a loss of function, leading to a decreased protection of dopaminergic neurons against toxic events. The most common mutated genes are the parkin 2 gene (*PARK2*) and PTEN-induced putative kinase 1 gene (*PINK1*) [40].

Recently, a deleterious variation in DNAJ heat shock protein family (Hsp40) member C6 (*DNAJC6*) by WES in juvenile parkinsonism was identified [41].

Mano et al. (2016) investigated three patients with apparent autosomal dominant PD and dementia. A WES technology-based study led to identify heterozygous mutation in prion protein gene (*PRNP*) [42]. Furthermore, another functional mutation in the same gene was described [43], thus expanding the spectrums of the parkinsonism phenotype and DNAJC6 variants.

Several common polymorphisms in *SNCA* and *LRRK2* were found, by NGS studies, to be highly associated with the risk to develop the disease.

Recently, mutations in transmembrane protein 230 gene (*TMEM230*) were claimed to be causative of Mendelian form of late-onset PD with typical Lewy body pathology in a large Canadian pedigree and seven smaller Chinese families [44]. The protein encoded by *TMEM230* remains largely uncharacterized, but initial evidence suggests a role in the trafficking of recycling vesicles, retromers, and endosomes, suggesting interesting links to the pathways targeted by other PD-causing genes.

Nevertheless, subsequent replication studies in different population were largely negative suggesting that more evaluation of genetic data from other different populations is needed to clearly understand the genetic role of this gene in PD pathology. Recently, a lack of evidence for a role of genetic variation in TMEM230 in the risk for Parkinson's disease in the Chinese population was reported [45].

Progress in FTLD Genetics

Frontotemporal dementia (FTD) is the second most common young-onset dementia affecting people of 45–65 years. The term frontotemporal lobar degeneration (FTLD) is used to describe the pathology associated with all kind of clinical FTD, and it is pathologically associated with degeneration of frontal and temporal lobes.

FTLD is heterogeneous with patients being classified into different subtypes basing on main components of pathologic protein aggregated: FTLD-tau, FTLD-TDP, FTLD-FET [46].

Genetic investigation over the past two decades in FTLD with Mendelian inheritance led to the identification of three common FTLD genes: microtubule-associated protein tau (*MAPT*), progranulin (*GRN*), and chromosome 9 open reading frame 72 (*C9ORF72*), and also to an additional small number of rare FTLD genes. All together the mutations in the abovementioned genes explain almost all autosomal dominant FTLD families.

Among the three known causal genes for the Mendelian forms of FTLD, *C9ORF72* was detected through GWAS in 2011 by two groups independently [47, 48].

GWAS focused on the identification of risk genes for ALS, and FTLD-TDP led initially to identify a significant susceptibility locus at chromosome 9p [49–51].

However, despite all these efforts, the research of the exact gene locus remained elusive until 2011 when it was determined a GGGGCC hexanucleotide repeat expansion within the noncoding region of the *C9ORF72* gene. Regarding the pathogenicity related to the number of expansions, there is not a clear consensus and a reliable cutoff so far. In healthy controls, most individuals carry between 2 and 20 copies of repeats, whereas ALS and FTLD patients carry from approximately 100 to several 1000s of copies of repeat. Anyway, up to now *C9ORF72* repeat expansion is the most common cause of FTLD and ALS worldwide [52]. A whole-genome sequencing study carried out by Pottier and colleagues in 2015 led to find rare variants in TANK binding kinase 1 gene (*TBK1*) [53], gene previously found to be implicated in ALS [54].

Moreover, in the same study, also heterozygous mutation in optineurin gene (*OPTN*) was found, suggesting that both *TBK1* and *OPTN* contribute to the etiology of FTLD-TDP.

Recently, a pathogenic mutation in the coiled-coil-helix-coiled-coil-helix domain containing 10 (*CHCHD10*) was identified by WES in an atypical family with late-onset motor neuron disease, cognitive decline resembling behavioral FTD [55]. Further studies carried out in FTLD Asian population led to establish *CHCHD10* as the most common FTLD gene in Asia [56].

A number of other genes have been investigated in relation to FTLD. Among those *TREM2* association with FTLD deserves particular interest. Recessive mutations in *TREM2* gene cause Nasu-Hakola disease. Guerreiro et al. (2013) found, through NGS techniques, mutations in *TREM2* in patients with atypical FTLD characterized by very young onset age, the presence of seizures and parkinsonism, and extensive white matter lesions on brain imaging as well [57]. Regarding the presence of risk genes or genetic modifiers, relatively few genes were identified. The most relevant and replicated risk factor involved in FTLD is the transmembrane protein 106B gene (*TMEM106B*). It was firstly identified as a risk factor for FTLD-TDP in a GWAS which enrolled more than 500 FTLD-TDP patients [51]. A GWAS in pathologically confirmed FTLD-TDP patients was also carried out and did not identify other common variants associated with the disease in addition to *TMEM106B*.

Recently, a further GWAS performed in 2154 clinical FTLD patients and 4308 controls identified for the first time the HLA locus and a further locus on chromosome 14q11 encompassing *RAB38* and cathepsin D (*CTSC*) [58].

The *HLA* locus association importantly suggests a link between FTLD and the immune system, as Ferrari et al. (2015) carried on a refined analysis of GWAS data and found two potential additional loci for FTLD susceptibility in chromosome 2p16.3 within the intronic region of a not characterized protein LOC30100 and on chromosome 17q25.3 within a region harboring *CEP131*, *ENTHD2*, and *C17ORF89* [59].

Progress in ALS Genetics

Amyotrophic lateral sclerosis (ALS) is characterized by a progressive degeneration of motor neurons in the brain and spinal cord, leading ultimately to paralysis and death within 1–5 years. Neuropathologic features include deposition of pathologicTDP-43 aggregates. As TDP-43 is also a pathologic hallmark in determinate forms of FTD, it is now ascertained that ALS and FTD belong to the same clinicopathologic spectrum of diseases.

Mendelian forms of ALS (familial ALS, FALS) account for about 10% of all ALS cases and show predominantly autosomal dominant inheritance [60].

Several genes are responsible for Mendelian forms of the disease: superoxide dismutase 1 (*SOD1*), TAR DNA-binding protein (*TARDBP*), FUS RNA-binding protein (*FUS*), and valosin-containing protein gene (*VCP*). Mutations in *SOD1* are responsible for 20% of familial ALS [60]. Recently, Wu and colleagues carried out a WES study in Chinese kindred and found a novel mutation in *SOD1* [61].

Mutations in *VCP* gene account for 1.5% of FALS cases. Johnson et al. (2010) identified a novel variant in the VCP gene by WES technology in an Italian family with autosomal dominant ALS [62].

Recently, a new NGS approach based on haloplex target enriched system was used to test 18 causative genes in ALS probands. Haloplex target enrichment system is a new targeted sequencing approach that enables to detect already known mutations or candidate genes. By using this approach, a novel dynactin 1 (*DCTN1*) pathogenic variant and three already known pathogenic mutations in *SOD1* gene were found [63]. This multigene panel NGS-based approach was also successfully employed in 4 index ALS patients and 148 sporadic cases from Korea. They tested 18 causative genes and identified 4 known mutations in *SOD1*, *ALS2*, *MAPT*, and *SQSTM1* genes and several variants in 9 genes potentially deleterious by in silico analyses [64].

These results suggest that multigene panel testing could be an effective approach for mutation screening in ALS and in other neurodegenerative disease-related genes.

A very recent paper by Morgan et al. (2017) tested a panel of known ALS genes in a wide ALS population consisting of 1126 patients and 613 controls. They found an increased burden of rare variants in patients within the untranslated regions of known disease-causing genes, driven by *SOD1, TARDBP, FUS, VCP, OPTN*, and *UBQLN2* [65].

Moreover, the hexanucleotide repeat expansion in the noncoding region of *C9ORF72* gene was shown to be the cause of 20–50% of familial ALS [66].

Lastly, four pathogenic mutations were found by WES in two large ALS families within the profiling 1 gene (*PFN1*) [67, 68] and a novel missense mutation in *UBQLN2* detected by WES in two ALS families [69].

Epigenetics: Basic Concepts

Epigenetics is focused on the investigation of mechanisms able to alter the expression of genes without altering the DNA sequence [70]. DNA methylation, chromatin remodeling, and noncoding RNAs (ncRNAs) are the three most investigated epigenetic modifications especially in relation to aging and neurodegenerative diseases. Although epigenetic changes are passed from parent to offspring through the germline and are retained through successive cell divisions, they can be reversed and are highly sensitive to environmental influences [71, 72].

Epigenetic processes are able to regulate DNA replication and repair, RNA transcription, and chromatin conformation that influence in turn transcriptional regulation and protein translation.

Methylation

DNA methylation is the best characterized epigenetic modification that involves the addition of a methyl group to the carbon-5 of a cytosine residue in DNA and is carried out by one of the several DNA methyltransferase (DNMT) enzymes. DNMT1 is the enzyme responsible for the maintenance of DNA methylation patterns during DNA replication. DNMT1 localizes to the DNA replication fork, where it methylates nascent DNA strands at the same locations as in the template strand [73]. DNMT3a and DNMT3b are involved in the de novo methylation of unmethylated and hemimethylated sites in nuclear and mitochondrial DNA, respectively [73, 74]. In mammals, DNA methylation occurs predominantly at CpG sites—locations where a cytosine nucleotide is followed by a guanine nucleotide. CpG sites can occur in concentrations of up to several hundred dinucleotide repeats, called CpG islands, which are frequently found in gene promoter regions. The methylation or hypermethylation of CpG islands in promoter regions usually prevents the expression of the associated gene [75]. DNA methylation is currently the best understood epigenetic mechanisms and is known to have a crucial role in normal development, cell proliferation, and genome stability [76]. Recently, however, non-CpG methylation has received increased attention [77]. The design and development of techniques for the identification, quantification, and positioning of individual CpG methylation across the genome is a milestone that needs to be accomplished in order to provide a reliable characterization of the human epigenome.

Early epigenetic investigations related to AD focused on DNA methylation finding non-AD-specific hypomethylation of the APP gene promoter region in one patient [78].

More recent studies support an overall reduction in DNA methylation in AD patients, thus highlighting the importance of DNA methylation in AD [79]. Interestingly Aβ has also been implicated as a trigger of epigenetic changes as it was found that Aβ induces global DNA hypomethylation [80].

Tau gene expression is also subject to complex epigenetic regulation, involving differentially methylated binding sites for transcription factors [81].

Recently, Bollati et al. (2011) investigated the methylation status of repetitive elements in blood, including *Arthrobacter luteus* elements (Alu) blood, long interspersed element 1 (LINE-1), and satellite-α (SAT-α), that comprise a wide portion of the human genome and are known to contain large numbers of CpG sites. Interestingly, they found that LINE-1 methylation was increased in AD patients and that enhanced LINE-1 methylation was associated with a better cognitive performance within the AD group [82].

DNA methylation was also reported to be altered in ALS postmortem brains by a comparison between sporadic ALS and control carried out using Affymetrix GeneChip Human Tiling 2.0R arrays [83].

Regarding FTD, two studies showed that the *GRN* promoter methylation is able to regulate progranulin expression, as they found that increased methylation in FTD subjects negatively correlates with GRN mRNA levels [84, 85].

A recent genome-wide study on DNA methylation pattern in peripheral blood of FTD and progressive supranuclear palsy (PSP) compared to healthy subjects found a specific methylation signature associated pathologically with tauopathy [86].

Interestingly, regarding the *C9ORF72*expansion, it was suggested that the length of the repeat might influence the level of DNA methylation at the *C9ORF72* promoter. This process was found in a family from Canada with father carrying an intermediate length allele, about 70 repeats with an unmethylated *C9ORF72* promoter that expanded to about 1750 repeats in his children. The expanded allele carried by the four children, two of them have developed ALS symptoms, was characterized by *C9ORF72* promoter hypermethylation and associated with reduced c9orf72 expression [87].

ncRNAs

It was widely believed in the past that most of the human genome consisted in "non-functional" DNA. It was later discovered that almost the whole genome is transcribed but that just about 2% in translated into proteins [88]. It is now ascertained that most of this "junk" is instead functional and composed by noncoding RNA (ncRNA), whose signaling and editing is able to play a crucial role in chromatin and nuclear structure. In particular, ncRNAs are involved in epigenetic regulation by recruiting chromatin-modifying complexes. ncRNAs operate through repressive control but have also the potential to act as gene activators [89].

ncRNAs comprise small RNAs (sRNAs) of less than 200 nucleotides and long noncoding RNA (lncRNAs) of more than 200 nucleotides. sRNAs are further subdivided as microRNAs (miRNAs), short interfering RNAs (siRNAs), and PIWI-associated RNAs (piRNAs), whereas lncRNAs are categorized according to their direction and position of their transcription in antisense, intergenic, exonic, intronic, and overlapping [90].

miRNAs are single-stranded, noncoding small RNAs that are abundant in plants and animals and are conserved across species [91]. The raw transcripts undergo several nuclear and cytoplasmic post-translational processing steps to generate mature, functional miRNAs. In the cytoplasm, mature miRNAs associate with other proteins to form the RNA-induced silencing complex (RISC), enabling the miRNA to imperfectly pair with cognate miRNA transcripts. The target mRNA is then degraded by the RISC, preventing its translation into protein [92, 93]. miRNA-mediated repression of translation is involved in many cellular processes, such as differentiation, proliferation, and apoptosis, as well as other key cellular mechanisms [94, 95].

It is now well established that altered RNA processing could act as contributing factor to several neurological conditions including aging-related neurodegenerative diseases as AD, PD, FTD, and ALS [96–98].

Impairments at all levels of gene regulation, from RNA synthesis, processing, function, and degradation, are associated with ncRNA.

Regarding miRNAs contribution in neurodegenerations, there are two ways in which they can drive it: alterations to miRNA biogenesis or to miRNA expression. In addition to this aspect, specific miRNAs affecting specific disease-linked genes are also associated with neurodegenerative diseases as AD. In particular, it has been demonstrated the implication miRNAs in $A\beta$ production via BACE1 modulation and in tau phosphorylation that leads to hyperphosphorylated neurofibrillary tangle formation. Moreover, several miRNAs have been also involved in ALS pathogenesis or as biomarkers of disease. MiR-23a was found overexpressed in skeletal muscle biopsies from ALS patients [99].

Furthermore, several miRNAs were found to be dysregulated also in spinal cord from individuals with ALS, including miR155-5p, miR-142-5p, let-7e, miR-148-5p, miR-133b, miR-140-3p, and miR-57. These are all miRNAs able to regulate neuronal homeostasis, pathogenesis of ALS, and other neurodegenerative-related transcripts [100].

In PD, several miRNAs dysregulations have been found, but one of the most involved miRNAs in the pathology appeared to be miR-133b that is particularly enriched in the midbrain region of normal subjects but was found to be deficient in samples from PD patients [101].

Altered miRNA signatures were also identified in AD and FTD. In particular, several miRNAs have been identified differentially expressed in postmortem tissue, blood, and cerebrospinal fluid that differ also by disease stage [102–105].

Interestingly, miRNA-based therapeutics, such as miRNA mimics or miRNA antagonists (antagomirs), have been designed to either reverse the downregulation or upregulation of disease-associated miRNAs, respectively.

Regarding lncRNAs they also have been involved in neurodegenerative diseases [106]. These ncRNAs are involved in different functions; they act as scaffolds for chromatin modifiers and nuclear paraspeckles, as transcriptional co-regulators, and even as decoys for other RNAs [107]. Dysregulations in lncRNAs can influence any one of these processes, thus contributing to neurodegeneration. lncRNAs associated with disease condition can post-transcriptionally increase gene expression, as it happens with the lncRNA BACE1-antisense (AS) whose expression is selectively increased in AD brains and competes with miR-545-5p binding to stabilize BACE1 mRNA. This will finally result in increased expression of BACE1 that contributes to the formation of the toxic Aβ peptides that is a major hallmark for AD [108].

Another lncRNA, BC200, was also found to be inked to AD. In particular, BC200 levels were found to be increased in specific brain regions mostly affected by AD as the Brodmann's area 9 [109]. MALAT1 and NEAT1 are other two lncRNAs very important for splicing and synapse formation [110, 111].

Chromatin Remodeling

In mammalian cells, histone proteins interact with DNA to form chromatin, the packaged form of DNA. Histones are octamer consisting of two copies of each of the four histone proteins: H2A, H2B, H3, and H4. Each histone octamer constitutes in 146 bp of the DNA strand wound around it to make up one nucleosome, which is the basic unit of chromatin. Histone proteins can be modified by post-translational changes, among those there are acetylation, methylation, phosphorylation, ubiquitination, and citrullination. These histone modifications induce changes to the structure of chromatin and thereby affect the accessibility of the DNA strand to transcriptional enzymes, resulting in activation or repression of genes associated with the modified histone [112]. The best-understood histone modification is acetylation, which is mediated by histone acetyltransferases and deacetylases. Acetylation of histones is usually associated with upregulated transcriptional activity of the associated gene, whereas deacetylation of histones to transcriptional silencing [113].

Therapeutic strategies are designed to target epigenetic modifiers such as histone deacetylases. Affecting the activity of this enzyme has been shown to be effective in myelodysplastic syndrome and acute myelogenous leukemia [114]. Thus, the analysis of histone acetylation levels on specific genes by chromatin immunoprecipitation (ChIP)-based technologies may be an interesting approach to monitor potential therapeutic strategies or follow the response of the patients to this therapy. Another recent interesting study has shown that histones released into the plasma enhance thrombin generation, a process that may contribute to microvascular thrombosis at sites of severe inflammation [115]. Under this point of view, the analysis of circulating histones in plasma may offer reliable information about the inflammation process. It has also been previously described that histones produce damage in endothelial cells and organ failure when injected into mice [116].

Epigenetic dysregulation in terms of chromatin remodeling has been found in neurodegenerative conditions. In particular, histone acetylation was found to be largely decreased in the temporal lobe of AD patients when compared to aged controls [117] and in mouse models of AD [118]. Moreover, increased H3 acetylation at the promoter region of the BACE1 gene in AD patients was found [119].

The involvement of histone modification in other neurodegenerative diseases is less and needs to be replicated; however, there is a study of a familial PD case with heterozygous A53T mutation in *SNCA* gene, in which the affected allele was epigenetically silenced through histone modifications and the normal allele showed expression levels exceeding those of the two normal alleles in controls [120].

Technologies Used in Epigenetic Studies

Most of the innovative technologies used for epigenetic studies have been developed from conventional assays. For example, the classical method of DNA methylation analysis was based on the capability of two restriction enzyme pairs (*Hpa*II-*Msp*I and *Sma*I-*Xma*I) to recognize or discriminate methylated regions. However, this method has some weak points that depend on the efficiency of the enzymes, the step of southern blot hybridization, and the expertise of users.

A major advance in DNA methylation analysis was the development of a method for sodium bisulfate modification of DNA to convert unmethylated cytosines to uracils, leaving methylated cytosines unchanged. This method was the precursor of most of the new technologies to analyze DNA methylation. In the case of the classical method after bisulfite conversion, PCR amplification is performed followed by determination of the sequences of amplification. However, it is too difficult to process large amounts of samples manually, and the critical step of bisulfate treatment could be not well performed, thus affecting the final results [121]. Automated methods offer several advantages versus classical procedures. Among them, automated analysis allows processing a large number of samples by a single technician. On the other hand, the automated technologies standardize the procedures, the results, and the analysis of the data. Moreover, automation and the use of these technologies deliver in high-throughput experiments, fast assays, and high reproducibility. Finally, the software of these systems offer high amount of information easy to interpret and analyze by the user.

It is essential that the epigenetic biomarkers that are applied to preclinical testing, diagnosis, disease progression, or treatment monitoring exhibit good sensitivity and reproducibility. Clearly, these technologies will allow us to discover epigenetic biomarkers for disease in the forthcoming years. They will also help identify or classify diseases, and finally monitoring disease progression or the efficacy of a drug in those diseases in which genetics alone cannot give definitive answers.

References

1. Nakamura Y. DNA variations in human and medical genetics: 25 years of my experience. J Hum Genet. 2009;54:1–8.
2. Frazer KA, Murray SS, Schork NJ, Topol EJ. Human genetic variation and its contribution to complex traits. Nat Rev Genet. 2009;10:241–51.
3. Ku CS, Loy EY, Salim A, Pawitan Y, Chia KS. The discovery of human genetic variations and their use as disease markers: past, present and future. J Hum Genet. 2010;55:403–15.
4. Kimura M. Evolutionary rate at the molecular level. Nature. 1968;217:624–6.
5. Dawn Teare M, Barrett JH. Genetic linkage studies. Lancet. 2005;366(9490):1036–44.
6. Hattersley AT, McCarthy MI. A question of standards: what makes a good genetic association study? Lancet. 2005;366(9493):1315–23.
7. Simón-Sánchez J, Singleton A. Genome-wide association studies in neurological disorders. Lancet Neurol. 2008;7(11):1067–72.
8. Pritchard JK. Are rare variants responsible for susceptibility to complex diseases? Am J Hum Genet. 2001;69:124–37.
9. Bamshad MJ, Ng SB, Bigham AW, et al. Exome sequencing as a tool for Mendelian disease gene discovery. Nat Rev Genet. 2011;12(11):745–55.
10. Boycott KM, Vanstone MR, Bulman DE, MacKenzie AE. Rare-disease genetics in the era of next-generation sequencing: discovery to translation. Nat Rev Genet. 2013;14(10):681–91.
11. Yang Y, Muzny DM, Reid JG, et al. Clinical whole-exome sequencing for the diagnosis of Mendelian disorders. N Engl J Med. 2013;369(16):1502–11.
12. Yang Y, Muzny DM, Xia F, et al. Molecular findings among patients referred for clinical whole-exome sequencing. JAMA. 2014;312(18):1870–9.
13. Lill CM, Bertram L. Towards unveiling the genetics of neurodegenerative diseases. Semin Neurol. 2001;31(5):531–41.
14. Blennow K, de Leon MJ, Zetterberg H. Alzheimer's disease. Lancet. 2006;368(9533):387–403.
15. Jarmolowicz AI, Chen HY, Panegyres PK. The patterns of inheritance in early-onset dementia: Alzheimer's disease and frontotemporal dementia. Am J Alzheimers Dis Other Dement. 2014;30:299–306.
16. Wallon D, Rousseau S, Rovelet-Lecrux A, et al. The French series of autosomal dominant early onset Alzheimer's disease cases: mutation spectrum and cerebrospinal fluid biomarkers. J Alzheimers Dis. 2012;30:847–56.
17. Guerreiro RJ, Lohmann E, Kinsella E, et al. Exome sequencing reveals an unexpected genetic cause of disease: NOTCH3 mutation in a Turkish family with Alzheimer's disease. Neurobiol Aging. 2012;33:1008.e17–23.
18. Pottier C, Hannequin D, Coutant S, et al. High frequency of potentially pathogenic SORL1 mutations in autosomal dominant early-onset Alzheimer disease. Mol Psychiatry. 2012;17:875–9.
19. Jiang T, Yu JT, Zhu XC, Tan L. TREM2 in Alzheimer's disease. Mol Neurobiol. 2013;48:180–5.
20. Pottier C, Wallon D, Rousseau S, et al. TREM2 R47H variant as a risk factor for early-onset Alzheimer's disease. J Alzheimers Dis. 2013;35:45–9.
21. Slattery CF, Beck JA, Harper L, et al. R47H *TREM2* variant increases risk of typical early-onset Alzheimer's disease but not of prion or frontotemporal dementia. Alzheimers Dement. 2014;10(6):602–608.e4.
22. Cacace R, Van den Bossche T, Engelborghs S, et al. Rare variants in PLD3 do not affect risk for early-onset Alzheimer disease in a European Consortium cohort. Hum Mutat. 2015;36(12):1226–35.
23. Piccoli E, Rossi G, Rossi T, et al. Novel PSEN1 mutations (H214N and R220P) associated with familial Alzheimer's disease identified by targeted exome sequencing. Neurobiol Aging. 2016;40:192.e7–11.
24. Farrer LA, Cupples LA, Haines JL, et al. Effects of age, sex, and ethnicity on the association between apolipoprotein E genotype and Alzheimer disease. A meta-analysis. APOE and Alzheimer Disease Meta Analysis Consortium. JAMA. 1997;278(16):1349–56.

25. Harold D, Abraham R, Hollingworth P, et al. Genomewide association study identifies variants at CLU and PICALM associated with Alzheimer's disease. Nat Genet. 2009;41:1088–93.
26. Hollingworth P, Harold D, Sims R, et al. Common variants at ABCA7, MS4A6A/MS4A4E, EPHA1, CD33 and CD2AP are associated with Alzheimer's disease. Nat Genet. 2011;43:429–35.
27. Lupton MK, Proitsi P, Danillidou M, et al. Deep sequencing of the Nicastrin gene in pooled DNA, the identification of genetic variants that affect risk of Alzheimer's disease. PLoS One. 2011;6:e17298.
28. Jonsson T, Stefansson H, Steinberg S, et al. Variant of *TREM2* associated with the risk of Alzheimer's disease. N Engl J Med. 2013;368:107–16.
29. Guerreiro R, Wojtas A, Bras J, et al. TREM2 variants in Alzheimer's disease. N Engl J Med. 2013;368:117–27.
30. Sims R, van der Lee SJ, Naj AC, et al. Rare coding variants in PLCG2, ABI3, and TREM2 implicate microglial-mediated innate immunity in Alzheimer's disease. Nat Genet. 2017;49(9):1373–84.
31. Kohli MA, Cukier HN, Hamilton-Nelson KL, et al. Segregation of a rare TTC3 variant in an extended family with late-onset Alzheimer disease. Neurol Genet. 2016;2:e41.
32. Valente EM, Abou-Sleiman PM, Caputo V, et al. Hereditary early-onset Parkinson's disease caused by mutations in PINK1. Science. 2004;304:1158–60.
33. Paisán-Ruíz C, Jain S, Evans EW, et al. Cloning of the gene containing mutations that cause PARK8-linked Parkinson's disease. Neuron. 2004;44:595–600.
34. Bonifati V, Rizzu P, van Baren MJ, et al. Mutations in the DJ-1 gene associated with autosomal recessive early-onset parkinsonism. Science. 2003;299:256–9.
35. Kitada T, Asakawa S, Hattori N, et al. Mutations in the parkin gene cause autosomal recessive juvenile parkinsonism. Nature. 1998;392:605–8.
36. Polymeropoulos MH, Lavedan C, Leroy E, et al. Mutation in the alpha-synuclein gene identified in families with Parkinson's disease. Science. 1997;276:2045–7.
37. Vilarino-Guell C, Wider C, Ross OA, et al. VPS35 mutations in Parkinson disease. Am J Hum Genet. 2011;89(1):162–7.
38. Chartier-Harlin MC, Dachsel JC, Vilarino-Guell C, et al. Translation initiator EIF4G1 mutations in familial Parkinson disease. Am J Hum Genet. 2011;89(3):398–406.
39. Nuytemans K, Bademci G, Inchausti V, et al. Whole exome sequencing of rare variants in EIF4G1 and VPS35 in Parkinson disease. Neurology. 2013;80:982–9.
40. Gasser T. Molecular pathogenesis of Parkinson disease: insights from genetic studies. Expert Rev Mol Med. 2009;11:e22.
41. Edvardson S, Cinnamon Y, Ta-Shma A, et al. A deleterious mutation in DNAJC6 encoding the neuronal-specific clathrin-uncoating co-chaperone auxilin, is associated with juvenile parkinsonism. PLoS One. 2012;7:e36458.
42. Mano KK, Matsukawa T, Mitsui J, et al. Atypical parkinsonism caused by Pro105Leu mutation of prion protein: a broad clinical spectrum. Neurol Genet. 2016;2:e48.
43. Köroğlu Ç, Baysal L, Cetinkaya M, et al. DNAJC6 is responsible for juvenile parkinsonism with phenotypic variability. Parkinsonism Relat Disord. 2013;19:320–4.
44. Deng HX, Shi Y, Yang Y, et al. Identification of TMEM230 mutations in familial Parkinson's disease. Nat Genet. 2016;48(7):733–9.
45. Giri A, Mok KY, Jansen I, et al. TMEM230 mutation analysis in Parkinson's disease in a Chinese population. Neurobiol Aging. 2017;49:219.e1–3.
46. Irwin DJ, Cairns NJ, Grossman M, et al. Frontotemporal lobar degeneration: defining phenotypic diversity through personalized medicine. Acta Neuropathol. 2015;129:469–91.
47. DeJesus-Hernandez M, Mackenzie IR, Boeve BF, et al. Expanded GGGGCC hexanucleotide repeat in noncoding region of C9ORF72 causes chromosome 9p-linked FTD and ALS. Neuron. 2011;72:245–56.
48. Renton AE, Majounie E, Waite A, et al. A hexanucleotide repeat expansion in C9ORF72 is the cause of chromosome 9p21-linked ALS-FTD. Neuron. 2011;72:257–68.

49. Laaksovirta H, Peuralinna T, Schymick JC, et al. Chromosome 9p21 in amyotrophic lateral sclerosis in Finland: a genome-wide association study. Lancet Neurol. 2010;9:978–85.
50. Boxer AL, Mackenzie IR, Boeve BF, et al. Clinical, neuroimaging and neuropathological features of a new chromosome 9p-linked FTD-ALS family. J Neurol Neurosurg Psychiatry. 2011;82:196–203.
51. Van Deerlin VM, Sleiman PM, Martinez-Lage M, et al. Common variants at 7p21 are associated with frontotemporal lobar degeneration with TDP-43 inclusions. Nat Genet. 2010;42:234–9.
52. Majounie E, Renton AE, Mok K, et al. Frequency of the C9orf72 hexanucleotide repeat expansion in patients with amyotrophic lateral sclerosis and frontotemporal dementia: a cross-sectional study. Lancet Neurol. 2012;11:23–330.
53. Pottier C, Bieniek KF, Finch N, et al. Whole-genome sequencing reveals important role for TBK1 and OPTN mutations in frontotemporal lobar degeneration without motor neuron disease. Acta Neuropathol. 2015;130:77–92.
54. Cirulli ET, Lasseigne BN, Petrovski S, et al. Exome sequencing in amyotrophic lateral sclerosis identifies risk genes and pathways. Science. 2015;347:1436–41.
55. Bannwarth S, Ait-El-Mkadem S, Chaussenot A, et al. A mitochondrial origin for frontotemporal dementia and amyotrophic lateral sclerosis through CHCHD10 involvement. Brain. 2014;137:2329–45.
56. Jiao B, Xiao T, Hou L, et al. High prevalence of CHCHD10 mutation in patients with frontotemporal dementia from China. Brain. 2015;139:1–4.
57. Guerreiro RJ, Lohmann E, Bras JM, et al. Using exome sequencing to reveal mutations in TREM2 presenting as a frontotemporal dementia-like syndrome without bone involvement. JAMA Neurol. 2013;70:78.
58. Ferrari R, Hernandez DG, Nalls MA, et al. Frontotemporal dementia and its subtypes: a genome-wide association study. Lancet Neurol. 2014;13:686–99.
59. Ferrari R, Grassi M, Salvi E, et al. A genome-wide screening and SNPs-to-genes approach to identify novel genetic risk factors associated with frontotemporal dementia. Neurobiol Aging. 2015;36(2904):e2913–26.
60. Dion PA, Daoud H, Rouleau GA. Genetics of motor neuron disorders: new insights into pathogenic mechanisms. Nat Rev Genet. 2009;10:769–82.
61. Wu J, Shen E, Shi D, et al. Identification of a novel Cys146X mutation of SOD1 in familial amyotrophic lateral sclerosis by whole-exome sequencing. Genet Med. 2012;14:823–6.
62. Johnson JO, Mandrioli J, Benatar M, et al. Exome sequencing reveals VCP mutations as a cause of familial ALS. Neuron. 2010;68:857–64.
63. Liu ZJ, Li HF, Tan GH, et al. Identify mutation in amyotrophic lateral sclerosis cases using HaloPlex target enrichment system. Neurobiol Aging. 2014;35(12):2881.e11–5.
64. Kim HJ, Oh KW, Kwon MJ, et al. Identification of mutations in Korean patients with amyotrophic lateral sclerosis using multigene panel testing. Neurobiol Aging. 2016;37:209.e9–16.
65. Morgan S, Shatunov A, Sproviero W, et al. A comprehensive analysis of rare genetic variation in amyotrophic lateral sclerosis in the UK. Brain. 2017;140(6):1611–8.
66. Herdewyn S, Zhao H, Moisse M, et al. Whole-genome sequencing reveals a coding non-pathogenic variant tagging a non-coding pathogenic hexanucleotide repeat expansion in C9orf72 as cause of amyotrophic lateral sclerosis. Hum Mol Genet. 2012;21:2412–9.
67. Wu CH, Fallini C, Ticozzi N, et al. Mutations in the profilin 1 gene cause familial amyotrophic lateral sclerosis. Nature. 2012;488:499–503.
68. Smith BN, Vance C, Scotter EL, et al. Novel mutations support a role for Profilin 1 in the pathogenesis of ALS. Neurobiol Aging. 2015;36(3):1602.e17–27.
69. Williams KL, Warraich ST, Yang S, et al. UBQLN2/ubiquilin 2 mutation and pathology in familial amyotrophic lateral sclerosis. Neurobiol Aging. 2012;33:2527.e3–10.
70. Egger G, Liang G, Aparicio A, Jones PA. Epigenetics in human disease and prospects for epigenetic therapy. Nature. 2004;429:57–463.
71. Jaenisch R, Bird A. Epigenetic regulation of gene expression: how the genome integrates intrinsic and environmental signals. Nat Genet. 2003;33(Suppl):245–54.

72. Skinner MK, Manikkam M, Guerrero-Bosagna C. Epigenetic transgenerational actions of environmental factors in disease etiology. Trends Endocrinol Metab. 2010;21:214–22.
73. Goll MG, Bestor TH. Eukaryotic cytosine methyltransferases. Annu Rev Biochem. 2005;74:481–514.
74. Okano M, Bell DW, et al. DNA methyltransferases Dnmt3a and Dnmt3b are essential for de novo methylation and mammalian development. Cell. 1999;99:247–57.
75. Klose RJ, Bird AP. Genomic DNA methylation: the mark and its mediators. Trends Biochem Sci. 2006;31:89–97.
76. Weber M, Schübeler D. Genomic patterns of DNA methylation: targets and function of an epigenetic mark. Curr Opin Cell Biol. 2007;19:273–80.
77. Guo JU, Su Y, Shin JH, et al. Distribution, recognition and regulation of non-CpG methylation in the adult mammalian brain. Nat Neurosci. 2013;17:215–22.
78. West RL, Lee JM, Maroun LE. Hypomethylation of the amyloid precursor protein gene in the brain of an Alzheimer's disease patient. J Mol Neurosci. 1995;6:141–6.
79. Mastroeni D, Grover A, Delvaux E, et al. Epigenetic changes in Alzheimer's disease: decrements in DNA methylation. Neurobiol Aging. 2010;31:2025–37.
80. Chen KL, Wang SS, Yang YY, et al. The epigenetic effects of amyloid-beta(1–40) on global DNA and neprilysin genes in murine cerebral endothelial cells. Biochem Biophys Res Commun. 2009;378:57–61.
81. Tohgi H, Utsugisawa K, Nagane Y, et al. The methylation status of cytosines in a tau gene promoter region alters with age to downregulate transcriptional activity in human cerebral cortex. Neurosci Lett. 1999;275:89–92.
82. Bollati V, Galimberti D, Pergoli L, et al. DNA methylation in repetitive elements and Alzheimer disease. Brain Behav Immun. 2011;25:1078–83.
83. Morahan JM, Yu B, Trent RJ, Pamphlett R. A genome-wide analysis of brain DNA methylation identifies new candidate genes for sporadic amyotrophic lateral sclerosis. Amyotroph Lateral Scler. 2009;10(5–6):418–29.
84. Banzhaf-Strathmann J, Claus R, Mucke O, et al. Promoter DNA methylation regulates progranulin expression and is altered in FTLD. Acta Neuropathol Commun. 2013;1:16.
85. Galimberti D, D'Addario C, Dell'osso B, et al. Progranulin gene (GRN) promoter methylation is increased in patients with sporadic frontotemporal lobar degeneration. Neurol Sci. 2013;34(6):899–903.
86. Li Y, Chen JA, Sears RL, et al. An epigenetic signature in peripheral blood associated with the haplotype on 17q21.31, a risk factor for neurodegenerative tauopathy. PLoS Genet. 2014;10(3):e1004211.
87. Xi Z, van Blitterswijk M, Zhang M, et al. Jump from pre-mutation to pathologic expansion in C9orf72. Am J Hum Genet. 2015;96(6):962–70.
88. Amaral PP, Dinger ME, Mercer TR, Mattick JS. The eukaryotic genome as an RNA machine. Science. 2008;319:1787–9.
89. Sanchez-Elsner T, Gou D, Kremmer E, Sauer F. Noncoding RNAs of trithorax response elements recruit Drosophila Ash1 to Ultrabithorax. Science. 2006;311(5764):1118–23.
90. Derrien T, Johnson R, Bussotti G, et al. The GENCODE v7 catalog of human long noncoding RNAs: analysis of their gene structure, evolution, and expression. Genome Res. 2012;22(9):1775–89.
91. Bernstein E, Allis CD. RNA meets chromatin. Genes Dev. 2005;19:1635–55.
92. Hwang H-W, Mendell JT. MicroRNAs in cell proliferation, cell death, and tumorigenesis. Br J Cancer. 2006;94:776–80.
93. Sevignani C, Calin GA, Siracusa LD, Croce CM. Mammalian microRNAs: a small world for fine-tuning gene expression. Mamm Genome. 2006;17:189–202.
94. Chang T-C, Mendell JT. MicroRNAs in vertebrate physiology and human disease. Annu Rev Genomics Hum Genet. 2007;8:215–39.
95. Fabbri M, Ivan M, Cimmino A, et al. Regulatory mechanisms of microRNAs involvement in cancer. Expert Opin Biol Ther. 2007;7:1009–19.

96. Anderson P, Ivanov P. tRNA fragments in human health and disease. FEBS Lett. 2014;588:4297–304.
97. Belzil VV, Gendron TF, Petrucelli L. RNA-mediated toxicity in neurodegenerative disease. Mol Cell Neurosci. 2013;56:406–19.
98. Ling S-C, Albuquerque CP, Han JS, et al. ALS-associated mutations in TDP-43 increase its stability and promote TDP-43 complexes with FUS/TLS. Proc Natl Acad Sci U S A. 2010;107:13318–23.
99. Liu EY, Cali CP, Lee EB. RNA metabolism in neurodegenerative disease. Dis Model Mech. 2017;10(5):509–18.
100. Figueroa-Romero C, Hur J, Lunn JS, et al. Expression of microRNAs in human post-mortem amyotrophic lateral sclerosis spinal cords provides insight into disease mechanisms. Mol Cell Neurosci. 2016;71:34–45.
101. Kim J, Inoue K, Ishii J, et al. A Micro RNA feedback circuit in midbrain dopamine neurons. Science. 2007;317:1220–4.
102. Cogswell JP, Ward J, Taylor IA, et al. Identification of miRNA changes in Alzheimer's disease brain and CSF yields putative biomarkers and insights into disease pathways. J Alzheimers Dis. 2008;14:7–41.
103. Lau P, Bossers K, Janky R, et al. Alteration of the microRNA network during the progression of Alzheimer's disease. EMBO Mol Med. 2013;5:1613–34.
104. Wang W-X, Huang Q, Hu Y, et al. Patterns of microRNA expression in normal and early Alzheimer's disease human temporal cortex: white matter versus gray matter. Acta Neuropathol. 2011;121:193–205.
105. Galimberti D, Villa C, Fenoglio C, et al. Circulating miRNAs as potential biomarkers in Alzheimer's disease. J Alzheimers Dis. 2014;42(4):1261–7.
106. Fenoglio C, Ridolfi E, Galimberti D, Scarpini E. An emerging role for long non-coding RNA dysregulation in neurological disorders. Int J Mol Sci. 2013;14(10):20427–42.
107. Prensner JR, Chinnaiyan AM. The emergence of lncRNAs in cancer biology. Cancer Discov. 2011;1:391–407.
108. Faghihi MA, Modarresi F, Khalil AM, et al. Expression of a noncoding RNA is elevated in Alzheimer's disease and drives rapid feed-forward regulation of beta-secretase. Nat Med. 2008;14:723–30.
109. Mus E, Hof PR, Tiedge H. Dendritic BC200 RNA in aging and in Alzheimer's disease. Proc Natl Acad Sci U S A. 2007;104:10679–84.
110. Bernard D, Prasanth KV, Tripathi V, et al. A long nuclear-retained non-coding RNA regulates synaptogenesis by modulating gene expression. EMBO J. 2010;29:3082–93.
111. Tripathi V, Ellis JD, Shen Z, et al. The nuclear-retained noncoding RNA MALAT1 regulates alternative splicing by modulating SR splicing factor phosphorylation. Mol Cell. 2010;39:925–38.
112. Dieker J, Muller S. Epigenetic histone code and autoimmunity. Clin Rev Allergy Immunol. 2010;39:78–84.
113. Brooks WH, Le Dantec C, Pers JO, et al. Epigenetics and autoimmunity. J Autoimmun. 2010;34:J207–19.
114. McDevitt MA. Clinical applications of epigenetic markers and epigenetic profiling in myeloid malignancies. Semin Oncol. 2012;39:109–22.
115. Ammollo CT, Semeraro F, Xu J, et al. Extracellular histones increase plasma thrombin generation by impairing thrombomodulin-dependent protein C activation. J Thromb Haemost. 2011;9:1795–803.
116. Xu J, Zhang X, Pelayo R, et al. Extracellular histones are major mediators of death in sepsis. Nat Med. 2009;15:1318–21.
117. Zhang K, Schrag M, Crofton A, et al. Targeted proteomics for quantification of histone acetylation in Alzheimer's disease. Proteomics. 2012;12:1261–8.
118. Graff J, Rei D, Guan JS, et al. An epigenetic blockade of cognitive functions in the neurodegenerating brain. Nature. 2012;483:222–6.

119. Marques S, Lemos R, Ferreiro E, et al. Epigenetic regulation of BACE1 in Alzheimer's disease patients and in transgenic mice. Neuroscience. 2012;220:256–66.
120. Voutsinas GE, Stavrou EF, Karousos G, et al. Allelic imbalance of expression and epigenetic regulation within the alpha-synuclein wild-type and p. Ala53Thr alleles in Parkinson disease. Hum Mutat. 2010;31:685–91.
121. Grunau C, Clark SJ, Rosenthal A. Bisulfite genomic sequencing: systematic investigation of critical experimental parameters. Nucleic Acids Res. 2001;29:e65.

Diagnosis of Alzheimer's Disease Typical and Atypical Forms

2

Laura Ghezzi

Abstract

Alzheimer's disease (AD) is the most common cause of dementia, with aging as the main risk factor for the development of the disease.

The classical form of AD presents with short-term memory loss and atrophy of the hippocampus and medial temporal lobe. With the progression of the disease, other cortical areas and cognitive domains are involved.

The diagnostic criteria for AD, previously based solely on the clinical and neuropsychological presentation, are currently implemented by the use of biomarkers and neuroimaging data. Moreover, the possibility of atypical forms, presenting with the involvement of different cortical areas, is taken into consideration.

Herein, the main diagnostic tools for AD are revised; atypical AD presentations and possible diagnostic pitfalls are also discussed.

Keywords

Alzheimer's disease (AD) · Biomarkers · Posterior cortical atrophy (PCA) · Primary progressive aphasia (PPA)

Introduction

Alzheimer's disease (AD) is an age-dependent neurodegenerative disorder and the most common cause of dementia with aging.

The early stages of AD are characterized by short-term memory loss. Once the disease progresses, patients experience difficulties in sense of direction, oral communication, calculation, ability to learn, and cognitive thinking. In addition, patients

L. Ghezzi
University of Milan, Centro Dino Ferrari, Fondazione Ca' Granda,
IRCCS Fondazione Policlinico, Milan, Italy

© Springer International Publishing AG 2018
D. Galimberti, E. Scarpini (eds.), *Neurodegenerative Diseases*,
https://doi.org/10.1007/978-3-319-72938-1_2

may develop language deficits, depression, aggressive behavior, and psychosis during the late stages, and eventually they need total care from caregivers.

Currently, diagnosis of AD is based on clinical presentation and on biological biomarkers, in particular radiological and cerebrospinal fluid amyloid, tau and phospho-tau levels.

Clinical Presentation of Typical AD

The disease onset is usually characterized by memory loss for recent events, associated with repetitive questioning and loss of ability to learn. Past memories are usually conserved, instead recent information, such as daily agenda or objects location, are lost (Ribot's law: recent memories are more likely to be lost than the more remote memories). Patient's awareness of memory loss generates depression and anxiety, but consciousness is quickly replaced by anosognosia and the patient loses his critical abilities. With disease progression, visuospatial deficits and dyscalculia appear. Caregivers report episodes of disorientation in known places, such as the patient's neighborhood or even his own home. Dressing apraxia thwarts patients' ability to dress themselves: they are neither able to choose the correct cloth nor to wear it; they need assistance even to put on a pair of trousers. In the late stages of the disease, apraxia affects every task of daily life, making impossible even the simplest action, such as taking a shower. Dyscalculia causes troubles with money, in particular, cash. Patients can't distinguish between 50 and 500; in this phase, they often lose money and are victims of cheaters. Prosopagnosia completes the clinical picture at the late stages of the disease. The patient is unable to recognize his friends or relatives' faces, making coping with their disease even more difficult. Communication also becomes a problem as vocabulary shrinks and fluency falters. Neuropsychiatric symptoms might appear too, such as wandering, irritability, disinhibition, apathy, psychosis, and affective and hyperactive behaviors (Fig. 2.1). These symptoms are collectively defined as behavioral and psychological symptoms of dementia (BPSD) [1]. Different from other dementia syndromes, such as frontotemporal dementia (FTD) and primary progressive aphasia (PPA), language and/or behavioral symptoms are rarely present at the beginning of the disease. Unfortunately, with disease progression, agitation and aggressiveness are frequent. BPSDs are a major source of distress and a reason for internalization of patients with AD [2].

Diagnostic Criteria for Typical AD

The NINCDS-ADRDA Criteria

The first diagnostic criteria for Alzheimer's disease (AD) were published in 1984 by the National Institute of Neurological and Communicative Disorders and Stroke-Alzheimer's Disease and Related Disorders Association (NINCDS-ADRDA) working group [3]. These widely accepted criteria supported a probabilistic diagnosis of AD within a clinical context where there was no definitive diagnostic biomarker. A definite diagnosis of AD was made possible only by histopathological confirmation [3].

Since the publication of the NINCDS-ADRDA criteria in 1984, the biological and pathogenic basis of the disease has been further elucidated. During the following decades, the histopathological and macroscopic changes occurring to the cerebral gray matter at different stages of the disease have been better described.

On these bases, biological markers have been included in the most recent diagnostic criteria. Neuroimages obtained by magnetic resonance imaging (MRI), positron emission tomography (PET) using fluorodeoxyglucose (FDG) or β-amyloid tracers, and cerebrospinal fluid (CSF) analysis of β-amyloid and tau proteins must be taken into consideration in the diagnostic process. Biological markers mirror the two degenerative processes characteristic of AD pathology: the deposition of β-amyloid in neuritic plaques and the tau path to neurofibrillary tangles. Moreover, pathological levels of CSF biomarkers (low β-amyloid, high tau and phospho-tau protein or, even more specifically, an abnormal ratio of tau to β-amyloid) are associated with very high rates of progression from amnestic mild cognitive impairment (MCI) to AD [4].

Revisions of the NINCDS-ADRDA Criteria

The first revision of the NINCDS-ADRDA criteria for research purposes was published in 2007 [5]. In order to satisfy a diagnosis of probable AD, patients must present an objective impairment of episodic memory with evidence of progression over more than 6 months plus medial temporal lobe atrophy at MRI images or pathological CSF markers or abnormally reduced glucose metabolism in bilateral parietal temporal regions at FDG-PET scan.

In the 2007 revision, a new terminology was introduced. The term prodromal AD was used to refer to the early pre-dementia phase of AD in which clinical symptoms are present, but not sufficiently severe to affect instrumental activities of daily living, and biomarker evidence from CSF or imaging is supportive of the presence of AD pathological changes. The state in which evidence of amyloidosis in the brain (with retention of specific PET amyloid tracers) or in the CSF (with changes in β-amyloid, tau, and phospho-tau concentrations) is not associated with any neuropsychological deficit was referred to as "preclinical AD" [5].

This "new lexicon" was further defined in 2010 by Dubois et al. [6]. The "preclinical" state of AD was split into two possible clinical entities: the "asymptomatic at-risk state for AD" and the "presymptomatic AD." The former refers to cognitive normal subjects with positive AD biomarkers; it's important to underline the "at-risk" state since we don't know much about the value of these biological changes to predict further development of the disease. Instead, the term "presymptomatic AD" applies to individuals who will develop AD. This can be ascertained only in families that are affected by rare autosomal dominant monogenic AD mutations (monogenic AD) [6].

The International Working Group (IWG-2) Criteria

In 2014, the last revision of the research criteria for AD was published [7]. According to these criteria, a diagnosis of AD can be made if the patient presents with a progressive (over more than 6 months) objective impairment of episodic memory plus one out of three in vivo evidence of AD pathology, meaning decreased $A\beta_{1-42}$ together with increased tau or phospho-tau in the CSF or increased tracer retention on amyloid PET or the presence of an AD autosomal dominant mutation [7]. Exclusion criteria include a sudden onset, the early occurrence of gait disturbances, seizures, and prevalent behavioral changes. Furthermore, the presence of focal neurological features, early extrapyramidal signs, early hallucinations, and cognitive fluctuations must be regarded as a "red flag," prompting for the research of an alternative diagnosis [7]. Obviously, other medical conditions responsible for cognitive impairment must be ruled out [6].

The 2014 revision of the criteria for typical AD includes a proposition of revision of the criteria for the diagnosis of atypical AD. Following these criteria, a diagnosis of atypical AD can be made in the presence of one of the three known atypical clinical presentations, including posterior cortical atrophy (PCA), logopenic aphasia or frontal variant, and at least one biomarker positive for AD [7].

NIA-AA Criteria

In 2011, the National Institute on Aging (NIA) and the Alzheimer's Association (AA) criteria for AD were published. These criteria maintain the distinction between different levels of diagnostic certainty with the clinical assessment as the core features of the diagnostic algorithm [8]. Positive biomarkers, such as characteristic changes in the CSF analysis, evocative atrophy at brain MRI, and carriers status for a known pathogenic mutation, were considered ancillary elements, adding degrees of certainty to the diagnosis but not mandatory [8].

The NIA-AA criteria proposed a diagnostic classification for patients with dementia caused by AD. Following these criteria, a diagnosis of "all-cause dementia" is possible when a decline from a previous level of functioning in any cognitive domain is present. The deficit must be documented through history and cognitive assessment and must interfere with daily life activities. Probable AD dementia can be diagnosed when the patient meets the criteria for general dementia with an insidious and gradual onset. In this case, onset symptoms can be either typical, characterized by an amnestic presentation, or atypical, with language, visuospatial, or behavioral impairment [8]. In the presence of atypical clinical course, cerebrovascular disease or features of other neurological or non-neurological condition influencing the cognitive status, a diagnosis of "possible" AD should be made. In patients who meet the core clinical criteria for probable AD dementia, evidence of a causative genetic mutation (in *APP*, *PSEN1*, or *PSEN2*) increases the certainty that the condition is caused by AD pathology [8].

DSM-V Criteria

Lastly, the *Diagnostic and Statistical Manual for Mental Disorders* also provides diagnostic criteria for Alzheimer's disease [9]. In the DSM-V, dementia has been newly named major cognitive disorder (MCD). The DSM-V also recognizes a less severe level of cognitive impairment, mild neurocognitive disorder (NCD), which provides a diagnosis for less disabling syndromes that may nonetheless be causing concern and could benefit from treatment. NCDs are characterized by cognitive impairment as the most prominent and defining feature of the condition. Six cognitive domains, which may be affected in NCD, are detailed in the manual, including complex attention, executive function, learning and memory, language, perceptual-motor function, and social cognition. Diagnosis of major NCD requires evidence of significant cognitive decline from a previous level of performance in one or more of the cognitive domains outlined above. Additionally, the cognitive deficit must be sufficient to interfere with independence in activities of daily living. The cognitive deficits must not be attributable to another mental disorder. Major NCD due to AD is then diagnosed if there is evidence of a causative genetic mutation or a steady, progressive decline in memory, learning, and at least one other cognitive domain without evidence of a mixed etiology [9].

Atypical Forms of AD

The initial presentation of AD can be atypical, with non-amnestic focal cortical cognitive symptoms. These syndromes are rare and often underestimated. The most common is PCA, also known as Benson's syndrome [10]. The prevalence and incidence of PCA are currently unknown; age of onset is 50–69 years old, much younger than typical amnestic AD. Patients often face considerable delays in diagnosis owing to the young age at onset and unusual symptoms at presentation. The neuropsychological deficits cited most frequently in individuals with PCA are visuospatial and visuoperceptual impairments, with individuals describing difficulties reading lines of text, judging distances, identifying static objects within the visual field, or having problems with stairs and escalators. Visual symptoms such as light sensitivity or visual distortions can be mistaken for migraine. Alexia, features of Balint's and/or Gerstmann's syndrome, can be part of the picture, but they are rarely reported spontaneously by the patient. Although higher order visual problems are reported more often than basic visual impairments, a recent study by Lehmann et al. (2011) demonstrates that such deficits are due to deficits in more basic visual processing (form, motion, color, and point localization) [11]. Many patients with PCA also present positive perceptual phenomena, such as prolonged color after-images, reverse size phenomena, and perception of movement of static stimuli [9]. Deficits in working memory and limb apraxia have also been described [9]. Moreover, Snowden et al. (2007) reported extrapyramidal signs and myoclonus with a frequency of 41 and 24% in their case histories. Indeed physical examination in most cases of PCA is unremarkable [12].

Voxel-based morphometry has shown the most widespread gray matter reduction in regions of the occipital and parietal lobes followed by areas in the temporal lobe. By 5 years of symptom duration, atrophy is widespread through the cortex. FDG-PET identifies areas of hypometabolism in the parieto-occipital areas and in the frontal eye fields. Data from single photon emission computed tomography (SPECT) usually confirm these findings [13].

Several studies confirm that AD is the most common pathology underlying PCA. However, some cases are attributable to other causes, such as corticobasal degeneration (CBD), dementia with Lewy body disease (LBD), and prion disease (PrD). Renner et al. (2004) reported pathological studies from 21 cases of PCA; of these 14 had AD, 3 had LBD, 2 had CBD, and 2 had PrD [14]. As for the distribution of pathological changes, unfortunately there are only a small number of studies on very few patients, so results are not consistent [14]. It's reasonable to think that there are differences between PCA and typical AD as some of these studies show, but results have to be confirmed by larger studies. All studies report higher density of neurofibrillary tangles and senile plaques in the occipital lobe, but findings in other cortical regions are discordant [13].

In 2017, Crutch et al., on behalf of the Alzheimer's Association ISTAART Atypical Alzheimer's Disease and Associated Syndromes Professional Interest Area, have proposed a three-level classification for the diagnosis of PCA [15]. Level 1 establishes the clinical and cognitive presentations compatible with a posterior cortical syndrome. Core clinical features are an insidious onset, a gradual progression, and a prominent early disturbance of visual and/or other posterior cognitive functions. Three or more posterior cognitive domains must be involved at disease presentation, with an evident impact on daily life activities. Typical "posterior" cognitive deficits, which are included in this classification, are space perception deficit, simultanagnosia, object perception deficit, constructional dyspraxia, environmental agnosia, oculomotor apraxia, dressing apraxia, optic ataxia, alexia, left/right disorientation, acalculia, limb apraxia, perceptive prosopagnosia, agraphia, homonymous visual field defects, and finger agnosia. Anterograde memory, speech, nonvisual language functions, executive functions, behavior, and personality must be spared. Neuroimaging can support the diagnosis with the evidence of a predominant occipito-parietal or occipito-temporal atrophy/hypometabolism/hypoperfusion on MRI/FDG-PET/SPECT. Obviously, other causes of cognitive impairment or visual deficits must be excluded. Level 2 distinguishes between patients meeting solely the criteria for PCA (pure PCA) and patients meeting also the criteria for other neurodegenerative diseases (PCA-plus). It's important to underline that all the criteria for level 1 must be fulfilled in order to diagnose a PCA-plus syndrome. Level 3 reflects current available evidence of the underlying pathology. PCA can be attributable to AD (AD-PCA), LBD (LBD-PCA), CBD (CBD-PCA), and PrP (PrP-PCA). Due to the fact that in vivo biomarkers are currently available only for AD and PrP, the diagnosis in vivo of CBD-PCA and LBD-PCA is pending the development of suitable biomarkers [15].

Frontal variant of AD (fvAD) is even more rare than PCA. It's characterized by prominent behavioral symptoms and executive dysfunction from disease onset and frontal lobe atrophy at the neuroimaging [16]. The few studies published in literature on fvAD suggest an underlying AD pathology and the presence of a CSF biomarkers profile consistent with AD [16].

Logopenic aphasia is the most recently described variant of primary progressive aphasia (PPA) [17]. Like the other variant of PPA, the core clinical feature is difficulty with language, with impairment in daily life activities requiring speech (i.e., using a telephone, asking for information, etc.). In Mesulam et al.'s diagnostic criteria for PPA, language deficit must be the symptom at onset and for the initial phases of the disease [18]. Word retrieval and sentence repetition deficits are the core features of the logopenic variant. Spontaneous speech is characterized by slow rate, with frequent pauses due to significant word-finding problems, but there is no frank agrammatism. Other diagnostic features include phonologic paraphasias in spontaneous speech and naming. The sound substitutions that result in phonologic paraphasias in logopenic patients are usually well articulated, without distortions. Lack of frank agrammatic errors and preservation of articulation and prosody help distinguish the logopenic from the nonfluent variants. Imaging abnormalities in the left temporo-parietal junction area are necessary to make a diagnosis of imaging-supported logopenic variant. Postmortem and in vivo studies using the β-amyloid ligand Pittsburgh compound B (PiB) indicate that lv-PPA is predominantly associated with AD pathology [19, 20]. In 2014, Leyton et al. (2014) demonstrated how the presence of phonologic errors at the neurological examination is the best predictive factor for the presence of an underlying amyloid pathology, whereas motor speech impairment and/or agrammatism are the best negative predictors [21].

References

1. Finkel SI, de Silva C, et al. Behavioral and psychological signs and symptoms of dementia: a consensus statement on current knowledge and implications for research and treatment. Int Psychogeriatr. 1996;8:497–500.
2. Steele C, Rovner B, Chase GA, et al. Psychiatric symptoms and nursing home placement of patients with Alzheimer's disease. Am J Psychiatry. 1990;147:1049–51.
3. McKhann G, Drachman DA, Folstein M, et al. Clinical diagnosis of Alzheimer's disease report of the NINCDS–ADRDA work group under the auspices of Department of Health and Human Services Task Force on Alzheimer's disease. Neurology. 1984;34:939–44.
4. Petersen RC, Roberts RO, Knopman DS, et al. Mild cognitive impairment: ten years later. Arch Neurol. 2009;66:1447–55.
5. Dubois B, Feldman HH, Jacova C, et al. Research criteria for the diagnosis of Alzheimer's disease: revising the NINCDS–ADRDA criteria. Lancet Neurol. 2007;6:734–46.
6. Dubois B, Feldman HH, Jacova C, et al. Revising the definition of Alzheimer's disease: a new lexicon. Lancet Neurol. 2010;9:1118–27.
7. Dubois B, Feldman HF, Jacova C, et al. Advancing research diagnostic criteria for Alzheimer's disease: the IWG-2 criteria. Lancet Neurol. 2014;13:614–29.
8. McKhann GM, Knopman DS, Chertkow H, et al. The diagnosis of dementia due to Alzheimer's disease: recommendations from the National Institute on Aging-Alzheimer's Association workgroups on diagnostic guidelines for Alzheimer's disease. Alzheimers Dement. 2011;7:263–9.
9. American Psychiatric Association. Diagnostic and statistical manual of mental disorders. 5th ed. Arlington: American Psychiatric Publishing; 2013.

10. Benson DF, Davis RJ, Snyder BD. Posterior cortical atrophy. Arch Neurol. 1988;45:789–93.
11. Lehmann M, Crutch SJ, Ridgway GR, et al. Cortical thickness and voxel-based morphometry in posterior cortical atrophy and typical Alzheimer's disease. Neurobiol Aging. 2011;32:1466–76.
12. Snowden JS, Stopford CL, Julien CL, et al. Cognitive phenotypes in Alzheimer's disease and genetic risk. Cortex. 2007;43:835–45.
13. Crutch SJ, Lehmann M, Schott JM, et al. Posterior cortical atrophy. Lancet Neurol. 2012;11:170–8.
14. Renner JA, Burns JM, Hou CE, et al. Progressive posterior cortical dysfunction: a clinicopathologic series. Neurology. 2004;63:1175–80.
15. Crutch SJ, Schott JM, Rabinovici GD, et al. Consensus classification of posterior cortical atrophy. Alzheimers Dement. 2017;13(8):870–84.
16. Blennerhassett R, Lillo P, Halliday GM, et al. Distribution of pathology in frontal variant Alzheimer's disease. J Alzheimers Dis. 2014;39:63–70.
17. Gorno-Tempini ML, Hillis AE, Weintraub S, et al. Classification of primary progressive aphasia and its variants. Neurology. 2011;76:1006–14.
18. Mesulam MM. Primary progressive aphasia. Ann Neurol. 2001;49:425–32.
19. Rohrer JD, Rossor MN, Warren JD. Alzheimer's pathology in primary progressive aphasia. Neurobiol Aging. 2012;33:744–52.
20. Rabinovici GD, Jagust WJ, Furst AJ, et al. Abeta amyloid and glucose metabolism in three variants of primary progressive aphasia. Ann Neurol. 2008;64:388–401.
21. Leyton CE, Ballard KJ, Piguet O, Hodges JR. Phonologic errors as a clinical marker of the logopenic variant of PPA. Neurology. 2014;82:1620–7.

Genetic Complexity of Early-Onset Alzheimer's Disease

Mahdi Ghani, Christiane Reitz, Peter St George-Hyslop, and Ekaterina Rogaeva

Abstract

The recent advances in "omics" technologies (e.g., next-generation sequencing) have made the precision medicine possible. Knowledge about genetics of Alzheimer's disease (AD), the most prevalent form of dementia, is important to manage the challenges of aging populations. So far, genetic analyses of families with autosomal dominant AD, presenting with early-onset dementia (<65 years of age), have found three causal genes: *APP*, *PSEN1*, and *PSEN2*. Genetics is now widely applied to AD diagnosis, monitoring, and the search for a potential treatment. The ability to detect carriers of causal mutations could help to evaluate

---

M. Ghani
Tanz Centre for Research in Neurodegenerative Diseases, University of Toronto, Toronto, ON, Canada

C. Reitz
Departments of Neurology, and Epidemiology, Taub Institute for Research on Alzheimer's Disease and the Aging Brain, Gertrude H. Sergievsky Center, Columbia University, New York, NY, USA

P.S. George-Hyslop
Tanz Centre for Research in Neurodegenerative Diseases, University of Toronto, Toronto, ON, Canada

Department of Medicine, University of Toronto, Toronto, ON, Canada

Department of Clinical Neurosciences, Cambridge Institute for Medical Research, University of Cambridge, Cambridge, UK

E. Rogaeva (✉)
Tanz Centre for Research in Neurodegenerative Diseases, University of Toronto, Toronto, ON, Canada

Department of Medicine, University of Toronto, Toronto, ON, Canada
e-mail: ekaterina.rogaeva@utoronto.ca

© Springer International Publishing AG 2018
D. Galimberti, E. Scarpini (eds.), *Neurodegenerative Diseases*,
https://doi.org/10.1007/978-3-319-72938-1_3

the efficacy of AD therapies in the longitudinal clinical trials of the individuals at either pre-symptomatic or early stages of dementia. We provide an overview for the molecular genetic findings available for early-onset AD; discuss how this knowledge can be applied in clinical practice and highlight strategies to detect novel AD genes.

Keywords

Gene · Alzheimer's disease · *APP* · *PSEN1* · *PSEN2*

Introduction

By 2050, 22% of the global population is predicted to be above 60 years old; hence, a dramatic increase in the prevalence of aging-associated diseases is expected. There were 35.6 million individuals with dementia worldwide in 2010. This number is expected to double every 20 years reaching an estimated 66 million people by 2030 [1]. However, recently demographic studies have shown promising changes in some developed countries. For instance, a ~20% drop in the incidence of dementia was recently reported in the UK population aged over 65 (mainly driven by men) [2], which might be attributable to lifestyle improvement and/or commonly used therapies (e.g., statins).

Dementia can be associated with environmental factors, such as head injury and alcohol intoxication, or can be a clinical manifestation of certain genetic defects. In general, dementia and cognitive impairment have been described for more than 100 genetic diseases recorded in the OMIM database (https://www.omim.org). For instance, early-onset dementia is observed in neuronal ceroid lipofuscinosis, spastic paraplegia, or spinocerebellar ataxia. Furthermore, some genes causing amyotrophic lateral sclerosis (e.g., *C9orf72*) show pleiotropy, in which the same mutation can result in different phenotypes, including frontotemporal dementia (FTD) [3]. However, almost two-thirds of dementia patients over age 80 are diagnosed with the Alzheimer's disease (AD) type of dementia (MIM: #104300) [4, 5].

The yearly cost of care for AD patients in the USA alone is expected to increase to about one trillion dollars by 2050 [6]. Therefore, the more we learn about the genetic factors influencing the risk of AD, the better we will manage the challenges of aging populations. The certainty of disease development in carriers of causal mutations makes it even more important to have a long-term plan for patient care, because early genetic diagnosis of AD, at a time when minimum comorbidities are present, is critical for slowing down disease progression once effective therapies become available.

The brain pathology of familial early-onset AD (<65 years of age) is similar to the sporadic late-onset form of the disease (≥65 years of age). It is characterized by progressive neuronal loss, inflammation, neurofibrillary tangles (consisting of hyper-phosphorylated tau protein encoded by *MAPT*), and amyloid plaques mainly consisting of amyloid beta (Aβ(beta))40/42 peptides generated by the cleavage of

Table 3.1 Major pathways identified by genomic studies of Alzheimer's disease, with most of the genes linked to more than one pathway

Pathway	Gene
Amyloid pathway	APOE, SORL1, CLU, CR1, PICALM, BIN1, ABCA7, CASS4
Immune system/inflammation	CLU, CR1, EPHA1, ABCA7, MS4A4A/MS4A6E, CD33, CD2AP, HLA-DRB5/DRB1, INPP5D, MEF2C, TREM2/TREML2, PLCG2, ABI3
Lipid transport and metabolism	APOE, CLU, ABCA7, SORL1
Synaptic cell functioning/endocytosis	CLU, PICALM, BIN1, EPHA1, MS4A4A/MS4A6E, CD33, CD2AP, PTK2B, SORL1, SLC24A4/RIN3, MEF2C
Tau pathology	BIN1, CASS4, FERMT2
Cell migration	PTK2B
Hippocampal synaptic function	MEF2C, PTK2B
Cytoskeletal function and axonal transport	CELF1, NME8, CASS4
Microglial and myeloid cell function	INPPD5

amyloid precursor protein (APP) [7]. The accumulation of Aβ(beta) peptides appears to be an early event that triggers a series of downstream events (e.g., the misprocessing of the tau protein and brain inflammation) [8, 9]. Of note, there is evidence that cases with early-onset AD have a greater tau accumulation in the parieto-occipital brain cortex than cases with late-onset AD [10].

In this chapter, we focus on the current state of knowledge of the early-onset type of AD. In contrast to late-onset AD, which is a more heterogeneous complex disorder with a heritability of up to 70% [11, 12], early-onset AD has a heritability between 92 and 100% and is almost entirely genetically determined [12]. Up to 60% of early-onset AD patients have at least one affected first-degree relative [13–15], and in 10–15% of these families, the mode of inheritance is autosomal dominant, in which the disease results from one copy of a mutant allele on an autosomal chromosome and patients have a 50% chance of passing the disease-causing mutation to children. So far, genetic analyses of extended pedigrees with an autosomal dominant mode of inheritance of early-onset AD have identified three causal genes, *APP* (MIM: *104760) [16], *PSEN1* (MIM: *104311) [17], and *PSEN2* (MIM: *600759) [18, 19]. *PSEN1* mutations are the most frequent defects in autosomal dominant AD (76%) followed by *APP* (19%) and *PSEN2* mutations (5%) (http://www.molgen.ua.ac.be/ADmutations/).

In addition to the causative genes, multiple genetic variants have been identified in association studies including candidate-gene and genome-wide tests. The major pathways identified by genomic studies of AD include amyloid pathway, inflammation, synaptic function, and lipid metabolism (Table 3.1). The variants in these genes are known to increase or decrease the risk of sporadic AD or modify the disease age at onset, among them the e4-allele of *APOE* (MIM: *107741) has the largest risk effect [20]. For instance, a

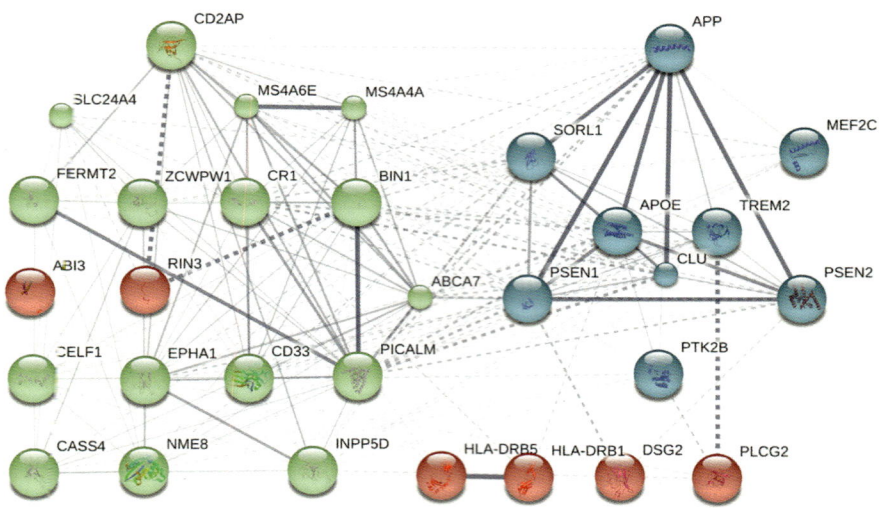

Fig. 3.1 Network of 31 well-confirmed AD genes using the STRING database of known and predicted protein-protein interactions (https://string-db.org/). Each node represents all the proteins produced by a single gene locus (e.g., splice isoforms). Line thickness indicates the strength of data supporting the connections between the genes according to the STRING database (e.g., gene co-expression). Large and small nodules represent proteins with known and unknown structures, respectively. STRING's k-means clustering function was applied to cluster the network, in which every color corresponds to a cluster and intercluster edges are represented by dashed lines

reproducible association between *SORL1* and late-onset AD was discovered by a candidate-gene approach [21]; and large genome-wide association studies (GWASs) have identified a link between sporadic late-onset AD and common polymorphisms in *CLU, PICALM, BIN1, MS4A4A/MS4A6E, CR1, CD2AP, CD33, EPHA1,* and *ABCA7* [22–26]. Furthermore, the largest meta-analysis of 74,538 individuals from published GWASs has confirmed AD risk associated with *SORL1* and identified 11 additional AD loci (*HLA-DRB5/HLA-DRB1, PTK2B, SLC24A4/RIN3, DSG2, INPP5D, MEF2C, NME8, ZCWPW1, CELF1, FERMT2,* and *CASS4*) [27]. Finally, a recent case-control study of ~85,000 subjects using a whole-exome microarray identified AD-associated rare missense variants in *PLCG2, ABI3,* and *TREM2*, strongly implicating microglial-mediated innate immunity in AD [28]. Predicted network of 31 well-confirmed AD genes is presented in Fig. 3.1.

The goal of this chapter is to provide an overview of the genetics of early-onset AD and explore how this knowledge can be applied in clinical practice. *APP, PSEN1,* and *PSEN2* are responsible for most reported large families with early-onset AD. However, up to 90% of early-onset AD cases, in particular cases without strong family history or families with a mix of early- and late-onset cases, are still genetically unexplained and currently investigated by whole-genome or whole-exome sequencing for discovery of novel causative variants.

APP

APP is a ~300 Kb gene located on the minus strand of chromosome 21 at position chr21:27,252,861-27,543,138 according to the February 2009 human reference sequence (GRCh37). It is composed of 18 exons encoding the APP, processing of which generates Aβ(beta) peptides. In addition to the major APP isoform that consists of 770 amino acids (AA), at least ten other isoforms are predicted as a result of alternative splicing (UniProt # P05067).

Knowledge about APP biology is critical for understanding the genetic basis of AD. In part, APP is expressed as a cell surface receptor and involved in neurite growth, neuronal adhesion, and axonogenesis [29]. In addition, APP accumulates in secretory transport vesicles leaving the late Golgi compartment and returning to the cell surface. APP is also implicated in cell mobility and transcriptional regulation [21, 29]. During maturation, APP (N-glycosylated in the endoplasmic reticulum) moves to the Golgi complex where complete maturation occurs (O-glycosylated and sulfated) [29]. The major secretory pathway is in fact non-amyloidogenic. Cleavage by either α(alpha)-, β(beta)-, or θ(theta)-secretase leads to the generation and extracellular release of soluble APP fragments, S-APP-alpha (18-687 AA) and S-APP-beta (18-671 AA), and retention of the corresponding membrane-anchored C-terminal fragments C83 (688-770 AA) and C99 (672-770 AA). Subsequent processing of C83 by γ(gamma)-secretase yields P3 peptides (688-713 or 688-711 AA) (http://www.uniprot.org). Several neuroprotective properties were reported for the soluble APP ectodomain fragments released from the cell surface by α(alpha)-secretases [30], which are zinc metalloproteinases that include members of the ADAM protein family [31].

The key component of amyloid plaques (Aβ(beta)) is derived from the sequential cleavage of APP by β(beta)- and γ(gamma)-secretase. Amyloidogenic beta-site APP-cleaving enzyme (BACE), also known as β(beta)-secretase, creates the C99 fragment, which is subjected to γ(gamma)-secretase, releasing Aβ(beta)40 (672-711 AA) or Aβ(beta)42 (672-713 AA), in addition to the cytotoxic C-terminal fragments γ(gamma)-CTF50 (721-770 AA), γ(gamma)-CTF57 (714-770 AA), and γ(gamma)-CTF59 (712-770 AA). Notably, in contrast to γ(gamma)-secretase, BACE is not genetically involved in AD, including the early-onset form of the disease [32, 33]. Yet, pathological *APP* mutations adjacent to the β(beta)-secretase cleavage site (671-672 AA) upregulate cleavage by β(beta)-secretase, increasing the generation of both Aβ(beta)40 and Aβ(beta)42 [34].

Aβ(beta)42 is not an abundant peptide, occurring in about a tenth of the amount of Aβ(beta)40, but is known to be more pathogenic as it aggregates faster and has been shown in cell culture assays to be more toxic than Aβ(beta)40 [35, 36]. Of note, *APP* mutations around the γ(gamma)-secretase cleavage site (711-721 AA) result in a modification of enzyme activity enhancing the production of Aβ(beta)42 [34]. Notably, the different fibril structures of the Aβ(beta) peptides could be responsible for different clinical AD subtypes [37]. Many other minor Aβ(beta) peptides, including the Aβ(beta)X-15 peptides, are produced by α(alpha)-secretase cleavage at 687-688 AA. In addition, cleavage at Asp739 by either caspase-6, -8, or -9 results

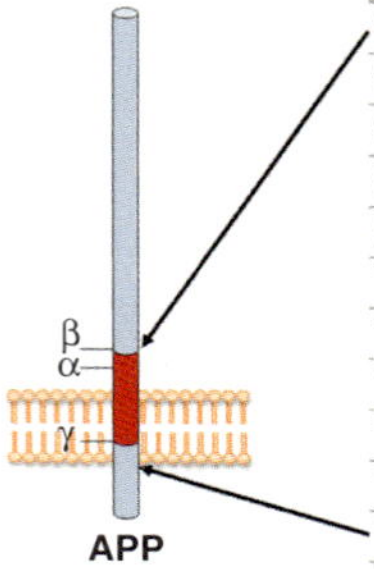

Fig. 3.2 Representative Alzheimer's disease mutations in *APP* are localized in or around the Aβ(beta) peptide domain (exons 16 and 17)

Codon	Mutation	Phenotype
670;671	Lys→Asn; Met→Leu	AD
673	Ala→Thr	Protective
692	Ala→Gly	AD
693	Glu→Gln	Haemorrhage
694	Asp→Asn	Haemorrhage
713	Ala→Thr	AD
714	Thr→Ile	AD
715	Val→Met	AD
716	Ile→Val	AD
717	Val→Ile/Phe/Gly	AD
723	Leu→Pro	AD
Entire Locus	Duplication	AD/Stroke

in the production of the neurotoxic C31 peptide and increased production of the Aβ(beta) peptides [38, 39].

The first AD mutation in *APP* (Val717Ile) affecting the transmembrane (TM) domain of the protein (700-723 AA) was reported in 1991 [16]. It segregated with AD in a large autosomal dominant UK family. Currently, there are 51 different pathological *APP* mutations observed in 121 families (http://www.molgen.ua.ac.be/ADMutations). In addition to missense *APP* mutations, there are 26 duplications overlapping *APP* with up to five neighboring genes. Mutant APP protein serves as an improved substrate for γ(gamma)- or β(beta)-secretase, leading to the overproduction of Aβ(beta) peptides [40], explaining why all pathologic missense mutations are clustered within *APP* exon 16 or 17 where the cleavage sites are located (Fig. 3.2).

The Mutation Database available at http://www.molgen.ua.ac.be/ADMutations [41] allows a quick evaluation of known mutations for evidence of co-segregation with AD, as well as their biological consequences in cell culture. Of note, most of the mutations (61%) implicated in major neurodegenerative disorders are reported in a single family, and only ~6% of them were described in more than ten families [41]. However, the number of affected families is underestimated, since the literature is biased toward novel findings.

In addition to the amyloid plaques in brain parenchyma, amyloid deposition is also often observed in brain blood vessels, called amyloid angiopathy. In particular, several AD families with duplications at the APP locus have been reported to have cerebral amyloid angiopathy (http://www.molgen.ua.ac.be/ADMutations) that frequently gets complicated by stroke [42, 43]. In fact, prior to the discovery of the first AD mutation [16], independent studies had suggested the *APP* Dutch mutation (Glu693Gln) to be responsible for an alternative phenotype described as "hereditary cerebral hemorrhage with amyloidosis-Dutch type" [44–46], which clinically manifested with stroke and was pathologically characterized by the deposition of Aβ(beta) in the leptomeningeal arteries and cortical arterioles, in addition to amyloid plaques [47]. Notably, codon Glu693 is located in a mutation hot spot of APP. For instance, the Glu693Lys substitution has been reported in Italian families with multiple strokes followed by epilepsy and cognitive decline [48], while the Arctic mutation at the same residue (Glu693Gly) is known to enhance Aβ(beta) proto-fibril formation and was reported in patients with typical AD without the severe cerebral amyloid angiopathy [49, 50]. Finally, a deletion of Glu693 has been reported in Japanese AD pedigrees leading to increased Aβ(beta) oligomerization rather than Aβ(beta) fibrilization [51]. Notably, missense substitutions at residues adjacent to Glu693, such as the Flemish (Ala692Gly) and Iowa (Asp694Asn) mutations, are also associated with cerebral amyloid angiopathy [52–56]. However, *APP* mutations are not a common cause of amyloid angiopathy in sporadic AD patients [57].

The pathogenic consequences of different *APP* mutations often depend on their position relative to the secretase sites (Fig. 3.2). For example, 11 *APP* mutations that occur between residues 714 and 717 (http://www.molgen.ua.ac.be/ADMutations) affect the site of γ(gamma)-secretase cleavage and cause AD by increasing the level of Aβ(beta)42 [58]. Indeed, the Thr714Ile, Val715Met, Val715Ala, Ile716Val, Ile716Phe, and Val717Ile mutations all increase Aβ(beta)42 levels and decrease Aβ(beta)40, raising the Aβ(beta)42/Aβ(beta)40 ratio [59–63]. In contrast, this ratio is not changed for the Swedish mutation (Lys670Asn/Met671Leu), which is located at the β(beta)-secretase site [50], since it elevates the level of both Aβ(beta)40 and Aβ(beta)42 [64, 65] and increases the rate of Aβ(beta) fibrilization [66]. Described 25 years ago, the Lys670Asn/Met671Leu mutation has only been observed in two large AD families from Sweden [67]. Yet, despite the rare frequency of this variant, mouse models generated based on the Swedish mutation have been extensively studied worldwide, including the search for AD treatments.

Of note, the *APP* Ala713Val variant located at the γ(gamma)-secretase cleavage site is not considered to be pathogenic. Hence, only specific conformational changes in the protein structure of APP lead to AD. This could explain the rarity of AD cases explained by *APP* mutations. Another important observation is that the Ala673Thr substitution, adjacent to the β(beta)-secretase site, appears to have a protective effect against late-onset AD (Fig. 3.2), with a ~40% reduction in the formation of amyloidogenic peptides in vitro. This suggests that reducing β(beta)-secretase cleavage of APP may protect against the disease [68]. This protective variant was detected by whole-genome sequencing of 1795 Icelanders in which the Thr-allele was significantly more frequent in controls (0.6%) than in AD patients (0.1%) (p-value = 5×10^{-7}). However, the study of the Ala673Thr frequency in a very large independent dataset (~10,000 AD cases and ~10,000 controls) concluded that it does not play a substantial role in AD risk in the North American population and may be restricted to Icelandic and Scandinavian populations [69].

Intriguingly, another substitution at the same codon (Ala673Val) has a dominant-negative effect on amyloidogenesis and causes AD only in a homozygous state [70]. Intriguingly, it was reported that an increased genome-wide average length of runs of homozygosity (ROH), which could harbor novel recessive mutations, is significantly associated with AD among an inbred Caribbean Hispanic population [71]. Importantly, the frequency of AD in Caribbean Hispanics is three times greater compared to non-Hispanics in the same community. The ongoing deep sequencing of the significant loci affected by ROH could detect novel recessive AD genes, which can then be tested in different ethnic groups.

In several neuropsychiatric disorders, genome-wide global burden measurements of copy number variations (CNVs), defined as genomic deletions or duplications ranging from 1 Kb to several Mb, are known to be important disease contributors. As mentioned earlier, the AD inheritance in some families is explained by duplications overlapping the entire *APP* locus (ranging from 0.6 to 6.4 Mb) [42, 43]. However, the role of CNVs has only recently begun to be systematically explored in AD [72]. For instance, an 18 Kb insertion in *CR1* (responsible for the CR1-S isoform) increases AD risk by twofold and explains the GWAS signals at the *CR1* locus [73]. A study of an early-onset AD dataset (including 261 families) revealed five deletions and five duplications that segregated with dementia [74], with two CNVs encompassing FTD genes (deletion of *CHMP2B* and duplication of *MAPT*); however, such findings could reflect the presence of FTD cases that are clinically misdiagnosed with AD. For 6 of the 10 CNVs, the *APOE* e4-allele also co-segregated with AD, suggesting that the genes affected by the CNVs and *APOE* could act together to modify AD risk. There is also some evidence that *APOE* alleles can modify the severity of AD in cases with *APP* mutations [75]. Indeed, cultured cells with an e4/e4 genotype are more vulnerable to Aβ(beta) than cells with the e3/e3 or e3/e4 genotypes [76].

PSEN1

In 1992, the 14q24 locus was linked to autosomal dominant AD [77, 78] and 3 years later was explained by mutations in *PSEN1* [17] that spans ~87 Kb on chr14:73,603,143-73,690,399 (GRCh37), including ten coding and two noncoding exons. The PSEN1 protein (467 AA for the longest isoform) (UniProt # P49768) is ubiquitously expressed [79], with the highest expression in the brain, whole blood, skin, and intestines, according to the database of Genotype-Tissue Expression (GTEx). PSEN1 has several highly conserved TM domains and a less conserved cytoplasmic hydrophilic loop domain.

PSEN1 is the catalytic subunit of the γ(gamma)-secretase complex [80] that cleaves multiple type-1 membrane proteins, including APP and Notch receptors. It is incorporated into the γ(gamma)-secretase complex together with three other critical components: NCSTN, APH1, and PEN2 [81]. Several endogenous proteins have been reported to selectively modulate the function of this complex, including TMP21, CD147 antigen (basigin), and gSAP [82]. AD-associated *PSEN1* mutations modify the conformation of the γ(gamma)-secretase complex, which increase production of Aβ(beta)42 [83]. Brain pathology of AD cases revealed that *PSEN1* and *PSEN2* mutations are associated with higher levels of insoluble Aβ(beta)42 (and to a lesser extent insoluble Aβ(beta)40) in comparison to sporadic AD cases [84–86].

PSEN1, PSEN2, and the signal peptide peptidases (SPPs) are proteases with a highly conserved GlyXGlyAsp motif including membrane-embedded Asp residues that are critical for their enzymatic activity [87]. In contrast to PSEN1 and PSEN2, SPPs cleave the TM region of type-2 but not type-1 membrane proteins. An AD-related Gly384Ala mutation in *PSEN1* (adjacent to the critical Asp in the

GlyXGlyAsp motif) causes a selective loss-of-function by slowing Aβ(beta)40 production, while the generation of Aβ(beta)42 remains unaffected [88], which is consistent with the effect of several *APP* mutations that result in an increased Aβ(beta)42/Aβ(beta)40 ratio [59–63].

Thus far, 219 different *PSEN1* mutations causing AD have been reported in ~480 families (http://www.molgen.ua.ac.be/ADMutations/); a majority of them are missense substitutions with only a few in-frame deletions or insertions [89–91]. Notably, sequencing of the DNA isolated from the histopathological slides of the first AD patient (Auguste Deter) reported by Dr. Alzheimer a century ago has revealed a novel Phe176Leu mutation in *PSEN1* explaining her early onset of AD (at age 51) [92].

Most of the reported coding *PSEN1* variations are pathogenic with just a few exceptions (e.g., Arg35Gln, Phe175Ser, and Val191Ala). Furthermore, the *PSEN1* variations that are recognized as benign polymorphisms could potentially increase AD risk in the presence of other AD risk factors. For instance, the Glu318Gly variation used to be categorized as nonpathogenic due to its presence in ~3% of controls and the absence of co-segregation with AD. However, an investigation of AD patients with extreme levels of cerebrospinal fluid biomarkers (Aβ(beta)-42, tau, and p-tau) revealed that the Glu318Gly variation increases the risk for AD through a gene-gene interaction with *APOE* [93]. Glu318Gly carriers who are also heterozygous for the *APOE* e4-allele have an AD risk similar to *APOE* e4 homozygotes. Such individuals have higher levels of neuronal degeneration and Aβ(beta) deposition and faster cognitive decline.

On average, *PSEN1* mutation carriers have an earlier onset and shorter duration of AD compared to carriers of *APP* or *PSEN2* mutations. AD symptoms in *PSEN1* mutation carriers could appear as early as the third decade of life (e.g., Ser170Phe) [94]. In contrast to *APP*, *PSEN1* mutations are broadly distributed throughout the gene, with the exception of the first three exons. In both *PSEN1* and *PSEN2*, exons 5–7 are the most frequently affected by missense mutations. In addition, genomic deletions of exon 9 and intronic mutations leading to its aberrant splicing are relatively frequent *PSEN1* variations (http://www.molgen.ua.ac.be/ADMutations/). Another splicing aberration in *PSEN1* is caused by a mutation in intron 4 [90, 91].

Identification of a common founder *PSEN1* mutation (Ala431Glu) has helped provide genetic counseling advice to Mexican AD patients [95, 96]. Also, *PSEN1* mutation Glu206Ala was responsible for 42% of early-onset AD families of Caribbean Hispanic origin [97]. This mutation shows a wide range of age at onset (22–77 years) which is neither explained by *APOE* e4-allele nor any antecedent environmental, health-related, or social factors, suggesting that the age of onset might be modified by other genetic factors [98]. Also, the high incidence of late-onset AD among Caribbean Hispanics is not explained by *PSEN1* [97, 99] and might be related to recessive mutations hidden within genomic ROH [71].

The phenotypic heterogeneity in *PSEN1* patients includes variant AD (vAD), in which dementia is accompanied by spastic paraparesis (MIM: *607822), a progressive spastic weakness of the limbs associated with axonal degeneration in the corticospinal tract and dorsal columns [100]. The brain pathology of these cases is

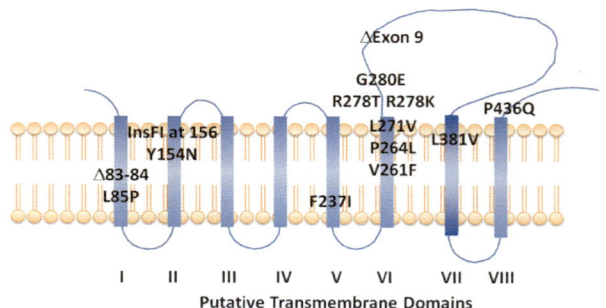

Fig. 3.3 *PSEN1* mutations reported to be associated with variant Alzheimer's disease presenting with spastic paraparesis

characterized by large, abundant, diffuse, Aβ(beta)-positive "cotton wool" plaques without the features of mature plaques (e.g., a congophilic core, neuritic pathology, and signs of inflammation) [89]. The "cotton wool" plaques can also be observed in rare cases of late-onset AD [101]. Furthermore, in a Japanese patient with early-onset familial AD, spastic paraparesis, and apraxia (inability to voluntarily perform certain movements), the PSEN1 Gly266Ser was identified [102]. Neuropathologic data was not available for this patient.

Mutations associated with vAD (Fig. 3.3) are broadly distributed within *PSEN1* (codons 83–436) and reported in families with variable ages of onset (24–51 years) [100]. Importantly, identical *PSEN1* mutations have been reported in families with either classical AD or vAD (e.g., the deletion of exon 9 in Finnish families [89, 103]). Surprisingly, even within a single family, a spectrum of disease phenotypes have been reported (e.g., AD, vAD, and pure spastic paraparesis) [104]. Cumulatively, these observations argue in favor of a genetic modifier responsible for vAD, the search for which is still ongoing. Currently, a modifier effect for coding variations in several spastic paraparesis genes has been excluded, including *ZFYVE26* which was selected as a promising gene candidate because it maps nearby *PSEN1* on chromosome 14 [105].

In addition to AD, *PSEN1* mutations have been implicated in several other disorders including the clinically overlapping phenotypes of FTD (MIM: *600274) and Pick disease (MIM: *172700), as well as phenotypically nonoverlapping disorders such as familial acne inversa (MIM: *613737) and dilated cardiomyopathy (MIM: *613694). For instance, the Asp333Gly mutation was found in a familial form of dilated cardiomyopathy, a disorder characterized by heart ventricular dilation resulting in congestive heart failure and arrhythmia [106]. The Gly183Val mutation was reported in patients with Pick-type tauopathy, while Ile211Met was associated with posterior cortical atrophy (a dementia with cortical visual dysfunction) [107, 108]. Furthermore, some *PSEN1* mutations have been reported in primary progressive aphasia and spinocerebellar ataxia [109, 110]. However, autopsy results and additional genetic analyses are important for clarification of these observations.

For example, the insArg352 in *PSEN1* was first detected in an AD patient [90], but the clinical diagnosis was later changed to FTD. In vitro study demonstrated that the insArg352 variation inhibits γ(gamma)-secretase cleavage of APP and Notch, suggesting that the loss-of-function of γ(gamma)-secretase may result in neurodegeneration [111]. However, subsequent analysis of *GRN* in this patient identified a pathogenic frameshift mutation responsible for FTD; and the autopsy results confirmed the diagnosis of FTD [112]. Hence, the insArg352 variation is most likely a benign polymorphism with regard to neurodegeneration. Another *PSEN1* variant (Leu113Pro) was described in a French family with six members affected with early-onset dementia across four generations, and three of them fulfilled the criteria for a clinical diagnosis of FTD [113]. There was no neuropathological data from this family, but neuroimaging data from two mutation carriers was consistent with a diagnosis of FTD.

PSEN1 is considered very intolerant to loss-of-function variants, and *Psen1*$^{-/-}$ mice die shortly after birth [114]. This notion is in agreement with genotype data from >60,000 unrelated individuals sequenced at The Exome Aggregation Consortium in which the frequency of *PSEN1* loss-of-function variants is significantly less than expected based on gene length (*PSEN1* pLI score 1.00; http://exac.broadinstitute.org/gene/ENSG00000080815).

Nevertheless, heterozygous loss-of-function mutations in *PSEN1* and other genes encoding the components of the γ(gamma)-secretase complex (*NCSTN* and *PEN2*) were reported in autosomal dominant Chinese families with acne inversa (an inflammatory disorder of the hair follicles), which develops after puberty (MIM:613737) [115]. Notably, patients with such mutations (15–81 years old) had no symptoms of dementia. Acne inversa is likely the result of compromised Notch signaling due to a 50% decrease in γ(gamma)-secretase activity in carriers of heterozygous mutations. The worldwide prevalence of acne inversa is up to 4% including 40% familial cases. Importantly, follow-up reports of γ(gamma)-secretase loss-of-function mutations in different ethnic groups confirmed that acne inversa is an allelic disorder of early-onset AD.

PSEN2

PSEN2 is a ~26 Kb gene with ten coding and three noncoding exons. It is located on chr1:227,058,273-227,083,804 (GRCh37) at the 1q42 locus and encodes the PSEN2 protein (448 AA for the longest isoform) (UniProt# P49810) that shares substantial structural and sequence similarities with PSEN1, apart from the cytoplasmic loop domain. PSEN2 also has the conserved GlyXGlyAsp motif with the Asp366 critical for γ(gamma)-secretase activity [116]. The second critical Asp is located at codon 263.

PSEN2 acts as a catalytic subunit of the γ(gamma)-secretase complex independent of PSEN1, but is contingent on the three other critical components (PEN2, NCSTN, and APH1), similar to PSEN1 complex. Notably, PSEN2-dependent γ(gamma)-secretase activity is predominant in microglia and modulates the release of proinflammatory cytokines [117]. Furthermore, both PSEN1 and PSEN2 are localized to the nuclear membrane and are considered to be involved in cell cycle regulation and mitosis [118].

PSEN2 was cloned right after *PSEN1* in 1995 [18, 19]; however, currently only 16 different mutations have been reported among 34 families, affecting *PSEN2* exons

4–7 and 12 (http://www.molgen.ua.ac.be/ADMutations). Nevertheless, in some populations *PSEN2* variants could be as frequent as *PSEN1* (e.g., in Volga Germans [119]). *PSEN2* mutation carriers have a less severe AD phenotype compared to *PSEN1* mutation carriers [17, 120], which in part could be explained by the lower brain expression of *PSEN2* vs. *PSEN1* [19]. Some *PSEN2* mutations are found in patients with late-onset AD (e.g., Val148Ile) [121], and some (e.g., Thr430Met) were observed in patients with a wide range of age at onset suggesting the action of unknown modifier gene(s) [122]. A genome-wide search for loci influencing the age at onset within nine families affected by the most common *PSEN2* substitution (Asn141Ile; the Volga German founder mutation) has revealed several candidate modifier loci in addition to *APOE* (1q23.3, 17p13.2, 7q33, and 11p14.2) [123].

Several reports have documented the phenotypic heterogeneity of *PSEN2* patients. For instance, the Ser130Leu substitution segregated with dilated cardiomyopathy in two families (similar to the Asp333Gly in *PSEN1*) [106]. In addition, the Ala85Val mutation was detected in a patient with a clinical and neuropathological phenotype indicative of dementia with Lewy bodies (MIM: *127750) [124], which is closely associated with both AD and Parkinson's disease. Lewy bodies are neuronal inclusions mainly consisting of α(alpha)-synuclein. Intriguingly, Lewy bodies were present in brains of 12 out of 19 (63%) AD cases affected with *PSEN1* or *APP* mutations [125], a neuropathological finding that remains to be explained.

Unexplained Early-Onset AD

Together, mutations in *APP*, *PSEN1*, and *PSEN2* account for less than 10% of early-onset AD cases. Most families enriched for early-onset AD do not suggest Mendelian inheritance and often have a mix of early- and late-onset cases. The wide range of age of onset likely indicates genetic modifier(s) modulating AD susceptibility. The few studies that have assessed this early-onset AD subtype have suggested a genetic architecture partially overlapping the late-onset form [14, 68, 126–129]. Thus, studying early-onset AD in subjects without *APP*, *PSEN1*, and *PSEN2* mutations would cover a critical gap and provide a unique opportunity for discovery of novel targets and disease pathways. Furthermore, the availability of new genetic markers will help refine different genetic signatures of clinical AD, allowing for a more accurate stratification of preclinical or clinical patient cohorts for medical research and clinical trials. In the long term, the ability to identify different underlying molecular pathologies of AD patients or at-risk individuals will pave the way for personalized medicine and health care (e.g., precision medicine).

Genetic Testing and Search for Novel AD Genes or Disease Modifiers

Mutation analyses for AD-causing genes are usually performed by Sanger sequencing of either the entire coding region (for *PSEN1* and *PSEN2*) or selected exons (mutation hot spot) and gene-dosage assessment (for *APP*). Diagnostic testing for the three causal AD genes is currently established in >20 certified laboratories in Europe and the USA (http://www.ncbi.nlm.nih.gov/gtr/).

Considering the variable clinical presentation associated with mutations in causal AD genes, there is still a need to improve understanding of the genotype-phenotype correlation in order to provide the best medical advice to mutation carriers. For instance, there is a concern for genetic counseling when novel mutations are detected. Without evidence of co-segregation of the mutation with AD, the analysis of Aβ(beta) levels in cell cultures from patients could provide functional support for the pathological impact of the mutation [130]. In addition, genotyping biological parents of a patient bearing a novel variant to confirm a de novo variant could support the pathogenic nature of the variant, as was shown for the *PSEN1* Val391Gly mutation [130, 131].

The detection of causal AD mutations could help evaluate the efficacy of therapies at either the asymptomatic phase or early stages of dementia. Since 2012, about 210 carriers of pathogenic *APP*, *PSEN1*, or *PSEN2* mutations have been enrolled in a longitudinal clinical trial, named the Dominantly Inherited Alzheimer Network (DIAN) (http://clinicaltrials.gov/ct2/show/NCT01760005). This trial is estimated to be completed in 2019 and has the goal of assessing the tolerability and biomarker efficacy of two potential modifying AD treatments, which are based on monoclonal antibodies that either bind to aggregated Aβ(beta) (gantenerumab) or soluble Aβ(beta) (solanezumab). It enrolls individuals 18–80 years of age (within 10–15 years of the anticipated age of onset) who either know or are unaware of their genetic status but have a 50% chance of inheriting such a mutation (first-degree relatives of a mutation carrier). Importantly, the outcome of the early intervention in the DIAN study and/or the discovery of AD biomarkers could have important implications for the treatment of the common sporadic form of AD.

The DIAN cohort allows for the search of genetic modifiers affecting AD phenotype. For instance, it was recently reported that the presence of the common Val66Met polymorphism in *BDNF* (rs6265), known to be responsible for a 30% reduction in BDNF secretion, modulates AD-related endophenotypes in individuals with *APP*, *PSEN1*, or *PSEN2* mutations [132]. The BDNF Met66 allele was associated with increased tau and p-tau levels in the CSF, as well as impairments in hippocampal metabolism and episodic memory. However, currently, no conclusions can be drawn on the relationship between the Val66Met substitution and subtypes of familial AD (e.g., *PSEN1* vs. *APP* mutation carriers). Also, the precise mechanism by which the Val66Met polymorphism causes differences in AD pathophysiology remains to be identified.

Another significant contribution of the DIAN cohort has been the functional investigation among carriers of mutations in *TREM2* (microglial gene associated with late-onset AD [133]). It was shown that the soluble ectodomain of TREM2 is increased in the CSF of mutation carriers 5 years before the expected symptom onset compared to noncarriers [134]. This study suggests that microglial activation occurs several years before disease onset, but after amyloidosis and neuronal injury.

The AD genes identified by GWASs could potentially modify the phenotype of patients with autosomal dominant early-onset AD. Therefore, new sequencing technologies are applied to identify rare deleterious variants within these genes.

Alternatively, known variants in genes associated with AD can be analyzed using cost-effective exome arrays, such as NeuroX [135] or the recently developed NeuroChip [136]. The cellular consequences of these variants can be further investigated in functional studies of patient cells harboring deleterious mutations or alternatively on animal models made by mutagenesis using data from sequencing projects [137–139]. The functional connections between known AD genes are currently ambiguous; however, these genes could be subdivided into a few categories (e.g., Aβ(beta) production, lipid/cholesterol metabolism, inflammation, vesicular trafficking, and synaptic function), with some genes fitting into several categories (e.g., *CLU*, which was implicated in both cholesterol metabolism and inflammation). Determining which genes or gene networks contribute to AD risk could reveal basic pathogenic mechanisms important for potential treatment (Fig. 3.1).

However, drug development is timely, costly, and burdened by a low success rate, while an attractive alternative strategy is drug repositioning. For instance, we recently analyzed publicly available "omics" data, including genomics, and generated a list of 524 anti-AD protein targets, 18 of which are targets for 75 existing drugs, including drugs modulating neuroinflammation that are particularly promising for AD intervention [140].

The source of the missing heritability could also be rare variants (allele frequency < 1%) that are not captured by GWAS. Rare variants are broadly distributed throughout the human genome; hence, genome-wide sequencing approaches are needed to determine which of them are related to AD. Of note, rare variants are not limited to Mendelian-type inheritance and along with common variants may affect risk for AD.

Recent advances in next-generation sequencing technology have provided a cost-effective approach to large-scale resequencing of the entire genome as a strategy for identifying risk or protective alleles [141, 142]. Notably, coding variations that constitute ~85% of known disease-causing mutations [143] can be captured with exome sequencing, which covers ~1% of the human genome (30 Mb; 180,000 exons), therefore reducing the time and cost of searching for highly penetrant variants in noncoding regions [144, 145]. The price of whole-genome sequencing however has also significantly dropped over the past few years enabling cost-effective analysis of large datasets (i.e., Illumina NovaSeq). Whole-exome sequencing identified AD-associated coding variants in *TREM2* [133] and revealed nonsense or missense mutations in *SORL1* in 7 of 29 unrelated cases from early-onset autosomal dominant AD families [127]. Whole-genome sequencing studies of large multiplex families are currently ongoing [146, 147].

Importantly, the amount of raw data produced by next-generation sequencing is enormous, and many computational steps are required to translate this output into reliable variant calls. Surprisingly, a large number of loss-of-function variants (~100 per genome) are identified in apparently healthy individuals [148]. Hence, the identified variants must be systematically filtered using high-quality catalogues of variants identified in the genomes of healthy individuals (e.g., from the gnomAD database http://gnomad.broadinstitute.org/). However, the age of individuals included in public databases is an important consideration for diseases with adult onset (e.g., AD). Some additional tools such as SIFT and Polyphen are also used to annotate the variants and predict their functional consequences, which can help to prioritize the variants selected for follow-up association or segregation studies. Further, in large extended families, linkage analyses can be employed prior to next-generation sequencing to identify chromosomal regions that most likely contain causative variants, which can then be prioritized in the analysis of the sequence data. Next-generation sequencing of two, distantly related, affected members of a family can be sufficient to identify shared, causative genetic variants residing in the linkage regions; additional availability of older unaffected family members allows for the validation of variants through demonstration of co-segregation. The significant advantage of combining linkage and next-generation sequencing analysis is the decrease in complexity of sequencing data to be analyzed.

Importantly, reports on the detection of rare variants with a large effect size need to be validated by independent studies, which might take time. For instance, whole-exome sequencing of large AD families with follow-up analyses of candidate variants in several large AD case-control datasets discovered a rare Val232Met mutation in *PLD3* segregating with AD in two autosomal dominant families [149]. This report suggested that the PLD3 Val232Met variant doubled the risk for AD and occurred more frequently in familial (~3%) than sporadic cases (~1%). Furthermore, gene-based burden analyses revealed that 14 *PLD3* variants could increase AD risk. However, so far only negative reports have been published on the association between AD and *PLD3*.

There are several study designs that can increase power for identifying AD variants. For example, an "extreme phenotype" study design could focus on AD families with a very early age of onset (e.g., <40 years), which are expected to be enriched for causative variants, although a limitation of the "extreme phenotype" approach is the difficulty of sampling a large enough number of families with a very early age of onset.

The study of trios or nuclear families (e.g., pedigrees consisting of non-demented parents and an affected offspring who is preferably negative for the *APOE* e4-allele) can allow for the identification of recessive or de novo mutations. In this study design, sequencing data of the affected offspring can be used to identify regions with ROH; subsequently these ROH regions can be analyzed together with the sequencing data of the parents to identify recessive mutations. In addition, the sequencing data can be analyzed for de novo mutations that are present in the offspring but absent in both parents. A recent study conducting exome sequencing in regions with long ROH in a consanguineous family with two siblings with early-onset AD identified a homozygous *CTSF* mutation [150]. Variants in this gene are associated with a type of adult-onset neuronal ceroid lipofuscinosis with some cases resembling the impairment seen in AD. Commonly encountered limitations in the analysis of trios or nuclear families for the identification of recessive or de novo mutations are the unavailability of both parents and the possibility that the early-onset AD in the offspring resulted from a dominant mutation that was undetected in the parent due to incomplete penetrance. Sporadic early-onset AD cases can also be explained by germ line or somatic mosaicism with the degree of mosaicism affecting onset age and clinical presentation (an example for this is the *PSEN1* Pro436Gln mutation) [151]. Finally, epigenetic mechanisms can affect gene expression and thereby the clinical presentation of AD.

Conclusion

Precision medicine could be possible in the near future as a result of the expansion in knowledge provided by the human genome project and the major advances in "omics" technologies made in the past decade including the application of next-generation sequencing. Genetics is currently widely applied to AD diagnosis, monitoring, and the search for a potential treatment. The detection of mutations in *APP*, *PSEN1*, and *PSEN2* is used for both pre-symptomatic and symptomatic AD diagnosis. In the near future, next-generation sequencing will likely uncover novel AD genes. The successful application of precision medicine to AD will demand extensive additional work to identify risk groups, the underlying pathological processes, and the development of new interventions. It will also require the significant involvement of biologists, physicians, technology developers, data scientists, and patient groups. We are only at the beginning of a broad precision medicine approach targeting the clinical and biological complexity of AD and building the evidence base needed to more effectively guide clinical practice.

Acknowledgements We are grateful for the support from the Canadian Institutes of Health Research, Wellcome Trust, Medical Research Council, National Institutes of Health, and the Canadian Consortium on Neurodegeneration in Aging.

References

1. Jonsson T, et al. Variant of TREM2 associated with the risk of Alzheimer's disease. N Engl J Med. 2013;368(2):107–16.
2. Matthews FE, et al. A two decade dementia incidence comparison from the Cognitive Function and Ageing Studies I and II. Nat Commun. 2016;7:11398.
3. Hardy J, Rogaeva E. Motor neuron disease and frontotemporal dementia: sometimes related, sometimes not. Exp Neurol. 2014;262(Pt B):75–83.
4. Mayeux R. Epidemiology of neurodegeneration. Annu Rev Neurosci. 2003;26:81–104.
5. Hebert LE, et al. Alzheimer disease in the United States (2010-2050) estimated using the 2010 census. Neurology. 2013;80(19):1778–83.
6. Stefanacci RG. The costs of Alzheimer's disease and the value of effective therapies. Am J Manag Care. 2011;17(Suppl 13):S356–62.
7. Hardy JA, Higgins GA. Alzheimer's disease: the amyloid cascade hypothesis. Science. 1992;256(5054):184–5.
8. St George-Hyslop PH, Petit A. Molecular biology and genetics of Alzheimer's disease. C R Biol. 2005;328(2):119–30.
9. Lippa CF, et al. Abeta-42 deposition precedes other changes in PS-1 Alzheimer's disease. Lancet. 1998;352(9134):1117–8.
10. Cho H, et al. Excessive tau accumulation in the parieto-occipital cortex characterizes early-onset Alzheimer's disease. Neurobiol Aging. 2017;53:103–11.

11. Wingo TS, et al. Autosomal recessive causes likely in early-onset Alzheimer disease. Arch Neurol. 2012;69(1):59–64.
12. Gatz M, et al. Role of genes and environments for explaining Alzheimer disease. Arch Gen Psychiatry. 2006;63(2):168–74.
13. Jarmolowicz AI, Chen HY, Panegyres PK. The patterns of inheritance in early-onset dementia: Alzheimer's disease and frontotemporal dementia. Am J Alzheimers Dis Other Demen. 2015;30(3):299–306.
14. van Duijn CM, et al. Apolipoprotein E4 allele in a population-based study of early-onset Alzheimer's disease. Nat Genet. 1994;7(1):74–8.
15. Campion D, et al. Early-onset autosomal dominant Alzheimer disease: prevalence, genetic heterogeneity, and mutation spectrum. Am J Hum Genet. 1999;65(3):664–70.
16. Goate A, et al. Segregation of a missense mutation in the amyloid precursor protein gene with familial Alzheimer's disease. Nature. 1991;349(6311):704–6.
17. Sherrington R, et al. Cloning of a gene bearing missense mutations in early-onset familial Alzheimer's disease. Nature. 1995;375(6534):754–60.
18. Levy-Lahad E, et al. Candidate gene for the chromosome 1 familial Alzheimer's disease locus. Science. 1995;269(5226):973–7.
19. Rogaev EI, et al. Familial Alzheimer's disease in kindreds with missense mutations in a gene on chromosome 1 related to the Alzheimer's disease type 3 gene. Nature. 1995;376(6543):775–8.
20. Raber J, Huang Y, Ashford JW. ApoE genotype accounts for the vast majority of AD risk and AD pathology. Neurobiol Aging. 2004;25(5):641–50.
21. Rogaeva E, et al. The neuronal sortilin-related receptor SORL1 is genetically associated with Alzheimer disease. Nat Genet. 2007;39(2):168–77.
22. Harold D, et al. Genome-wide association study identifies variants at CLU and PICALM associated with Alzheimer's disease. Nat Genet. 2009;41(10):1088–93.
23. Lambert JC, et al. Genome-wide association study identifies variants at CLU and CR1 associated with Alzheimer's disease. Nat Genet. 2009;41(10):1094–9.
24. Carrasquillo MM, et al. Replication of CLU, CR1, and PICALM associations with Alzheimer disease. Arch Neurol. 2010;67(8):961–4.
25. Naj AC, et al. Common variants at MS4A4/MS4A6E, CD2AP, CD33 and EPHA1 are associated with late-onset Alzheimer's disease. Nat Genet. 2011;43(5):436–41.
26. Hollingworth P, et al. Common variants at ABCA7, MS4A6A/MS4A4E EPHA1, CD33 and CD2AP are associated with Alzheimer's disease. Nat Genet. 2011;43(5) 429–35.
27. Lambert JC, et al. Meta-analysis of 74,046 individuals identifies 11 new susceptibility loci for Alzheimer's disease. Nat Genet. 2013;45(12):1452–8.
28. Sims R, et al. Rare coding variants in PLCG2, ABI3 and TREM2 implicate microglial-mediated innate immunity in Alzheimer's disease. Nat Genet. 2017;49(9):1373–84.
29. Bignante EA, et al. Amyloid beta precursor protein as a molecular target for amyloid beta-induced neuronal degeneration in Alzheimer's disease. Neurobiol Aging. 2013;34(11):2525–37.
30. Postina R. Activation of alpha-secretase cleavage. J Neurochem. 2012;120(Suppl 1):46–54.
31. Allinson TM, et al. ADAMs family members as amyloid precursor protein alpha-secretases. J Neurosci Res. 2003;74(3):342–52.
32. Nicolaou M, et al. Mutations in the open reading frame of the beta-site APP cleaving enzyme (BACE) locus are not a common cause of Alzheimer's disease. Neurogenetics. 2001;3(4):203–6.
33. Cruts M, et al. Amyloid beta secretase gene (BACE) is neither mutated in nor associated with early-onset Alzheimer's disease. Neurosci Lett. 2001;313(1–2):105–7.
34. Citron M, et al. Mutation of the beta-amyloid precursor protein in familial Alzheimer's disease increases beta-protein production. Nature. 1992;360(6405):672–4.
35. Wolfe MS. The gamma-secretase complex: membrane-embedded proteolytic ensemble. Biochemistry. 2006;45(26):7931–9.
36. Selkoe DJ, Wolfe MS. Presenilin: running with scissors in the membrane. Cell. 2007;131(2):215–21.

37. Qiang W, et al. Structural variation in amyloid-beta fibrils from Alzheimer's disease clinical subtypes. Nature. 2017;541(7636):217–21.
38. Gervais FG, et al. Involvement of caspases in proteolytic cleavage of Alzheimer's amyloid-beta precursor protein and amyloidogenic A beta peptide formation. Cell. 1999;97(3):395–406.
39. Lu DC, et al. A second cytotoxic proteolytic peptide derived from amyloid beta-protein precursor. Nat Med. 2000;6(4):397–404.
40. Sisodia SS, St George-Hyslop PH. gamma-Secretase, Notch, Abeta and Alzheimer's disease: where do the presenilins fit in? Nat Rev Neurosci. 2002;3(4):281–90.
41. Cruts M, Theuns J, Van Broeckhoven C. Locus-specific mutation databases for neurodegenerative brain diseases. Hum Mutat. 2012;33(9):1340–4.
42. Rovelet-Lecrux A, et al. APP locus duplication causes autosomal dominant early-onset Alzheimer disease with cerebral amyloid angiopathy. Nat Genet. 2006;38(1):24–6.
43. Guyant-Marechal I, et al. Intrafamilial diversity of phenotype associated with app duplication. Neurology. 2008;71(23):1925–6.
44. Levy E, et al. Mutation of the Alzheimer's disease amyloid gene in hereditary cerebral hemorrhage, Dutch type. Science. 1990;248(4959):1124–6.
45. Van Broeckhoven C, et al. Amyloid beta protein precursor gene and hereditary cerebral hemorrhage with amyloidosis (Dutch). Science. 1990;248(4959):1120–2.
46. Fernandez-Madrid I, et al. Codon 618 variant of Alzheimer amyloid gene associated with inherited cerebral hemorrhage. Ann Neurol. 1991;30(5):730–3.
47. Bornebroek M, et al. Hereditary cerebral hemorrhage with amyloidosis-Dutch type (HCHWA-D): I—a review of clinical, radiologic and genetic aspects. Brain Pathol. 1996;6(2):111–4.
48. Bugiani O, et al. Hereditary cerebral hemorrhage with amyloidosis associated with the E693K mutation of APP. Arch Neurol. 2010;67(8):987–95.
49. Kamino K, et al. Linkage and mutational analysis of familial Alzheimer disease kindreds for the APP gene region. Am J Hum Genet. 1992;51(5):998–1014.
50. Nilsberth C, et al. The 'Arctic' APP mutation (E693G) causes Alzheimer's disease by enhanced Abeta protofibril formation. Nat Neurosci. 2001;4(9):887–93.
51. Tomiyama T, et al. A new amyloid beta variant favoring oligomerization in Alzheimer's-type dementia. Ann Neurol. 2008;63(3):377–87.
52. Grabowski TJ, et al. Novel amyloid precursor protein mutation in an Iowa family with dementia and severe cerebral amyloid angiopathy. Ann Neurol. 2001;49(6):697–705.
53. Greenberg SM, et al. Hemorrhagic stroke associated with the Iowa amyloid precursor protein mutation. Neurology. 2003;60(6):1020–2.
54. Hendriks L, et al. Presenile dementia and cerebral haemorrhage linked to a mutation at codon 692 of the beta-amyloid precursor protein gene. Nat Genet. 1992;1(3):218–21.
55. Roks G, et al. Presentation of amyloidosis in carriers of the codon 692 mutation in the amyloid precursor protein gene (APP692). Brain. 2000;123(Pt 10):2130–40.
56. Kumar-Singh S, et al. Dense-core senile plaques in the Flemish variant of Alzheimer's disease are vasocentric. Am J Pathol. 2002;161(2):507–20.
57. Domingues-Montanari S, et al. No evidence of APP point mutation and locus duplication in individuals with cerebral amyloid angiopathy. Eur J Neurol. 2011;18(10):1279–81.
58. Haass C, et al. Mutations associated with a locus for familial Alzheimer's disease result in alternative processing of amyloid beta-protein precursor. J Biol Chem. 1994;269(26):17741–8.
59. Kumar-Singh S, et al. Nonfibrillar diffuse amyloid deposition due to a gamma(42)-secretase site mutation points to an essential role for N-truncated A beta(42) in Alzheimer's disease. Hum Mol Genet. 2000;9(18):2589–98.
60. De Jonghe C, et al. Pathogenic APP mutations near the gamma-secretase cleavage site differentially affect Abeta secretion and APP C-terminal fragment stability. Hum Mol Genet. 2001;10(16):1665–71.
61. Cruts M, et al. Novel APP mutation V715A associated with presenile Alzheimer's disease in a German family. J Neurol. 2003;250(11):1374–5.

62. Eckman CB, et al. A new pathogenic mutation in the APP gene (I716V) increases the relative proportion of A beta 42(43). Hum Mol Genet. 1997;6(12):2087–9.
63. Herl L, et al. Mutations in amyloid precursor protein affect its interactions with presenilin/gamma-secretase. Mol Cell Neurosci. 2009;41(2):166–74.
64. Citron M, et al. Excessive production of amyloid beta-protein by peripheral cells of symptomatic and presymptomatic patients carrying the Swedish familial Alzheimer disease mutation. Proc Natl Acad Sci U S A. 1994;91(25):11993–7.
65. Perez RG, Squazzo SL, Koo EH. Enhanced release of amyloid beta-protein from codon 670/671 "Swedish" mutant beta-amyloid precursor protein occurs in both secretory and endocytic pathways. J Biol Chem. 1996;271(15):9100–7.
66. Kirkitadze MD, Condron MM, Teplow DB. Identification and characterization of key kinetic intermediates in amyloid beta-protein fibrillogenesis. J Mol Biol. 2001;312(5):1103–19.
67. Mullan M, et al. A pathogenic mutation for probable Alzheimer's disease in the APP gene at the N-terminus of beta-amyloid. Nat Genet. 1992;1(5):345–7.
68. Jonsson T, et al. A mutation in APP protects against Alzheimer's disease and age-related cognitive decline. Nature. 2012;488(7409):96–9.
69. Wang LS, et al. Rarity of the Alzheimer disease-protective APP A673T variant in the United States. JAMA Neurol. 2015;72(2):209–16.
70. Di Fede G, et al. A recessive mutation in the APP gene with dominant-negative effect on amyloidogenesis. Science. 2009;323(5920):1473–7.
71. Ghani M, et al. Evidence of recessive Alzheimer's disease loci in Caribbean Hispanics: genome-wide survey of runs of homozygosity. JAMA Neurol. 2013;70:1261–7.
72. Ghani M, et al. Genome-wide survey of large rare copy number variants in Alzheimer's disease among Caribbean hispanics. G3 (Bethesda). 2012;2(1):71–8.
73. Hazrati LN, et al. Genetic association of CR1 with Alzheimer's disease: a tentative disease mechanism. Neurobiol Aging. 2012;33(12):2949.e5–12.
74. Hooli BV, et al. Rare autosomal copy number variations in early-onset familial Alzheimer's disease. Mol Psychiatry. 2014;19(6):676–81.
75. St George-Hyslop P, et al. Alzheimer's disease and possible gene interaction. Science. 1994;263(5146):537.
76. Wilhelmus MM, et al. Apolipoprotein E genotype regulates amyloid-beta cytotoxicity. J Neurosci. 2005;25(14):3621–7.
77. Schellenberg GD, et al. Genetic linkage evidence for a familial Alzheimer's disease locus on chromosome 14. Science. 1992;258(5082):668–71.
78. St George-Hyslop P, et al. Genetic evidence for a novel familial Alzheimer's disease locus on chromosome 14. Nat Genet. 1992;2(4):330–4.
79. Hruz T, et al. Genevestigator v3: a reference expression database for the meta-analysis of transcriptomes. Adv Bioinforma. 2008;2008:420747.
80. Haass C, De Strooper B. The presenilins in Alzheimer's disease—proteolysis holds the key. Science. 1999;286(5441):916–9.
81. Yu G, et al. Nicastrin modulates presenilin-mediated notch/glp-1 signal transduction and betaAPP processing. Nature. 2000;407(6800):48–54.
82. St George-Hyslop P, Fraser PE. Assembly of the presenilin gamma-/epsilon-secretase complex. J Neurochem. 2012;120(Suppl 1):84–8.
83. Borchelt DR, et al. Familial Alzheimer's disease-linked presenilin 1 variants elevate Abeta1-42/1-40 ratio in vitro and in vivo. Neuron. 1996;17(5):1005–13.
84. Lemere CA, et al. The E280A presenilin 1 Alzheimer mutation produces increased A beta 42 deposition and severe cerebellar pathology. Nat Med. 1996;2(10):1146–50.
85. Mann DM, et al. Amyloid beta protein (Abeta) deposition in chromosome 14-linked Alzheimer's disease: predominance of Abeta42(43). Ann Neurol. 1996;40(2):149–56.
86. Mann DM, et al. Amyloid (Abeta) deposition in chromosome 1-linked Alzheimer's disease: the Volga German families. Ann Neurol. 1997;41(1):52–7.
87. Wolfe MS, et al. Two transmembrane aspartates in presenilin-1 required for presenilin endoproteolysis and gamma-secretase activity. Nature. 1999;398(6727):513–7.

88. Fluhrer R, et al. Intramembrane proteolysis of GXGD-type aspartyl proteases is slowed by a familial Alzheimer disease-like mutation. J Biol Chem. 2008;283(44):30121–8.
89. Crook R, et al. A variant of Alzheimer's disease with spastic paraparesis and unusual plaques due to deletion of exon 9 of presenilin 1. Nat Med. 1998;4(4):452–5.
90. Rogaeva EA, et al. Screening for PS1 mutations in a referral-based series of AD cases: 21 novel mutations. Neurology. 2001;57(4):621–5.
91. De Jonghe C, et al. Aberrant splicing in the presenilin-1 intron 4 mutation causes presenile Alzheimer's disease by increased Abeta42 secretion. Hum Mol Genet. 1999;8(8):1529–40.
92. Muller U, Winter P, Graeber MB. A presenilin 1 mutation in the first case of Alzheimer's disease. Lancet Neurol. 2013;12(2):129–30.
93. Benitez BA, et al. The PSEN1, p.E318G variant increases the risk of Alzheimer's disease in APOE-epsilon4 carriers. PLoS Genet. 2013;9(8):e1003685.
94. Snider BJ, et al. Novel presenilin 1 mutation (S170F) causing Alzheimer disease with Lewy bodies in the third decade of life. Arch Neurol. 2005;62(12):1821–30.
95. Yescas P, et al. Founder effect for the Ala431Glu mutation of the presenilin 1 gene causing early-onset Alzheimer's disease in Mexican families. Neurogenetics. 2006;7(3):195–200.
96. Murrell J, et al. The A431E mutation in PSEN1 causing familial Alzheimer's disease originating in Jalisco State, Mexico: an additional fifteen families. Neurogenetics. 2006;7(4):277–9.
97. Athan ES, et al. A founder mutation in presenilin 1 causing early-onset Alzheimer disease in unrelated Caribbean Hispanic families. JAMA. 2001;286(18):2257–63.
98. Lee JH, et al. Genetic modifiers of age at onset in carriers of the G206A mutation in PSEN1 with familial Alzheimer disease among Caribbean Hispanics. JAMA Neurol. 2015;72(9):1043–51.
99. Tang MX, et al. Incidence of AD in African-Americans, Caribbean Hispanics, and Caucasians in northern Manhattan. Neurology. 2001;56(1):49–56.
100. Rogaeva E. The solved and unsolved mysteries of the genetics of early-onset Alzheimer's disease. Neuromolecular Med. 2002;2(1):1–10.
101. Le TV, et al. Cotton wool plaques in non-familial late-onset Alzheimer disease. J Neuropathol Exp Neurol. 2001;60(11):1051–61.
102. Matsubara-Tsutsui M, et al. Molecular evidence of presenilin 1 mutation in familial early onset dementia. Am J Med Genet. 2002;114(3):292–8.
103. Hiltunen M, et al. Identification of a novel 4.6-kb genomic deletion in presenilin-1 gene which results in exclusion of exon 9 in a Finnish early onset Alzheimer's disease family: an Alu core sequence-stimulated recombination? Eur J Hum Genet. 2000;8(4):259–66.
104. Smith MJ, et al. Variable phenotype of Alzheimer's disease with spastic paraparesis. Ann Neurol. 2001;49(1):125–9.
105. Sinha N, et al. Variant Alzheimer's disease with spastic paraparesis and supranuclear gaze palsy. Can J Neurol Sci. 2013;40(2):249–51.
106. Li D, et al. Mutations of presenilin genes in dilated cardiomyopathy and heart failure. Am J Hum Genet. 2006;79(6):1030–9.
107. Dermaut B, et al. A novel presenilin 1 mutation associated with Pick's disease but not beta-amyloid plaques. Ann Neurol. 2004;55(5):617–26.
108. Sitek EJ, et al. A patient with posterior cortical atrophy possesses a novel mutation in the presenilin 1 gene. PLoS One. 2013;8(4):e61074.
109. Mahoney CJ, et al. The presenilin 1 P264L mutation presenting as non-fluent/agrammatic primary progressive aphasia. J Alzheimers Dis. 2013;36(2):239–43.
110. Braga-Neto P, et al. Early-onset familial Alzheimer's disease related to presenilin 1 mutation resembling autosomal dominant spinocerebellar ataxia. J Neurol. 2013;260(4):1177–9.
111. Amtul Z, et al. A presenilin 1 mutation associated with familial frontotemporal dementia inhibits gamma-secretase cleavage of APP and notch. Neurobiol Dis. 2002;9(2):269–73.
112. Pickering-Brown SM, et al. Mutations in progranulin explain atypical phenotypes with variants in MAPT. Brain. 2006;129(Pt 11):3124–6.
113. Raux G, et al. Dementia with prominent frontotemporal features associated with L113P presenilin 1 mutation. Neurology. 2000;55(10):1577–8.

114. Shen J, et al. Skeletal and CNS defects in Presenilin-1-deficient mice. Cell. 1997;89(4):629–39.
115. Wang B, et al. Gamma-secretase gene mutations in familial acne inversa. Science. 2010;330(6007):1065.
116. Kimberly WT, et al. The transmembrane aspartates in presenilin 1 and 2 are obligatory for gamma-secretase activity and amyloid beta-protein generation. J Biol Chem. 2000;275(5):3173–8.
117. Jayadev S, et al. Presenilin 2 is the predominant gamma-secretase in microglia and modulates cytokine release. PLoS One. 2010;5(12):e15743.
118. Li J, et al. Alzheimer presenilins in the nuclear membrane, interphase kinetochores, and centrosomes suggest a role in chromosome segregation. Cell. 1997;90(5):917–27.
119. Blauwendraat C, et al. Pilot whole-exome sequencing of a German early-onset Alzheimer's disease cohort reveals a substantial frequency of PSEN2 variants. Neurobiol Aging. 2016;37:208.e11–7.
120. Bird TD, et al. Wide range in age of onset for chromosome 1—related familial Alzheimer's disease. Ann Neurol. 1996;40(6):932–6.
121. Lao JI, et al. A novel mutation in the predicted TM2 domain of the presenilin 2 gene in a Spanish patient with late-onset Alzheimer's disease. Neurogenetics. 1998;1(4):293–6.
122. Ezquerra M, et al. A novel mutation in the PSEN2 gene (T430M) associated with variable expression in a family with early-onset Alzheimer disease. Arch Neurol. 2003;60(8):1149–51.
123. Marchani EE, et al. Evidence for three loci modifying age-at-onset of Alzheimer's disease in early-onset PSEN2 families. Am J Med Genet B Neuropsychiatr Genet. 2010;153B(5):1031–41.
124. Piscopo P, et al. A novel PSEN2 mutation associated with a peculiar phenotype. Neurology. 2008;70(17):1549–54.
125. Lippa CF, et al. Lewy bodies contain altered alpha-synuclein in brains of many familial Alzheimer's disease patients with mutations in presenilin and amyloid precursor protein genes. Am J Pathol. 1998;153(5):1365–70.
126. Cacace R, Sleegers K, Van Broeckhoven C. Molecular genetics of early-onset Alzheimer's disease revisited. Alzheimers Dement. 2016;12(6):733–48.
127. Pottier C, et al. High frequency of potentially pathogenic SORL1 mutations in autosomal dominant early-onset Alzheimer disease. Mol Psychiatry. 2012;17(9):875–9.
128. Pastor P, et al. Apolipoprotein Eepsilon4 modifies Alzheimer's disease onset in an E280A PS1 kindred. Ann Neurol. 2003;54(2):163–9.
129. Dermaut B, et al. PRNP Val129 homozygosity increases risk for early-onset Alzheimer's disease. Ann Neurol. 2003;53(3):409–12.
130. Richards S, et al. Standards and guidelines for the interpretation of sequence variants: a joint consensus recommendation of the American College of Medical Genetics and Genomics and the Association for Molecular Pathology. Genet Med. 2015;17(5):405–24.
131. Lou F, et al. Very early-onset sporadic Alzheimer's disease with a de novo mutation in the PSEN1 gene. Neurobiol Aging. 2017;53:193.e1–5.
132. Rogaeva E, Schmitt-Ulms G. Does BDNF Val66Met contribute to preclinical Alzheimer's disease? Brain. 2016;139(Pt 10):2586–9.
133. Guerreiro R, et al. TREM2 variants in Alzheimer's disease. N Engl J Med. 2013;368(2):117–27.
134. Suarez-Calvet M, et al. Early changes in CSF sTREM2 in dominantly inherited Alzheimer's disease occur after amyloid deposition and neuronal injury. Sci Transl Med. 2016;8(369):369ra178.
135. Ghani M, et al. Mutation analysis of patients with neurodegenerative disorders using NeuroX array. Neurobiol Aging. 2015;36(1):545.e9–14.
136. Blauwendraat C, et al. NeuroChip, an updated version of the NeuroX genotyping platform to rapidly screen for variants associated with neurological diseases. Neurobiol Aging. 2017;57:247.e9–13.
137. Ghani M, et al. Mutation analysis of the MS4A and TREM gene clusters in a case-control Alzheimer's disease data set. Neurobiol Aging. 2016;42:217.e7–13.

138. Vardarajan BN, et al. Rare coding mutations identified by sequencing of Alzheimer disease genome-wide association studies loci. Ann Neurol. 2015;78(3):487–98.
139. Vardarajan BN, et al. Coding mutations in SORL1 and Alzheimer disease. Ann Neurol. 2015;77(2):215–27.
140. Zhang M, et al. Drug repositioning for Alzheimer's disease based on systematic 'omics' data mining. PLoS One. 2016;11(12):e0168812.
141. Bertram L, Lill CM, Tanzi RE. The genetics of Alzheimer disease: back to the future. Neuron. 2010;68(2):270–81.
142. Kuwano R, Hara N. Personal genomics for Alzheimer's disease. Brain Nerve (Shinkei kenkyu no shinpo). 2013;65(3):235–46.
143. Choi M, et al. Genetic diagnosis by whole exome capture and massively parallel DNA sequencing. Proc Natl Acad Sci U S A. 2009;106(45):19096–101.
144. Ng SB, et al. Targeted capture and massively parallel sequencing of 12 human exomes. Nature. 2009;461(7261):272–6.
145. Pruitt KD, et al. The consensus coding sequence (CCDS) project: identifying a common protein-coding gene set for the human and mouse genomes. Genome Res. 2009;19(7):1316–23.
146. Barral S, et al. Linkage analyses in Caribbean Hispanic families identify novel loci associated with familial late-onset Alzheimer's disease. Alzheimers Dement. 2015;11(12):1397–406.
147. Kunkle BW, et al. Genome-wide linkage analyses of non-Hispanic white families identify novel loci for familial late-onset Alzheimer's disease. Alzheimers Dement. 2016;12(1):2–10.
148. MacArthur DG, Tyler-Smith C. Loss-of-function variants in the genomes of healthy humans. Hum Mol Genet. 2011;19(R2):R125–30.
149. Cruchaga C, et al. Rare coding variants in the phospholipase D3 gene confer risk for Alzheimer's disease. Nature. 2014;505(7484):550–4.
150. Bras J, et al. Exome sequencing in a consanguineous family clinically diagnosed with early-onset Alzheimer's disease identifies a homozygous CTSF mutation. Neurobiol Aging. 2016;46:236.e1–6.
151. Beck JA, et al. Somatic and germline mosaicism in sporadic early-onset Alzheimer's disease. Hum Mol Genet. 2004;13(12):1219–24.

Genetic Risk Factors for Complex Forms of Alzheimer's Disease

Céline Bellenguez and Jean-Charles Lambert

Abstract

The most frequent forms of Alzheimer's disease (AD) are complex, and their distribution within families cannot be explained by a Mendelian model of inheritance. In fact, these forms of AD result from a combination of genetic and environmental factors, with the estimated heritability ranging from 58 to 79%. This chapter reviews the large body of research on genetic risk factors in AD. Linkage analyses and candidate gene association studies have notably identified *APOE* (the major genetic risk factor for AD) and *SORL1*. Most of the other loci known to be associated with AD have been identified in genome-wide association studies and (more recently) analyses of rare variants. These AD-associated loci and genes have highlighted a number of underlying biological mechanisms, which will be discussed briefly. Although some of these pathways (e.g., amyloid precursor protein (APP) metabolism and tau pathology) fit with the amyloid cascade hypothesis, others point out innate immunity and microglia.

Keywords

Alzheimer's disease · Genetic · Genome-wide association study · GWAS · Rare variant · Sequencing · Amyloid · Tau · Immunity · Microglia

C. Bellenguez (✉) • J.-C. Lambert
Inserm, U1167, RID-AGE—Risk Factors and Molecular Determinants of Aging-Related Diseases, F-59000 Lille, France

Institut Pasteur de Lille, F-59000 Lille, France

Univ. Lille, U1167—Excellence Laboratory LabEx DISTALZ, F-59000 Lille, France
e-mail: celine.bellenguez@pasteur-lille.fr; jean-charles.lambert@pasteur-lille.fr

© Springer International Publishing AG 2018
D. Galimberti, E. Scarpini (eds.), *Neurodegenerative Diseases*,
https://doi.org/10.1007/978-3-319-72938-1_4

Introduction

Less than 2% of cases of Alzheimer's disease (AD) correspond to monogenic, early-onset, familial forms of the disease. The most frequent forms of AD are complex, and their distribution in families cannot be explained by a Mendelian model of inheritance. In fact, these forms of AD result from a combination of genetic and environmental factors and are conventionally categorized into early-onset AD (EOAD) and late-onset AD (LOAD) by using an age at onset cutoff of 65. LOAD accounts for around 95% of all cases of AD [1]. A large twin study confirmed earlier reports whereby monozygotic twins are more likely to both have AD than dizygotic twins, with an estimated heritability of between 58 and 79% [2]. More generally, a positive family history of AD is a risk factor for the disease. For example, one study estimated that first-degree relatives of AD patients have more than twice the risk of developing the disease by the age of 85 years than individuals who merely share the same environment as AD patients [3]. These data suggest that complex forms of AD have a substantial genetic component.

For many years, the *APOE* gene (encoding for apolipoprotein E) was the only known AD susceptibility locus. However, many other loci have now been identified in genome-wide association studies (GWASs) and (more recently) studies of rare variants. This chapter first presents the findings of the various types of genetic studies performed in the field of AD, i.e., linkage analyses (designed to identify genomic regions that are preferentially transmitted with the disease in families) and association studies (designed to compare the allelic or genotypic distribution of genetic markers in AD cases and controls). These studies differ in the number of genetic markers or loci analyzed (e.g., candidate approaches vs. genome-wide approaches), the number of individuals considered (from less than a hundred to more than 50,000), or the technique used to assess genetic variations (e.g., genotyping vs. sequencing). Schematically, we have considered separately the periods before and after the year in which the results of the first large-scale GWASs were published (2009). In the last part of this chapter, we briefly review the underlying biological mechanisms highlighted by the discovery of AD-linked loci and genes. Unless otherwise specified, the reported results refer to populations of European ancestry, since most of the large-scale genetic studies of AD have been performed in this setting.

Early Linkage Studies and Candidate Gene Association Studies

The *APOE* Gene

The *APOE* gene is located on chromosome 19, which was first reported as being involved in AD in 1987, following the identification of an association between familial AD and the *APOC2* gene [4]. This discovery was followed by the identification of a linkage between AD and the 19q13.2–13.3 region, which also contains the *APOC2* gene [5]. In parallel with these genetic studies, biochemical analyses revealed the presence of APOE in senile plaques—one of the two main pathological features of AD [6]. Lastly, it has been shown that APOE binds β-amyloid (Aβ) peptides with high affinity and is likely to act as a chaperone protein [7]. By performing an association analysis in 30 AD cases and 91 controls, Strittmatter et al. [7] were the first to identify an association between AD and the *APOE* gene (located close to the *APOC2* gene in the 19q13.2–13.3 region).

The *APOE* gene's three major alleles (ε2, ε3, and ε4) are defined by the single nucleotide polymorphisms (SNPs) rs429358 and rs7412, which encode amino acids change at codons 112 and 158, respectively. The ε3 allele is characterized by a cysteine at codon 112 and an arginine at codon 158. Relative to ε3, the ε4 allele has an arginine at codon 112, and the ε2 allele has a cysteine at codon 158. These three alleles generate six different *APOE* genotypes, namely, ε2ε2, ε2ε3, ε2ε4, ε3ε3, ε3ε4, and ε4ε4 [8, 9]. Even though the ε4 allele seems to be the ancestral allele [10, 11], the ε3 allele is the most common, with frequencies of around 79%, 13%, and 8% for ε3, ε4, and ε2, respectively [12]. Strittmatter et al.'s study showed that the frequency of ε4 was abnormally high in patients with familial LOAD. This association was then extended to sporadic LOAD [13] and EOAD [14, 15]. The disease-promoting effect of the *APOE* ε4 allele was then confirmed in many studies (as summarized in a meta-analysis by Farrer et al. [16]), whereas the ε2 allele has a protective effect [14, 16–18].

There is a dose-response relationship between the number of ε4 alleles carried by an individual and both the risk and age at onset of AD [19]. Compared with homozygous ε3 individuals, the AD risk is about threefold higher for heterozygous ε4 carriers and 12- to 15-fold higher for homozygous ε4 carriers [16, 20]. The risk associated with the ε3ε4 and ε4ε4 genotypes is age-dependent, and the highest risk occurs at an intermediate age (between 60 and 79) [16, 20].

Although the ε3 allele is the most frequent in all ethnic groups, the *APOE* allele frequencies vary with ethnicity. For example, the frequency of the ε4 allele is estimated to be 21%, 13%, and 9% in African, European, and Asian populations, respectively [12]. Similarly, the strength of the association between the *APOE* ε4 allele and AD seems to vary from one population to another, and the results have initially been inconsistent in African-American and Caribbean Hispanic populations [16, 21, 22]. However, the association between AD and the *APOE* gene has been emphasized in recent studies of larger numbers of African-Americans [23], Caribbean Hispanics [24], Japanese people [25, 26], or Chinese people [27].

At present, the *APOE* gene is still the strongest known genetic factor for AD and accounts for 17–25% of the disease's heritability [28–31]. It has been estimated that ε4ε4 genotype carriers (who account for around 2% of individuals with European ancestry [16]), have an AD lifetime risk of around 30% by the age of 75 and over 50% by the age of 85; this compares with values of around 3% and 12% for the general population by the ages of 75 and 85, respectively [20]. Hence, Genin et al. suggested that *APOE* should be considered as a major gene for AD, since its effect on AD is more similar to that of causative genes in monogenic diseases than that of genetic risk factors for susceptibility to complex diseases.

In GWASs, the association signal for the *APOE* gene is quite large. This raises the question of whether another genetic risk factor for AD is located in the same region or whether this observation is entirely due to the extended correlation between variants (i.e., linkage disequilibrium (LD)) in the region. However, no other genes from this region have been unambiguously identified as being associated with AD [32, 33].

Early Genome-Wide Linkage Analyses

Following the success of linkage analyses in identifying the *APP*, *PSEN1*, and *PSEN2* genes in familial EOAD forms and the *APOE* gene in complex forms of AD, several genome-wide linkage studies (GWLSs) of families or affected sib-pairs were performed between 1997 and 2009 with a view to detecting other contributory genes [34–55]. Most of these studies were performed using quite sparse microsatellite maps (generally fewer than 450 microsatellites, with an average inter-marker distance of 7–16 cM) on a small number of affected individuals (for a review, see [56]). The largest GWLS performed before 2009 resulted from the meta-analysis of five GWLSs, totaling 2206 AD individuals from 785 families [37]. Even though some regions were found to been linked to AD, the results have not always been replicated, and no new, robust genetic factors for AD were immediately identified. However, other types of genetic analysis have subsequently identified loci (such as *CLU*, *CD33*, and *CD2AP*) in some of the AD linkage regions reported by Butler et al. (see section "Large-Scale Genome-Wide Association Studies"). However, these linkage regions are quite large, and it is not known whether the subsequently identified loci explain the detected linkage signal. As is the case for other complex diseases, the difficulty of identifying new genetic factors for AD in linkage analyses has been attributed to various methodological issues and to the characteristics of the genetic component in such complex phenotypes. In particular, linkage analyses are not well powered to identify common variants with small effects that might be involved in these diseases [57, 58].

Candidate Gene Association Studies

Many candidate gene association studies have been performed in the field of AD although the results have also been disappointing. These candidate genes were selected on the basis of their biological functions (and thus potential disease involvement) or their location in AD linkage regions. Up until August 2006, a total of 875 AD candidate association studies had been performed on 1055 polymorphisms in 355 genes [59]. To distinguish between true positives and false positives, Bertram et al. [59] created the Alzgene database (http://www.alzgene.org/). Alzgene catalogued all the association studies performed on AD and provided meta-analysis results for frequent polymorphisms (i.e., with a minor allele frequency (MAF) >1%) considered in at least four independent studies. As of December 2005, 13 genes other than *APOE* had significant associations with AD according to the Alzgene database [59]. However, none of those genes is currently considered to be a confirmed genetic risk factor for AD (with the exception of *PSEN1* for monogenic forms only), thus illustrating the limitations of candidate gene approaches. However, this strategy allowed to identify the *SORL1* gene as an AD-associated gene. The association with *SORL1* was first identified in a study of seven genes involved in the retromer system [60]. Although the results of subsequent candidate gene association studies of *SORL1* were inconsistent, the association was then confirmed in a meta-analysis [61], and a genome-wide significant association signal has been reported in large-scale genome-wide association studies (see sections "Large-Scale Genome-Wide Association Studies" and "Genome-Wide Association Studies in Non-European Populations"). Interestingly, rare variants in *SORL1* are also associated with AD (see section "The Search for Associations with Rare Variants").

Genome-Wide Association Studies

The lack of robust findings in candidate gene association studies of most complex diseases has been attributed to a variety of factors, including small sample sizes [62]. One other major limitation is the studies' reliance on regions identified by linkage analyses or on prior knowledge of the gene's function. In the field of AD, a number of hypothesis-free association studies have first been performed using sparse microsatellite maps [63–65]. However, those maps can only capture a small proportion of the genetic information through the LD between genotyped and non-genotyped genetic variations, unless there is extended LD in the population. In an intermediate hypothesis-free approach, a genome-wide panel of around 17,000 potentially causative SNPs was also tested for an association with AD [66]. Over the years, the scientific community has generated data on common genetic variations and genome-wide patterns of LD [67]. Thanks to technical progress, SNP genotyping has become more reliable and less expensive. Consequently, the development of very dense SNP maps (comprising around 500,000 SNPs and covering the entire genome) has enabled use of the GWAS approach in the identification of new genetic risk factors for complex diseases, including AD. These GWASs test the association between AD and each of the SNPs present on a microarray. To deal with the issue of false positives, most GWAS studies adopt a two-stage approach. Firstly, the analysis of a discovery sample allows to select SNPs showing the most promising association signals. Secondly, the selected SNPs are tested on a second, independent sample at the replication stage. A signal is considered to have been replicated if (1) the effect's direction is the same in both stages and (2) the associated *p*-value in the replication stage is below the nominal threshold of 0.05. Furthermore, a very low *p*-value (typically $<5 \times 10^{-8}$) is required for considering a signal as genome-wide significant in a meta-analysis of discovery and replication stages [68]. When describing the results of a GWAS, it is common practice to report the gene that is situated closest to the SNP with the lowest *p*-value; accordingly, this is also what will be done in the following sections. In theory, however, functional variants could be located in any of the genes included in the LD block surrounding the best SNP hit. Furthermore, the association signal might even be generated by variants that regulate genes located far away from the detected locus. Hence, further studies are required to identify functional genes and variants.

Early Small-Scale Genome-Wide Association Studies

As with the candidate gene studies, the first AD GWASs suffered from small sample sizes (with discovery samples of less than 1100 cases and 1300 controls) [69–77]. Again, only a robust association with *APOE* was identified. These results suggested that there were no common disease-associated variants (other than *APOE*) with a large effect. One of these GWASs [71] detected a significant association (based on a study-specific significance threshold that was less strict than the genome-wide significance threshold of 5×10^{-8}) for SNP rs3826656 close to the *CD33* gene. An association with this locus was later reported to meet the genome-wide significance threshold in a meta-analysis of two large-scale GWASs, although the association with the initially reported SNP itself could not be replicated [78, 79]. However, the association with *CD33* could not be replicated in the second stage of the largest meta-analysis of AD GWASs performed to date [80]. Further studies are required to confirm or exclude an association between AD and the *CD33* locus.

Large-Scale Genome-Wide Association Studies

The first large-scale GWASs were published in 2009 by the European Alzheimer's Disease Initiative (EADI) and the Genetic and Environmental Risk in Alzheimer's Disease (GERAD) consortia [81, 82]. They enabled more than 500,000 SNPs to be screened for an association with AD. For each of these GWASs, the discovery stage included more than 2000 cases and 5000 controls, and the replication stage included more than 2000 cases and 2000 controls. The EADI study detected genome-wide significant association signals for the *CLU* gene and the *CR1* gene. The GERAD study also reported a genome-wide significant association signal for *CLU*, and detected an additional one near the *PICALM* gene. The associations with the *PICALM* and *CR1* loci were cross-replicated in the EADI and GERAD samples, respectively. As expected, the effect sizes of the genes' associations were smaller than that of *APOE* effect (Table 4.1); this highlights the need for large samples and thus sufficient power for detecting genetic associations with AD.

These discoveries were followed closely by the detection of a genome-wide significant association for a SNP near the *BIN1* gene by the Cohorts for Heart and Aging Research in Genomic Epidemiology (CHARGE) consortium [83]. The four-stage CHARGE study covered 9511 AD cases and 28,174 controls and included the EADI and GERAD samples in its second and third stages. It is noteworthy that the GERAD consortium had already reported a suggestive association signal for the *BIN1* locus [81]. The discovery stage (stage 1) of the CHARGE GWAS combined results from studies that had used different genotyping microarrays. It was therefore

Table 4.1 Genetic loci associated at the genome-wide significance level with complex forms of Alzheimer's disease in large meta-analyses

Locus	Index variant	Chr	Position[a]	EA/OA[b]	EAF[c]	OR (95% CI)	References	References for index variant/OR
CR1 (complement C3b/C4b receptor 1)	rs6656401	1	207,692,049	A/G	0.174	1.18 (1.14–1.22)	[82]	[80]
BIN1 (bridging integrator 1)	rs6733839	2	127,892,810	T/C	0.380	1.22 (1.18–1.25)	[83]	[80]
INPP5D (inositol polyphosphate-5-phosphatase D)	rs35349669	2	234,068,476	T/C	0.460	1.08 (1.05–1.11)	[80]	
HS3ST1 (heparan sulfate-glucosamine 3-sulfotransferase 1)	rs13113697	4	11,711,232	T/G	0.289	1.07 (1.05–1.11)	[84]	
MEF2C (myocyte enhancer factor 2C)	rs190982	5	88,223,420	G/A	0.371	0.93 (0.90–0.95)	[80]	
HBEGF (heparin-binding EGF-like growth factor)	rs2074612	5	139,714,690	T/C	0.422	1.08 (1.05–1.11)	[85, 86]	[86]
HLA-DRB5/HLA-DRB1 (major histocompatibility complex, class II, DR beta 5/beta 1)	rs9271192	6	32,578,530	C/A	0.265	1.11 (1.08–1.15)	[80]	
TREM2 (triggering receptor expressed on myeloid cells 2)	p.R47H (rs75932628)	6	41,129,252	T/C	0.005	2.46 (2.06–2.92)	[87, 88]	[89]
	Rare variants	6	/	/	/	Up to 6		[90, 91]
CD2AP (CD2 associated protein)	rs10948363	6	47,487,762	G/A	0.252	1.10 (1.07–1.13)	[78, 79]	[80]
NME8 (NME/NM23 family member 8)	rs2718058	7	37,841,534	G/A	0.372	0.93 (0.90–0.95)	[80]	

(continued)

Table 4.1 (continued)

Locus	Index variant	Chr	Position[a]	EA/OA[b]	EAF[c]	OR (95% CI)	References	References for index variant/OR
ZCWPW1 (zinc finger CW-type and PWWP domain containing 1)	rs1476679	7	100,004,446	C/T	0.299	0.91 (0.89–0.94)	[80]	
EPHA1 (EPH receptor A1)	rs11771145	7	143,110,762	A/G	0.363	0.90 (0.88–0.93)	[78, 79]	[80]
PTK2B (protein tyrosine kinase 2 beta)	rs28834970	8	27,195,121	C/T	0.344	1.10 (1.08–1.13)	[80]	
CLU (clusterin)	rs9331896	8	27,467,686	C/T	0.396	0.86 (0.84–0.89)	[81, 82]	[80]
ECHDC3 (enoyl-CoA hydratase domain containing 3)	rs7920721	10	11,720,308	G/A	0.368	1.07 (1.04–1.11)	[84]	
CELF1 (CUGBP Elav-like family member 1)	rs10838725	11	47,557,871	C/T	0.279	1.08 (1.05–1.11)	[80]	
MS4A cluster (membrane spanning 4-domains A)	rs983392	11	59,923,508	G/A	0.415	0.90 (0.87–0.92)	[78, 79]	[80]
PICALM (phosphatidylinositol-binding clathrin assembly protein)	rs10792832	11	85,867,875	A/G	0.372	0.87 (0.85–0.89)	[81]	[80]
SORL1 (sortilin-related receptor 1)	rs11218343	11	121,435,587	C/T	0.043	0.77 (0.72–0.82)	[60]	[80]
	Rare variants	11	/	/	/	Up to 12	[92, 93]	[90, 94]
FERMT2 (fermitin family member 2)	rs17125944	14	53,400,629	C/T	0.081	1.14 (1.09–1.19)	[80]	
SLC24A4/RIN3 (solute carrier family 24 member 4/ Ras and Rab interactor 3)	rs10498633	14	92,926,952	T/G	0.215	0.91 (0.88–0.94)	[80]	
SPPL2A (signal peptide peptidase like 2A)	rs59685680	15	51,001,534	G/T	0.190	0.92 (0.89–0.95)	[86]	

Gene	SNP	Chr	Position	EA/OA	EAF	OR (95% CI)	Refs
TRIP4 (thyroid hormone receptor interactor 4)	rs74615166	15	64,725,490	C/T	0.016	1.31 (1.17–1.42)	[95]
PLCG2 (phospholipase C gamma 2)	rs72824905	16	81,942,028	G/C	0.011	0.68 (0.60–0.77)	[89]
SCIMP (SLP adaptor and CSK interacting membrane protein)	rs77493189	17	5,118,951	G/T	0.140	1.11 (1.07–1.15)	[86]
ABI3 (ABI family member 3)	rs616338	17	47,297,297	T/C	0.007	1.43 (1.28–1.60)	[89]
TSPOAP1-AS1 (TSPO-associated protein 1 antisense RNA 1)	rs2632516	17	56,409,089	C/G	0.467	0.92 (0.91–0.94)	[85]
ABCA7 (ATP-binding cassette subfamily A member 7)	rs4147929	19	1,063,443	A/G	0.185	1.15 (1.11–1.19)	[78]
	Rare variants	19	/	/	/	Up to 4	[96]
APOE (apolipoprotein E)	rs429358/ rs7412	19	45,411,941 45,412,079	ε2ε2 + ε2ε3/ ε3ε3 ε2ε4/ε3ε3 ε3ε4/ε3ε3 ε4ε4/ε3ε3	0.135 0.026 0.213 0.018	0.56 (0.49–0.64) 2.64 (2.13–3.27) 3.63 (3.37–3.90) 14.49 (11.91–17.64)	[7, 19] [90, 96–98] [20]
CASS4 (Cas scaffolding protein family member 4)	rs7274581	20	55,018,260	C/T	0.080	0.88 (0.84–0.92)	[80]

Chr chromosome, *EA* effect allele, *OA* other allele, *EAF* effect allele frequency, *OR* odds ratio, *CI* confidence interval

[a] Build 37, assembly hg19
[b] For *APOE*, effect genotype/reference genotype is provided
[c] Effect allele frequency in 1000G Phase 3 reference panel [99], restricted to the European populations. For *APOE*, the effect genotype frequency reported in [16] for the controls of the Caucasian ethnic group is provided

necessary to use an imputation procedure so that all the studies had results for the same SNPs. Imputation enables the prediction of genotypes at non-typed SNPs for chip-genotyped individuals by taking into account LD between the SNPs. It depends on the availability of reference panels (e.g., HapMap [67] or 1000 Genomes [99]) that have been genotyped on millions of genetic variants. In the CHARGE GWAS, imputation was performed with the HapMap reference panel; this enabled about 2.5 million SNPs to be tested for an association with AD.

The same imputation strategy was considered by the Alzheimer Disease Genetics Consortium (ADGC) in a GWAS whose discovery stage involved a meta-analysis of ten GWAS datasets totaling 8309 AD cases and 7366 controls. Following a replication stage in 3531 AD cases and 3565 controls, this study led to the identification of a genome-wide significant association signal for the *MS4A* gene cluster [79]. This locus was also detected in a companion study [78], whose discovery stage covered 6688 AD cases and 13,685 controls by meta-analyzing four GWAS datasets (including the GERAD and EADI data). A total of 13,182 cases and 26,161 controls were considered in the two-stage replication analysis. Along with the *MS4A* locus, Hollingworth et al.'s study identified a genome-wide significant association signal for the *ABCA7* gene. This signal was replicated in the ADGC study, and other studies have since reported an association between AD and rare variants in the *ABCA7* gene [96] (see section "The Search for Associations with Rare Variants"). Combining the results from the ADGC study and Hollingworth et al.'s study further identified genome-wide association signals at or near the *CD2AP*, *CD33*, and *EPHA1* genes. As had been the case for *CD33*, suggestive association signals had already been reported for the *EPHA1* and *MS4A* loci [71, 81, 83].

These GWASs and their meta-analyses showed that increasing the sample size and the number of tested variants by imputation can identify new genetic risk factors for AD. Based on this observation, the International Genomics of Alzheimer's Project (IGAP) consortium was created to perform a meta-analysis of the ADGC, CHARGE, EADI, and GERAD GWAS datasets [80]. Following imputation with 1000 Genomes data, the IGAP discovery stage was performed on more than seven million SNPs, 17,008 AD cases, and 37,154 controls. The replication stage included 8572 AD cases and 11,312 controls. Genome-wide association signals were detected at 19 loci other than *APOE*. Eight of these had been identified in earlier analyses of the same GWAS datasets (*ABCA7*, *BIN1*, *CD2AP*, *CLU*, *CR1*, *EPHA1*, the *MS4A* cluster, and *PICALM*). *SORL1* had initially been identified in candidate gene association studies [60, 61], and a genome-wide significant association signal was then reported in a multi-ancestry GWAS [26]. The ten novel loci were *CASS4*, *CELF1*, *FERMT2*, *HLA-DRB5/HLA-DRB1*, *INPP5D*, *MEF2C*, *NME8*, *PTK2B*, *SLC24A4/RIN3*, and *ZCWPW1*. As mentioned above, the association signal at the *CD33* locus was not replicated in the second stage of the IGAP study and did not reach genome-wide significance in the meta-analysis. Furthermore, the IGAP consortium provided a list of 13 loci showing suggestive association signals (i.e., with $p < 1 \times 10^{-6}$), namely, the chr1q31.2, *HS3ST1*, *SQSTM1*, *TREML2*, *NDUFAF6*, *ECHDC3*, *AP2A2*, *ADAMTS20*, the *IGH* cluster, *SPPL2A*, *TRIP4*, *SCIMP*, and *ACE* loci. These loci were deemed worthy of further investigation, since experience had shown that many initially suggestive signals become significant by increasing the sample size. It is noteworthy that the *ACE* gene was among the best candidate genes for AD in meta-analyses of candidate gene association studies [59, 100, 101] and that the *TREML2*-associated variant is close to the *TREM2* gene (rare variants of which are associated with AD; see section "The Search for Associations with Rare Variants") [87, 88]. However, the association signals for the rare variants in *TREM2* gene and the frequent variant in *TREML2* loci might be independent [102].

The IGAP GWAS merged samples from many of the previous GWASs and candidate gene studies performed on individuals of European ancestry. The IGAP report thus did not provide replication evidence for the previously identified loci, but it provided an overview (for individuals of European ancestry) of the evidence for associations with AD. With regard to the newly identified loci, the signal at the *ZCWPW1* locus has been independently replicated [95]. Furthermore, a gene-based analysis of the IGAP data identified a significant signal (on the basis of a gene-based p-value threshold) at the *TP53INP1* and *IGHV1-67* genes (located in the *NDUFAF6* and *IGH* cluster loci, respectively, where a suggestive signal had been reported in the IGAP single-variant analysis) [103]. Lastly, meta-analyses of the IGAP results and those of other GWASs identified genome-wide significant association signals for the *ECHDC3, HS3ST1, TRIP4, SCIMP*, and *SPPL2A* loci (for which suggestive associations had been reported in the IGAP study) and for two new loci: *HBEGF* and *TSPOAP1-AS1* (previously named *BZRAP1-AS1*) [84–86, 95]. However, further assessment of the associations with loci identified in or after the IGAP study is required.

Table 4.1 provides an overview of the AD-associated loci. The data shows that the association signals identified in GWASs arise from rather common alleles with modest effect sizes.

Genome-Wide Association Studies in Non-European Populations

The IGAP GWAS meta-analysis focused on populations of European ancestry. Genome-wide genetic studies have also been performed for African-American, Caribbean Hispanic, Israeli-Arab, and Japanese populations for example [23–26, 104–106], albeit with much smaller sample sizes (no more than 3000 cases in the discovery stage). As mentioned above, a genome-wide significant association signal for *SORL1* was identified in a study whose discovery stage was performed in Japanese individuals [26], and genome-wide significant association signals for *ECHDC3, HBEGF* and *TSPOAP1-AS1* were detected by a meta-analysis of results from several non-European datasets and from the IGAP European dataset [85]. The latter meta-analysis further reported the *TPBG* gene as being associated with AD on the basis of a gene-based p-value threshold. A SNP in the *FBXL7* gene was found to be significantly associated with AD in Caribbean Hispanics [24]. The signal for this SNP could not be replicated in datasets from other ethnicities, although other SNPs in LD showed some evidence of association in these samples. The *COBL* and

SLC10A2 loci have been identified in African Americans, although replication studies have not been performed [107].

For loci initially identified in European populations, the replication of signals in other ethnicities is a complicated task. Firstly, studies of non-European populations have quite small sample sizes, which limits the replication power. Secondly, the LD structure differs from one ethnic group to another; hence, the top SNP identified in GWASs in populations of European ancestry might not tag as well functional variants in other populations—again leading to a loss of power if only the top SNP is tested. After correction for multiple testing, a gene-based analysis performed in African-Americans replicated the associations with the *ABCA7*, *BIN1*, *CD33*, *CR1* and *EPHA1* loci [23], while the associations with the *PICALM* and *BIN1* loci have been replicated in a Japanese population [26]. Many studies have been performed in Asian populations; although the results are divergent, a meta-analysis has provided some evidence of associations for *BIN1*, *CLU* and *PICALM* [108].

Moving Beyond Conventional Genome-Wide Association Studies

The abovementioned GWASs mainly looked at the association between AD status and each variant. However, other strategies have been considered. For example, Herold et al.'s family-based GWAS identified genome-wide significant associations for the *PTPRG*, *OSBPL6* and *PDCL3* loci by (1) taking account of both linkage and association information and (2) considering a multivariate phenotype combining AD status and AD age at onset [109]. By performing a genome-wide haplotype analysis, Lambert et al. identified a genome-wide significant association between AD and the *FRMD4A* gene [110]. This association was subsequently replicated in a gene-based analysis performed in Caribbean Hispanics [24]. Gene–gene interactions have also been studied, and some significant associations have been claimed [111]. In particular, large-scale studies have identified interactions between *CRYL1* and *KHDRBS2*, between *SIRT1* and *ABCB1*, between *PSAP* and *PEBP4*, and between *GRIN2B* and *ADRA1A* [112, 113]. However, these interactions have yet to be extensively replicated. Furthermore, several reports have suggested that some SNPs can interact with the *APOE* genotype [114]. In particular, some SNPs in the 17q21.31 region reportedly decrease the AD risk in individuals who do not carry the *APOE* ε4 allele although the interaction between the SNPs and the *APOE* genotype was not statistically significant. The 17q21.31 region includes the *MAPT* gene, which encodes the tau protein. It has been suggested that rare and/or structural variants of the *MAPT* gene have an impact on AD [115, 116]. However, the role of the *MAPT* gene in AD remains unclear, and further studies are required. Some studies have focused on the impact of structural variants (and particularly copy number variants (CNVs)) on the AD risk, but the results have been inconsistent [117, 118]. It is noteworthy that the association signal observed at the *CR1* locus has been ascribed to a CNV [119]. Lastly, it was recently reported that mosaic loss of chromosome Y is associated with AD [120].

The Search for Associations with Rare Variants

The *index variants at the APOE* gene and the loci identified in GWASs account for around 30% of AD's heritability [28, 30]; this figure suggests that other genetic factors are yet to be discovered. Although the abovementioned GWASs focused on frequent variants (i.e., a MAF >1%), it has been hypothesized that the analysis of rare variants could enable a more exhaustive characterization of the genetic background of complex diseases, including AD [121]. In addition to the identification of new genetic loci, the analysis of rare variants might highlight functional variants in known AD-associated loci. Several strategies can be considered for this purpose. The cheapest approach is to impute rare variants by using available GWAS data and public reference panels. The latter continues to grow; the largest reference panel (the Haplotype Reference Consortium panel) provides data on 39,235,157 SNPs in 32,488 individuals, and this enables researchers to improve the quality of imputation for rare variants [122]. Another strategy consists in genotyping rare variants by using a human exome array. This array can genotype both rare and frequent coding variants detected in genetic data on the exome—the coding part of the genome. These genotyped variants are most likely to be functional and thus involved in disease. Lastly, the most exhaustive information on rare variants is provided by whole-exome or whole-genome sequencing. Thanks to technical progress and a fall in the cost of sequencing, a sufficiently large sample can be sequenced, ensuring enough power to detect associations between AD and rare variants.

Indeed, a sequencing-based study in an Icelandic population reported a protective effect against AD of the rare coding A673T *APP* mutation [123]. Jonsson et al. performed whole-genome sequencing on a subset of 1795 Icelanders and imputed the discovered variants in 71,743 Icelanders with chip genotyping data and (thanks to genealogical records) in more than 290,000 Icelanders lacking any genotyping data. Thanks to this strategy, the study included 3048 AD cases and 79,248 controls. The rarity of the A673T *APP* mutation in other populations means that replication is difficult [124], but a study performed in a Finnish population reported decreased plasma Aβ levels in carriers of the A673T *APP* mutation [125].

Sequencing-based studies also identified an association between AD and the rare missense mutation p.R47H (with a frequency of 0.26% in the ExAC database [126]) in the *TREM2* gene [87, 88]. Jonsson et al.'s study was performed in the Icelandic population using the same strategy as for the *APP* gene's study. The discovery stage included 3550 AD cases and 110,050 controls, and the replication stage was performed on 2037 cases and 9727 controls of European ancestry. Guerreiro et al. applied a candidate gene approach to 1092 AD cases and 1107 controls with sequencing data and then performed the replication stage on genotyped or imputed data from 7428 AD cases and 17,469 controls. This association has been robustly replicated in several studies (including studies of EOAD), with an estimated 2.5-fold increase in the risk of AD for carriers of the p.R47H mutation [89, 127, 128]. Other *TREM2* variants (particularly the p.R62H variant) reportedly contribute to the AD risk [89–91]. Lastly, the results have been less conclusive in African-American and Asian populations, in which the p.R47H variant is more rare [129–131].

However, other variants in the *TREM2* gene might be associated with AD in the African-American population [130, 131].

In an updated version of the Icelandic study, Steinberg et al. searched for rare variant associations within 104 genes located in 18 IGAP-listed AD GWAS loci (*HLA* and *APOE* were not considered) [96]. The researchers reported that rare loss-of-function variants in *ABCA7* are associated with a twofold increase in the AD risk. This association seems to be independent of the common variant signal identified in GWASs and has been replicated in several independent studies (including EOAD studies) [90, 97, 98].

Rare mutations in *SORL1* have been identified in autosomal dominant cases of EOAD [93]. Rare coding variants of *SORL1* are also enriched in cases of EOAD [92, 132] and may not highly contribute to the risk of LOAD [90, 94, 133, 134]. The effect size varies greatly with the pathogenicity class of the rare *SORL1* variants considered, with some classes increasing the AD risk by a factor of 12 or more [90, 94].

Recently, the IGAP consortium performed a large-scale association study on 37,022 cases and 48,402 controls. The discovery stage was performed on individuals genotyped with an exome chip. In addition to the *TREM2* p.R47H and p.R62H variants, variants in the *ABI3* and *PLCG2* genes (each with a MAF of around 1%) were found to be associated with AD at the genome-wide level of significance [89]. However, no independent replication studies have been published yet.

Other association signals between rare variants and AD have been detected but have less supporting evidence. For example, rare coding variants in the *PLD3* gene reportedly increase the AD risk [135], although many replication studies were negative [136–141]. An increased risk of AD was identified for carriers of the rare missense P155L mutation in the *TM2D3* gene in an exome chip analysis of a small Icelandic cohort [142]. Replication of this finding is difficult due to the rarity of the mutation in other populations. Another study reported an association between AD and the rare T835M mutation in the *UNC5C* gene [143]. Rare variants in many other genes (including GWAS genes and dementia genes with Mendelian inheritance) may contribute to complex forms of AD, albeit with lower levels of evidence; further studies (particularly in larger samples) are needed to validate these findings [115, 144–147].

The Post-GWAS Era: A New Frontier

While the GWAS and sequencing data represent a major breakthrough in our understanding of the genetic risk underlying AD, it is often difficult to identify the functional genes and variants within the associated loci and to determine how (in mechanistic terms) they contribute to the pathogenesis of AD. These are important challenges in the post-GWAS era, and characterization of the link between AD genetics and pathogenesis will require major research efforts. This is even more critical for genomic approaches that select genes/loci on a hypothesis-free basis, i.e., without predetermined ideas about their respective functions. Accordingly, new insights on the genetics of AD might oblige us to significantly revise our understanding of the underlying pathophysiological pathways.

As is the case for other multifactorial diseases, gene-set enrichment analyses have been developed from the AD genetic data in order to characterize specific pathways where genes associated with AD risk are overrepresented and which are thus potentially involved in the disease process. Due to the particularly large genetic component of AD (relative to other multifactorial diseases), the genetic determinants of AD might be distributed in one or more specific biological processes, rather than randomly. Although this type of statistical analysis is strongly dependent on the inherent limitations of gene-set enrichment approaches (e.g., the non-optimal quality, annotation, and exhaustiveness of the biological databases), the involvement of the innate immune system has been evidenced in the initial reports (in 2010) [148, 149] through to the most recent studies (based on the IGAP dataset) [150]. More precisely, microglial-mediated innate immunity was highlighted (1) in 2013 by the finding that *TREM2* variants are major genetic risk factors of AD [87, 88] and (2) more recently by the identification of three other AD-risk genes in the same protein–protein interaction network as *TREM2* (*SPI1*—located in the *CELF1* locus—*PLCG2* and *ABI3*, which are almost exclusively expressed in microglia in the brain) [89, 151]. It has been suggested that TREM2 is required for the early expansion of microglia around Aβ plaques, which thus slows the latter from spreading [152]. Loss-of-function mutations in *TREM2* may thus facilitate amyloid-related neuronal damage [153]. Furthermore, potential other defects in Aβ clearance have been highlighted [154]. If we first consider APOE, there is a self-consistent body of evidence to suggest that this protein regulates Aβ clearance in the brain [155]. Furthermore, it was recently reported that the GWAS-defined gene *PICALM* is a central factor in the transcytosis and clearance of Aβ at the blood–brain barrier [156]. As previously suggested, these observations suggest that defects in clearance/degradation/sequestration may lead to the progressive, harmful accumulation of Aβ peptides (at or near synapses, probably).

Interestingly, the new genetic landscape of AD (apart from the monogenic forms of AD) does not rule out a role of APP metabolism and Aβ production in the pathophysiology of AD. As a relevant genetic risk factor for both EOAD and LOAD, *SORL1* has been characterized as a major player in the APP metabolism. Both under-expression and loss-of-function variants associated with AD risk systematically lead to an increase in Aβ secretion [157, 158]. The results of in vitro and in vivo experiments have shown that ABCA7 deficiency is correlated with higher levels of Aβ secretion [159, 160], which is in line with an association between rare loss-of-functions variants and the AD risk [96, 161]. Lastly, a systematic, high-content screening study showed that *FERMT2* was also likely to modulate the APP metabolism because under-expression of this gene led to an increase in Aβ secretion [162].

In conclusion, a dozen GWAS-defined genes have been successfully linked (mainly through defects in Aβ peptide clearance/degradation) to the conventional amyloid cascade hypothesis. However, a large number of GWAS-defined genes do not fit with this hypothesis—suggesting that some of them contribute to AD through mechanisms unrelated to Aβ peptides.

To address this specific question, several studies have sought to determine whether some GWAS-defined genes can modulate tau pathology. To this end, *Drosophila* models have been used to over-express tau in the eyes—leading to the "rough eyes" phenotype. Consequences of the expression modulation of potential human genetic modifiers on this tau-dependent external phenotype can then be easily assessed. The first tauopathy-linked genetic risk factor for AD to be identified in a *Drosophila* model was *BIN1* [163]. Neuropathologic and biochemical studies subsequently confirmed the interactions between tau and BIN1 in vitro and in vivo [163]. Lastly, a detailed characterization of the protein–protein interaction highlighted a direct interaction between the BIN1 SH3 domains and the tau proline-rich domain. It has also been shown that this interaction depends on tau's phosphorylation status [164]. Systematic screening in *Drosophila* has then been developed to characterize other GWAS-defined genes likely to interact with tau. In addition to *BIN1*, orthologs of *FERMT2, CD2AP, CASS4, MADD* (located in the *CELF1* locus), *EPHA1*, and *PTK2B* appear to modify tau toxicity in *Drosophila* models [165, 166]. Further investigations have shown that PTK2B co-localizes with hyperphosphorylated and oligomeric tau in progressive disease stages in the brains of AD patients and transgenic tau mice. These data indicate that PTK2B acts as an early marker and in vivo modulator of tau toxicity [165]. Taken as a whole, the data in *Drosophila* clearly indicate that the core of the focal adhesion pathway is involved in the pathophysiology of AD. This observation is of interest because the focal adhesion pathway is also involved in axon maturation and synaptic plasticity.

It is also important to bear in mind that in addition to the main results described above, other types of analysis can also indicate putative pleiotropic functions for some of the GWAS-defined genes. For example, *PICALM* has also been linked to tau pathology and autophagy [167–169], whereas *BIN1* may potentially interfere with APP metabolism and Aβ peptide production [170, 171].

Conclusion

Genes identified in GWASs are already opening up new perspectives for research on the pathophysiology of AD. The characterization of new genes will not only provide more details of known disease processes but may also enable the discovery of hitherto unsuspected mechanisms. This new genetic landscape and the next discoveries in this field will likely highly influence our way to define the AD pathophysiological process and thus relevant therapeutic targets to develop potential drugs. However, it is likely that it will take several years before seeing the first applications. Indeed, whereas the first GWASs were published in 2009, the number of publications related to the pathophysiological functions of the GWAS-defined genes is still limited, and stronger efforts are needed.

Since AD has a significant polygenic component, several studies have also evaluated the relevance of this new genetic landscape as a diagnosis tool, by assessing the predictive utility of AD genetic risk scores using the most recent GWAS data. Using the IGAP dataset, it has been shown that polygenic scores improve risk prediction with increased prediction at polygenic extremes [172]. Polygenic scores may also help to evaluate individual differences in age-specific genetic risk for AD [173]. Finally, when restricting such analyses to incident cases, a risk score incorporating common genetic variations associated with AD outside the APOEε4 locus also improved slightly AD-risk prediction [174]. However, there is a consensus to say that these approaches cannot be used at the clinical level even if they might be useful for stratifying AD risk in enrichment strategy for therapeutic trials.

References

1. Winblad B, Amouyel P, Andrieu S, Ballard C, Brayne C, Brodaty H, et al. Defeating Alzheimer's disease and other dementias: a priority for European science and society. Lancet Neurol. 2016;15(5):455–532.
2. Gatz M, Reynolds CA, Fratiglioni L, Johansson B, Mortimer JA, Berg S, et al. Role of genes and environments for explaining Alzheimer disease. Arch Gen Psychiatry. 2006;63(2):168–74.
3. Green RC, Cupples LA, Go R, Benke KS, Edeki T, Griffith PA, et al. Risk of dementia among white and African American relatives of patients with Alzheimer disease. JAMA. 2002;287(3):329.
4. Schellenberg GD, Deeb SS, Boehnke M, Bryant EM, Martin GM, Lampe TH, et al. Association of an apolipoprotein CII allele with familial dementia of the Alzheimer type. J Neurogenet. 1987;4(2–3):97–108.
5. Pericak-Vance MA, Bebout JL, Gaskell PC, Yamaoka LH, Hung WY, Alberts MJ, et al. Linkage studies in familial Alzheimer disease: evidence for chromosome 19 linkage. Am J Hum Genet. 1991;48(6):1034–50.
6. Namba Y, Tomonaga M, Kawasaki H, Otomo E, Ikeda K. Apolipoprotein E immunoreactivity in cerebral amyloid deposits and neurofibrillary tangles in Alzheimer's disease and kuru plaque amyloid in Creutzfeldt-Jakob disease. Brain Res. 1991;541(1):163–6.
7. Strittmatter WJ, Saunders AM, Schmechel D, Pericak-Vance M, Enghild J, Salvesen GS, et al. Apolipoprotein E: high-avidity binding to beta-amyloid and increased frequency of type 4 allele in late-onset familial Alzheimer disease. Proc Natl Acad Sci U S A. 1993;90(5):1977–81.
8. Weisgraber KH, Rall SC, Mahley RW. Human E apoprotein heterogeneity. Cysteine-arginine interchanges in the amino acid sequence of the apo-E isoforms. J Biol Chem. 1981;256(17):9077–83.
9. Zannis VI, Breslow JL, Utermann G, Mahley RW, Weisgraber KH, Havel RJ, et al. Proposed nomenclature of apoE isoproteins, apoE genotypes, and phenotypes. J Lipid Res. 1982;23(6):911–4.
10. Fullerton SM, Clark AG, Weiss KM, Nickerson DA, Taylor SL, Stengård JH, et al. Apolipoprotein E variation at the sequence haplotype level: implications for the origin and maintenance of a major human polymorphism. Am J Hum Genet. 2000;67(4):881–900.
11. Hanlon CS, Rubinsztein DC. Arginine residues at codons 112 and 158 in the apolipoprotein E gene correspond to the ancestral state in humans. Atherosclerosis. 1995;112(1):85–90.
12. Singh PP, Singh M, Mastana SS. APOE distribution in world populations with new data from India and the UK. Ann Hum Biol. 2006;33(3):279–308. Taylor & Francis.
13. Saunders AM, Strittmatter WJ, Schmechel D, George-Hyslop PH, Pericak-Vance MA, Joo SH, et al. Association of apolipoprotein E allele epsilon 4 with late-onset familial and sporadic Alzheimer's disease. Neurology. 1993;43(8):1467–72.
14. Chartier-Harlin MC, Parfitt M, Legrain S, Pérez-Tur J, Brousseau T, Evans A, et al. Apolipoprotein E, epsilon 4 allele as a major risk factor for sporadic early and late-onset forms of Alzheimer's disease: analysis of the 19q13.2 chromosomal region. Hum Mol Genet. 1994;3(4):569–74.

15. van Duijn CM, de Knijff P, Cruts M, Wehnert A, Havekes LM, Hofman A, et al. Apolipoprotein E4 allele in a population-based study of early-onset Alzheimer's disease. Nat Genet. 1994;7(1):74–8.
16. Farrer LA, Cupples LA, Haines JL, Hyman B, Kukull WA, Mayeux R, et al. Effects of age, sex, and ethnicity on the association between apolipoprotein E genotype and Alzheimer disease. A meta-analysis. APOE and Alzheimer Disease Meta Analysis Consortium. JAMA. 1997;278(16):1349–56.
17. Corder EH, Saunders AM, Risch NJ, Strittmatter WJ, Schmechel DE, Gaskell PC, et al. Protective effect of apolipoprotein E type 2 allele for late onset Alzheimer disease. Nat Genet. 1994;7(2):180–4.
18. Talbot C, Lendon C, Craddock N, Shears S, Morris JC, Goate A. Protection against Alzheimer's disease with apoE epsilon 2. Lancet (London, England). 1994;343(8910):1432–3.
19. Corder EH, Saunders AM, Strittmatter WJ, Schmechel DE, Gaskell PC, Small GW, et al. Gene dose of apolipoprotein E type 4 allele and the risk of Alzheimer's disease in late onset families. Science. 1993;261(5123):921–3.
20. Genin E, Hannequin D, Wallon D, Sleegers K, Hiltunen M, Combarros O, et al. APOE and Alzheimer disease: a major gene with semi-dominant inheritance. Mol Psychiatry. 2011;16(9):903–7. Nature Publishing Group.
21. Reitz C, Mayeux R. Genetics of Alzheimer's disease in Caribbean Hispanic and African American populations. Biol Psychiatry. 2014;75(7):534–41.
22. Tang MX, Stern Y, Marder K, Bell K, Gurland B, Lantigua R, et al. The APOE-epsilon4 allele and the risk of Alzheimer disease among African Americans, whites, and Hispanics. JAMA. 1998;279(10):751–5.
23. Reitz C, Jun G, Naj A, Rajbhandary R, Vardarajan BN, Wang L-S, et al. Variants in the ATP-binding cassette transporter (ABCA7), apolipoprotein E ϵ4, and the risk of late-onset Alzheimer disease in African Americans. JAMA. 2013;309(14):1483–92.
24. Tosto G, Fu H, Vardarajan BN, Lee JH, Cheng R, Reyes-Dumeyer D, et al. F-box/LRR-repeat protein 7 is genetically associated with Alzheimer's disease. Ann Clin Transl Neurol. 2015;2(8):810–20. Wiley-Blackwell.
25. Hirano A, Ohara T, Takahashi A, Aoki M, Fuyuno Y, Ashikawa K, et al. A genome-wide association study of late-onset Alzheimer's disease in a Japanese population. Psychiatr Genet. 2015;25(4):139–46.
26. Miyashita A, Koike A, Jun G, Wang L-S, Takahashi S, Matsubara E, et al. SORL1 is genetically associated with late-onset Alzheimer's disease in Japanese, Koreans and Caucasians. Toft M, editor. PLoS One. 2013;8(4):e58618. Public Library of Science.
27. Wu P, Li H-L, Liu Z-J, Tao Q-Q, Xu M, Guo Q-H, et al. Associations between apolipoprotein E gene polymorphisms and Alzheimer's disease risk in a large Chinese Han population. Clin Interv Aging. 2015;10:371–8. Dove Press.
28. Cuyvers E, Sleegers K. Genetic variations underlying Alzheimer's disease: evidence from genome-wide association studies and beyond. Lancet Neurol. 2016;15(8):857–68.
29. Lee SH, Harold D, Nyholt DR, Goddard ME, Zondervan KT, Williams J, et al. Estimation and partitioning of polygenic variation captured by common SNPs for Alzheimer's disease, multiple sclerosis and endometriosis. Hum Mol Genet. 2013;22(4):832–41.
30. Ridge PG, Hoyt KB, Boehme K, Mukherjee S, Crane PK, Haines JL, et al. Assessment of the genetic variance of late-onset Alzheimer's disease. Neurobiol Aging. 2016;41:200.e13–20.
31. Ridge PG, Mukherjee S, Crane PK, Kauwe JSK. Alzheimer's disease: analyzing the missing heritability. PLoS One. 2013;8(11):e79771. Public Library of Science.
32. Guerreiro RJ, Hardy J. TOMM40 association with Alzheimer disease: tales of APOE and linkage disequilibrium. Arch Neurol. 2012;69(10):1243–4.
33. Jun G, Vardarajan BN, Buros J, Yu C-E, Hawk MV, Dombroski BA, et al. Comprehensive search for Alzheimer disease susceptibility loci in the APOE region. Arch Neurol. 2012;69(10):1270. American Medical Association.

34. Ashley-Koch AE, Shao Y, Rimmler JB, Gaskell PC, Welsh-Bohmer KA, Jackson CE, et al. An autosomal genomic screen for dementia in an extended Amish family. Neurosci Lett. 2005;379(3):199–204.
35. Avramopoulos D, Fallin MD, Bassett SS. Linkage to chromosome 14q in Alzheimer's disease (AD) patients without psychotic symptoms. Am J Med Genet B Neuropsychiatr Genet. 2005;132B(1):9–13.
36. Blacker D, Bertram L, Saunders AJ, Moscarillo TJ, Albert MS, Wiener H, et al. Results of a high-resolution genome screen of 437 Alzheimer's disease families. Hum Mol Genet. 2003;12(1):23–32.
37. Butler AW, Ng MYM, Hamshere ML, Forabosco P, Wroe R, Al-Chalabi A, et al. Meta-analysis of linkage studies for Alzheimer's disease—a web resource. Neurobiol Aging. 2009;30(7):1037–47.
38. Curtis D, North BV, Sham PC. A novel method of two-locus linkage analysis applied to a genome scan for late onset Alzheimer's disease. Ann Hum Genet. 2001;65(Pt 5):473–81.
39. Giedraitis V, Hedlund M, Skoglund L, Blom E, Ingvast S, Brundin R, et al. New Alzheimer's disease locus on chromosome 8. J Med Genet. 2006;43(12):931–5. BMJ Group.
40. Hahs DW, McCauley JL, Crunk AE, McFarland LL, Gaskell PC, Jiang L, et al. A genome-wide linkage analysis of dementia in the Amish. Am J Med Genet B Neuropsychiatr Genet. 2006;141B(2):160–6.
41. Hamshere ML, Holmans PA, Avramopoulos D, Bassett SS, Blacker D, Bertram L, et al. Genome-wide linkage analysis of 723 affected relative pairs with late-onset Alzheimer's disease. Hum Mol Genet. 2007;16(22):2703–12.
42. Holmans P, Hamshere M, Hollingworth P, Rice F, Tunstall N, Jones S, et al. Genome screen for loci influencing age at onset and rate of decline in late onset Alzheimer's disease. Am J Med Genet B Neuropsychiatr Genet. 2005;135B(1):24–32.
43. Kehoe P, Wavrant-De Vrieze F, Crook R, Wu WS, Holmans P, Fenton I, et al. A full genome scan for late onset Alzheimer's disease. Hum Mol Genet. 1999;8(2):237–45.
44. Lee JH, Cheng R, Graff-Radford N, Foroud T, Mayeux R, National Institute on Aging Late-Onset Alzheimer's Disease Family Study Group. Analyses of the National Institute on Aging Late-Onset Alzheimer's Disease Family Study: implication of additional loci. Arch Neurol. 2008;65(11):1518–26.
45. Lee JH, Cheng R, Santana V, Williamson J, Lantigua R, Medrano M, et al. Expanded genomewide scan implicates a novel locus at 3q28 among Caribbean Hispanics with familial Alzheimer disease. Arch Neurol. 2006;63(11):1591. American Medical Association.
46. Lee JH, Mayeux R, Mayo D, Mo J, Santana V, Williamson J, et al. Fine mapping of 10q and 18q for familial Alzheimer's disease in Caribbean Hispanics. Mol Psychiatry. 2004;9(11):1042–51.
47. Liu F, Arias-Vásquez A, Sleegers K, Aulchenko YS, Kayser M, Sanchez-Juan P, et al. A genomewide screen for late-onset Alzheimer disease in a genetically isolated Dutch population. Am J Hum Genet. 2007;81(1):17–31. Elsevier.
48. Myers A, Wavrant De-Vrieze F, Holmans P, Hamshere M, Crook R, Compton D, et al. Full genome screen for Alzheimer disease: stage II analysis. Am J Med Genet. 2002;114(2):235–44.
49. Olson JM, Goddard KAB, Dudek DM. A second locus for very-late-onset Alzheimer disease: a genome scan reveals linkage to 20p and epistasis between 20p and the amyloid precursor protein region. Am J Hum Genet. 2002;71(1):154–61.
50. Pericak-Vance MA, Bass MP, Yamaoka LH, Gaskell PC, Scott WK, Terwedow HA, et al. Complete genomic screen in late-onset familial Alzheimer disease. Evidence for a new locus on chromosome 12. JAMA. 1997;278(15):1237–41.
51. Pericak-Vance MA, Grubber J, Bailey LR, Hedges D, West S, Santoro L, et al. Identification of novel genes in late-onset Alzheimer's disease. Exp Gerontol. 2000;35(9–10):1343–52.
52. Rademakers R, Cruts M, Sleegers K, Dermaut B, Theuns J, Aulchenko Y, et al. Linkage and association studies identify a novel locus for Alzheimer disease at 7q36 in a Dutch population-based sample. Am J Hum Genet. 2005;77(4):643–52.

53. Scott WK, Hauser ER, Schmechel DE, Welsh-Bohmer KA, Small GW, Roses AD, et al. Ordered-subsets linkage analysis detects novel Alzheimer disease loci on chromosomes 2q34 and 15q22. Am J Hum Genet. 2003;73(5):1041–51.
54. Sillén A, Andrade J, Lilius L, Forsell C, Axelman K, Odeberg J, et al. Expanded high-resolution genetic study of 109 Swedish families with Alzheimer's disease. Eur J Hum Genet. 2008;16(2):202–8. Nature Publishing Group.
55. Sillén A, Forsell C, Lilius L, Axelman K, Björk BF, Onkamo P, et al. Genome scan on Swedish Alzheimer's disease families. Mol Psychiatry. 2006;11(2):182–6. Nature Publishing Group.
56. Ertekin-Taner N. Genetics of Alzheimer's disease: a centennial review. Neurol Clin. 2007;25(3):611–67.
57. Hirschhorn JN, Daly MJ. Genome-wide association studies for common diseases and complex traits. Nat Rev Genet. 2005;6(2):95–108. Nature Publishing Group.
58. Risch NJ. Searching for genetic determinants in the new millennium. Nature. 2000;405(6788):847–56. Nature Publishing Group.
59. Bertram L, McQueen MB, Mullin K, Blacker D, Tanzi RE. Systematic meta-analyses of Alzheimer disease genetic association studies: the AlzGene database. Nat Genet. 2007;39(1):17–23.
60. Rogaeva E, Meng Y, Lee JH, Gu Y, Kawarai T, Zou F, et al. The neuronal sortilin-related receptor SORL1 is genetically associated with Alzheimer disease. Nat Genet. Nature Publishing Group. 2007;39(2):168–77.
61. Reitz C, Cheng R, Rogaeva E, Lee JH, Tokuhiro S, Zou F, et al. Meta-analysis of the association between variants in SORL1 and Alzheimer disease. Arch Neurol. 2011;68(1):99.
62. Cardon LR, Bell JI. Association study designs for complex diseases. Nat Rev Genet. 2001;2(2):91–9. Nature Publishing Group.
63. Farrer LA, Bowirrat A, Friedland RP, Waraska K, Korczyn AD, Baldwin CT. Identification of multiple loci for Alzheimer disease in a consanguineous Israeli-Arab community. Hum Mol Genet. 2003;12(4):415–22.
64. Hiltunen M, Mannermaa A, Thompson D, Easton D, Pirskanen M, Helisalmi S, et al. Genome-wide linkage disequilibrium mapping of late-onset Alzheimer's disease in Finland. Neurology. 2001;57(9):1663–8.
65. Zubenko GS, Hughes HB, Stiffler JS, Hurtt MR, Kaplan BB. A genome survey for novel Alzheimer disease risk loci: results at 10-cM resolution. Genomics. 1998;50(2):121–8.
66. Grupe A, Abraham R, Li Y, Rowland C, Hollingworth P, Morgan A, et al. Evidence for novel susceptibility genes for late-onset Alzheimer's disease from a genome-wide association study of putative functional variants. Hum Mol Genet. 2007;16(8):865–73. Oxford University Press.
67. The International HapMap Consortium. A haplotype map of the human genome. Nature. 2005;437(7063):1299–320.
68. McCarthy MI, Abecasis GR, Cardon LR, Goldstein DB, Little J, Ioannidis JPA, et al. Genome-wide association studies for complex traits: consensus, uncertainty and challenges. Nat Rev Genet. 2008;9(5):356–69. Nature Publishing Group.
69. Abraham R, Moskvina V, Sims R, Hollingworth P, Morgan A, Georgieva L, et al. A genome-wide association study for late-onset Alzheimer's disease using DNA pooling. BMC Med Genet. 2008;1(1):44.
70. Beecham GW, Martin ER, Li Y-J, Slifer MA, Gilbert JR, Haines JL, et al. Genome-wide association study implicates a chromosome 12 risk locus for late-onset Alzheimer disease. Am J Hum Genet. 2009;84(1):35–43.
71. Bertram L, Lange C, Mullin K, Parkinson M, Hsiao M, Hogan MF, et al. Genome-wide association analysis reveals putative Alzheimer's disease susceptibility loci in addition to APOE. Am J Hum Genet. 2008;83(5):623–32.
72. Carrasquillo MM, Zou F, Pankratz VS, Wilcox SL, Ma L, Walker LP, et al. Genetic variation in PCDH11X is associated with susceptibility to late-onset Alzheimer's disease. Nat Genet. 2009;41(2):192–8.

73. Coon KD, Myers AJ, Craig DW, Webster JA, Pearson JV, Lince DH, et al. A high-density whole-genome association study reveals that APOE is the major susceptibility gene for sporadic late-onset Alzheimer's disease. J Clin Psychiatry. 2007;68(4):613–8.
74. Li H, Wetten S, Li L, St Jean PL, Upmanyu R, Surh L, et al. Candidate single-nucleotide polymorphisms from a genomewide association study of Alzheimer disease. Arch Neurol. 2008;65(1):45–53.
75. Poduslo SE, Huang R, Huang J, Smith S. Genome screen of late-onset Alzheimer's extended pedigrees identifies TRPC4AP by haplotype analysis. Am J Med Genet B Neuropsychiatr Genet. 2009;150B(1):50–5.
76. Potkin SG, Guffanti G, Lakatos A, Turner JA, Kruggel F, Fallon JH, et al. Hippocampal atrophy as a quantitative trait in a genome-wide association study identifying novel susceptibility genes for Alzheimer's disease. Domschke K, editor. PLoS One. 2009;4(8):e6501.
77. Reiman EM, Webster JA, Myers AJ, Hardy J, Dunckley T, Zismann VL, et al. GAB2 alleles modify Alzheimer's risk in APOE ε4 carriers. Neuron. 2007;54(5):713–20.
78. Hollingworth P, Harold D, Sims R, Gerrish A. Common variants at ABCA7, MS4A6A, MS4A4E, EPHA1, CD33 and CD2AP are associated with Alzheimer's disease. Nat Genet. 2011;43(5):429–35.
79. Naj AC, Jun G, Beecham GW, Wang L-S, Vardarajan BN, Buros J, et al. Common variants at MS4A4/MS4A6E, CD2AP, CD33 and EPHA1 are associated with late-onset Alzheimer's disease. Nat Genet. 2011;43(5):436–41.
80. Lambert JC, Ibrahim-Verbaas CA, Harold D, Naj AC, Sims R, Bellenguez C, et al. Meta-analysis of 74,046 individuals identifies 11 new susceptibility loci for Alzheimer's disease. Nat Genet. 2013;45(12):1452–8.
81. Harold D, Abraham R, Hollingworth P, Sims R, Gerrish A, Hamshere ML, et al. Genome-wide association study identifies variants at CLU and PICALM associated with Alzheimer's disease. Nat Genet. 2009;41(10):1088–93.
82. Lambert J-C, Heath S, Even G, Campion D, Sleegers K, Hiltunen M, et al. Genome-wide association study identifies variants at CLU and CR1 associated with Alzheimer's disease. Nat Genet. 2009;41(10):1094–9.
83. Seshadri S, Fitzpatrick AL, Ikram MA, DeStefano AL, Gudnason V, Boada M, et al. Genome-wide analysis of genetic loci associated with Alzheimer disease. JAMA. 2010;303(18):1832–40.
84. Desikan RS, Schork AJ, Wang Y, Thompson WK, Dehghan A, Ridker PM, et al. Polygenic overlap between C-reactive protein, plasma lipids, and Alzheimer disease. Circulation. 2015;131(23):2061–9. NIH Public Access.
85. Jun GR, Chung J, Mez J, Barber R, Beecham GW, Bennett DA, et al. Transethnic genome-wide scan identifies novel Alzheimer's disease loci. Alzheimers Dement. 2017;13(7):727–38.
86. Liu JZ, Erlich Y, Pickrell JK. Case-control association mapping by proxy using family history of disease. Nat Genet. 2017;49(3):325–31.
87. Guerreiro R, Wojtas A, Bras J, Carrasquillo M, Rogaeva E, Majounie E, et al. TREM2 variants in Alzheimer's disease. N Engl J Med. 2013;368(2):117–27. Massachusetts Medical Society.
88. Jonsson T, Stefansson H, Steinberg S, Jonsdottir I, Jonsson PV, Snaedal J, et al. Variant of *TREM2* associated with the risk of Alzheimer's disease. N Engl J Med. 2013;368(2):107–16. Massachusetts Medical Society.
89. Sims R, van der Lee SJ, Naj AC, Bellenguez C, Badarinarayan N, Jakobsdottir J, et al. Rare coding variants in PLCG2, ABI3, and TREM2 implicate microglial-mediated innate immunity in Alzheimer's disease. Nat Genet. 2017;49(9):1373–84.
90. Bellenguez C, Charbonnier C, Grenier-Boley B, Quenez O, Le Guennec K, Nicolas G, et al. Contribution to Alzheimer's disease risk of rare variants in TREM2, SORL1 and ABCA7 in 1,779 cases and 1,273 controls. Neurobiol Aging. 2017;59:220.e1–9.
91. Jin SC, Benitez BA, Karch CM, Cooper B, Skorupa T, Carrell D, et al. Coding variants in TREM2 increase risk for Alzheimer's disease. Hum Mol Genet. 2014;23(21):5838–46.

92. Nicolas G, Charbonnier C, Wallon D, Quenez O, Bellenguez C, Grenier-Boley B, et al. SORL1 rare variants: a major risk factor for familial early-onset Alzheimer's disease. Mol Psychiatry. 2016;21(6):831–6. Nature Publishing Group.
93. Pottier C, Hannequin D, Coutant S, Rovelet-Lecrux A, Wallon D, Rousseau S, et al. High frequency of potentially pathogenic SORL1 mutations in autosomal dominant early-onset Alzheimer disease. Mol Psychiatry. 2012;17(9):875–9.
94. Holstege H, van der Lee SJ, Hulsman M, Wong TH, van Rooij JG, Weiss M, et al. Characterization of pathogenic SORL1 genetic variants for association with Alzheimer's disease: a clinical interpretation strategy. Eur J Hum Genet. 2017;25(8):973–81.
95. Ruiz A, Heilmann S, Becker T, Hernández I, Wagner H, Thelen M, et al. Follow-up of loci from the International Genomics of Alzheimer's Disease Project identifies TRIP4 as a novel susceptibility gene. Transl Psychiatry. 2014;4(2):e358. Nature Publishing Group.
96. Steinberg S, Stefansson H, Jonsson T, Johannsdottir H, Ingason A, Helgason H, et al. Loss-of-function variants in ABCA7 confer risk of Alzheimer's disease. Nat Genet. 2015;47(5):445–7. Nature Publishing Group, a division of Macmillan Publishers Limited. All Rights Reserved.
97. Allen M, Lincoln SJ, Corda M, Watzlawik JO, Carrasquillo MM, Reddy JS, et al. ABCA7 loss-of-function variants, expression, and neurologic disease risk. Neurol Genet. 2017;3(1):e126.
98. Cuyvers E, De Roeck A, Van den Bossche T, Van Cauwenberghe C, Bettens K, Vermeulen S, et al. Mutations in ABCA7 in a Belgian cohort of Alzheimer's disease patients: a targeted resequencing study. Lancet Neurol. 2015;14(8):814–22.
99. The 1000 Genomes Project Consortium. A global reference for human genetic variation. Nature. 2015;526(7571):68–74.
100. Kehoe PG, Russ C, McIlroy S, Williams H, Holmans P, Holmes C, et al. Variation in DCP1, encoding ACE, is associated with susceptibility to Alzheimer disease. Nat Genet. 1999;21(1):71–2. Nature Publishing Group.
101. Lehmann DJ, Cortina-Borja M, Warden DR, Smith AD, Sleegers K, Prince JA, et al. Large meta-analysis establishes the ACE insertion-deletion polymorphism as a marker of Alzheimer's disease. Am J Epidemiol. 2005;162(4):305–17. Oxford University Press.
102. Benitez BA, Jin SC, Guerreiro R, Graham R, Lord J, Harold D, et al. Missense variant in TREML2 protects against Alzheimer's disease. Neurobiol Aging. 2014;35(6):1510.e19–26.
103. Escott-Price V, Bellenguez C, Wang L-S, Choi S-H, Harold D, Jones L, et al. Gene-wide analysis detects two new susceptibility genes for Alzheimer's disease. PLoS One. Public Library of Science. 2014;9(6):e94661.
104. Lee JH, Cheng R, Barral S, Reitz C, Medrano M, Lantigua R, et al. Identification of novel loci for Alzheimer disease and replication of CLU, PICALM, and BIN1 in Caribbean Hispanic individuals. Arch Neurol. 2011;68(3):320–8.
105. Logue MW, Schu M, Vardarajan BN, Buros J, Green RC, Go RCP, et al. A comprehensive genetic association study of Alzheimer disease in African Americans. Arch Neurol. 2011;68(12):1569.
106. Sherva R, Baldwin CT, Inzelberg R, Vardarajan B, Cupples LA, Lunetta K, et al. Identification of novel candidate genes for Alzheimer's disease by autozygosity mapping using genome wide SNP data. J Alzheimers Dis. 2011;23(2):349–59. NIH Public Access.
107. Mez J, Chung J, Jun G, Kriegel J, Bourlas AP, Sherva R, et al. Two novel loci, COBL and SLC10A2, for Alzheimer's disease in African Americans. Alzheimers Dement. 2017;13(2):119–29. NIH Public Access.
108. Wang H-Z, Bi R, Hu Q-X, Xiang Q, Zhang C, Zhang D-F, et al. Validating GWAS-identified risk loci for Alzheimer's disease in Han Chinese populations. Mol Neurobiol. 2016;53(1):379–90. Springer US.
109. Herold C, Hooli BV, Mullin K, Liu T, Roehr JT, Mattheisen M, et al. Family-based association analyses of imputed genotypes reveal genome-wide significant association of Alzheimer's disease with OSBPL6, PTPRG, and PDCL3. Mol Psychiatry. 2016;21(11):1608–12.
110 Lambert J-C, Grenier-Boley B, Harold D, Zelenika D, Chouraki V, Kamatani Y, et al. Genome-wide haplotype association study identifies the FRMD4A gene as a risk locus for Alzheimer's disease. Mol Psychiatry. 2013;18(4):461–70.

111. Ebbert MTW, Ridge PG, Kauwe JSK. Bridging the gap between statistical and biological epistasis in Alzheimer's disease. Biomed Res Int. 2015;2015:870123.
112. Gusareva ES, Carrasquillo MM, Bellenguez C, Cuyvers E, Colon S, Graff-Radford NR, et al. Genome-wide association interaction analysis for Alzheimer's disease. Neurobiol Aging. 2014;35(11):2436–43.
113. Hohman TJ, Bush WS, Jiang L, Brown-Gentry KD, Torstenson ES, Dudek SM, et al. Discovery of gene-gene interactions across multiple independent data sets of late onset Alzheimer disease from the Alzheimer Disease Genetics Consortium. Neurobiol Aging. 2016;38:141–50.
114. Wijsman EM, Pankratz ND, Choi Y, Rothstein JH, Faber KM, Cheng R, et al. Genome-wide association of familial late-onset Alzheimer's disease replicates BIN1 and CLU and nominates CUGBP2 in interaction with APOE. PLoS Genet. 2011;7(2):e1001308. Public Library of Science.
115. Coppola G, Chinnathambi S, Lee JJ, Dombroski BA, Baker MC, Soto-Ortolaza AI, et al. Evidence for a role of the rare p.A152T variant in MAPT in increasing the risk for FTD-spectrum and Alzheimer's diseases. Hum Mol Genet. 2012;21(15):3500–12. Oxford University Press.
116. Le Guennec K, Quenez O, Nicolas G, Wallon D, Rousseau S, Richard A-C, et al. 17q21.31 duplication causes prominent tau-related dementia with increased MAPT expression. Mol Psychiatry. 2017;22(8):1119–25.
117. Chapman J, Rees E, Harold D, Ivanov D, Gerrish A, Sims R, et al. A genome-wide study shows a limited contribution of rare copy number variants to Alzheimer's disease risk. Hum Mol Genet. 2013;22(4):816–24.
118. Cuccaro D, De Marco EV, Cittadella R, Cavallaro S. Copy number variants in Alzheimer's disease. Campion D, editor. J Alzheimers Dis. 2016;55(1):37–52. IOS Press.
119. Brouwers N, Van Cauwenberghe C, Engelborghs S, Lambert J-C, Bettens K, Le Bastard N, et al. Alzheimer risk associated with a copy number variation in the complement receptor 1 increasing C3b/C4b binding sites. Mol Psychiatry. 2012;17(2):223–33.
120. Dumanski JP, Lambert J-C, Rasi C, Giedraitis V, Davies H, Grenier-Boley B, et al. Mosaic loss of chromosome Y in blood is associated with Alzheimer disease. Am J Hum Genet. 2016;98(6):1208–19.
121. Manolio TA, Collins FS, Cox NJ, Goldstein DB, Hindorff LA, Hunter DJ, et al. Finding the missing heritability of complex diseases. Nature. 2009;461(7265):747–53. Nature Publishing Group.
122. McCarthy S, Das S, Kretzschmar W, Delaneau O, Wood AR, Teumer A, et al. A reference panel of 64,976 haplotypes for genotype imputation. Nat Genet. 2016;48(10):1279–83.
123. Jonsson T, Atwal JK, Steinberg S, Snaedal J, Jonsson PV, Bjornsson S, et al. A mutation in APP protects against Alzheimer's disease and age-related cognitive decline. Nature. 2012;488(7409):96–9.
124. Mengel-From J, Jeune B, Pentti T, McGue M, Christensen K, Christiansen L. The APP A673T frequency differs between Nordic countries. Neurobiol Aging. 2015;36(10):2909. e1–4.
125. Martiskainen H, Herukka S-K, Stančáková A, Paananen J, Soininen H, Kuusisto J, et al. Decreased plasma β-amyloid in the Alzheimer's disease APP A673T variant carriers. Ann Neurol. 2017;82(1):128–32.
126. Lek M, Karczewski KJ, Minikel EV, Samocha KE, Banks E, Fennell T, et al. Analysis of protein-coding genetic variation in 60,706 humans. Nature. 2016;536(7616):285–91.
127. Lill CM, Rengmark A, Pihlstrøm L, Fogh I, Shatunov A, Sleiman PM, et al. The role of TREM2 R47H as a risk factor for Alzheimer's disease, frontotemporal lobar degeneration, amyotrophic lateral sclerosis, and Parkinson's disease. Alzheimers Dement. 2015;11(12):1407–16.
128. Pottier C, Wallon D, Rousseau S, Rovelet-Lecrux A, Richard A-C, Rollin-Sillaire A, et al. TREM2 R47H variant as a risk factor for early-onset Alzheimer's disease. J Alzheimers Dis. 2013;35(1):45–9.

129. Huang M, Wang D, Xu Z, Xu Y, Xu X, Ma Y, et al. Lack of genetic association between TREM2 and Alzheimer's disease in East Asian population: a systematic review and meta-analysis. Am J Alzheimers Dis Other Demen. 2015;30(6):541–6.
130. Jin SC, Carrasquillo MM, Benitez BA, Skorupa T, Carrell D, Patel D, et al. TREM2 is associated with increased risk for Alzheimer's disease in African Americans. Mol Neurodegener. 2015;10:19.
131. Reitz C, Mayeux R, Alzheimer's Disease Genetics Consortium. TREM2 and neurodegenerative disease. N Engl J Med. 2013;369(16):1564–5.
132. Verheijen J, Van den Bossche T, van der Zee J, Engelborghs S, Sanchez-Valle R, Lladó A, et al. A comprehensive study of the genetic impact of rare variants in SORL1 in European early-onset Alzheimer's disease. Acta Neuropathol. 2016;132(2):213–24.
133. Fernández MV, Black K, Carrell D, Saef B, Budde J, Deming Y, et al. SORL1 variants across Alzheimer's disease European American cohorts. Eur J Hum Genet. 2016;24(12):1828–30. Nature Publishing Group.
134. Vardarajan BN, Zhang Y, Lee JH, Cheng R, Bohm C, Ghani M, et al. Coding mutations in SORL1 and Alzheimer disease. Ann Neurol. 2015;77(2):215–27.
135. Cruchaga C, Karch CM, Jin SC, Benitez BA, Cai Y, Guerreiro R, et al. Rare coding variants in the phospholipase D3 gene confer risk for Alzheimer's disease. Nature. 2013;505(7484):550–4. Nature Publishing Group.
136. Cacace R, Van den Bossche T, Engelborghs S, Geerts N, Laureys A, Dillen L, et al. Rare variants in *PLD3* do not affect risk for early-onset Alzheimer disease in a European Consortium Cohort. Hum Mutat. 2015;36(12):1226–35.
137. Heilmann S, Drichel D, Clarimon J, Fernández V, Lacour A, Wagner H, et al. PLD3 in non-familial Alzheimer's disease. Nature. 2015;520(7545):E3–5.
138. Hooli BV, Lill CM, Mullin K, Qiao D, Lange C, Bertram L, et al. PLD3 gene variants and Alzheimer's disease. Nature. 2015;520(7545):E7–8.
139. Lambert J-C, Grenier-Boley B, Bellenguez C, Pasquier F, Campion D, Dartigues J-F, et al. PLD3 and sporadic Alzheimer's disease risk. Nature. 2015;520(7545):E1.
140. van der Lee SJ, Holstege H, Wong TH, Jakobsdottir J, Bis JC, Chouraki V, et al. PLD3 variants in population studies. Nature. 2015;520(7545):E2–3.
141. Schulte EC, Kurz A, Alexopoulos P, Hampel H, Peters A, Gieger C, et al. Excess of rare coding variants in PLD3 in late- but not early-onset Alzheimer's disease. Hum Genome Var. 2015;2:14028.
142. Jakobsdottir J, van der Lee SJ, Bis JC, Chouraki V, Li-Kroeger D, Yamamoto S, et al. Rare functional variant in TM2D3 is associated with late-onset Alzheimer's disease. Haines JL, editor. PLoS Genet. 2016;12(10):e1006327. Public Library of Science.
143. Wetzel-Smith MK, Hunkapiller J, Bhangale TR, Srinivasan K, Maloney JA, Atwal JK, et al. A rare mutation in UNC5C predisposes to late-onset Alzheimer's disease and increases neuronal cell death. Nat Med. 2014;20(12):1452–7.
144. Bettens K, Brouwers N, Engelborghs S, Lambert J-C, Rogaeva E, Vandenberghe R, et al. Both common variations and rare non-synonymous substitutions and small insertion/deletions in CLU are associated with increased Alzheimer risk. Mol Neurodegener. 2012;7(1):3.
145. Kim JH, Song P, Lim H, Lee J-H, Lee JH, Park SA, et al. Gene-based rare allele analysis identified a risk gene of Alzheimer's disease. PLoS One. 2014;9(10):e107983.
146. Logue MW, Schu M, Vardarajan BN, Farrell J, Bennett DA, Buxbaum JD, et al. Two rare AKAP9 variants are associated with Alzheimer's disease in African Americans. Alzheimers Dement. 2014;10(6):609–618.e11.
147. Vardarajan BN, Ghani M, Kahn A, Sheikh S, Sato C, Barral S, et al. Rare coding mutations identified by sequencing of Alzheimer disease genome-wide association studies loci. Ann Neurol. 2015;78(3):487–98.
148. Jones L, Holmans PA, Hamshere ML, Harold D, Moskvina V, Ivanov D, et al. Genetic evidence implicates the immune system and cholesterol metabolism in the aetiology of Alzheimer's disease. El Khoury J, editor. PLoS One. 2010;5(11):e13950.

149. Lambert J-C, Grenier-Boley B, Chouraki V, Heath S, Zelenika D, Fievet N, et al. Implication of the immune system in Alzheimer's disease: evidence from genome-wide pathway analysis. J Alzheimers Dis. 2010;20(4):1107–18.
150. Jones L, Lambert J-C, Wang L-S, Choi S-H, Harold D, Vedernikov A, et al. Convergent genetic and expression data implicate immunity in Alzheimer's disease. Alzheimers Dement. 2015;11(6):658–71.
151. Huang K-L, Marcora E, Pimenova AA, Di Narzo AF, Kapoor M, Jin SC, et al. A common haplotype lowers PU.1 expression in myeloid cells and delays onset of Alzheimer's disease. Nat Neurosci. 2017;20(8):1052–61.
152. Wang Y, Ulland TK, Ulrich JD, Song W, Tzaferis JA, Hole JT, et al. TREM2-mediated early microglial response limits diffusion and toxicity of amyloid plaques. J Exp Med. 2016;213(5):667–75.
153. Yuan P, Condello C, Keene CD, Wang Y, Bird TD, Paul SM, et al. TREM2 haplodeficiency in mice and humans impairs the microglia barrier function leading to decreased amyloid compaction and severe axonal dystrophy. Neuron. 2016;90(4):724–39.
154. Lambert J-C, Amouyel P. Genetics of Alzheimer's disease: new evidences for an old hypothesis? Curr Opin Genet Dev. 2011;21(3):295–301.
155. Castellano JM, Kim J, Stewart FR, Jiang H, DeMattos RB, Patterson BW, et al. Human apoE isoforms differentially regulate brain amyloid-β peptide clearance. Sci Transl Med. 2011;3(89):89ra57.
156. Zhao Z, Sagare AP, Ma Q, Halliday MR, Kong P, Kisler K, et al. Central role for PICALM in amyloid-β blood-brain barrier transcytosis and clearance. Nat Neurosci. 2015;18(7):978–87.
157. Caglayan S, Takagi-Niidome S, Liao F, Carlo A-S, Schmidt V, Burgert T, et al. Lysosomal sorting of amyloid-β by the SORLA receptor is impaired by a familial Alzheimer's disease mutation. Sci Transl Med. 2014;6(223):223ra20.
158. Young JE, Boulanger-Weill J, Williams DA, Woodruff G, Buen F, Revilla AC, et al. Elucidating molecular phenotypes caused by the SORL1 Alzheimer's disease genetic risk factor using human induced pluripotent stem cells. Cell Stem Cell. 2015;16(4):373–85.
159. Sakae N, Liu C-C, Shinohara M, Frisch-Daiello J, Ma L, Yamazaki Y, et al. ABCA7 deficiency accelerates amyloid-β generation and Alzheimer's neuronal pathology. J Neurosci. 2016;36(13):3848–59.
160. Satoh K, Abe-Dohmae S, Yokoyama S, St George-Hyslop P, Fraser PE. ATP-binding cassette transporter A7 (ABCA7) loss of function alters Alzheimer amyloid processing. J Biol Chem. 2015;290(40):24152–65.
161. Le Guennec K, Nicolas G, Quenez O, Charbonnier C, Wallon D, Bellenguez C, et al. ABCA7 rare variants and Alzheimer disease risk. Neurology. 2016;86(23):2134–7.
162. Chapuis J, Flaig A, Grenier-Boley B, Eysert F, Pottiez V, Deloison G, et al. Genome-wide, high-content siRNA screening identifies the Alzheimer's genetic risk factor FERMT2 as a major modulator of APP metabolism. Acta Neuropathol. 2017;133(6):955–66.
163. Chapuis J, Hansmannel F, Gistelinck M, Mounier A, Van Cauwenberghe C, Kolen KV, et al. Increased expression of BIN1 mediates Alzheimer genetic risk by modulating tau pathology. Mol Psychiatry. 2013;18(11):1225–34.
164. Sottejeau Y, Bretteville A, Cantrelle F-X, Malmanche N, Demiaute F, Mendes T, et al. Tau phosphorylation regulates the interaction between BIN1's SH3 domain and Tau's proline-rich domain. Acta Neuropathol Commun. 2015;3(1):58.
165. Dourlen P, Fernandez-Gomez FJ, Dupont C, Grenier-Boley B, Bellenguez C, Obriot H, et al. Functional screening of Alzheimer risk loci identifies PTK2B as an in vivo modulator and early marker of Tau pathology. Mol Psychiatry. 2017;22(6):874–83.
166. Shulman JM, Imboywa S, Giagtzoglou N, Powers MP, Hu Y, Devenport D, et al. Functional screening in Drosophila identifies Alzheimer's disease susceptibility genes and implicates Tau-mediated mechanisms. Hum Mol Genet. 2014;23(4):870–7.
167. Ando K, Brion J-P, Stygelbout V, Suain V, Authelet M, Dedecker R, et al. Clathrin adaptor CALM/PICALM is associated with neurofibrillary tangles and is cleaved in Alzheimer's brains. Acta Neuropathol. 2013;125(6):861–78.

168. Ando K, Tomimura K, Sazdovitch V, Suain V, Yilmaz Z, Authelet M, et al. Level of PICALM, a key component of clathrin-mediated endocytosis, is correlated with levels of phosphotau and autophagy-related proteins and is associated with tau inclusions in AD, PSP and Pick disease. Neurobiol Dis. 2016;94:32–43.
169. Tian Y, Chang JC, Fan EY, Flajolet M, Greengard P. Adaptor complex AP2/PICALM, through interaction with LC3, targets Alzheimer's APP-CTF for terminal degradation via autophagy. Proc Natl Acad Sci U S A. 2013;110(42):17071–6.
170. Miyagawa T, Ebinuma I, Morohashi Y, Hori Y, Young Chang M, Hattori H, et al. BIN1 regulates BACE1 intracellular trafficking and amyloid-β production. Hum Mol Genet. 2016;25(14):2948–58.
171. Ubelmann F, Burrinha T, Salavessa L, Gomes R, Ferreira C, Moreno N, et al. Bin1 and CD2AP polarise the endocytic generation of beta-amyloid. EMBO Rep. 2017;18(1):102–22.
172. Escott-Price V, Sims R, Bannister C, Harold D, Vronskaya M, Majounie E, et al. Common polygenic variation enhances risk prediction for Alzheimer's disease. Brain. 2015;138(Pt 12):3673–84.
173. Desikan RS, Fan CC, Wang Y, Schork AJ, Cabral HJ, Cupples LA, et al. Genetic assessment of age-associated Alzheimer disease risk: development and validation of a polygenic hazard score. Brayne C, editor. PLoS Med. 2017;14(3):e1002258.
174. Chouraki V, Reitz C, Maury F, Bis JC, Bellenguez C, Yu L, et al. Evaluation of a genetic risk score to improve risk prediction for Alzheimer's disease. Hall A, editor. J Alzheimers Dis. 2016;53(3):921–32.

Role of Cerebrovascular Disease in Cognition

5

Ana Verdelho

Abstract

Vascular risk factors and cerebrovascular disease are recognized factors implicated in the evolution towards dementia, not only of vascular origin, but also of degenerative dementia as Alzheimer's disease. Even among nondemented subjects, hypertension, diabetes, and stroke are associated with worse performance in attention, speed and motor control, and executive functions. Influence of vascular risk factors in cognition starts early in life. Recently, several publications expressed that intervention in potential modifiable risk factors should receive special attention in order to delay or prevent dementia. Current scientific evidence sustains that policy actions should be conducted in order to reduce vascular risk factors in middle life, with population and community-level measures. Cerebral small vessel disease, which can be expressed by white matter changes, lacunes, and microbleeds, has gained clinical relevance in the last decades. Intervention in prevention of this previously overlooked disease can represent a potential outcome in experimental studies aiming to reduce cerebrovascular burden.

Keywords

Vascular risk factors · Hypertension · Diabetes · Stroke · Cerebral small vessel disease · White matter changes · Lacunes · Microbleeds

A. Verdelho, M.D., Ph.D.
Department of Neurosciences and Mental Health, Centro Hospitalar Lisboa Norte-Hospital de Santa Maria, Instituto de Medicina Molecular—IMM e Instituto de Saúde Ambiental-ISAMB Medical School, University of Lisbon, Lisbon, Portugal
e-mail: averdelho@medicina.ulisboa.pt

© Springer International Publishing AG 2018
D. Galimberti, E. Scarpini (eds.), *Neurodegenerative Diseases*,
https://doi.org/10.1007/978-3-319-72938-1_5

Introduction

Vascular risk factors and cerebrovascular disease of the brain influence cognition and are implicated in the evolution towards dementia, not only of vascular origin, but also of degenerative dementia as Alzheimer's disease (AD).

In the last few years several publications have stressed acknowledge of an overlap between risk factors for vascular disease and neurodegeneration and dementia [1]. On this behalf, efforts should be done in order to improve research and population recognition of vascular risk factors [1–3]. Recently, in an initiative funded by the Joint Programme for Neurodegenerative Disease Research, a survey was conducted within a group of international experts. The results and recommendations for the study of vascular disease and its contribution to cognitive decline and neurodegeneration retrieved from that survey were published, and the interested reader may find it in a quite fine comprehensive review [4]. There are other publications made upon this approach, but this chapter does not aim to be an exhaustive bibliographic review under the topic. The author proposes a reflection based on selected bibliographic references and his own clinical experience, aiming to share the concern about a (although) frequent, sometimes neglected topic in daily practice.

This chapter has two different sections. The first section covers the impact of main vascular factors in cognition and in the risk of dementia. As small vessel disease is closely linked to vascular risk factors, and represents one of the consequences of several vascular risk factors measured in the brain, we approach, in the second section, the impact of cerebral small vessel disease in cognition and in dementia.

Role of Vascular Risk Factors in Cognition

Vascular risk factors have been implicated in cognitive decline and dementia. A recent review that considered the most frequent vascular risk factors (diabetes, midlife hypertension, midlife obesity, physical inactivity, and smoking) plus depression and educational level, concluded that even among degenerative dementia, around a third of Alzheimer's disease cases might be attributable to potentially modifiable risk factors [5]. Moreover, midlife vascular risk factors were associated with higher amyloid deposition in the brain [6]. Among the whole spectrum of vascular risk factors, hypertension, stroke, and diabetes seem to play the most important role [7–18]. Before exploring evidence that support the relationship between some of the major risk factors and cognitive impairment, we present two concepts that have evolved in the last years. The first is that cognitive decline is insidious and slowly developing starting early in life, around the fourth decade [19]. This is probably one of the explanations for many of the controversial data concerning some of the vascular risk factors, namely, cholesterol blood levels and body mass index [20–24]. It is likely that these pathologies contribute to cognitive decline mainly when present in midlife.

The second concept is that the interaction between several cardiovascular risk factors contributes more strongly for cognitive decline than isolated risk factors [10, 22]. A systematic review stressed that the risk of dementia in diabetes is increased when associated with other vascular risk factors, a phenomenon that was also identified for other risk factors [10, 22, 25], mainly if they are concomitantly present in midlife [10, 26].

Role of Diabetes in Cognition

Diabetes has increasingly been identified as a risk factor for cognitive impairment and dementia [18, 27–30], including AD [31]. Among nondemented subjects, diabetics have worse cognitive performance when compared to nondiabetics [13, 28, 32] in global tests of cognition [33], attention, executive functions, processing speed, and motor control, and also memory, praxis, and language [33, 34], independently of other confounders. Diabetic subjects have a twofold increase in risk of mild cognitive impairment and dementia comparing to nondiabetics [13, 18, 35], an effect that stands long time after diabetes diagnosis [30].

Diabetes has several pathways to be implicated in the progression of dementia: not only due to the higher risk of vascular disease, but also mediated through metabolic changes due to the insulin and glycemia pathways, interfering with imbalance of glucagon/insulin homeostasis [36] that is implicated in the metabolic production of beta-amyloid protein and tau protein [27], promoting neuronal degeneration [37] and thus implicated in pathogenesis of AD [13, 38, 39]. Moreover, recent data suggest a genetic link between diabetes and the pathogenesis of AD [40, 41] and that insulin may modulate distribution of amyloid beta 40 and 42 in the brain [42].

Role of Stroke in Cognition

Stroke is a well-recognized risk factor for cognitive impairment in prospective community studies [7, 14, 35, 43, 44] and is associated with a twofold risk of dementia [44], not only for vascular dementia and vascular cognitive impairment, but also for degenerative dementias such as AD [44].

The higher risk of dementia in stroke survivors can be partially explained by concomitant vascular factors [45] and by pre-stroke dementia, but this is not the only explanation [44–46]. Nondemented stroke survivors have worse performance in tasks of attention and executive functions [33] comparing to subjects without stroke. On the other hand, small vessel disease predicts vascular dementia [47], even without clinical stroke.

The clear impact of stroke on the development of degenerative types of dementia is not well established. Although a higher risk of AD is associated with stroke, the pathological association between the two diseases is not clear. Neuropathological data suggested that vascular disease could affect cognition, not only through the effects on subcortical connections and white matter disease, but also exacerbating cortical atrophy [48–50]. One of the likely explanations could be that vascular acute events anticipate incipient cognitive impairment due to concomitant amyloid pathology or otherwise have a synergistic or additive effect to develop degenerative dementia. In line with this hypothesis, amyloid pathology was associated with more severe and rapid post-stroke/TIA cognitive decline in a recent publication [51]. However, so far, no evidence exists that stroke per se leads to increase of amyloid deposits [52]. On the other hand, in the DEDEMAS study [53], the majority of post-stroke cognitively impaired patients were not due to amyloid pathology, as deficits developed in the absence of amyloid pathology [53]. These findings suggest an alternative explanation implicating stroke as the direct cause of cognitive decline. In the same line, in a mouse model of recurrent photothrombotic stroke, recurrent infarcts (parietal cortex) were recently associated with progressive cognitive decline, with histopathologic evaluation showing remote astrogliosis of the hippocampus [54].

Role of Hypertension in Cognition

There is a considerable controversy between studies approaching some of the vascular risk factors and cognitive decline. One of the examples is the effect of hypertension. One of the most important variables that explain differences between studies considering hypertension is age of included subjects in those studies, with midlife hypertension being the cue for the explanation of the impact in cognition [55]. Hypertension in midlife has been consistently associated with later development of cognitive decline and dementia, with a higher effect in non-treated hypertensive subjects [56]. Sustained midlife hypertension was also associated with brain atrophy [57]. Although the strongest association is with vascular dementia, there is also an increased risk of degenerative dementia as Alzheimer's disease [7, 10, 17, 56, 58–60]. It was indeed suggested that hypertension was associated with greater amyloid burden not only in middle aged but also among older adults [61]. Treatment with antihypertensive treatment was associated with reduced hippocampus atrophy in hypertensive subjects [62] and with less AD neuropathology [63].

However, the relationship between late-onset hypertension and cognitive decline and dementia is less clear: some studies were negative for this association [11, 12, 64] or sustain that a very low systolic and/or diastolic value was associated with higher risk of cognitive decline [58, 59].

In cross-sectional studies among nondemented subjects, hypertension in late life was associated with worse performance in several cognitive tests mainly related with executive functions and attention, digit symbol test, and word fluency [33] but also difficulties in some global cognitive functioning tests [65, 66]. The most likely explanation for these discrepancies is that the deleterious effect of hypertension is due to chronic vascular damage starting in midlife that later originates cognitive impairment [60]. Results from trials focusing on the prevention of dementia using antihypertensive medication have failed to show a consistent protective effect, sustaining this explanation [67–69] and precluding a recommendation [69]. From the six main randomized placebo-controlled studies, four were negative for a protective effect [70–73], one found a small effect on the prevention of dementia [74], and the other [75] found a protective effect only for post-stroke dementia. Other studies, with concomitant treatments other than hypertension therapy, failed to show an effect in cognition [76], and from three recent studies approaching multifactorial intervention including hypertension control risk, in different settings, only one had a positive outcome [77–79]. In fact those studies were probably performed in older ages than what was desirable to prevent dementia and, additionally, the follow-up was short.

Role of Alcohol Intake and Smoking in Cognition

Influence of alcohol intake on brain structure and cognition has been a focus of interest in the two last decades. In the LADIS study [33], among subjects with white matter changes free of dementia and living independently, mild and moderate alcohol consumption was associated with better performance on global measures of cognition compared to non-drinkers (included never drinkers), but this relation was lost over time [33, 47]. Low or moderate alcohol intake was associated with reduced risk of AD in a systematic review with meta-analysis, compared to the risk of dementia in non-drinkers [80]. In this review, non-drinkers had a small higher risk compared also to excessive drinkers. However, non-drinkers could include former excessive drinkers that stopped consuming due to health problems [80]. These favorable results were replicated in a recent overview of systematic reviews under the topic [81]. However, a study conducted among older subjects could not find evidence that moderate alcohol intake could prevent cognitive decline [82]. Moreover, higher alcohol consumption and drinking have been associated with increased risk of dementia (both for vascular and Alzheimer's dementia) [83]. A recent review approached alcohol dose associated with a stratified risk of dementia and found that low dose (6 g/day for best association and 12.5 g/day maximum dosage for benefit) had the best association with low risk for dementia [84]. High risk of dementia was particularly found with dosages above 23 drinks/week or 38 g/day [84]. Considering imaging data, controversial data exists considering brain atrophy: brain atrophy was associated with alcohol intake even for low drinkers [85], but a recent study suggested that wine (among different types of alcohol beverages) was associated with larger total brain volume [86]. Direct effect of alcohol consumption on WMC and infarcts remains unclear [85].

Risk of dementia associated with smoking has also been studied. Smoking habits could have a theoretical beneficial effect in cognition, mediated through the stimulating effect of nicotine. In fact, the acute administration of nicotine in non-smoking young adults with attention deficit was associated with improvement in attention, executive functions, and working memory, probably mediated through the activation of the cholinergic system [87]. In a pilot study, an improvement in measures of attention, memory, and mental processing was found after 6 months of transdermal nicotine in non-smoking subjects with amnestic mild cognitive impairment, in a double-blind randomized trial [88]. Nevertheless, the deleterious effect of smoking, mediated through oxidative stress, triggering atherogenesis and inflammation could, even indirectly, mediate increased risk for cognitive decline. In a meta-analysis of

19 observational prospective studies, smoking increased the risk for dementia, not only vascular dementia, but also for degenerative dementias, an effect found mainly comparing active smokers against never-smokers [89]. This risk could potentially be more pronounced among persons without the APOE4 allele than among APOE4 carriers [90]. In a small study using estimates of relative risk, an increased relative risk was found between cigarette smoking and AD [91].

Role of Small Vessel Disease in Cognition

Small vessel disease is a broad concept used in several contexts and involves the cognitive, clinical, and imaging consequences of the pathological changes of the small vessels of the brain [92]. As small vessels are not visualized in vivo, visible imaging consequences of small vessel disease are usually considered as the marker of the disease. Clinical expression of small vessel disease is not uniform; to make it more complex, definition of small vessel disease varies between the different studies. Expression of small vessel disease includes lacunar infarcts, white matter changes, or hemorrhagic events, as microbleeds (Fig. 5.1). More recently, perivascular spaces that are mostly visible through MRI gained attention as an additional marker of small vessel disease. In a recent study, using genome-wide association study data from two different large sets of cases and controls, Traylor et al. found results supporting a shared pathophysiological process between AD and specifically small vessel disease strokes [93]. Location of MRI-visible perivascular space may potentially be different in these two pathologies [94].

In this section we will focus on the cognitive implications of small vessel disease.

White matter changes designate the changes of the radiological appearance of the white matter of the brain, detected through CT or MRI, of probable vascular etiology, that are frequently described in older subjects with or without cognitive deficit [95–106]. White matter changes do not follow specific vascular territories and are usually described as periventricular and subcortical but can also appear infratentorial in the pons. Age is the most frequent risk factor, but white matter changes are increased in subjects with hypertension and stroke [107]. Traditional clinical manifestations of white matter changes include cognitive decline, gait disturbances, urinary dysfunction, personality, and mood changes [92]. The knowledge of an implication of white matter changes in cognition has more than a century, but it was only after the advent of brain imaging that this concept gained interest, and the term leukoaraiosis was introduced [108]. Periventricular white matter changes are frequent in demented subjects, independently of the type of dementia [98]. White matter changes are associated with worse cognitive performance among nondemented older subjects, mainly in executive functions, attention, processing speed, and motor control [33, 99, 100, 109] but also in global measures of cognition [33, 99, 109], independently of other confounders. WMC severity is implicated in higher risk of cognitive impairment and dementia [47, 49, 102–105], and the relation is stronger with vascular dementia [47, 106–111]. Recently, Kandiah N et al. showed that white matter changes increased over the

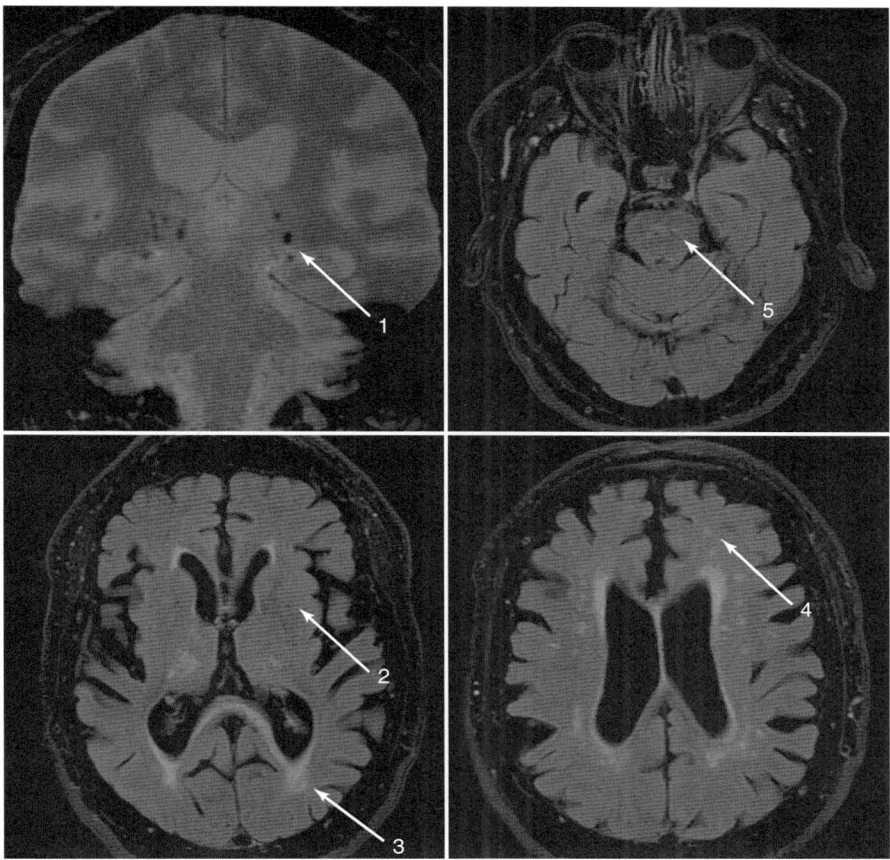

Fig. 5.1 Different expressions of small vessel disease, in the same patient. (1) Microbleeds. (2) Lacunes. (3) Periventricular white matter changes. (4) Subcortical white matter changes. (5) White matter changes in the pons

continuum of mild cognitive impairment and mild AD evolution, suggesting a synergistic effect between white matter changes and amyloid pathology [112]. Moreover, white matter changes were associated with cortical thickness [113], and effect found associated with other vascular lesions, as incident subcortical infarcts [114] and acute infarcts [50], and association eventually mediated through remote disconnecting phenomena. A nice summary of these effects is described in the METACOHORTS Consortium Statement [4].

Lacunes are frequently described in CT and MRI of elderly subjects and have been implicated in higher risk of dementia [115]. A recent systematic review and meta-analysis found an increased risk of mild cognitive impairment and dementia after lacunar stroke, the same risk described in other clinical non-lacunar strokes [116]. Similarly to white matter changes, lacunes have been implicated in worse executive functioning [117], processing speed, and motor control [118] among demented and nondemented subjects, with or without previous clinical stroke. The high frequency of lacunes in demented and nondemented subjects [119], and the coexistence to other small vessel disease types with lacunes [120] difficult the exact influence of lacunes in cognition. Specific locations, such as thalamic and basal ganglia lacunes, can have a specific impact in cognition [107], but further studies are needed to understand the individual effect of lacunes, even considering other concomitant confounders.

Cerebral microbleeds have been progressively described using specific susceptible MRI sequences. Prevalence data is highly variable, lower in community studies (7–36%), higher among demented subjects, mainly in subcortical vascular dementia (up to 85%) [121–124], but also in AD, where cerebral microbleeds are located more frequently in lobar areas [125].

Cerebral microbleeds have been associated with worse performance mainly in executive functions [122, 126–128], processing and motor speed [129–131], and attention [130]. Some recent evidence sustains a specific association between lobar microbleeds and memory deficit [132], and an association between cerebral microbleeds and cerebrospinal fluid biomarkers, emphasizing the link with amyloid pathology [131]. The increasing number of microbleeds seems to be associated with an increasing cognitive decline [127, 132], including AD [132].

Conclusions

Vascular risk factors are associated with an increased risk of cognitive decline and dementia, including degenerative dementia, and even among nondemented subjects, are associated with worse cognitive performance. Treatment and control of vascular risk factors in midlife has a key role in order to prevent cognitive impairment associated with aging. Nowadays, enough evidence sustains treatment of diabetes, prevention of stroke and stroke recurrence, and also treatment of hypertension in midlife, in order to prevent progression towards dementia. Further studies are needed to determine the type of intervention in each subject, considering other vascular risk factors [132]. Small vessel disease is increased in subjects with vascular risk factors, can be monitored with brain imaging, is associated with cognitive decline, and can be used as a hallmark of cerebral vascular disease. In future studies, small vessel disease, namely, white matter changes, represents a potential end point of experimental studies.

References

1. Lincoln P, Fenton K, Alessi C, Prince M, Brayne C, Wortmann M, Patel K, Deanfield J, Mwatsama M. The Blackfriars Consensus on brain health and dementia. Lancet. 2014;383:1805–6.
2. Smith D, Yaffe K. Dementia (including Alzheimer's disease) can be prevented: statement supported by international experts. J Alzheimers Dis. 2014;38:699–703.
3. Orrell M, Brayne C, INTERDEM (early detection and timely INTERvention in DEMentia); Alzheimer Europe; Alzheimer's Disease International; European Association of Geriatric Psychiatry. Dementia prevention: call to action. Lancet. 2015;386(10004):1625.

4. METACOHORTS Consortium. Electronic address: joanna.wardlaw@ed.ac.uk; METACOHORTS Consortium. METACOHORTS for the study of vascular disease and its contribution to cognitive decline and neurodegeneration: an initiative of the Joint Programme for Neurodegenerative Disease Research. Alzheimers Dement. 2016;12:1235–49.
5. Norton S, Matthews FE, Barnes DE, Yaffe K, Brayne C. Potential for primary prevention of Alzheimer's disease: an analysis of population-based data. Lancet Neurol. 2014;13:788–94.
6. Gottesman RF, Schneider AL, Zhou Y, Coresh J, Green E, Gupta N, Knopman DS, Mintz A, Rahmim A, Sharrett AR, Wagenknecht LE, Wong DF, Mosley TH. Association between midlife vascular risk factors and estimated brain amyloid deposition. JAMA. 2017;317:1443–50.
7. Hénon H, Pasquier F, Leys D. Poststroke dementia. Cerebrovasc Dis. 2006;22:61–70.
8. Troncoso JC, Zonderman AB, Resnick SM, Crain B, Pletnikova O, O'Brien RJ. Effect of infarcts on dementia in the Baltimore longitudinal study of aging. Ann Neurol. 2008;64:168–76.
9. Xu WL, Qiu CX, Wahlin A, Winblad B, Fratiglioni L. Diabetes mellitus and risk of dementia in the Kungsholmen project: a 6-year follow-up study. Neurology. 2004;63:1181–6.
10. Kivipelto M, Helkala EL, Laakso MP, Hänninen T, Hallikainen M, Alhainen K, et al. Midlife vascular risk factors and Alzheimer's disease in later life: longitudinal, population based study. BMJ. 2001;322:1447–51.
11. Hebert LE, Scherr PA, Bennett DA, Bienias JL, Wilson RS, Morris MC, Evans DA. Blood pressure and late-life cognitive function change. A biracial longitudinal population study. Neurology. 2004;62:2021–4.
12. Shah RC, Wilson RS, Bienias JL, Arvanitakis Z, Evans DA, Bennett DA. Relation of blood pressure to risk of incident Alzheimer's disease and change in global cognitive function in older persons. Neuroepidemiology. 2006;26:30–6.
13. Arvanitakis Z, Wilson RS, Bienias JL, Evans DA, Bennett DA. Diabetes Mellitus and risk of Alzheimer disease and decline in cognitive function. Arch Neurol. 2004;61:661–6.
14. Rastas S, Pirttilä T, Mattila K, Verkkoniemi A, Juva K, Niinistö L, et al. Vascular risk factors and dementia in the general population aged >85 years. Prospective population-based study. Neurobiol Aging. 2010;31:1–7.
15. Ruitenberg A, Skoog I, Ott A, Aevarsson O, Witteman JC, Lernfelt B, et al. Blood pressure and risk of dementia: results from the Rotterdam study and the Gothenburg H-70 Study. Dement Geriatr Cogn Disord. 2001;12:33–9.
16. Harrington F, Saxby BK, McKeith IG, Wesnes K, Ford GA. Cognitive performance in hypertensive and normotensive older subjects. Hypertension. 2000;36:1079–82.
17. Launer LJ, Ross GW, Petrovitch H, Masaki K, Foley D, White LR, Havlik RJ. Midlife blood pressure and dementia: the Honolulu-Asia aging study. Neurobiol Aging. 2000;21:49–55.
18. Biessels GJ, Staekenborg S, Brunner E, Brayne C, Scheltens P. Risk of dementia in diabetes mellitus: a systematic review. Lancet Neurol. 2006;5:64–74.
19. Singh-Manoux A, Kivimaki M, Glymour MM, Elbaz A, Berr C, Ebmeier KP, et al. Timing of onset of cognitive decline: results from Whitehall II prospective cohort study. BMJ. 2011;344:d7622.
20. Strand BH, Langballe EM, Hjellvik V, Handal M, Næss O, Knudsen GP, et al. Midlife vascular risk factors and their association with dementia deaths: results from a Norwegian prospective study followed up for 35 years. J Neurol Sci. 2013;324(1–2):124–30.
21. Alonso A, Jacobs DR Jr, Menotti A, Nissinen A, Dontas A, Kafatos A, Kromhout D. Cardiovascular risk factors and dementia mortality: 40 years of follow-up in the Seven Countries Study. J Neurol Sci. 2009;280(1–2):79–83.
22. Whitmer RA, Sidney S, Selby J, Johnston SC, Yaffe K. Midlife cardiovascular risk factors and risk of dementia in late life. Neurology. 2005;64(2):277–81.
23. Whitmer RA, Gustafson DR, Barrett-Connor E, Haan MN, Gunderson EP, Yaffe K. Central obesity and increased risk of dementia more than three decades later. Neurology. 2008;71:1057–64.
24. Anstey KJ, Cherbuin N, Budge M, Young J. Body mass index in midlife and late-life as a risk factor for dementia: a meta-analysis of prospective studies. Obes Rev. 2011;12(5):e426–37.

25. Purnell C, Gao S, Callahan CM, Hendrie HC. Cardiovascular risk factors and incident Alzheimer disease: a systematic review of the literature. Alzheimer Dis Assoc Disord. 2009;23:1–10.
26. Virta JJ, Heikkilä K, Perola M, Koskenvuo M, Räihä I, Rinne JO, Kaprio J. Midlife cardiovascular risk factors and late cognitive impairment. Eur J Epidemiol. 2013;28:405–16.
27. Luchsinger JA. Adiposity, hyperinsulinemia, diabetes and Alzheimer's disease: an epidemiological perspective. Eur J Pharmacol. 2008;585:119–29.
28. Euser SM, Sattar N, Witteman JC, Bollen EL, Sijbrands EJ, Hofman A, Perry IJ, Breteler MM, Westendorp RG, PROSPER and Rotterdam Study. A prospective analysis of elevated fasting glucose levels and cognitive function in older people: results from PROSPER and the Rotterdam Study. Diabetes. 2010;59:1601–7.
29. Rawlings AM, Sharrett AR, Mosley TH, Ballew SH, Deal JA, Selvin E. Glucose peaks and the risk of dementia and 20-year cognitive decline. Diabetes Care. 2017;40:879–86.
30. Rawlings AM, Sharrett AR, Schneider AL, Coresh J, Albert M, Couper D, Griswold M, Gottesman RF, Wagenknecht LE, Windham BG, Selvin E. Diabetes in midlife and cognitive change over 20 years: a cohort study. Ann Intern Med. 2014;161:785–93.
31. Vagelatos NT, Eslick GD. Type 2 diabetes as a risk factor for Alzheimer's disease: the confounders, interactions, and neuropathology associated with this relationship. Epidemiol Rev. 2013;2013(35):152–60.
32. Cukierman T, Gerstein HC, Williamson JD. Cognitive decline and dementia in diabetes-systematic overview of prospective observational studies. Diabetologia. 2005;48:2460–9.
33. Verdelho A, Madureira S, Ferro JM, Basile AM, Chabriat H, Erkinjuntti T, et al. Differential impact of cerebral white matter changes, diabetes, hypertension and stroke on cognitive performance among non-disabled elderly. The LADIS study. J Neurol Neurosurg Psychiatry. 2007;78:1325–30.
34. Manschot SM, Brands AM, van der Grond J, Kessels RP, Algra A, Kappelle LJ, et al. Brain magnetic resonance imaging correlates of impaired cognition in patients with type 2 diabetes. Diabetes. 2006;55:1106–13.
35. Yip AG, Brayne C, Matthews FE, MRC Cognitive Function and Ageing Study. Risk factors for incident dementia in England and Wales: the Medical Research Council Cognitive Function and Ageing Study. A population-based nested case–control study. Age Ageing. 2006;35:154–60.
36. Morsi M, Maher A, Metwally A, Abo-Elmagd O, Johar D, Bernstein L. A shared comparison of diabetes mellitus and neurodegenerative disorders. J Cell Biochem. 2017. https://doi.org/10.1002/jcb.26261. [Epub ahead of print].
37. Folch J, Pedrós I, Patraca I, Martínez N, Sureda F, Camins A. Metabolic basis of sporadic Alzheimer's disease. Role of hormones related to energy metabolism. Curr Pharm Des. 2013;19(38):6739–48.
38. Liu F, Shi J, Tanimukai H, Gu J, Gu J, Grundke-Iqbal I, Iqbal K, Gong CX. Reduced O-GlcNAcylation links lower brain glucose metabolism and tau pathology in Alzheimer's disease. Brain. 2009;132:1820–32.
39. de la Monte SM, Wands JR. Alzheimer's disease is type 3 diabetes-evidence reviewed. J Diabetes Sci Technol. 2008;2:1101–13.
40. Mirza Z, Kamal MA, Abuzenadah AM, Al-Qahtani MH, Karim S. Establishing genomic/transcriptomic links between Alzheimer's disease and type II diabetes mellitus by meta-analysis approach. CNS Neurol Disord Drug Targets. 2014;13:501–16.
41. Abdul-Rahman O, Sasvari-Szekely M, Ver A, Rosta K, Szasz BK, KeresztURI E, Keszler G. Altered gene expression profiles in the hippocampus and prefrontal cortex of type 2 diabetic rats. BMC Genomics. 2012;13:81.
42. Swaminathan SK, Ahlschwede KM, Sarma V, Curran GL, Omtri RS, Decklever T, Lowe VJ, Poduslu JF, Kandimalla KK. Insulin differentially affects the distribution kinetics of amyloid beta 40 and 42 in plasma and brain. J Cereb Blood Flow Metab. 2017:271678X17709709. https://doi.org/10.1177/0271678X17709709. [Epub ahead of print].

43. Reitz C, Bos MJ, Hofman A, Koudstaal PJ, Breteler MM. Prestroke cognitive performance, incident stroke, and risk of dementia: the Rotterdam Study. Stroke. 2008;39:36–41.
44. Savva GM, Stephan BC, Alzheimer's Society Vascular Dementia Systematic Review Group. Epidemiological studies of the effect of stroke on incident dementia: a systematic review. Stroke. 2010;41:e41–6.
45. Allan LM, Rowan EN, Firbank MJ, Thomas AJ, Parry SW, Polvikoski TM, et al. Long term incidence of dementia, predictors of mortality and pathological diagnosis in older stroke survivors. Brain. 2011;134(Pt 12):3716–27.
46. Pendlebury ST, Rothwell PM. Prevalence, incidence, and factors associated with pre-stroke and post-stroke dementia: a systematic review and meta-analysis. Lancet Neurol. 2009;8(11):1006–18.
47. Verdelho A, Madureira S, Moleiro C, Ferro JM, Santos CO, Erkinjuntti T, et al. White matter changes and diabetes predict cognitive decline in the elderly: the LADIS study. Neurology. 2010;75:160–7.
48. Jagust WJ, Zheng L, Harvey DJ, Mack WJ, Vinters HV, Weiner MW, et al. Neuropathological basis of magnetic resonance images in aging and dementia. Ann Neurol. 2008;63:72–80.
49. Capizzano AA, Ación L, Bekinschtein T, Furman M, Gomila H, Martínez A, et al. White matter hyperintensities are significantly associated with cortical atrophy in Alzheimer's disease. J Neurol Neurosurg Psychiatry. 2004;75:822–7.
50. Duering M, Righart R, Wollenweber FA, Zietemann V, Gesierich B, Dichgans M. Acute infarcts cause focal thinning in remote cortex via degeneration of connecting fiber tracts. Neurology. 2015;84:1685–92.
51. Liu W, Wong A, Au L, Yang J, Wang Z, Leung EY, Chen S, Ho CL, Mok VC. Influence of amyloid-β on cognitive decline after stroke/transient ischemic attack: three-year longitudinal study. Stroke. 2015;46:3074–80.
52. Sahathevan R, Linden T, Villemagne VL, Churilov L, Ly JV, Rowe C, Donnan G, Brodtmann A. Positron emission tomographic imaging in stroke: cross-sectional and follow-up assessment of amyloid in ischemic stroke. Stroke. 2016;47:113–9.
53. Wollenweber FA, Därr S, Müller C, Duering M, Buerger K, Zietemann V, Malik R, Brendel M, Ertl-Wagner B, Bartenstein P, Rominger A, Dichgans M. Prevalence of amyloid positron emission tomographic positivity in poststroke mild cognitive impairment. Stroke. 2016;47:2645–8.
54. Schmidt A, Diederich K, Strecker JK, Geng B, Hoppen M, Duning T, Schäbitz WR, Minnerup J. Progressive cognitive deficits in a mouse model of recurrent photothrombotic stroke. Stroke. 2015;46:1127–31.
55. Muller M, Sigurdsson S, Kjartansson O, Aspelund T, Lopez OL, Jonnson PV, et al. Joint effect of mid- and late-life blood pressure on the brain: the AGES-Reykjavik Study. Neurology. 2014;82:2187–95.
56. Gottesman RF, Schneider AL, Albert M, Alonso A, Bandeen-Roche K, Coker L, Coresh J, Knopman D, Power MC, Rawlings A, Sharrett AR, Wruck LM, Mosley TH. Midlife hypertension and 20-year cognitive change: the atherosclerosis risk in communities neurocognitive study. JAMA Neurol. 2014;71:1218–27.
57. Power MC, Schneider ALC, Wruck L, Griswold M, Coker LH, Alonso A, Jack CR Jr, Knopman D, Mosley TH, Gottesman RF. Life-course blood pressure in relation to brain volumes. Alzheimers Dement. 2016;12:890–9.
58. Qiu C, Winblad B, Fratiglioni L. Low diastolic pressure and risk of dementia in very old people: a longitudinal study. Dement Geriatr Cogn Disord. 2009;28:213–9.
59. Razay G, Williams J, King E, Smith AD, Wilcock G. Blood pressure, dementia and Alzheimer's disease: the OPTIMA longitudinal study. Dement Geriatr Cogn Disord. 2009;28:70–4.
60. Stewart R, Xue QL, Masaki K, Petrovitch H, Ross GW, White LR, Launer LJ. Change in blood pressure and incident dementia: a 32-year prospective study. Hypertension. 2009;54:233–40.
61. Rodrigue KM, Rieck JR, Kennedy KM, Devous MD Sr, Diaz-Arrastia R, Park DC. Risk factors for β-amyloid deposition in healthy aging: vascular and genetic effects. JAMA Neurol. 2013;70:600–6.

62. Korf ES, White LR, Scheltens P, Launer LJ. Midlife blood pressure and the risk of hippocampal atrophy: the Honolulu Asia Aging Study. Hypertension. 2004;44:29–34.
63. Hoffman LB, Schmeidler J, Lesser GT, Beeri MS, Purohit DP, Grossman HT, Haroutunian V. Less Alzheimer disease neuropathology in medicated hypertensive than nonhypertensive persons. Neurology. 2009;72:1720–6.
64. Di Carlo A, Baldereschi M, Amaducci L, Maggi S, Grigoletto F, Scarlato G, Inzitari D. Cognitive impairment without dementia in older people: prevalence, vascular risk factors, impact on disability. The Italian Longitudinal Study on Aging. J Am Geriatr Soc. 2000;48:775–82.
65. Cacciatore F, Abete P, Ferrara N, Paolisso G, Amato L, Canonico S, et al. The role of blood pressure in cognitive impairment in an elderly population. Osservatorio Geriatrico Campano Group. J Hypertens. 1997;15:135–42.
66. Budge MM, de Jager C, Hogervorst E, Smith AD. Total plasma homocysteine, age, systolic blood pressure, and cognitive performance in older people. J Am Geriatr Soc. 2002;50:2014–8.
67. Gorelick PB, Scuteri A, Black SE, Decarli C, Greenberg SM, Iadecola C, et al. Vascular contributions to cognitive impairment and dementia: a statement for healthcare professionals from the American Heart Association/American Stroke Association. Stroke. 2011;42:2672–713.
68. McGuinness B, Todd S, Passmore P, Bullock R. Blood pressure lowering in patients without prior cerebrovascular disease for prevention of cognitive impairment and dementia. Cochrane Database Syst Rev. 2009;(4):CD004034.
69. Iadecola C, Yaffe K, Biller J, Bratzke LC, Faraci FM, Gorelick PB, Gulati M, Kamel H, Knopman DS, Launer LJ, Saczynski JS, Seshadri S, Zeki Al Hazzouri A, American Heart Association Council on Hypertension; Council on Clinical Cardiology; Council on Cardiovascular Disease in the Young; Council on Cardiovascular and Stroke Nursing; Council on Quality of Care and Outcomes Research; and Stroke Council. Impact of hypertension on cognitive function: a scientific statement from the American Heart Association. Hypertension. 2016;68:e67–94.
70. SHEP Cooperative Research Group. Prevention of stroke by antihypertensive drug treatment in older persons with isolated systolic hypertension. Final results of the Systolic Hypertension in the Elderly Program (SHEP). JAMA. 1991;265:3255–64.
71. Lithell H, Hansson L, Skoog I, Elmfeldt D, Hofman A, Olofsson B, et al. The Study on Cognition and Prognosis in the Elderly (SCOPE): principal results of a randomized double-blind intervention trial. J Hypertens. 2003;21:875–86.
72. Peters R, Beckett N, Forette F, Tuomilehto J, Clarke R, Ritchie C, et al. Incident dementia and blood pressure lowering in the Hypertension in the Very Elderly Trial cognitive function assessment (HYVET-COG): a double-blind, placebo controlled trial. Lancet Neurol. 2008;7:683–9.
73. Yusuf S, Diener HC, Sacco RL, Cotton D, Ounpuu S, Lawton WA, et al. Telmisartan to prevent recurrent stroke and cardiovascular events. N Engl J Med. 2008;359:1225–37.
74. Forette F, Seux ML, Staessen JA, Thijs L, Birkenhager WH, Babarskiene MR, et al. Prevention of dementia in randomised double-blind placebocontrolled Systolic Hypertension in Europe (Syst-Eur) trial. Lancet. 1998;352:1347–51.
75. Tzourio C, Anderson C, Chapman N, Woodward M, Neal B, MacMahon S, et al. Effects of blood pressure lowering with perindopril and indapamide therapy on dementia and cognitive decline in patients with cerebrovascular disease. Arch Intern Med. 2003;163:1069–75.
76. Williamson JD, Launer LJ, Bryan RN, Coker LH, Lazar RM, Gerstein HC, Murray AM, Sullivan MD, Horowitz KR, Ding J, Marcovina S, Lovato L, Lovato J, Margolis KL, Davatzikos C, Barzilay J, Ginsberg HN, Linz PE, Miller ME, Action to Control Cardiovascular Risk in Diabetes Memory in Diabetes Investigators. Cognitive function and brain structure in persons with type 2 diabetes mellitus after intensive lowering of blood pressure and lipid levels: a randomized clinical trial. JAMA Intern Med. 2014;174:324–33.
77. Ihle-Hansen H, Thommessen B, Fagerland MW, Øksengård AR, Wyller TB, Engedal K, Fure B. Multifactorial vascular risk factor intervention to prevent cognitive impairment after stroke and TIA: a 12-month randomized controlled trial. Int J Stroke. 2014;9:932–8.

78. Matz K, Teuschl Y, Firlinger B, Dachenhausen A, Keindl M, Seyfang L, Tuomilehto J, Brainin M, ASPIS Study Group. Multidomain lifestyle interventions for the prevention of cognitive decline after ischemic stroke: randomized trial. Stroke. 2015;46(10):2874–80.
79. Ngandu T, Lehtisalo J, Solomon A, Levalahti E, Ahtiluoto S, Antikainen R, Bäckman L, Hänninen T, Jula A, Laatikainen T, Lindström J, Mangialasche F, Paajanen T, Pajala S, Peltonen M, Rauramaa R, Stigsdotter-Neely A, Strandberg T, Tuomilehto J, Soininen H, Kivipelto M. A 2 year multidomain intervention of diet, exercise, cognitive training, and vascular risk monitoring versus control to prevent cognitive decline in at-risk elderly people (FINGER): a randomised controlled trial. Lancet. 2015;385:2255–63.
80. Anstey KJ, Mack HA, Cherbuin N. Alcohol consumption as a risk factor for dementia and cognitive decline: meta-analysis of prospective studies. Am J Geriatr Psychiatry. 2009;17:542–55.
81. Ilomaki J, Jokanovic N, Tan EC, Lonnroos E. Alcohol consumption, dementia and cognitive decline: an overview of systematic reviews. Curr Clin Pharmacol. 2015;10:204–12.
82. Hogenkamp PS, Benedict C, Sjögren P, Kilander L, Lind L, Schiöth HB. Late-life alcohol consumption and cognitive function in elderly men. Age (Dordr). 2014;36:243–9.
83. Langballe EM, Ask H, Holmen J, Stordal E, Saltvedt I, Selbæk G, Fikseaunet A, Bergh S, Nafstad P, Tambs K. Alcohol consumption and risk of dementia up to 27 years later in a large, population-based sample: the HUNT study, Norway. Eur J Epidemiol. 2015;30:1049–56.
84. Xu W, Wang H, Wan Y, Tan C, Li J, Tan L, Yu JT. Alcohol consumption and dementia risk: a dose-response meta-analysis of prospective studies. Eur J Epidemiol. 2017;32:31–42.
85. Ding J, Eigenbrodt ML, Mosley TH Jr, Hutchinson RG, Folsom AR, Harris TB, Nieto FJ. Alcohol intake and cerebral abnormalities on magnetic resonance imaging in a community-based population of middle-aged adults: the Atherosclerosis Risk in Communities (ARIC) study. Stroke. 2004;35:16–21.
86. Gu Y, Scarmeas N, Short EE, Luchsinger JA, DeCarli C, Stern Y, Manly JJ, Schupf N, Mayeux R, Brickman AM. Alcohol intake and brain structure in a multiethnic elderly cohort. Clin Nutr. 2014;33:662–7.
87. Potter AS, Newhouse PA. Acute nicotine improves cognitive deficits in young adults with attention-deficit/hyperactivity disorder. Pharmacol Biochem Behav. 2008;88:407–17.
88. Newhouse P, Kellar K, Aisen P, White H, Wesnes K, Coderre E, et al. Nicotine treatment of mild cognitive impairment: a 6-month double-blind pilot clinical trial. Neurology. 2012;78:91–101.
89. Anstey KJ, von Sanden C, Salim A, O'Kearney R. Smoking as a risk factor for dementia and cognitive decline: a meta-analysis of prospective studies. Am J Epidemiol. 2007;166:367–78.
90. Reitz C, den Heijer T, van Duijn C, Hofman A, Breteler MM. Relation between smoking and risk of dementia and Alzheimer disease: the Rotterdam Study. Neurology. 2007;69:998–1005.
91. Saito EK, Diaz N, Chung J, McMurtray A. Smoking history and Alzheimer's disease risk in a community-based clinic population. J Educ Health Promot. 2017;6:24.
92. Pantoni L. Cerebral small vessel disease: from pathogenesis and clinical characteristics to therapeutic challenges. Lancet Neurol. 2010;9:689–701.
93. Traylor M, Adib-Samii P, Harold D, Alzheimer's Disease Neuroimaging Initiative, International Stroke Genetics Consortium (ISGC), UK Young Lacunar Stroke DNA Resource, Dichgans M, Williams J, Lewis CM, Markus HS, METASTROKE, International Genomics of Alzheimer's Project (IGAP), Investigators. Shared genetic contribution to ischaemic stroke and Alzheimer's disease. Ann Neurol. 2016;79(5):739–47. https://doi.org/10.1002/ana.24621.
94. Banerjee G, Kim HJ, Fox Z, Jäger HR, Wilson D, Charidimou A, Na HK, Na DL, Seo SW, Werring DJ. MRI-visible perivascular space location is associated with Alzheimer's disease independently of amyloid burden. Brain. 2017;140:1107–16.
95. de Leeuw FE, de Groot JC, Achten E, Oudkerk M, Ramos LM, Heijboer R, et al. Prevalence of cerebral white matter lesions in elderly people: a population based magnetic resonance imaging study. The Rotterdam scan study. J Neurol Neurosurg Psychiatry. 2001;70:9–14.
96. Longstreth WT, Manolio TA, Arnold A. Clinical correlates of white matter findings on cranial magnetic resonance imaging of 3301 elderly people: the cardiovascular health study. Stroke. 1996;27:1274–82.

97. Ylikoski A, Erkinjuntti T, Raininko R, Sarna S, Sulkava R, Tilvis R. White matter hyperintensities on mri in the neurologically nondiseased elderly. Analysis of cohorts of consecutive subjects aged 55 to 85 years living at home. Stroke. 1995;26:1171–7.
98. Schmidt R, Schmidt H, Haybaeck J, Loitfelder M, Weis S, Cavalieri M, Seiler S, Enzinger C, Ropele S, Erkinjuntti T, Pantoni L, Scheltens P, Fazekas F, Jellinger K. Heterogeneity in age-related white matter changes. Acta Neuropathol. 2011;122:171–85.
99. Skoog I, Berg S, Johansson B, Palmertz B, Andreasson LA. The influence of white matter lesions on neuropsychological functioning in demented and non-demented 85-yeras-olds. Acta Neurol Scand. 1996;93:142–8.
100. de Leeuw FE, de Groot JC, Oudkerk M, Witteman JC, Hofman A, van Gijn J, Breteler MM. Hypertension and cerebral white matter lesions in a prospective cohort study. Brain. 2002;125:765–72.
101. Ylikoski R, Ylikoski A, Raininko R, Keskivaara P, Sulkava R, Tilvis R, Erkinjuntti T. Cardiovascular diseases, health status, brain imaging findings and neuropsychological functioning in neurologically healthy elderly individuals. Arch Gerontol Geriatr. 2000;30:115–30.
102. Inaba M, White L, Bell C, Chen R, Petrovitch H, Launer L, Abbott RD, Ross GW, Masaki K. White matter lesions on brain magnetic resonance imaging scan and 5-year cognitive decline: the Honolulu-Asia aging study. J Am Geriatr Soc. 2011;59:1484–9.
103. Silbert LC, Howieson DB, Dodge H, Kaye JA. Cognitive impairment risk: white matter hyperintensity progression matters. Neurology. 2009;73:120–5.
104. Jokinen H, Kalska H, Ylikoski R, Madureira S, Verdelho A, van der Flier WM, Scheltens P, Barkhof F, Visser MC, Fazekas F, Schmidt R, O'Brien J, Waldemar G, Wallin A, Chabriat H, Pantoni L, Inzitari D, Erkinjuntti T, LADIS Group. Longitudinal cognitive decline in subcortical ischemic vascular disease—the LADIS study. Cerebrovasc Dis. 2009;27:384–91.
105. Steffens DC, Potter GG, McQuoid DR, MacFall JR, Payne ME, Burke JR, Plassman BL, Welsh-Bohmer KA. Longitudinal magnetic resonance imaging vascular changes, apolipoprotein e genotype, and development of dementia in the neurocognitive outcomes of depression in the elderly study. Am J Geriatr Psychiatry. 2007;15:839–49.
106. Kuller LH, Lopez OL, Newman A, Beauchamp NJ, Burke G, Dulberg C, Fitzpatrick A, Fried L, Haan MN. Risk factors for dementia in the cardiovascular health cognition study. Neuroepidemiology. 2003;22:13–22.
107. The LADIS Study Group. 2001–2011: a decade of the LADIS (Leukoaraiosis and DISability) study: what have we learned about white matter changes and small-vessel disease? Cerebrovasc Dis. 2011;32:577–88.
108. Hachinski VC, Potter P, Merskey H. Leuko-araiosis: an ancient term for a new problem. Can J Neurol Sci. 1986;13:533–4.
109. Madureira S, Verdelho A, Ferro J, Basile AM, Chabriat H, Erkinjuntti T, Fazekas F, Hennerici M, O'brien J, Pantoni L, Salvadori E, Scheltens P, Visser MC, Wahlund LO, Waldemar G, Wallin A, Inzitari D, LADIS Study Group. Development of a neuropsychological battery for a multinational study: the LADIS. Neuroepidemiology. 2006;27:101–16.
110. Bombois S, Debette S, Bruandet A, Delbeuck X, Delmaire C, Leys D, Pasquier F. Vascular subcortical hyperintensities predict conversion to vascular and mixed dementia in mci patients. Stroke. 2008;39:2046–51.
111. Meguro K, Ishii H, Kasuya M, Akanuma K, Meguro M, Kasai M, Lee E, Hashimoto R, Yamaguchi S, Asada T. Incidence of dementia and associated risk factors in japan: the osaki-tajiri project. J Neurol Sci. 2007;260:175–82.
112. Kandiah N, Chander RJ, Ng A, Wen MC, Cenina AR, Assam PN. Association between white matter hyperintensity and medial temporal atrophy at various stages of Alzheimer's disease. Eur J Neurol. 2015;22:150–5.

113. Tuladhar AM, Reid AT, Shumskaya E, de Laat KF, van Norden AG, van Dijk EJ, van Norden AG, van Dijk EJ, Norris DG, de Leeuw FE. Relationship between white matter hyperintensities, cortical thickness, and cognition. Stroke. 2015;46:425–32.
114. Duering M, Righart R, Csanadi E, Jouvent E, Herve D, Chabriat H, Dichgans M. Incident subcortical infarcts induce focal thinning in connected cortical regions. Neurology. 2012;79:2025–8.
115. Loeb C, Gandolfo C, Crose R, Conti M. Dementia associated with lacunar infarction. Stroke. 1992;23:1225–9.
116. Makin S, Turpin S, Dennis M, Wardlaw J. Cognitive impairment after lacunar stroke: systematic review and meta-analysis of incidence, prevalence and comparison with other stroke sub-types. J Neurol Neurosurg Psychiatry. 2013;84:893–900.
117. Carey CL, Kramer JH, Josephson SA, Mungas D, Reed BR, Schuff N, Weiner MW, Chui HC. Subcortical lacunes are associated with executive dysfunction in cognitively normal elderly. Stroke. 2008;39:397–402.
118. Benisty S, Gouw AA, Porcher R, Madureira S, Hernandez K, Poggesi A, van der Flier WM, Van Straaten EC, Verdelho A, Ferro J, Pantoni L, Inzitari D, Barkhof F, Fazekas F, Chabriat H, LADIS Study Group. Location of lacunar infarcts correlates with cognition in a sample of non-disabled subjects with age-related white-matter changes: the LADIS study. J Neurol Neurosurg Psychiatry. 2009;80:478–83.
119. Jellinger KA, Attems J. Incidence of cerebrovascular lesions in Alzheimer's disease: a postmortem study. Acta Neuropathol. 2003;105:14–7.
120. Miyao S, Takano A, Teramoto J, Takahashi A. Leukoaraiosis in relation to prognosis for patients with lacunar infarction. Stroke. 1992;23:1434–8.
121. Hanyu H, Tanaka Y, Shimizu S, Takasaki M, Fujita H, Kaneko N, Yamamoto Y, Harada M. Cerebral microbleeds in Binswanger's disease: a gradient-echo t2*-weighted magnetic resonance imaging study. Neurosci Lett. 2003;340:213–6.
122. Poels MM, Vernooij MW, Ikram MA, Hofman A, Krestin GP, van der Lugt A, Breteler MM. Prevalence and risk factors of cerebral microbleeds: an update of the Rotterdam scan study. Stroke. 2010;41:S103–6.
123. Seo SW, Hwa Lee B, Kim EJ, Chin J, Sun Cho Y, Yoon U, Na DL. Clinical significance of microbleeds in subcortical vascular dementia. Stroke. 2007;38:1949–51.
124. Ayaz M, Boikov AS, Haacke EM, Kido DK, Kirsch WM. Imaging cerebral microbleeds using susceptibility weighted imaging: one step toward detecting vascular dementia. J Magn Reson Imaging. 2010;31:142–8.
125. Shams S, Martola J, Granberg T, Li X, Shams M, Fereshtehnejad SM, Cavallin L, Aspelin P, Kristoffersen-Wiberg M, Wahlund LO. Cerebral microbleeds: different prevalence, topography, and risk factors depending on dementia diagnosis—the Karolinska Imaging Dementia Study. Am J Neuroradiol. 2015;36:661–6.
126. Gregoire SM, Smith K, Jager HR, Benjamin M, Kallis C, Brown MM, Cipolotti L, Werring DJ. Cerebral microbleeds and long-term cognitive outcome: longitudinal cohort study of stroke clinic patients. Cerebrovasc Dis. 2012;33:430–5.
127. Werring DJ, Frazer DW, Coward LJ, Losseff NA, Watt H, Cipolotti L, et al. Cognitive dysfunction in patients with cerebral microbleeds on t2*-weighted gradient-echo MRI. Brain. 2004;127:2265–75.
128. Qiu C, Cotch MF, Sigurdsson S, Jonsson PV, Jonsdottir MK, Sveinbjrnsdottir S, et al. Cerebral microbleeds, retinopathy, and dementia: the ages-Reykjavik study. Neurology. 2010;75:2221–8.
129. Poels MM, Ikram MA, van der Lugt A, Hofman A, Niessen WJ, Krestin GP, Breteler MM, Vernooij MW. Cerebral microbleeds are associated with worse cognitive function: the Rotterdam scan study. Neurology. 2012;78:326–33.

130. van Norden AG, van den Berg HA, de Laat KF, Gons RA, van Dijk EJ, de Leeuw FE. Frontal and temporal microbleeds are related to cognitive function: the Radboud University Nijmegen Diffusion Tensor and Magnetic Resonance Cohort (RUN DMC) Study. Stroke. 2011;42:3382–6.
131. Shams S, Granberg T, Martola J, Charidimou A, Li X, Shams M, Fereshtehnejad SM, Cavallin L, Aspelin P, Wiberg-Kristoffersen M, Wahlund LO. Cerebral microbleeds topography and cerebrospinal fluid biomarkers in cognitive impairment. J Cereb Blood Flow Metab. 2017;37:1006–13.
132. Akoudad S, Wolters FJ, Viswanathan A, de Bruijn RF, van der Lugt A, Hofman A, Koudstaal PJ, Ikram MA, Vernooij MW. Association of cerebral microbleeds with cognitive decline and dementia. JAMA Neurol. 2016;73:934–43.

Risk Factors and Prevention in Alzheimer's Disease and Dementia

6

Giulia Grande, Davide L. Vetrano, and Francesca Mangialasche

Abstract

Along with global aging, the number of people suffering from dementia and Alzheimer's disease (AD) will dramatically increase with burdensome consequences at both individual and societal levels. Since so far no effective curative drugs have been found, the identification of modifiable factors to reduce the risk of cognitive decline remains a public health priority. Up to one-third of AD cases worldwide can be attributable to the presence of seven potentially modifiable risk factors: physical inactivity, smoking, midlife hypertension and obesity, DM, depression, and low level of education. Therefore, it might be possible to substantially reduce AD occurrence through public health interventions promoting activities enhancing cognitive reserve and healthy lifestyles. In this chapter, we summarize the major findings concerning risk and protective factors for demen-

G. Grande, M.D. (✉)
Aging Research Center, Department of Neurobiology, Care Sciences, and Society (NVS), Karolinska Institutet and Stockholm University, Stockholm, Sweden

Center for Research and Treatment on Cognitive Dysfunctions, Biomedical and Clinical Sciences Department, "Luigi Sacco" Hospital, University of Milan, Milano, Italy
e-mail: giulia.grande@ki.se

D.L. Vetrano, M.D.
Aging Research Center, Department of Neurobiology, Care Sciences, and Society (NVS), Karolinska Institutet and Stockholm University, Stockholm, Sweden

Department of Geriatrics, Catholic University of Rome, Rome, Italy

F. Mangialasche, M.D., Ph.D.
Aging Research Center, Department of Neurobiology, Care Sciences, and Society (NVS), Karolinska Institutet and Stockholm University, Stockholm, Sweden

Section of Gerontology and Geriatrics, University of Perugia, Perugia, Italy

© Springer International Publishing AG 2018
D. Galimberti, E. Scarpini (eds.), *Neurodegenerative Diseases*,
https://doi.org/10.1007/978-3-319-72938-1_6

tia and AD, based on current epidemiological evidence from observational and interventional studies. We also discuss the impact of ongoing interventional studies testing the effect of preventive measures for dementia and AD.

Keywords

Alzheimer's disease · Dementia · Risk factors · Prevention · Multi-domain intervention

Introduction

Dementia represents a growing global challenge. The World Alzheimer Report estimated that in 2015 approximately 47 million people were living with dementia worldwide. These figures are expected to reach 75 million by 2030, and 131 million by 2050, with the greatest increase expected in low- and middle-income countries [1]. The World Health Organization (WHO) has described this trend in terms of a fast-growing epidemic, concluding that Alzheimer's disease (AD) and other dementias should be regarded as a public health priority [2, 3], with global economic costs exceeding $818 billion in 2015 [1].

AD is the leading cause of dementia, accounting for 60–70% of cases, although increasing evidence has shown that mixed brain pathologies (AD together with vascular lesions) account for the majority of dementia cases, especially in advanced age (85+ years) [4]. Dementia and AD are multifactorial disorders, where genetic susceptibility and environmental factors (e.g. psychosocial, lifestyle and biological factors), as well as their interaction over the lifespan, contribute to the pathological process and the clinical expression of the disease (Table 6.1). The frequent co-occurrence of AD and cerebrovascular disease is consistent with the evidence that both disorders share several risk and protective factors, supporting the validity of dementia syndrome as a target for prevention. Findings from projection studies have suggested that prevention is likely to delay the onset and therefore reduce the prevalence of AD, which is currently incurable [5]. For example, it has been estimated that an intervention that delays AD onset of 1 year would reduce the worldwide total number of AD cases by 11% [6]. Although population-based studies are not entirely consistent, it does appear that the incidence of all-cause dementia has been declining in high-income countries over the past decades [7–9], mainly due to the reduction in the prevalence of many vascular risk factors over time [7]. This supports the hypothesis that prevention is a strategy to halt or delay cognitive decline in older people. In line with these findings, a growing body of literature has reported that multi-domain interventions, aimed at reducing several risk factors at the same time, can benefit cognition in older adults [10].

In this chapter, we summarize the major findings concerning risk and protective factors for dementia and AD, based on current epidemiological evidence from observational and interventional studies. As many epidemiological findings here discussed apply to both dementia and AD, these terms are used interchangeably when appropriate. We also discuss the impact of ongoing interventional studies testing the effect of preventive measures for dementia/AD.

Table 6.1 Risk and protective factors for cognitive decline and dementia

Risk factors	Protective factors
Age	**Genetic**
Gender (female sex)	Different genes (e.g. *APP, APOE* ε2)
Genetic	have been proposed (see Chap. 4)
Familial aggregation	Psychosocial factors
APOE ε4	High levels of education and
Different genes have been proposed (see Chap. 4)	socioeconomic status
Vascular and metabolic	High level of occupational
Hypercholesterolaemia	complexity
Diabetes mellitus and pre-diabetes	Rich social network and social
Cerebrovascular disorders (stroke, clinically silent	engagement
brain infarcts and cerebral microvascular lesions)	Mentally stimulating activities
Cardiovascular diseases (myocardial infarction,	**Lifestyle**
coronary heart disease, atrial fibrillation)	Physical activity
Smoking	Moderate alcohol intake
Metabolic syndrome	**Diet**
Midlife positive association but late-life negative	Mediterranean, DASH and MIND
association	diets
Hypertension	PUFAs and fish-related fats
High BMI (overweight and obesity)	Vitamins B_6 and B_{12}, folate
High serum cholesterol	Antioxidant vitamins (A, C, and E)
Diet	Vitamin D
Saturated fats	**Drugs**
Homocysteine	Antihypertensive drugs
Others	Statins
Depression	HRT
Obstructive sleep apnea	NSAIDs
Hearing loss	
Traumatic brain injury	
Occupational exposure (heavy metals, ELF-EMFs)	
Air pollution	
Infective agents (herpes simplex virus type I, *Chlamydophila pneumoniae*, spirochetes)	
Drugs (benzodiazepines, anticholinergics, opioids, antipsychotics)	

APP amyloid precursor protein, *APOE* apolipoprotein E, *BMI* body mass index, *DASH* Dietary Approaches to Stop Hypertension, *ELF-EMF* extremely low-frequency electromagnetic field, *HRT* hormone replacement therapy, *MIND* Mediterranean-DASH Intervention for Neurodegenerative Delay, *NSAIDs* non-steroidal anti-inflammatory drug, *PUFA* polyunsaturated fatty acid

Non-modifiable Risk Factors

Age

Age remains the strongest risk factor for the onset of dementia, particularly of AD type [11]. The incidence of dementia/AD approximately doubles every 5 years after the age of 60, with up to two-thirds of nursing home residents estimated to suffer from dementia [12]. In Europe, approximately 2 of every 1000 person-years become demented among people aged 65–69 years, and the incidence increases from 70 to 80 of every 1000 person-years among people aged 90+. It is still unclear if the incidence of dementia continues to increase in very advanced age or reaches a plateau. The Cache County Study found that the incidence of dementia increased with age, peaked, and then started to decline at extreme old ages for both men and women [13]. However, some meta-analyses and large-scale studies in Europe provided no evidence for the potential decline in the incidence of dementia among the oldest-old adults [14, 15].

Familial Aggregation

Familial aggregation is an important risk factor for dementia, particularly of AD type. A first-degree relative history of dementia is associated with an approximately twofold increase in the relative risk of AD [16, 17]. Familial clustering might be explained by both shared disease susceptibility genes and similar lifestyle habits and environmental factors among family members [18]. Risk estimates gradually decline with advancing parental age at diagnosis of AD, with small or even no increased risk when the diagnosis is set after 80 years of age [19].

For a detailed discussion regarding the genetic factors in the occurrence of AD, refer to Chap. 4 of this book.

Gender

Several studies have suggested that women have higher incidence and prevalence of dementia than men [20]. This can be partly due to sex differences in survival, with women having a longer life expectancy than men. Other factors which might contribute to this gender difference include variations in risk factors exposure, namely, a lower cognitive reserve in women, due to a lower level of education with a less qualified occupational status. In addition, differences in sex-specific hormones (levels of oestrogens vs. testosterone) might explain in part the gender difference in dementia occurrence [21, 22].

Modifiable Risk Factors

Vascular and Metabolic Risk Factors

Vascular and metabolic risk factors have been consistently associated with cognitive decline and dementia, both vascular and AD subtype. These associations are much stronger when exposure to these factors occurs in midlife rather later in life [23–25].

Hypercholesterolaemia
Hypercholesterolaemia represents a major risk factor for atherosclerosis, and in both AD and vascular dementia a role of vascular damage has been highlighted. Hypercholesterolaemia during young and middle age (<65 years) has been

associated with an increased risk of dementia/AD, whereas low blood total cholesterol in late-life (age >75 years) has been associated with subsequent development of dementia/AD [4]. It is plausible that the age-dependent association is due to the decrease of blood total cholesterol in the early, asymptomatic stages of dementia, which most likely occurs as a consequence of the disease (concept identified as "reverse causality"). The apolipoprotein E (*APOE*) genotype, a well-established risk factor for AD, might modify the impact of hypercholesterolaemia on dementia risk. The biological plausibility of this interaction derives from the role of *APOE* in encoding a brain cholesterol-transporter protein [26].

Although some retrospective observational studies have suggested that statins (cholesterol-lowering medications) use might reduce the risk of AD [27, 28], a meta-analysis of these studies [29] concluded that statins did not protect against dementia. Moreover, two randomized controlled trials (RCTs) on pravastatin and simvastatin, as well as a recent systematic review, failed in demonstrating a protective effect of statins on dementia incidence in older populations with high cardiovascular risk [30–32]. It is however worth to mention that these studies were not primarily designed to detect changes in AD incidence. It is also possible that a treatment started in late-life is unlikely to halt an already established neurodegenerative process. In favour of a protective role of statins is the fact that these drugs reduce the risk of stroke and cerebrovascular diseases, which are established risk factors for cognitive decline and dementia [33]. On the other hand, the results of these studies might suffer from a confounding by indication bias, namely, clinicians might be less prone to prescribe statins to individuals with complex clinical conditions that already imply a high pharmacological burden and are associated with increased risk of dementia.

Firm conclusions of the use of statins on the occurrence of dementia or AD cannot be drawn, and RCTs with AD as a primary outcome are needed to disentangle this association [34].

Diabetes Mellitus

Robust evidence has related diabetes mellitus (DM) with cognitive decline and dementia/AD [35–39]. Both hyperglycaemia and insulin resistance seem to play a direct role in cognitive decline [40–42].

A systematic review of observational studies has found a 50–100% increased risk of AD and a 150% increased risk of vascular dementia in individuals with DM [43]. Typically, DM clusters with other cardiovascular risk factors or diseases and is consistently associated with the incidence of vascular dementia. On the other hand, Cherbuin and coworkers found an association between higher plasma glucose and hippocampal atrophy, suggesting a direct damage of hyperglycaemia in the anatomical structures implicated in AD [44].

Approximately 3% of AD cases worldwide have been attributed to the presence of DM in the general population, which means around one million cases [5].

If the association between DM and dementia/AD is quite well established, the role of antidiabetic drugs in dementia/AD occurrence is rather questionable, and the evidence is still limited [45]. In the Rotterdam Study, individuals with DM treated with insulin had a substantially greater risk of developing AD [35]. By contrast, a neuropathological study has reported that individuals treated with both insulin and oral antidiabetic drugs had a significantly lower neuritic plaque density than individuals without DM [46, 47]. However, some RCTs testing oral antidiabetic medications on people with mild to moderate AD have failed to demonstrate positive effects of these drugs on cognition, with the drawbacks of higher risk of adverse cardiovascular events of rosiglitazone [48]. Additionally, inconsistent data have been reported for metformin use and cognition in diabetic patients [49]. Finally, the cognitive benefits of an intensive glycaemic control in elderly subjects are not yet clear, due to the increased vulnerability of older adults to hypoglycaemia, which can in turn increase the risk of dementia [50].

Blood Pressure

The association between blood pressure and dementia/AD is complex and varies with age [51]. Hypertension in midlife has been consistently associated with increased risk of dementia/AD later in life. Conversely, hypertension later in life is not significantly related with AD; instead, hypotension appears to be a risk factor for dementia/AD in older populations [52]. In the CAIDE cohort, a population-based study, midlife high systolic blood pressure nearly doubled the risk of late-life AD [53]. In another survey—the Honolulu-Asia Aging Study (HAAS)—untreated high diastolic blood pressure increased the risk of AD by four times [54]. Both the HAAS and the CAIDE have a long follow-up: 27 and 21 years, respectively. Other prospective population-based studies reported a U-shaped relationship between blood pressure and dementia/AD [55, 56].

Norton and colleagues [5] pooled the findings coming from observational studies associating midlife hypertension and AD and calculated a weighted relative risk of 1.61 (95% confidence interval, CI: 1.16–2.24). Interestingly, the authors estimated that around 5% of AD cases worldwide could be related to the presence of hypertension in midlife.

Hypertension can raise the risk of dementia by boosting the risk of stroke and multi-lacunar cerebral infarcts. High blood pressure can also increase the burden of white matter lesions in the brain, thus lowering the threshold at which AD pathology produces clinically relevant symptoms. Finally, hypertension can also exacerbate AD-related neurodegeneration.

The effect of antihypertensive treatments in reducing the risk of dementia/AD is stronger in midlife than in late-life [57]. Findings from longitudinal studies associated a decreased risk of both cognitive decline and dementia in people on antihypertensive treatment [58], but RCTs have yielded mixed results. This heterogeneity may stem from the different length of the follow-up and the drugs considered. No consistency has been found to support the use of an antihypertensive molecule over another [59].

Additionally, there is insufficient knowledge on optimal therapeutic targets for blood pressure control among older adults and oldest-old adults (85+ years). Nevertheless, recommendations from the American Heart Association and American Stroke Associations (AHA/ASA) suggest that it is reasonable to lower blood pressure in midlife to reduce the risk of post-stroke and vascular dementia later in life [57]. Caution is recommended regarding the same treatment among adults older than 80 years of age [51].

Obesity

Midlife obesity—defined as body mass index (BMI) ≥30—has been often [53, 60–62], but not always [63], found to be associated with higher risk of incident dementia/AD. Not only obesity, but also being overweight (BMI ≥25) has been associated with an increased risk of dementia [60]. Moreover, a greater BMI has been associated with a higher risk of hospital or death certificate diagnosis of dementia during 20–30 years of follow-up in a population-based study [61].

Based on the available evidence, a pooled relative risk of 1.60 (95% CI: 1.34–1.92) has been estimated for midlife obesity and the risk of AD, with about 2% of all AD cases worldwide being potentially related to the presence of this risk factor [5].

Similar to hypertension and hypercholesterolaemia, findings suggest that the association between BMI and dementia/AD is age dependent, with both a higher BMI and obesity in late-life associated with a lower risk of developing dementia/AD [64, 65]. In the HAAS study, weight loss accelerated the time of dementia diagnosis [66]. Another cohort study on older adults reported an increased rate of cognitive decline in patients who lost one unit of BMI compared with those with stable BMI. Post-mortem data in a sub-sample of this cohort showed a higher degree of AD pathology in people with a lower BMI [67].

In conclusion, overweight and obesity in midlife can increase the risk of dementia/AD later in life, probably by promoting hormonal imbalance, accelerated brain aging, and by enhancing vascular and neurodegenerative pathways in individuals with high levels of adiposity [68]. Conversely, underweight later in life is associated with dementia/AD development, reflecting more an early manifestation of the disease rather than a true risk factor [69].

Smoking

Smoking is a renowned cardiovascular risk factor, and a meta-analysis of 19 studies showed that current smokers had higher dementia risk (relative risk: 1.27; 95% CI: 1.02–1.60) as compared to those who never smoked, with a relative risk peaking up to 1.79 (95% CI: 1.43–2.23) when considering AD [70]. The same study reported a greater cognitive decline in older people who were current smokers.

Recent evidence reviewed by the WHO confirms the strong link between smoking and the risk of dementia, reporting a dose-response effect. This means that the more a person smokes, the higher is the risk of cognitive decline and dementia [71]. It has been estimated that 14% of AD cases worldwide (about 4.7 million) are potentially attributable to smoking [5].

Less consistent results are available for the link between history of smoking and dementia/AD risk, possibly indicating that quitting smoking later in life is still beneficial and could reduce the risk of AD or other forms of dementia compared with continued smoking [70].

Potential mechanisms underlying the effect of smoking in the risk of dementia/AD include promotion of both vascular and neurodegenerative damage. Smoking is also an indicator of an unhealthy lifestyle, with concurrent factors which can contribute in increasing the risk of dementia/AD.

Metabolic Syndrome and Other Vascular Risk Disorders

The metabolic syndrome is defined by the co-occurrence of cardiovascular risk factors, including hypertension, obesity, insulin resistance, and dyslipidaemia, and has been strongly related with cardiovascular diseases and mortality [72, 73]. Nonetheless, evidence associating the metabolic syndrome with dementia or cognitive decline is somehow inconclusive, with longitudinal studies yielding mixed results [74, 75]. In the French Three-City Study [76], the metabolic syndrome was associated with cognitive decline, whereas in the HAAS study this association was weak and related to the incident risk of vascular dementia, but not AD.

Other vascular disorders related to increased risk of dementia/AD and cognitive decline include myocardial infarction, coronary heart disease, heart failure, carotid atherosclerosis, stroke, clinically silent brain infarcts and cerebral microvascular lesions [77–79]. Additionally, meta-analyses of longitudinal studies have reported a hazard ratio of 1.4 for atrial fibrillation and dementia [80, 81].

Lifestyle

Physical, mental, and social activities are among the main components of the lifestyle, and a growing body of evidence has linked these activities with dementia/AD occurrence.

So far, the majority of the findings on lifestyle and dementia comes from observational studies, and methodological challenges might affect the results. In this context, it is arduous to avoid a reverse causation bias, since prodromal dementia may manifest itself with decreased initiative and interest, as well as low mood. These can lead to a reduced mental and physical activity and social isolation.

Physical Activity

Physical activity is one of the lifestyle components collecting strong evidence as protective factor against dementia [82–84]. Exercise can indeed reduce the risk of cognitive decline, dementia, and AD when practised in midlife, but also when maintained or increased in late-life [4, 85]. Additionally, a systematic review reported an increased risk of cognitive impairment in physically inactive people in 20 out of the 24 longitudinal studies included [86].

In the analysis by Norton and colleagues, which took into account seven modifiable risk factors for AD, physical inactivity accounted for the largest proportion of AD cases in Europe and the USA. The same study reported that about 13% of AD cases worldwide are potentially attributable to this risk factor [5].

The neuroprotective effects of physical activity have been investigated in neuroimaging studies, which documented the reduction of age-related brain atrophy, and also the increase of grey and white matter in brain areas involved in dementia/AD (i.e. frontal and temporal lobes). Additionally, animal studies have shown that exercise can enhance neuroplasticity, including angiogenesis and upregulation of growth factors [87].

Education and Other Mentally Stimulating Activities
Higher levels of education have been consistently related to a reduced risk of AD [89, 90] and can also mitigate the risk of dementia due to the *APOEε4* allele [88]. The greatest proportion, one out of five, of AD cases worldwide is attributable to a low educational attainment, according to the analyses reported by Norton and colleagues [5]. The prevailing model to explain this association hypothesizes a positive contribution of education to the cognitive reserve able to counteract the burden of neurodegenerative pathology [89–91]. In persons with higher cognitive reserve, more cerebral lesions are needed to clinically express dementia [92]. In line with this statement, an analysis of large autopsy data found that subjects with higher educational levels were less likely to suffer from pre-mortem clinical dementia symptoms, among those individuals with similar amount of neuropathological lesions [93]. Interestingly, once subjects express cognitive symptoms of AD, people with higher education experience a faster cognitive decline. This can be partially explained by the fact that they accumulated a greater amount of AD pathology with respect to those people with a lower level of education [94–96].

Activities providing mental stimulation include not only formal education, but also occupational complexity (i.e. intellectually demanding job) and leisure activities. Several observational studies have shown that greater cognitive engagement across the lifespan is associated with a decreased risk of AD [94, 97, 98]. Wilson and co-authors [99] conducted the analyses using physical and social activity levels as covariates, reporting that the protective role of cognitive engagement is independent from both social and physical activities.

Social Network
One of the main challenges in identifying the role of the social network in dementia/AD occurrence is the heterogeneity in the definition of this component of the lifestyle. The assessment of this factor includes objective measures such as marital status, living situation, number of people in the social network, as well as subjective measures such as feelings of loneliness. Despite the methodological differences, several studies point at social stimulation as a protective factor for late-life dementia/AD, while poor social interaction has been associated with increased risk of dementia [4]. Some studies reported that living with a partner during midlife was associated with reduced risk of cognitive impairment and dementia later in life, suggesting that being in a relationship entails cognitive and social challenges that can increase the cognitive reserve [4]. Additionally, accumulating evidence suggests that a higher degree of loneliness and being single and not cohabiting with a partner in later life are risk factors for AD [100, 101]. Social relationships may influence health through several mechanisms. For instance, social integration may have a beneficial effect on health through influencing health behaviours, while social support can benefit health through stress reduction, by providing psychological and material support.

Diet

Several cohort studies have investigated the association between specific nutrients and dietary patterns and risk of dementia/AD and cognitive decline. Nevertheless, results are still conflicting and RCTs are required for the formulation of dietary recommendations.

As an example, higher rates of cognitive decline have been related with higher intake of saturated fats or cholesterol [102–104], but some studies did not confirm such an association [105]. Moreover, if the protective role of fish oil (dietary intake) or omega-3 fatty acid supplementation on death from coronary heart disease is well established, the potential benefits of these nutrients against dementia are still questionable [106, 107]. More consistent findings have related a diet rich in fruit and vegetables with a decreased risk of dementia and AD.

Dietary Patterns

Among dietary patterns, the Mediterranean diet is characterized by a high intake of legumes, cereals, fruits and vegetables and a moderately high intake of fish. It also contains high amounts of unsaturated fatty acids from vegetables oils (i.e. olive oil) and a low saturated fatty acid content. The intake of dairy products, such as cheese and yogurt, is low to moderate in this dietary pattern, and the intake of meat and poultry is limited. Finally, the Mediterranean diet is characterized by a regular but moderate amount of alcohol, primarily in the form of wine [108, 109].

Several longitudinal studies together with a large RCT (PREDIMED) [110] have suggested that adherence to the Mediterranean diet improves cardiovascular outcomes, and as consequence reduces the risk of dementia, especially of vascular type [111]. High adherence to the Mediterranean diet has been also associated to a reduced risk of AD [4].

Other dietary patterns which seem to benefit cognition are the DASH (Dietary Approaches to Stop Hypertension) and the hybrid MIND (Mediterranean-DASH Intervention for Neurodegenerative Delay) diet [108, 112]. They are both similar to the Mediterranean diet, since they include a high intake of vegetables, nuts and legumes; preference for whole grains and low consumptions of red meat and high saturated fat foods. Differences in these dietary profiles entail indications of the quantity/quality of fruit and vegetable oils (in general high consumption), fish, poultry and dairy products (low-moderate intake).

Antioxidant Vitamins

Oxidative and nitrosative stress are involved in AD pathophysiology, and some observational studies have reported a reduced risk of dementia/AD in relation to the dietary intake of antioxidant vitamins (A, C, E) [113]. These findings have led to test the neuroprotective effect of antioxidants (e.g. vitamin C, E, beta-carotene, flavonoids) in RCTs, with overall negative results in terms of dementia/AD prevention. Additionally, potential health risks have been reported with vitamin E supplementation [114]. Therefore, no recommendations can be given for supplementations with antioxidants in the prevention of dementia/AD [115–117].

Vitamin B6, B12, and Folate

Observational studies have related elevated serum homocysteine and low serum levels of folate, vitamin B6 and B12 with cognitive decline and dementia, but data on B vitamins supplementation studies are inconsistent [118]. The association between B vitamins and late-life dementia/AD can be mediated both by their effect on neurodegenerative pathways, and through influence on blood levels of homocysteine, which is a risk factor for atherosclerosis.

Alcohol Consumption

Alcohol abuse is unquestionably associated with poor cognitive performance and increased risk of dementia. Conversely, low to moderate alcohol intake has been found to be associated with a decreased risk of AD [119] in some studies, while others did not support this finding. In a meta-analysis, the pooled estimated relative risk for AD in light to moderate drinkers versus non-drinkers was 0.72 (95% CI: 0.61–0.86) [119]. The underlying mechanism by which low/moderate alcohol consumption might prevent AD is at present unknown. It might exert its protective effect via the reduction in vascular risk factors (e.g. lipid and lipoprotein levels, inflammatory and haemostatic factors) or through the antioxidant effect of polyphenols, which are richly represented in red wine [120].

Air Pollution

Air pollution has received increasing attention as a potential risk factor for cognitive decline and dementia. A recent systematic review has reported a consistent positive association between the exposure to at least one pollutant and dementia [121]. Nevertheless, several methodological limitations need to be taken into account when considering these findings. First, the studies might suffer from a misclassification bias, coming from the identification of dementia cases from healthcare system records. Second, the majority of the studies used data on recent exposure to air pollution as a proxy of long-term exposure. Third, it is extremely difficult to point out the specific putative causal agent across different pollutants. Despite these issues, the existing epidemiologic evidence is in favour of considering air pollution as a risk factor for dementia. Further studies with better designs are needed to better disentangle this association, in order to be a solid foundation for recommendations and possible interventions.

Other Modifiable Risk Factors

Depression

Although there is no doubt that depression and cognition are strictly linked, the direction of the association when investigating dementia occurrence is still unclear [122]. Depression has been associated with poor cognitive function, but it represents a psychiatric symptom of AD as well [123, 124]. Therefore, understanding the relation of depression and AD is complicated by the possibility that depression may be a prodromal symptom of AD rather than a risk factor for the disease.

In a meta-analysis including 23 longitudinal studies, late-life depression was associated with a higher risk of all-cause dementia (odd ratio, OR: 1.96; 95% CI: 1.64–2.34), AD (OR: 1.85; 95% CI: 1.45–2.37), and vascular dementia (OR: 2.53; 95% CI: 1.42–4.50) [125]. Overall, 8% of all AD cases worldwide are potentially attributable to depression, according to Norton and colleagues, with strong clinical and public health implications [5].

Traumatic Brain Injuries

Accumulating and consistent evidence has shown that mild but repeated traumatic brain injuries could result in chronic traumatic encephalopathy (CTE), manifested with a worsening in cognitive performance, neuropsychiatric symptoms, and parkinsonism [126]. Traumatic brain injuries can occur in motor vehicle crashes and falls, as well as in contact sports. In a recent large autopsy study in deceased players of American football, CTE was neuropathologically diagnosed in 87% of the players, across all levels of play, supporting the link between repeated mild traumatic brain injuries and neuropathological damage [127].

Obstructive Sleep Apnea Syndrome

Accumulating evidence relates obstructive sleep apnea (OSA) with mild cognitive impairment and dementia [128, 129]. This observation has led to the hypothesis that hypoxia may be implicated in the biological mechanism of dementia onset. The use of continuous positive airway pressure (CPAP) appears to be associated with a lower risk of impairment in cognition in subjects with OSA.

Dementia Risk Scores

Risk and protective factors often co-occur in the same person, leading to interactive effects, which can amplify or reduce the overall dementia/AD risk. Cumulative exposure to risk factors can be accounted for by using risk scores, which help to estimate the overall risk of dementia in individuals. Several risk scores have been proposed, accounting for midlife or late-life risk factors (e.g. age, gender, education, vascular disorders) [130]. Among others [131–134], the Cardiovascular Risk Factors, Aging, and Dementia (CAIDE) risk score [135] takes into account age, sex, education, and midlife risk factors, namely, blood pressure, BMI, total cholesterol and the level of physical activity. The CAIDE risk score provides an estimate of the risk to develop dementia over 20 years and has been validated in different populations [132, 136].

In general, several key elements need to be considered when developing a risk score. First, since dementia is a slowly progressive disease, it becomes challenging to differentiate a diagnostic model—assessing the risk of a disease already present—from a prognostic one estimating the risk of the disease that might occur later on. Second, the association between several risk factors and dementia/AD is age dependent; as a consequence, a score specifically developed for midlife risk factors can have a lower predictive value when applied in older populations.

Despite these challenges, risk scores can have several areas of application, either in the community—to help individuals in becoming aware of their risk profile—or in clinical settings—to improve patients' lifestyle and/or adherence to pharmacological treatments. Dementia risk scores can also be utilized for early identification of at-risk subjects, who can be target of preventative interventions.

Multi-domain Preventative Interventions

Based on the aforementioned epidemiologic evidence and biologic plausibility on the role of vascular care, lifestyle and dementia, it appears reasonable to promote preventative interventions based on the management of vascular and lifestyle-related risk factors. The multifactorial nature of late-life cognitive impairment and dementia/AD suggests that multi-domain interventions targeting several risk factors simultaneously are needed for optimum preventive effects. Three large European RCTs [137–139] on prevention of dementia recently tested pioneering multi-domain interventions in older adults, mainly based on improvement of lifestyle and adherence to medical treatments for vascular risk factors and vascular diseases. The crucial aspects of these studies are the multi-domain approach and the use of clinical evaluation and neuropsychological tests to detect cognitive changes and dementia incidence. The Finnish Geriatric Intervention Study to Prevent Cognitive Impairment and Disability (FINGER) tested a 2-year multi-domain intervention comprising nutritional guidance, exercise, cognitive training, social activity, and intensive monitoring/management of metabolic and vascular risk factors. The intervention targeted subjects with higher dementia risk, based on the CAIDE risk score, and was associated with improvement or stability of global cognition and specific cognitive domains (processing speed and executive function), which are important for carrying on activities of daily living [137]. Follow-up of participants is ongoing to determine the effect of the interventions on dementia/AD onset.

The Dutch Prevention of Dementia by Intensive Vascular Care (PreDIVA) study lasted 6 years and compared standard and intensive care of cardiovascular risk factors in an unselected population of older people. The multi-domain intervention did not result in an overall decrease of dementia incidence, but reduced occurrence of dementia was found in a subgroup of people with baseline untreated hypertension, for whom therapy was initiated. This highlights the importance of focusing preventive interventions in at-risk groups [139, 140].

In the French Multidomain Alzheimer Preventive Trial (MAPT), a 3-year multi-domain lifestyle intervention (cognitive training, advice on nutrition and physical activity), administered alone or in combination with omega-3 polyunsaturated fatty-acid supplementation, did not change the occurrence of cognitive decline among older persons with memory complaints [138, 141].

Even if divergent, the findings generated by these RCTs are crucial to model future trials based on multi-domain interventions [10]. Methodological challenges include the identification of risk profiles which can benefit from specific interventions. Duration and intensity of the interventions can also vary depending on the target group, and tailored interventions, shaped for different cultural and geographical settings are needed. International collaborations are pivotal to reach these goals, and a recent step in this direction has been the launch of the World Wide FINGERS network (http://wwfingers.com/), which aims to promote globally the collaboration in the dementia/AD prevention field. Within this initiative, the FINGER RCT model is tested in other countries, including China, Singapore, the USA, and the UK, with the aim to generate robust evidence to define effective preventive approaches for various at-risk groups and settings.

Conclusions

Given the increasing number of people with dementia and AD, together with the current lack of curative treatments, prevention has been highlighted by international bodies as a key strategy to halt this growing epidemic [2, 3, 142]. Up to one-third of AD cases worldwide can be attributable to the presence of seven potentially modifiable risk factors: physical inactivity, smoking, midlife hypertension and obesity, DM, depression, and low level of education [5]. Therefore, it might be possible to substantially reduce AD prevalence through public health interventions promoting activities enhancing the cognitive reserve and healthy lifestyles. To be successful, preventative interventions should be carried out in the framework of a life-course approach. Moreover, despite the many risk factors, the beneficial effect is probably higher when they are managed in midlife, and improvement of subjects' risk profile in older age can still prevent or postpone cognitive impairment and dementia, supporting the role of prevention in older adults.

Multi-domain interventions are deemed promising preventative strategies, which need to be further investigated, to define effective and feasible interventions for specific risk profiles. Ongoing international collaborations, such as World Wide FINGERS, are crucial to address this issue in the most comprehensive way.

It is also important to consider that the majority of dementia cases in the general population occurs in subjects with advanced age (75+ years), making dementia one of the most burdensome geriatric syndromes. As such, prevention of dementia in subjects with advanced age demands the multidimensional approach which defines the comprehensive geriatric assessment. Through such approach, all the factors contributing to cognitive and behavioural symptoms are evaluated and addressed, wherever possible.

Overall, the current available knowledge allows to identifying risk factors which can be managed to reduce the risk of dementia/AD in late-life. At the same time, ongoing coordinated international efforts will help to define evidence-based preventative approaches accessible and sustainable for populations with different geographical, economic and cultural settings.

References

1. Alzheimer Disease International, ADI, The Global Impact of Dementia. An analysis on prevalence, incidence, costs, and trends. 2015. Accessed 1 Aug 2017.
2. The Lancet. WHO has a dementia plan, now we need action. Lancet Neurol. 2017;16(8):571.
3. First World Health Organization ministerial conference on global action against dementia: meeting report. Geneva: World Health Organization; 2015. Accessed 1 Aug 2017.
4. Solomon A, et al. Advances in the prevention of Alzheimer's disease and dementia. J Intern Med. 2014;275(3):229–50.
5. Norton S, et al. Potential for primary prevention of Alzheimer's disease: an analysis of population-based data. Lancet Neurol. 2014;13(8):788–94.
6. Brookmeyer R, et al. Forecasting the global burden of Alzheimer's disease. Alzheimers Dement. 2007;3(3):186–91.
7. Satizabal C, Beiser AS, Seshadri S. Incidence of dementia over three decades in the Framingham Heart Study. N Engl J Med. 2016;375(1):93–4.
8. Langa KM, et al. A comparison of the prevalence of dementia in the United States in 2000 and 2012. JAMA Intern Med. 2017;177(1):51–8.
9. Winblad B, et al. Defeating Alzheimer's disease and other dementias: a priority for European Science and Society. Lancet Neurol. 2016;15(5):455–532.
10. Kivipelto M, Mangialasche F, Ngandu T. Can lifestyle changes prevent cognitive impairment? Lancet Neurol. 2017;16(5):338–9.
11. Kelley BJ, Boeve BF, Josephs KA. Young-onset dementia: demographic and etiologic characteristics of 235 patients. Arch Neurol. 2008;65(11):1502–8.
12. Larson EB, Langa KM. The rising tide of dementia worldwide. Lancet. 2008;372(9637):430–2.
13. Mercy L, et al. Incidence of early-onset dementias in Cambridgeshire, United Kingdom. Neurology. 2008;71(19):1496–9.
14. Corrada MM, et al. Prevalence of dementia after age 90: results from the 90+ study. Neurology. 2008;71(5):337–43.
15. Corrada MM, et al. Dementia incidence continues to increase with age in the oldest old: the 90+ study. Ann Neurol. 2010;67(1):114–21.
16. Green RC, et al. Risk of dementia among white and African American relatives of patients with Alzheimer disease. JAMA. 2002;287(3):329–36.
17. Fratiglioni L, et al. Risk factors for late-onset Alzheimer's disease: a population-based, case-control study. Ann Neurol. 1993;33(3):258–66.
18. Fratiglioni L, Qiu C. Prevention of common neurodegenerative disorders in the elderly. Exp Gerontol. 2009;44(1–2):46–50.
19. Wolters FJ, et al. Parental family history of dementia in relation to subclinical brain disease and dementia risk. Neurology. 2017;88(17):1642–9.
20. Mazure CM, Swendsen J. Sex differences in Alzheimer's disease and other dementias. Lancet Neurol. 2016;15(5):451–2.
21. Vina J, Lloret A. Why women have more Alzheimer's disease than men: gender and mitochondrial toxicity of amyloid-beta peptide. J Alzheimers Dis. 2010;20(Suppl 2):S527–33.
22. Schenck-Gustafsson K, Decola PR, Pfaff DW, Pisetsky DS, editors. Handbook of clinical gender medicine. Basel: Karger.
23. Debette S, et al. Midlife vascular risk factor exposure accelerates structural brain aging and cognitive decline. Neurology. 2011;77(5):461–8.
24. Yaffe K, et al. Early adult to midlife cardiovascular risk factors and cognitive function. Circulation. 2014;129(15):1560–7.
25. Gottesman RF, et al. Midlife hypertension and 20-year cognitive change: the atherosclerosis risk in communities neurocognitive study. JAMA Neurol. 2014;71(10):1218–27.
26. Romas SN, et al. APOE genotype, plasma lipids, lipoproteins, and AD in community elderly. Neurology. 1999;53(3):517–21.
27. Jick H, et al. Statins and the risk of dementia. Lancet. 2000;356(9242):1627–31.

28. Wolozin B, et al. Decreased prevalence of Alzheimer disease associated with 3-hydroxy-3-methyglutaryl coenzyme A reductase inhibitors. Arch Neurol. 2000;57(10):1439–43.
29. Zhou B, Teramukai S, Fukushima M. Prevention and treatment of dementia or Alzheimer's disease by statins: a meta-analysis. Dement Geriatr Cogn Disord. 2007;23(3):194–201.
30. Shepherd J, et al. Pravastatin in elderly individuals at risk of vascular disease (PROSPER): a randomised controlled trial. Lancet. 2002;360(9346):1623–30.
31. Heart Protection Study Collaborative Group. MRC/BHF Heart Protection Study of cholesterol lowering with simvastatin in 20,536 high-risk individuals: a randomised placebo-controlled trial. Lancet. 2002;360(9326):7–22.
32. McGuinness B, Cardwell CR, Passmore P. Statin withdrawal in people with dementia. Cochrane Database Syst Rev. 2016;(9):CD012050.
33. Snowdon DA, et al. Brain infarction and the clinical expression of Alzheimer disease. The Nun Study. JAMA. 1997;277(10):813–7.
34. Shepardson NE, Shankar GM, Selkoe DJ. Cholesterol level and statin use in Alzheimer disease: I. Review of epidemiological and preclinical studies. Arch Neurol. 2011;68(10):1239–44.
35. Ott A, et al. Diabetes mellitus and the risk of dementia: the Rotterdam Study. Neurology. 1999;53(9):1937–42.
36. Gregg EW, et al. Is diabetes associated with cognitive impairment and cognitive decline among older women? Study of Osteoporotic Fractures Research Group. Arch Intern Med. 2000;160(2):174–80.
37. Arvanitakis Z, et al. Diabetes mellitus and risk of Alzheimer disease and decline in cognitive function. Arch Neurol. 2004;61(5):661–6.
38. Schnaider Beeri M, et al. Diabetes mellitus in midlife and the risk of dementia three decades later. Neurology. 2004;63(10):1902–7.
39. Xu WL, et al. Diabetes mellitus and risk of dementia in the Kungsholmen project: a 6-year follow-up study. Neurology. 2004;63(7):1181–6.
40. Craft S. Insulin resistance syndrome and Alzheimer's disease: age- and obesity-related effects on memory, amyloid, and inflammation. Neurobiol Aging. 2005;26(Suppl 1):65–9.
41. Cukierman T, Gerstein HC, Williamson JD. Cognitive decline and dementia in diabetes—systematic overview of prospective observational studies. Diabetologia. 2005;48(12):2460–9.
42. Yaffe K, et al. Diabetes, impaired fasting glucose, and development of cognitive impairment in older women. Neurology. 2004;63(4):658–63.
43. Biessels GJ, et al. Risk of dementia in diabetes mellitus: a systematic review. Lancet Neurol. 2006;5(1):64–74.
44. Cherbuin N, Sachdev P, Anstey KJ. Higher normal fasting plasma glucose is associated with hippocampal atrophy: the PATH Study. Neurology. 2012;79(10):1019–26.
45. Launer LJ, et al. Effects of intensive glucose lowering on brain structure and function in people with type 2 diabetes (ACCORD MIND): a randomised open-label substudy. Lancet Neurol. 2011;10(11):969–77.
46. Pathan AR, et al. Rosiglitazone attenuates the cognitive deficits induced by high fat diet feeding in rats. Eur J Pharmacol. 2008;589(1–3):176–9.
47. Abbatecola AM, et al. Rosiglitazone and cognitive stability in older individuals with type 2 diabetes and mild cognitive impairment. Diabetes Care. 2010;33(8):1706–11.
48. Gold M, et al. Rosiglitazone monotherapy in mild-to-moderate Alzheimer's disease: results from a randomized, double-blind, placebo-controlled phase III study. Dement Geriatr Cogn Disord. 2010;30(2):131–46.
49. Imfeld P, et al. Metformin, other antidiabetic drugs, and risk of Alzheimer's disease: a population-based case-control study. J Am Geriatr Soc. 2012;60(5):916–21.
50. Whitmer RA, et al. Hypoglycemic episodes and risk of dementia in older patients with type 2 diabetes mellitus. JAMA. 2009;301(15):1565–72.
51. Qiu C, Winblad B, Fratiglioni L. The age-dependent relation of blood pressure to cognitive function and dementia. Lancet Neurol. 2005;4(8):487–99.
52. Ruitenberg A, et al. Cerebral hypoperfusion and clinical onset of dementia: the Rotterdam Study. Ann Neurol. 2005;57(6):789–94.

53. Kivipelto M, et al. Obesity and vascular risk factors at midlife and the risk of dementia and Alzheimer disease. Arch Neurol. 2005;62(10):1556–60.
54. Launer LJ, et al. Midlife blood pressure and dementia: the Honolulu-Asia aging study. Neurobiol Aging. 2000;21(1):49–55.
55. Elkins JS, et al. Pre-existing hypertension and the impact of stroke on cognitive function. Ann Neurol. 2005;58(1):68–74.
56. Glynn RJ, et al. Current and remote blood pressure and cognitive decline. JAMA. 1999;281(5):438–45.
57. Pantoni L, Gorelick P. Advances in vascular cognitive impairment 2010. Stroke. 2011;42(2):291–3.
58. Levi Marpillat N, et al. Antihypertensive classes, cognitive decline and incidence of dementia: a network meta-analysis. J Hypertens. 2013;31(6):1073–82.
59. Yasar S, et al. Relationship between antihypertensive medications and cognitive impairment: Part I. Review of human studies and clinical trials. Curr Hypertens Rep. 2016;18(8):67.
60. Whitmer RA, et al. Obesity in middle age and future risk of dementia: a 27 year longitudinal population based study. BMJ. 2005;330(7504):1360.
61. Rosengren A, et al. Body mass index, other cardiovascular risk factors, and hospitalization for dementia. Arch Intern Med. 2005;165(3):321–6.
62. Gustafson D, et al. An 18-year follow-up of overweight and risk of Alzheimer disease. Arch Intern Med. 2003;163(13):1524–8.
63. Albanese E, et al. Overweight and obesity in midlife and brain structure and dementia 26 years later: the AGES-Reykjavik Study. Am J Epidemiol. 2015;181(9):672–9.
64. Fitzpatrick AL, et al. Midlife and late-life obesity and the risk of dementia: cardiovascular health study. Arch Neurol. 2009;66(3):336–42.
65. Hughes TF, et al. Association between late-life body mass index and dementia: The Kame Project. Neurology. 2009;72(20):1741–6.
66. Stewart R, et al. A 32-year prospective study of change in body weight and incident dementia: the Honolulu-Asia Aging Study. Arch Neurol. 2005;62(1):55–60.
67. Buchman AS, et al. Body mass index in older persons is associated with Alzheimer disease pathology. Neurology. 2006;67(11):1949–54.
68. Gazdzinski S, et al. Body mass index and magnetic resonance markers of brain integrity in adults. Ann Neurol. 2008;63(5):652–7.
69. Burns JM, et al. Reduced lean mass in early Alzheimer disease and its association with brain atrophy. Arch Neurol. 2010;67(4):428–33.
70. Anstey KJ, et al. Smoking as a risk factor for dementia and cognitive decline: a meta-analysis of prospective studies. Am J Epidemiol. 2007;166(4):367–78.
71. McKenzie J, Bhatti L. Tursan d'Espaignet E World Health Organization: tobacco and dementia. Geneva 2014. Accessed 1 Aug 2017.
72. Lakka HM, et al. The metabolic syndrome and total and cardiovascular disease mortality in middle-aged men. JAMA. 2002;288(21):2709–16.
73. Sundstrom J, et al. Clinical value of the metabolic syndrome for long term prediction of total and cardiovascular mortality: prospective, population based cohort study. BMJ. 2006;332(7546):878–82.
74. Vanhanen M, et al. Association of metabolic syndrome with Alzheimer disease: a population-based study. Neurology. 2006;67(5):843–7.
75. Yaffe K, et al. The metabolic syndrome and development of cognitive impairment among older women. Arch Neurol. 2009;66(3):324–8.
76. Raffaitin C, et al. Metabolic syndrome and cognitive decline in French elders: the Three-City Study. Neurology. 2011;76(6):518–25.
77. Dolan H, et al. Atherosclerosis, dementia, and Alzheimer disease in the Baltimore Longitudinal Study of Aging cohort. Ann Neurol. 2010;68(2):231–40.
78. Vidal JS, et al. Coronary artery calcium, brain function and structure: the AGES-Reykjavik Study. Stroke. 2010;41(5):891–7.

79. Luchsinger JA, et al. Aggregation of vascular risk factors and risk of incident Alzheimer disease. Neurology. 2005;65(4):545–51.
80. Santangeli P, et al. Atrial fibrillation and the risk of incident dementia: a meta-analysis. Heart Rhythm. 2012;9(11):1761–8.
81. de Bruijn RF, et al. Association between atrial fibrillation and dementia in the general population. JAMA Neurol. 2015;72(11):1288–94.
82. Sofi F, et al. Physical activity and risk of cognitive decline: a meta-analysis of prospective studies. J Intern Med. 2011;269(1):107–17.
83. Grande G, et al. Physical activity reduces the risk of dementia in mild cognitive impairment subjects: a cohort study. J Alzheimers Dis. 2014;39(4):833–9.
84. Ravaglia S, et al. Changes in nutritional status and body composition during enzyme replacement therapy in adult-onset type II glycogenosis. Eur J Neurol. 2010;17(7):957–62.
85. Tolppanen AM, et al. Leisure-time physical activity from mid- to late life, body mass index, and risk of dementia. Alzheimers Dement. 2015;11(4):434–443 e6.
86. Rolland Y, Abellan van Kan G, Vellas B. Physical activity and Alzheimer's disease: from prevention to therapeutic perspectives. J Am Med Dir Assoc. 2008;9(6):390–405.
87. Chieffi S, et al. Neuroprotective effects of physical activity: evidence from human and animal studies. Front Neurol. 2017;8:188.
88. Wang HX, et al. Education halves the risk of dementia due to apolipoprotein epsilon4 allele: a collaborative study from the Swedish brain power initiative. Neurobiol Aging. 2012;33(5):1007 e1–7.
89. Prince M, et al. Dementia incidence and mortality in middle-income countries, and associations with indicators of cognitive reserve: a 10/66 Dementia Research Group population-based cohort study. Lancet. 2012;380(9836):50–8.
90. Ngandu T, et al. Education and dementia: what lies behind the association? Neurology. 2007;69(14):1442–50.
91. Wang HX, et al. Association of lifelong exposure to cognitive reserve-enhancing factors with dementia risk: a community-based cohort study. PLoS Med. 2017;14(3):e1002251.
92. Yaffe K, et al. Association of plasma beta-amyloid level and cognitive reserve with subsequent cognitive decline. JAMA. 2011;305(3):261–6.
93. Roe CM, et al. Education and Alzheimer disease without dementia: support for the cognitive reserve hypothesis. Neurology. 2007;68(3):223–8.
94. Wilson RS, et al. Participation in cognitively stimulating activities and risk of incident Alzheimer disease. JAMA. 2002;287(6):742–8.
95. Hall CB, et al. Education delays accelerated decline on a memory test in persons who develop dementia. Neurology. 2007;69(17):1657–64.
96. Singh-Manoux A, et al. Does cognitive reserve shape cognitive decline? Ann Neurol. 2011;70(2):296–304.
97. Verghese J, et al. Leisure activities and the risk of dementia in the elderly. N Engl J Med. 2003;348(25):2508–16.
98. Akbaraly TN, et al. Leisure activities and the risk of dementia in the elderly: results from the Three-City Study. Neurology. 2009;73(11):854–61.
99. Wilson RS, et al. Relation of cognitive activity to risk of developing Alzheimer disease. Neurology. 2007;69(20):1911–20.
100. Fratiglioni L, et al. Influence of social network on occurrence of dementia: a community-based longitudinal study. Lancet. 2000;355(9212):1315–9.
101. Hakansson K, et al. Association between midlife marital status and cognitive function in later life: population based cohort study. BMJ. 2009;339:b2462.
102. Morris MC, et al. Dietary fat intake and 6-year cognitive change in an older biracial community population. Neurology. 2004;62(9):1573–9.
103. Kalmijn S, et al. Dietary intake of fatty acids and fish in relation to cognitive performance at middle age. Neurology. 2004;62(2):275–80.
104. Okereke OI, et al. Dietary fat types and 4-year cognitive change in community-dwelling older women. Ann Neurol. 2012;72(1):124–34.

105. Engelhart MJ, et al. Diet and risk of dementia: does fat matter?: The Rotterdam Study. Neurology. 2002;59(12):1915–21.
106. Wu S, et al. Omega-3 fatty acids intake and risks of dementia and Alzheimer's disease: a meta-analysis. Neurosci Biobehav Rev. 2015;48:1–9.
107. Zhang Y, et al. Intakes of fish and polyunsaturated fatty acids and mild-to-severe cognitive impairment risks: a dose-response meta-analysis of 21 cohort studies. Am J Clin Nutr. 2016;103(2):330–40.
108. Morris MC, et al. MIND diet associated with reduced incidence of Alzheimer's disease. Alzheimers Dement. 2015;11(9):1007–14.
109. Valls-Pedret C, et al. Mediterranean diet and age-related cognitive decline: a randomized clinical trial. JAMA Intern Med. 2015;175(7):1094–103.
110. Martinez-Lapiscina EH, et al. Mediterranean diet improves cognition: the PREDIMED-NAVARRA randomised trial. J Neurol Neurosurg Psychiatry. 2013;84(12):1318–25.
111. Estruch R, et al. Primary prevention of cardiovascular disease with a Mediterranean diet. N Engl J Med. 2013;368(14):1279–90.
112. Smith PJ, et al. Effects of the dietary approaches to stop hypertension diet, exercise, and caloric restriction on neurocognition in overweight adults with high blood pressure. Hypertension. 2010;55(6):1331–8.
113. Mangialasche F, et al. Biomarkers of oxidative and nitrosative damage in Alzheimer's disease and mild cognitive impairment. Ageing Res Rev. 2009;8(4):285–305.
114. Mangialasche F, et al. Dementia prevention: current epidemiological evidence and future perspective. Alzheimers Res Ther. 2012;4(1):6.
115. Masaki KH, et al. Association of vitamin E and C supplement use with cognitive function and dementia in elderly men. Neurology. 2000;54(6):1265–72.
116. Engelhart MJ, et al. Dietary intake of antioxidants and risk of Alzheimer disease. JAMA. 2002;287(24):3223–9.
117. Kryscio RJ, et al. Association of antioxidant supplement use and dementia in the Prevention of Alzheimer's Disease by Vitamin E and Selenium Trial (PREADViSE). JAMA Neurol. 2017;74(5):567–73.
118. Clarke R, et al. Effects of homocysteine lowering with B vitamins on cognitive aging: meta-analysis of 11 trials with cognitive data on 22,000 individuals. Am J Clin Nutr. 2014;100(2):657–66.
119. Anstey KJ, Mack HA, Cherbuin N. Alcohol consumption as a risk factor for dementia and cognitive decline: meta-analysis of prospective studies. Am J Geriatr Psychiatry. 2009;17(7):542–55.
120. Ruitenberg A, et al. Alcohol consumption and risk of dementia: the Rotterdam Study. Lancet. 2002;359(9303):281–6.
121. Power MC, et al. Exposure to air pollution as a potential contributor to cognitive function, cognitive decline, brain imaging, and dementia: a systematic review of epidemiologic research. Neurotoxicology. 2016;56:235–53.
122. Mirza SS, et al. 10-year trajectories of depressive symptoms and risk of dementia: a population-based study. Lancet Psychiatry. 2016;3(7):628–35.
123. Ismail Z, et al. Prevalence of depression in patients with mild cognitive impairment: a systematic review and meta-analysis. JAMA Psychiatry. 2017;74(1):58–67.
124. Steffens DC. Late-life depression and the prodromes of dementia. JAMA Psychiatry. 2017;74(7):673–4.
125. Diniz BS, et al. Late-life depression and risk of vascular dementia and Alzheimer's disease: systematic review and meta-analysis of community-based cohort studies. Br J Psychiatry. 2013;202(5):329–35.
126. Washington PM, Villapol S, Burns MP. Polypathology and dementia after brain trauma: does brain injury trigger distinct neurodegenerative diseases, or should they be classified together as traumatic encephalopathy? Exp Neurol. 2016;275(Pt 3):381–8.
127. Mez J, et al. Clinicopathological evaluation of chronic traumatic encephalopathy in players of American Football. JAMA. 2017;318(4):360–70.

128. Ramos AR, et al. Obstructive sleep apnea and neurocognitive function in a Hispanic/Latino population. Neurology. 2015;84(4):391–8.
129. Lutsey PL, et al. Sleep characteristics and risk of dementia and Alzheimer's disease: the atherosclerosis risk in communities study. Alzheimers Dement. 2017;14(2):157–66. https://doi.org/10.1016/j.jalz.2017.06.2269.
130. Solomon A, Soininen H. Dementia: risk prediction models in dementia prevention. Nat Rev Neurol. 2015;11(7):375–7.
131. Barnes DE, et al. Predicting risk of dementia in older adults: the late-life dementia risk index. Neurology. 2009;73(3):173–9.
132. Anstey KJ, et al. A self-report risk index to predict occurrence of dementia in three independent cohorts of older adults: the ANU-ADRI. PLoS One. 2014;9(1):e86141.
133. Exalto LG, et al. Risk score for prediction of 10 year dementia risk in individuals with type 2 diabetes: a cohort study. Lancet Diabetes Endocrinol. 2013;1(3):183–90.
134. Mitnitski A, et al. A vascular risk factor index in relation to mortality and incident dementia. Eur J Neurol. 2006;13(5):514–21.
135. Kivipelto M, et al. Risk score for the prediction of dementia risk in 20 years among middle aged people: a longitudinal, population-based study. Lancet Neurol. 2006;5(9):735–41.
136. Exalto LG, et al. Midlife risk score for the prediction of dementia four decades later. Alzheimers Dement. 2014;10(5):562–70.
137. Ngandu T, et al. A 2 year multidomain intervention of diet, exercise, cognitive training, and vascular risk monitoring versus control to prevent cognitive decline in at-risk elderly people (FINGER): a randomised controlled trial. Lancet. 2015;385(9984):2255–63.
138. Andrieu S, et al. Effect of long-term omega 3 polyunsaturated fatty acid supplementation with or without multidomain intervention on cognitive function in elderly adults with memory complaints (MAPT): a randomised, placebo-controlled trial. Lancet Neurol. 2017;16(5):377–89.
139. Moll van Charante EP, et al. Effectiveness of a 6-year multidomain vascular care intervention to prevent dementia (preDIVA): a cluster-randomised controlled trial. Lancet. 2016;388(10046):797–805.
140. Richard E, et al. Prevention of dementia by intensive vascular care (PreDIVA): a cluster-randomized trial in progress. Alzheimer Dis Assoc Disord. 2009;23(3):198–204.
141. Vellas B, et al. Mapt study: a multidomain approach for preventing Alzheimer's disease: design and baseline data. J Prev Alzheimers Dis. 2014;1(1):13–22.
142. Livingston G, et al. Dementia prevention, intervention, and care. Lancet. 2017;390(10113):2673–734. https://doi.org/10.1016/S0140-6736(17)31363-6.

Diagnosis of Frontotemporal Dementia

Giorgio Giulio Fumagalli

Abstract

Frontotemporal dementia (FTD) is a progressive neurodegenerative disease that can present with three different clinical syndromes: behavioural-variant frontotemporal dementia (bvFTD), associated with behavioural and executive deficits; non-fluent variant primary progressive aphasia (nfPPA), with progressive deficits in speech, grammar, and word output; and semantic variant primary progressive aphasia (svPPA), which is a progressive disorder of semantic knowledge and naming.

The disease can mimic various psychiatric disorders and sometimes can be difficult to discriminate against other forms of dementia. Advances in clinical, imaging, and molecular characterisation have increased the accuracy of the diagnosis of frontotemporal dementia. Updated diagnostic criteria have been developed and are now widely used. Recognition and accurate diagnoses of FTD subtypes will aid the neurologist in the management of patients.

Keywords

Frontotemporal dementia · Semantic aphasia · Non-fluent aphasia · Diagnosis · Clinical

Introduction

The term frontotemporal dementia (FTD) refers to a heterogeneous group of syndromes caused by progressive and selective degeneration of the frontal and temporal lobes that cause changes in behaviour or language deficits.

G.G. Fumagalli
Department of Neurosciences, Psychology, Drug Research and Child Health (NEUROFARBA), University of Florence, Milan, Italy

Neurodegenerative Disease Unit,
Fondazione Ca' Granda, IRCCS Ospedale Policlinico, Milan, Italy

© Springer International Publishing AG 2018
D. Galimberti, E. Scarpini (eds.), *Neurodegenerative Diseases*,
https://doi.org/10.1007/978-3-319-72938-1_7

Arnold Pick in 1892 made the first description of a patient with frontotemporal dementia, presenting with aphasia, lobar atrophy, and presenile dementia [1]. In 1911, Alois Alzheimer recognised the characteristic association with Pick bodies and named the clinicopathological entity Pick's disease [2]. In 1982, Mesulam described a language subtype of the disorder, later defined as primary progressive aphasia [3]. New discoveries have raised the attention on the disease, and recently revised diagnostic criteria have been issued [4, 5].

FTD is the second most prevalent type of dementia in patients younger than 65 years [6] and the third in all age groups [7]. FTD affects both genders in roughly equal distribution. The estimated prevalence of FTD is highest in the 45–64 year age group and ranges from 15 to 22 per 100,000 persons [8], but it is probably underestimated due to lack of recognition and diagnosis of the FTD syndromes [9].

Clinical Features

There are three clinical variants of FTD: behavioural-variant frontotemporal dementia (bvFTD), which is associated with early behavioural and executive deficits; semantic variant primary progressive aphasia (svPPA), which is a progressive disorder of semantic knowledge and naming, and non-fluent variant primary progressive aphasia (nfPPA), with progressive deficits in speech, grammar, and word output.

As the disease progresses, the symptoms of the three clinical variants can converge, as an initially focal degeneration can become more diffuse. Over time, patients develop global cognitive impairment and can have motor deficits, including parkinsonism, and few of them develop motor neuron disease. Patients with end-stage disease have difficulty eating, moving, and swallowing. Death occurs on average 8 years after symptom onset and is typically caused by pneumonia or other secondary infections [10].

Behavioural-Variant Frontotemporal Dementia

The most common reported early symptoms of bvFTD include personality changes, with disinhibition and impulsivity or apathy. Behavioural disinhibition can result in tactless and socially inappropriate behaviour, impulsive or careless actions, and offensive personal remarks. Reduced inhibition often results in bad fiscal decisions and these patients, that can be overly trusting, may also become susceptible to financial scams. In some instances, patients are overly friendly and start conversations that are inappropriately explicit or personal. Patients may lose the ability to empathise with their families and friends with a decrease in social interest and responsiveness to the emotions and needs of other people. Patients may become more irritable and may commit antisocial or even criminal acts. Such acts, however, are usually not malevolent but rather poorly thought out or impulsive in nature. Although patients might make inappropriate sexual comments, they usually have decreased libido. Apathy manifests as reduced interest in work, hobbies, social interaction, and hygiene and can be mistaken for depression. Patients often show stereotyped behaviours that vary from simple repetitive movements (like foot tapping or pacing) or repetitive use of verbal phrases to compulsive ritualistic behaviours. Hoarding is common. Some patients become more mentally rigid and resistant to changes in scheduled routines or plans [11]. Some patients with bvFTD, particularly the carriers of a chromosome 9 open reading frame 72 (*C9ORF72*) expansion, may exhibit psychotic features early in the disease course, including visual or auditory hallucinations and bizarre or somatic delusions [12].

Eating habits frequently change in patients with bvFTD. Some patients develop a strong preference for sweets, binge eating, or even attempts to eat nonedible objects [13]. Family members may have to lock kitchen cabinets to stop the patient from stealing food. In addition, many patients with bvFTD demonstrate changes with language, including echolalia, progressive reduction of speech, and semantic deficits.

Patients have limited insight into their own behaviour making an informant critically important when collecting the patient history [14]. Neuropsychological testing in patients with bvFTD frequently reveals executive dysfunction. Although memory can be better than in Alzheimer's disease (AD), episodic memory can be impaired even in early stages of the disease, which may contribute to misdiagnoses of AD [15]. By contrast, drawing and other visuospatial functions are often remarkably spared.

Amyotrophic lateral sclerosis accompanies bvFTD in about 15% of cases and can be suggested by the presence of upper and lower motor neuron findings [16, 17].

Parkinsonism may also be present in patients with bvFTD [18], especially those signs typical of progressive supranuclear palsy (PSP) or corticobasal syndrome (CBS). The classical PSP findings of vertical gaze palsy and axial-predominant parkinsonism may occur early or emerge later in some patients with bvFTD. Similarly, asymmetric parkinsonism, alien limb phenomena, hemineglect and apraxia, as in the corticobasal syndrome, are often associated with FTD neuropathology. Both parkinsonism and motor neuron disease, however, can be caused by a variety of other disorders which must be taken into consideration.

Some individuals have a very slow disease course with slow progression of cognitive impairment and often normal MRI and PET studies. Their disease is classified as frontotemporal dementia phenocopy [19].

While patients may exhibit either increased or decreased activity in the beginning of their disease course, all eventually develop symptoms of apathy and inertia. This may progress to mutism and immobility in the end stages of the disease.

Primary Progressive Aphasia

Patients with primary progressive aphasia (PPA) present an insidious decline in linguistic skills during the initial phase of the disease, and language dysfunction remains the main symptom for the first years of the illness. Deficits include language production, object naming, syntax, or word comprehension and are apparent during conversation or through speech and language assessment.

Although the underlying cause is more often frontotemporal dementia, PPA can also be associated with AD in the form of logopenic aphasia.

Semantic Variant Primary Progressive Aphasia (svPPA)

Semantic loss causes anomia, word-finding difficulties, and impaired word comprehension. Anomia tends to be more pronounced for nouns than for verbs or pronouns, and typically patients ask what a word means. Comprehension of individual words is impaired, especially for words that are not routinely used by the patient. Patients may lose the normal give and take of conversation, talking incessantly and requiring interruption to conduct the examination. The deficits in recognition of objects and people go beyond the visual domain, and tactile, olfactory, or gustatory clues do not help. Patients have surface dyslexia and dysgraphia, impairments in which words with atypical spelling or pronunciation are regularised. Other language domains are spared, especially during the initial disease phase, and patients retain correct grammar and fluent speech [5]. Patients with svPPA also demonstrate abnormal behaviours, such as irritability, emotional withdrawal, insomnia, and strict or selective eating, often focused around one particular type of food; sometimes depression emerges [20]. As the disease progresses, speech becomes increasingly empty, with vague words or jargon phrases replacing specific nouns and verbs, and patients may also develop visual agnosia and prosopagnosia. Symptoms result from early asymmetrical (left more than right) degeneration of anterior temporal lobes. Patients with right-sided temporal atrophy may present with behavioural features and relatively preserved language but over time will also develop semantic deficits.

Non-fluent Variant Primary Progressive Aphasia

Non-fluent variant primary progressive aphasia is characterised by slow, hesitant, and halting speech production and by agrammatism. Patients often make inconsistent speech sound errors, including insertions, deletions, substitutions, transpositions, and distortions. Patients might have trouble understanding sentences with complex syntactic constructions but retain the ability to understand simpler sentences with the same semantic content. Grammatical errors are observed in spontaneous speech and frequently include omission of small words and conjuctions, dropping of verb endings, and errors in subject/verb agreement. Early in the disease, written language production and syntactic comprehension tests reveal mild grammatical errors. Some patients maintain intact writing despite the presence of marked deficits in spoken language. Single-word comprehension and object knowledge are not affected although patients can have a mild anomia that is usually more pronounced for verbs than for nouns [5]. Patients with nfPPA frequently also demonstrate apraxia of speech, defined as impaired motor speech planning, manifest by articulation deficits. Patients may also demonstrate or develop behavioural changes of bvFTD or features of CBS or PSP.

Diagnostic Criteria

In 2011, an international consortium developed revised criteria for the diagnosis of bvFTD [4]. The diagnostic criteria outline features that increase the likelihood that frontotemporal dementia-related neuropathology will be identified [4, 5].

The patients can be classified as possible bvFTD if they present at least three among disinhibition, apathy, loss of empathy, stereotyped or ritualistic behaviour, hyperorality, and typical neuropsychological profile (executive deficit with relative sparing of memory and visuospatial functions). The progressive deterioration of behaviour or cognition by observation or history must be present in all the categories and should not be accounted better by another psychiatric or medical disease. Biomarkers strongly indicative of AD or other neurodegenerative process are considered exclusionary criteria for bvFTD.

Regarding PPA, the Gorno-Tempini criteria have been developed to distinguish the different subtypes. svPPA must have impaired object naming and single-word comprehension with at least three among impaired object knowledge, surface dyslexia or dysgraphia, spared repetition and spared grammaticality, or motor aspect of the speech.

On the other hand, to be diagnosed with non-fluent PPA is mandatory to have agrammatism in language production or effortful, halting speech with at least two among impaired comprehension of syntactically complex sentences, spared single-word comprehension, and spared object knowledge [5].

If patients have a significant functional decline over time and show atrophy, hypometabolism, or hypoperfusion at neuroimaging in the typical areas (frontal and/or temporal lobes for bvFTD, predominant anterior temporal for semantic PPA, and predominant left posterior fronto-insular for non-fluent PPA), the diagnosis can be considered as probable.

The status of definite diagnosis is reserved only for those that are carriers of a known pathogenic mutation or have a histopathologic evidence on biopsy or at postmortem.

Imaging

Neuroimaging results are required by the current criteria [4, 5] to rank the diagnosis as "probable"; however early in the disease, imaging may show only subtle changes.

Structural imaging, with CT and structural MRI, allows to evaluate the regional atrophy, and visual rating scales can be useful in the clinical setting to guide and quantify the atrophy [21, 22].

bvFTD patients typically show atrophy in the frontal and temporal lobes, svPPA shows predominant temporal pole atrophy usually more pronounced on the left side whereas nfPPA more left-sided fronto-opercular atrophy. In the case of genetic cases, it is also possible to identify typical pattern of atrophy for mutation: in *MAPT* mutations, the involvement is anteromedial temporal simmetrically, in *GRN* is fronto-temporo-parietal asimmetrically, and in *C9ORF72* expansions atrophy is predominantly in the frontal lobes, with some atrophy also observed in the thalamus, the cerebellum, the anterior temporal, parietal, and occipital lobes [23].

Other MRI techniques such as functional MRI have shown promise in the differential diagnosis of FTD in research but are not yet part of the current clinical practice.

Fluorodeoxyglucose positron emission tomography (FDG-PET) imaging may be more sensitive than MRI in early stages and can be clinically useful in distinguishing FTD from AD [24]. FDG-PET reveals hypometabolism of frontal, anterior cingulate, and anterior temporal regions in FTD, in contrast to temporoparietal and posterior cingulate hypometabolism in AD.

Patterns of frontal or anterior temporal hypoperfusion with preserved parietal signal on single-photon emission computed tomography (SPECT) can also be useful in distinguishing FTD from AD [25].

PET amyloid imaging allows the identification of amyloid deposition in vivo. This can be particularly useful for the differential of PPA with patients with FTD pathology typically showing low levels of amyloid binding on PET (amyloid negative), while patients with AD pathology showing elevate amyloid binding (amyloid positive) [26].

Several PET tau ligands are currently under investigation in FTD and other form of dementia but are not validated to date.

Workout

The first step of the workout to diagnose FTD is a careful history collected by an informant that include symptoms onset, progression over time and family history of dementia, movement disorders, or psychosis.

Ascertainment of behavioural changes can be facilitated by standardised questionnaires such as the frontal behavioural inventory [27] or the frontotemporal dementia rating scale [28].

A full neurological examination should be performed, especially the assessment of vertical saccades, axial tone, the presence of parkinsonism, cortical sensory tests, apraxia testing, and frontal release signs.

Standard neuropsychological testing is useful to assess different cognitive domains and should show primarily executive dysfunction with relative sparing of visuospatial ability and memory although some patients with bvFTD have significant episodic memory deficits [29].

In patients with PPA, language should be tested, including spontaneous speech, picture naming, word and sentence comprehension and repetition, semantic association, reading, and writing.

Laboratory studies should be done in all patients, including liver and kidney function tests, complete blood count, vitamin B12, and thyroid function.

Neuroimaging studies should always be available, with MRI giving more information than CT scan, to rule out structural and vascular abnormalities and to assess patterns of focal atrophy.

When structural imaging is inconclusive, FDG-PET or SPECT imaging can support the diagnosis in case of a hypomethabolism or hypoperfusion in frontal and/or temporal lobes [24].

PET amyloid is helpful, particularly in young patients or with language problems, to rule out AD, although AD and FTD neuropathology can co-occur.

Lumbar puncture may help to distinguish between FTD and AD or Creutzfeldt–Jakob disease (CJD); in fact, low cerebrospinal fluid beta-amyloid concentrations are suggestive of Alzheimer's disease, whereas very high cerebrospinal tau concentrations could suggest CJD [30]. Unfortunately, there are no validated biomarkers that can reliably distinguish patients with FTD from controls or other dementias, but low serum progranulin levels can predict GRN mutation status in carriers and patients [31].

Genetic testing is available for several mutations that cause familial FTD. Such testing should be done selectively and guided by the family history, clinical syndrome, and imaging. Genetic counselling is advisable for those who wish to know the results [32]. Knowledge of genetic status can confirm a diagnosis and may aid referral of patients and carriers to current and future clinical trials targeting specific FTD mutations.

Differential Diagnosis

Among the many condition that can mimic FTD, reversible causes should always be considered such as neurological infections (syphilis, HIV), toxic-metabolic disorders (heavy metals, illicit drugs), and vascular and paraneoplastic diseases. Both normal pressure hydrocephalus and low intracranial pressure syndromes can sometimes be misdiagnosed as FTD [33].

Psychiatric disorders can mimic FTD; however, an onset of symptoms in patients in middle age should lead to consideration of bvFTD's inclusion in the differential diagnosis [34]. A misdiagnosis of obsessive-compulsive disorder can be made noting the repetitive and compulsive behaviours in patients with FTD. Often apathy and emotional withdrawal might lead to a misdiagnosis of depression, although FTD patients do not usually have other symptoms typical of depression and often deny sadness. Frontotemporal dementia can cause delusions and euphoria, which are features of bipolar disorder and schizophrenia. Personality disorders can be the heralding sign of the behavioural-variant and borderline, antisocial, schizoid, and schizotypal personality changes, and addictive disorders are common features of patients in the early stages of bvFTD. A high rate of late-onset psychosis is a characteristic feature of FTD associated with *C9ORF72* mutations which have also been found in patients with schizophrenia and bipolar disorder [35–37].

The main differential is with other types of dementia particularly AD, vascular dementia, and dementia with Lewy bodies (DLB). AD pathology could be suggested by predominance of memory and visuospatial deficits, social appropriateness, normal neurological examination, and evidence of generalised brain atrophy on imaging. Among the PPA, a form with prominent anomia, acalculia, and word-finding pauses is the logopenic variant primary progressive aphasia, which is usually caused by AD neuropathology [5].

Executive dysfunction, parkinsonism, and hallucinations can be seen in both DLB and FTD; however, patients with DLB have more pronounced parkinsonism, visuospatial deficits, and cognitive fluctuations compared with patients with FTD [38, 39].

The movement abnormalities in FTD, PSP, and CBS are typically less responsive to levodopa than those in classic Parkinson's disease. A diagnosis of PSP is suggested with predominant postural imbalance, slowing of saccadic velocities, a history of early falls, dysphagia, and pseudobulbar affect. PSP and CBS can initially present as either bvFTD or nfPPA [40].

References

1. Pick A. Uber die Beziehungen der senilen Hirnatrophie zur Aphasie. Prager Med Wochenschr. 1892;17:165–7.
2. Alzheimer A. Uber eigenartige Krankheitsfalle der spateren Alters. Z Gesamte Neurol Psychiatr. 1911;4:356–85.
3. Mesulam MM. Primary progressive aphasia. Ann Neurol. 2001;49:425–32.
4. Rascovsky K, Hodges JR, Knopman D, et al. Sensitivity of revised diagnostic criteria for the behavioural variant of frontotemporal dementia. Brain. 2011;134:2456–77.
5. Gorno-Tempini ML, Hillis AE, Weintraub S, et al. Classification of primary progressive aphasia and its variants. Neurology. 2011;76:1006–14.
6. Vieira RT, Caixeta L, Machado S, et al. Epidemiology of early-onset dementia: a review of the literature. Clin Pract Epidemiol Ment Health. 2013;9:88–95.
7. Ratnavalli E, Brayne C, Dawson K, Hodges JR. The prevalence of frontotemporal dementia. Neurology. 2002;58:1615–21.
8. Knopman DS, Roberts RO. Estimating the number of persons with frontotemporal lobar degeneration in the US population. J Mol Neurosci. 2011;45(3):330–5.
9. Onyike CU, Diehl-Schmid J. The epidemiology of frontotemporal dementia. Int Rev Psychiatry. 2013;25(2):130–7.
10. Roberson ED, Hesse JH, Rose KD, et al. Frontotemporal dementia progresses to death faster than Alzheimer disease. Neurology. 2005;65(5):719–25.
11. Perry DC, Whitwell JL, Boeve BF, et al. Voxel-based morphometry in patients with obsessive-compulsive behaviors in behavioral variant frontotemporal dementia. Eur J Neurol. 2012;19:911–7.
12. Snowden JS, Rollinson S, Thompson JC, et al. Distinct clinical and pathological characteristics of frontotemporal dementia associated with C9ORF72 mutations. Brain. 2012;135(pt 3):693Y708.
13. Wooley J, Gorno-Tempini M, Seeley W, et al. Binge eating is associated with right orbitofrontal-insular-striatal atrophy in frontotemporal dementia. Neurology. 2007;69:1424–33.
14. Mendez M, Shapira JS. Loss of insight and functional neuroimaging in frontotemporal dementia. J Neuropsych Clini Neurosci. 2005;17:413–6.
15. Hodges JR, Davies RR, Xuereb JH, et al. Clinicopathological, correlates in frontotemporal dementia. Ann Neurol. 2004;56:399–406.
16. Gustafson L. Frontal lobe degeneration of non-Alzheimer type. II. Clinical picture and differential diagnosis. Arch Gerontol Geriatr. 1987;6:209–23.
17. Brun A. Frontal lobe degeneration of non-Alzheimer type. I. Neuropathology. Arch Gerontol Geriatr. 1987;6:193–208.
18. Le Ber I, Guedj E, Gabelle A, et al. Demographic, neurological and behavioural characteristics and brain perfusion SPECT in frontal variant of frontotemporal dementia. Brain. 2006;129:3051–65.
19. Kipps CM, Hodges JR, Hornberger M. Nonprogressive behavioural frontotemporal dementia: recent developments and clinical implications of the 'bvFTD phenocopy syndrome'. Curr Opin Neurol. 2010;23:628–32.
20. Seeley WW, Bauer AM, Miller BL, et al. The natural history of temporal variant frontotemporal dementia. Neurology. 2005;64:1384–90.
21. Harper L, Fumagalli GG, Barkhof F, et al. MRI visual rating scales in the diagnosis of dementia: evaluation in 184 post-mortem confirmed cases. Brain. 2016;139(Pt 4):1211–25.
22. Kipps CM, Davies RR, Mitchell J, et al. Clinical significance of lobar atrophy in frontotemporal dementia: application of an MRI visual rating scale. Dement Geriatr Cogn Disord. 2007;23(5):334–42.
23. Whitwell JL, Weigand SD, Boeve BF, et al. Neuroimaging signatures of frontotemporal dementia genetics: C9ORF72, tau, progranulin and sporadics. Brain. 2012;135(Pt 3):794–806.

24. Womack KB, Diaz-Arrastia R, Aizenstein HJ, et al. Temporoparietal hypometabolism in frontotemporal lobar degeneration and associated imaging diagnostic errors. Arch Neurol. 2011;68:329–37.
25. McNeill R, Sare GM, Manoharan M, et al. Accuracy of single-photon emission computed tomography in differentiating frontotemporal dementia from Alzheimer's disease. J Neurol Neurosurg Psychiatry. 2007;78(4):350–5.
26. Rabinovici GD, Rosen HJ, Alkalay A, et al. Amyloid vs FDG-PET in the differential diagnosis of AD and FTLD. Neurology. 2011;77(23):2034–42.
27. Kertesz A, Davidson W, Fox H. Frontal behavioral inventory: diagnostic criteria for frontal lobe dementia. Can J Neurol Sci. 1997;24(1):29–36.
28. Mioshi E, Hsieh S, Savage S, et al. Clinical staging and disease progression in frontotemporal dementia. Neurology. 2010;74(20):1591–7.
29. Pennington C, Hodges JR, Hornberger M. Neural correlates of episodic memory in behavioral variant frontotemporal dementia. J Alzheimers Dis. 2011;24(2):261–8.
30. Bian H, Swieten JV, Leight S, et al. CSF biomarkers in frontotemporal lobar degeneration with known pathology. Neurology. 2008;70:1827–35.
31. Ghidoni R, Stoppani E, Rossi G, et al. Optimal plasma progranulin cutoff value for predicting null progranulin mutations in neurodegenerative diseases: a multicenter Italian study. Neurodegener Dis. 2012;9(3):121–7.
32. Goldman JS, Adamson J, Karydas A, et al. New genes, new dilemmas: FTLD genetics and its implications for families. Am J Alz Dis Other Dem. 2007;22:507–15.
33. Wicklund MR, Mokri B, Drubach DA, et al. Frontotemporal brain sagging syndrome: an SIH-like presentation mimicking FTD. Neurology. 2011;76:1377–82.
34. Woolley JD, Khan BK, Murthy NK, et al. The diagnostic challenge of psychiatric symptoms in neurodegenerative disease: rates of and risk factors for prior psychiatric diagnosis in patients with early neurodegenerative disease. J Clini Psych. 2011;72:126–33.
35. Galimberti D, Reif A, Dell'Osso B, et al. C9orf72 hexanucleotide repeat expansion is a rare cause of schizophrenia. Neurobiol Aging. 2014;35:1214.e7–10.
36. Galimberti D, Reif A, Dell'Osso B, et al. C9orf72 hexanucleotide repeat expansion as a rare cause of bipolar disorder. Bipolar Disord. 2014;16:448–9.
37. Galimberti D, Fenoglio C, Serpente M, et al. Autosomal dominant frontotemporal lobar degeneration due to the C9orf72 hexanucleotide repeat expansion: late-onset psychotic clinical presentation. Biol Psychiatry. 2013;74:384–91.
38. Claassen DO, Parisi JE, Giannini C, et al. Frontotemporal dementia mimicking dementia with Lewy bodies. Cogn Behav Neurol. 2008;21:157–63.
39. Perri R, Monaco M, Fadda L, et al. Neuropsychological correlates of behavioral symptoms in Alzheimer's disease, frontal variant of frontotemporal, subcortical vascular, and lewy body dementias: a comparative study. J Alzheimers Dis. 2014;39:669–77.
40. Litvan I, Agid Y, Calne D, et al. Clinical research criteria for the diagnosis of progressive supranuclear palsy (Steele-Richardson-Olszewski syndrome): report of the NINDS-SPSP international workshop. Neurology. 1996;47:1–9.

Autosomal Dominant Frontotemporal Lobar Degeneration: From Genotype to Phenotype

Maria Serpente and Daniela Galimberti

Abstract

Frontotemporal lobar degeneration (FTLD) is the most frequent dementia in presenile population. It presents with different syndromes, including frontotemporal dementia (FTD), primary non-fluent aphasia (PNFA), and semantic dementia (SD). Motor neuron disease often co-occur with FTLD. In the last few years, different autosomal dominant mutations have been demonstrated to be the cause of the familial aggregation frequently reported in FTLD. Major causal genes so far discovered include microtubule-associated protein tau (*MAPT*), progranulin (*GRN*), and chromosome 9 open reading frame (*C9ORF*) *72*. Mutations in *MAPT* are generally associated with early onset and with the FTD phenotype, whereas mutations in *GRN* and *C9ORF72* are associated with high clinical heterogeneity and age at disease onset. In addition, other genes are linked to rare cases of familial FTLD. Moreover, the use of next-generation sequencing approach allowed the identification of disease modifier (risk) genes such as common variants in the transmembrane protein 106b.

Keywords

Frontotemporal lobar degeneration · Tau · Progranulin (*GRN*) · *C9ORF72* · Genetics · Risk factor

---

M. Serpente (✉) • D. Galimberti
Neurodegenerative Disease Unit, Department of Pathophysiology and Transplantation, University of Milan, Fondazione Cà Granda, IRCCS Ospedale Maggiore Policlinico, Milan, Italy
e-mail: maria.serpente@unimi.it; daniela.galimberti@unimi.it; daniela.galimberti@policlinico.mi.it

© Springer International Publishing AG 2018
D. Galimberti, E. Scarpini (eds.), *Neurodegenerative Diseases*,
https://doi.org/10.1007/978-3-319-72938-1_8

Introduction

Frontotemporal lobar degeneration (FTLD) refers to a clinically, pathologically, and genetically heterogeneous group of disorders that affect principally the frontal and temporal lobes of the brain. After Alzheimer's disease (AD), FTLD is the second most common form of dementia with presenile onset (<65 years) and a mean average in the 50s. Its prevalence has been estimated at 10.8 per 100,000 and lifetime risk at 1 in 742 [1]. There is, however, a wide variation in the age at onset as well as in age disease duration. The mean duration of symptoms from onset until death is around 8 years, and the progression may be rapid or very slow.

FTLD encompasses three main clinical syndromes: frontotemporal dementia (FTD), progressive non-fluent aphasia (PNFA), and semantic dementia (SD).

Clinical Syndromes

The most common clinical syndrome, which accounts for more than half of cases [2], is FTD, which is characterized by behavioral changes and progressive deterioration of personality. Main behavioral features are social disinhibition, apathy, loss of empathy for others, repetitive, obsessive, and stereotyped behaviors, and dietary changes. These key characteristics form the basis for contemporary clinical diagnostic criteria [3]. Moreover, it is now recognized that some FTD patients experience psychotic symptoms of delusions and hallucination [4]. Behavioral changes are accompanied by cognitive impairments in frontal executive functions, with relative sparing of memory.

A second, less frequent, clinical syndrome is termed PNFA [5], or non-fluent variant primary progressive aphasia [6]. It is a disorder of expressive language, associated with asymmetric atrophy of left hemisphere, typically characterized by effortful speech and impaired use of grammar, but there are no uniform and precise language impairments across patients [7].

The third syndrome is SD, which is characterized by impaired understanding of a meaning of words, faces, objects, and other sensory stimuli. SD is also known as semantic variant primary progressive aphasia [6] because of the prominence of language-related problems. Neuroimaging data shows bilateral but often asymmetric atrophy of the temporal lobes [8]. The symptoms depend on the side of atrophy; for example, patients with left-sided atrophy have difficulties with the comprehension of words, whereas right-predominant patients show impairment in face recognition [9]. Nevertheless, there is a gradual loss of conceptual understanding that finally affects all sensory domains.

FTLD can also be associated with an extrapyramidal movement disorder, such as parkinsonism or corticobasal syndrome (CBS), or with amyotrophic lateral sclerosis (ALS). The presence of ALS accelerates the disease course [10] and accounts for patients with very short disease duration. Moreover, the overlap with ALS influences also the gender distribution of the disease; in fact, FTLD affects male and female equally, but the presence of ALS produces a significant male bias [11].

Moreover, ALS is most often seen in combination with FTD, whereas an association with PNFA or SD is rare [11]. In these frameworks, to date, ALS and FTLD are considered part of a disease spectrum based on clinical, pathological, and genetic evidence [12, 13].

Neuropathology of FTLD

The first histopathological description was done in 1911 by Alois Alzheimer, who observed, in the brain of a patient with language and behavior disturbances, the presence of ballooned neurons containing tau protein and argyrophilic intracytoplasmic inclusions. He named them "Pick cells" and "Pick bodies", respectively, after Arnold Pick, who reported that case in 1892.

Nevertheless, the Pick pathology is not always present. From the neuropathological point of view, FTLD patients are classified according to the main components of pathological protein aggregates. Currently, there are three neuropathological categories: FTLD-tau, FTLD-TAR DNA-binding protein (TDP)-43, and FTLD-FET (FUS and its related proteins, EWS and TAF15, namely, FET family) [14]. In about 45% of neuropathological confirmed FTLD cases, neuronal and glial inclusions of the microtubule-(MT)-binding protein tau (FTLD-tau) have been observed. Microtubule-associated protein tau (*MAPT*) gene may generate six different tau isoforms that are expressed in the adult brain and consist of either three (3R) or four (4R) MT-binding repeats. FTLD patients are classified according to predominant accumulation of tau species [13]. Accumulation of 3R tau forms produces typical rounded bodies, known as Pick bodies, whereas 4R tau can be found also in patients with CBS.

However, the majority of FTLD patients present pathological inclusion negative for tau protein but positive for ubiquitin. In fact, in about 50% of FTLD patients, aggregates of RNA- and DNA-binding protein (TDP-43) are present in those inclusions with subsequently creation of a new subtype called FTLD-TDP. A harmonized nomenclature defines four different subtypes of FTLD-TDP (A-D) based on the morphology and anatomical distribution of TDP-43 inclusions [15]. The remaining 5% of FTLD patients present ubiquitinated inclusion bodies positive for the fused in sarcoma (FUS) protein [15]. Recently, it was shown that in addition to FUS two other members of FET family proteins, such as Transportin-1, TATA-binding protein-associated factor 15 (TAF15), and Ewing's sarcoma protein (EWS), are found in the inclusions. Thus, this group of FTLD patients was renamed to FTLD-FET [15].

The clinical phenotypes do not easily predict the underlying type of FTLD pathology. For example, FTD can be associated with FTLD-tau, FTLD-TDP, and also with FTD-FUS pathology. However, when behavioral disorder occurs in combination with ALS, FTD is often associated with FTLD-TDP rather than tau pathology. Moreover, a very early onset of disease is a strong predictor of FUS pathology.

Although this heterogeneity is a bias for genetic studies of FTLD, significant progresses in causal gene identification have been made thank to the study of individual FTLD families and the use of pathologically homogenous confirmed subpopulation of patients.

Genetics of FTLD: Causal Genes and Disease Genetic Modifiers

The majority of FTLD cases are sporadic and likely caused by the interaction between genetic and environmental factors. A number of cases, however, present familial aggregation and are inherited in an autosomal dominant fashion, suggesting a genetic cause [16–18]. Up to 40% of patients have a family history, suggesting FTLD in at least one extra family member [17, 19]. A clearly autosomal dominant pattern of inheritance is documented in about 10% of patients.

The current knowledge about genetics of FTLD has been recently enlarged by the identification of multiple novel genetic defects and chromosomal loci involved in hereditary forms. At present, three major causal genes have been identified: *MAPT*, *GRN*, and *C9ORF72*. Mutations in other genes have also, less commonly, associated with FTLD. Of these, the most notable is charged multivesicular body 2 protein gene (*CHMP2B*), and the others are valosin-containing protein gene (*VCP*), sequestosome 1 gene (*SQSTM2*, also known as p62), coiled-coil-helix-coiled-coil-helix domain containing 10 (*CHCHD10*), TANK-binding protein gene (*TBK1*), *TARDBP*, and *FUS*.

Major Causal Genes

MAPT

The first evidence of a genetic cause for familial FTLD came from the demonstration of a linkage with chromosome 17q21.2 in autosomal dominantly inherited form of FTD with parkinsonism [20], named FTDP-17. A comparison of the linked regions in each analyzed family localized the disease gene to a 3-cM region at chromosome 17q21-22 [21]. The gene responsible for such association, *MAPT*, was discovered few years later [22]. *MAPT* encodes the microtubule-associated protein tau, which is involved in microtubule stabilization, assembly, and cytoskeletal dynamics [12]. It is composed by 15 exons and transcribed, by alternative splicing, in six different isoforms ranging from 353 to 441 amino acids. All six isoforms play a role in the maintenance of microtubular structure. If one or more fails, microtubule formation will become more difficult or the stability compromised. Any excess of tau protein can be bundled into protein aggregates that fill the cells and induce neurotoxicity. Tau has four repeat domains in the C-terminus, which mediate the interaction with microtubules. These domains are encoded by exons 9, 10, 11, and 12 in which the majority of pathogenic mutations have been reported. Alternative splicing of exon 10 leads to two different isoforms that contain either three (3R) or four (4R) 31-amino acids repeats [23].

All patients with *MAPT* mutations are characterized by the deposition of insoluble aggregated tau proteins within neurons and glial cells in the cerebral cortex and in other brain regions.

To date, more than 40 pathogenic *MAPT* mutations have been described and classified according to their position in the gene [24], their effects on *MAPT* transcription, and the type of tauopathy. *MAPT* mutations include missense mutations, deletions, or intronic mutations located close to the splice-donor site of the intron after the alternatively spliced exon 10. They are mainly clustered in exons 9–13, which contain the microtubule-binding regions, except for two mutations in exon 1 [25]. The frequency of *MAPT* mutations is highly variable, but in general *MAPT* mutations are very rare in sporadic patients, whereas in most familial cases the frequency ranges between 5% and 20% depending on the geographic distribution [26].

The pathogenic mechanism of each different mutation depends on the type and location of the genetic defect and affects the normal function of tau, i.e., the stabilization of microtubules promoting their assembly by binding tubulin. Some mutations increase the free cytoplasmic portion of the protein promoting tau aggregation, while others lead to an aberrant phosphorylation of tau protein, which damages microtubule stabilization [26]. The mechanism is clearest for mutations localized in the donor splicing site following exon 10. It was shown that these intronic mutants increase the inclusion of *MAPT* exon 10 by destabilizing the stem-loop structure that spans the splice site of exon 10 resulting in an increased production of 4R protein tau. Mutations in the acceptor splicing site following exon 10 lead to an enhanced inclusion of this exon [27].

Alternatively, other mutations affect the alternative splicing, thus producing altered ratios of the different isoforms (3R/4R tau). Most of missense mutations, such as the p.P301L mutation, reduce the ability of tau to binding microtubules leading to a decreased tau capacity to promote microtubules assembly [28]. Moreover, it was observed in vitro studies that several coding mutations accelerate the aggregation of tau [29]. So far, five mutations have been identified outside the tau repeat domains: p.R5H, p.R5L (N-terminal), p.K369I, p.K389R, and p.R406W (C-terminal) [25].

Grisart et al. observed a microduplication on chromosome 17g21.31 that was associated with behavioral problems and skills impairments [30]. The authors suggested that the overexpression of *MAPT* in neurons could contribute to the behavioral changes and the duplication of the corticotropin-releasing hormone receptor 1 gene (*CRHR1*), located 59 kb centromeric from *MAPT*, could explain the impaired motor skills. The presence of structural changes at the *MAPT* locus in the presence of behavioral changes led the authors to believe that rearrangements at this locus might be associated with FTLD [31]. Several subsequently studies failed to identify abnormal copy number variations (CNVs) at the genetic region encompassing *GRN* and *MAPT* [31]. However, in 2009, Rovelet-Lecrux et al. identified a heterozygous 17.3 kb deletion responsible for the removal of exons 6–9 of *MAPT* in one FTD patient [32]. This deletion caused the loss of the first microtubule-binding domain and a decrease in the binding abilities of tau to the microtubules. The same group reported a 439-kb duplication in the region encompassing *CRHR1*, *MAPT*, and saithoin (*STH*) in one patient affected by behavioral and amnestic disorders [33]. These are the first evidence of a possible link between rearrangements at the *MAPT* locus and the FTLD.

Rossi et al. recently suggested that tau plays a role in genome and chromosome stability that can be ascribed to its function as a microtubule-associated protein as well as a protein protecting chromatin integrity through interaction with DNA [34]. At autopsy, patients with *MAPT* mutations show tau-positive inclusions [26].

The clinical presentation in *MAPT* mutation carriers is heterogeneous, but behavioral changes, semantic impairment, episodic memory decline, and parkinsonism have been proposed as key clinical features [35]. From a pathological point of view, patients present atrophy of the frontotemporal lobes and basal ganglia and variable presences of tau-positive inclusions, typical of FTLD-tau [14].

Although MAPT mutations are considered to have complete penetrance, in 2015 Rossi and Tagliavini observed the presence of unaffected carriers in the same family suggesting the possibility of incomplete or delayed penetrance [36].

GRN

After the discovery of *MAPT* as causal gene for FTDP-17, there were still numerous autosomal dominant FTLD cases genetically linked to the same chromosomal region of *MAPT* (chr17q21), without any mutation in *MAPT*, in spite of an extensive fine mapping of the gene. A small region rich of genes, localized approximately 6.2 Mb in physical distance to *MAPT* locus, had been recognized as that one containing the gene responsible for the disease in these families. The first identified mutation in *GRN*, identified in 2006, consisted of a 4-bp insertion of *CTGC* between coding nucleotides 90 and 91, causing a frameshift and premature termination in progranulin (C31LfsX34) [37]. In another parallel study, Cruts et al., analyzing other families with a FTLD pathology without *MAPT* mutation, found at the same time another mutation of five base pairs into the intron following the first noncoding exon of *GRN* (IVS1 + 5G > C) [38]. This mutation causes the splicing out of the intron 0, leading the retention of mRNA within the nucleus and its degradation.

GRN mutations were subsequently found to account for 5–20% of FTLD patients with positive family history and 1–5% of apparently sporadic patients [39].

GRN gene encodes for the growth regulation factor named progranulin. GRN is an 88-kDa secreted glycoprotein, which in the brain is expressed by neurons and microglia [40]. Its expression is low in early development and increases with age, and it is composed by seven and one half cysteine-rich granulin domains and can be cleaved by several proteases into 6 kDa units called granulins. It belongs to a family of proteins involved in multiple biological functions, including development, wound repair, and inflammation, by activating signaling cascades that control cell cycle progression and cell motility.

Since the original identification of null mutations in FTLD, more than 70 different mutations have been described so far in 231 families. Most of the known pathogenic *GRN* mutations, particularly frameshift, splice-site, and nonsense mutations, are predicted to result in a premature stop codon. The resulting aberrant mRNA is degraded through the process of nonsense-mediated decay, leading to haploinsufficiency [41] and to a 50% loss in GRN protein levels. Moreover, a number of missense *GRN* mutations have been described, but only one, p.A9D, has been confirmed as pathogenic. It was observed that in vitro p.A9D mutant GRN was not secreted but sequestered in the Golgi network [42]. Other potential pathogenic missense mutations are p.P248L, p.R432C (alteration of GRN secretion), p.C139R, and p.C521Y (alteration of cysteine residues leading to impaired physiological processing of GRN) [43].

At the neuropathological examination, *GRN*-mutated FTLD cases displayed ubiquitin-positive, tau-negative inclusions (FTLD-U) similar to the microvacuolar type still observed in a large proportion of apparently sporadic FTLD that were different from the tau-positive inclusions typical of *MAPT* mutated cases. According to the novel neuropathological classification of FTLD-TDP pathology in FTLD, TDP-43 neuropathological subtype A is consistently found in association with *GRN*-mutated cases [15]. Truncated and hyperphosphorylated isoforms of the TDP-43 were recognized as main components of the ubiquitin-positive inclusions typical of the *GRN*-mutated families, as well as of idiopathic FTLD and of a proportion of ALS cases [44]. Nevertheless, at present, the linkage between progranulin haploinsufficiency and TDP-43 accumulation in the cytoplasm has not been clarified.

As mentioned above, *GRN* mutation accounts for about 5–10% of all FTD cases, markedly varying depending on the population considered. A collaborative study [45] analyzing *GRN* mutations in 434 FTLD patients, clinically ranging from bvFTD to PNFA, FTLD associated with parkinsonism or MND, estimates a frequency of 6.9% of all included FTLD-spectrum cases. About 56% of such cases were represented by FTLD subjects with ubiquitinated inclusions at the neuropathology (FTLD-U) with a positive family history of FTLD. The most common phenotype was bvFTD, but a few patients were diagnosed with PNFA, AD, or CBS. As expected, the majority of *GRN* mutations introduced a premature termination codon, suggesting that their corresponding mRNA has been degraded through nonsense-mediated decay, thus supporting the hypothesis that most *GRN* mutations create a functionally null allele [37]. Accumulation of ubiquitinated proteins, p62, and lysosomal proteases was observed probably resulting from impairment in proteasomal or lysosomal activity. Intriguingly, this is particular relevant for homozygous GRN mutations that are, recently, associated with the lysosomal storage disorder, neuronal ceroid lipofuscinosis (NCL) [46].

From a clinical point of view, mutations in *GRN* are associated with extremely heterogeneous phenotypes, but the main clinical diagnosis is bvFTD followed by diagnosis of PPA [47]. Language impairment seems to be more relevant as the disease progresses. About 40% of patients who have parkinsonism and episodic memory impairment are frequently observed leading to a clinical diagnosis of AD in some cases [48]. Although rarely, an overlap between psychiatric disorders and genetically determined FTLD can occur, as shown by Rainero et al. [49], who described a patient with heterosexual pedophilia who was a carrier of a *GRN* mutation and developed bvFTD over time, and by Cerami et al. [50], who reported two clinically different, apparently sporadic FTLD cases sharing the Thr272fs *GRN* mutation, who had had a premorbid bipolar disorder history.

The penetrance for *GRN* mutations is age dependent with only 50% of *GRN* mutation carriers affected at the age of 60s and 90% of mutation carriers affected at 70 years of age.

Age at disease onset is extremely wide, even in the same family, ranging from 47 to 79 years [51].

For example, in a large Calabrian family harboring a heterozygous c.1145insA mutation, the age at onset ranged from 35 to 87 years whereas the age of death from 56 to 87 years [52]. In that family the clinical presentation is homogenous; all affected members had clinical diagnosis of FTD with subsequent language impairment.

A major contribution to achieve a correct diagnosis independent of the phenotypic presentation is the demonstration that progranulin plasma levels are extremely low in *GRN* mutation carriers, even in asymptomatic subjects [51, 53].

Regarding the function of progranulin, Pickford et al. [54] demonstrated, in an in vitro model, that it has chemotactic properties towards cultured mouse neurons. In addition, progranulin-treated primary neurons secrete a number of cytokines and chemokines, particularly those involved in proliferation (i.e., IL-4), and, importantly, induce microglia to switch from a pro-inflammatory to an anti-inflammatory phenotype [54]. Another recent observation is that progranulin binds the TNFR2 that is expressed specifically in neuronal subtypes and glial cells in the brain, leading to an anti-inflammatory cascade [55].

Yin et al. [56] generated conditional *GRN* knockout mice. They observed that *GRN*-deficient macrophages produced more pro-inflammatory cytokines and chemokines, including CCL2, CXCL1, IL-6, IL-12p40, and TNF, but less anti-inflammatory cytokine IL-10 compared to wild-type (wt) macrophages, when exposed to bacterial lipopolysaccharide. However, *GRN*-deficient mice failed to clear bacterial infection as fast as wt mice and were characterized by an exaggerated inflammatory tissue damage. Immunostaining of brain sections for CD68 revealed greater activation of microglia with age in *GRN*-deficient than wt mice. Moreover, *GRN*-deficient microglia responded to inflammatory stimuli by becoming more cytotoxic than wt microglia, and *GRN*-deficient neurons were more susceptible than wt to damage by activated microglia and by certain cytotoxic stresses, such as depletion of glucose and oxygen. They also showed enhanced hippocampal ubiquitin immunostaining and increased phosphorylation of TDP-43 in the hippocampus and thalamus of old *GRN*-deficient mice. In light of these observations, authors hypothesized that FTLD may arise from the congruence of two independent phenotypes of *GRN* insufficiency: deregulated inflammation and increased neuronal vulnerability to damage [56].

In vivo studies in progranulin heterozygous mice (Grn+/−), that mimic progranulin haploinsufficiency, were carried out as well. These mice developed age-dependent social and emotional deficits potentially relevant to bvFTD. Nevertheless, no gliosis or neuroinflammation was observed, suggesting that microglial activation is independent from functional deficits, and thus progranulin deficiency could have effects directly on neurons [57]. It is important to underlie that the understanding of GRN physiological roles and the identification of factors that are able to upregulate GRN may be an important therapeutic area. For example, several studies identified sortilin and prosaponin as key regulators of GRN lysosomal trafficking and modulators of GRN secretion in human [56]. Moreover, it was observed that, in mouse models of PD and AD, the addiction of GRN had neuroprotective effect [57].

C9ORF72

One of the most intriguing discoveries in the genetics of FTLD has been the investigation of FTD/MND families linked to a locus on chromosome 9q21-22. The first evidence of linkage with this locus comes from a study carried out in families with autosomal dominant FTD-MND [58]. Additional data confirmed the linkage to chr9q21-22 in FTD-MND families [59], until, in 2011, two international groups of researchers identified the gene responsible for the disease in this locus, *C9ORF72* [60, 61]. The mutation consists of a large hexanucleotide (GGCGCC) repeat expansion in the first intron of a gene named Open Reading Frame 72 (*C9ORF72*) that segregates with ALS or combined FTD-MND phenotype and TDP-43-based pathology.

In healthy subjects, most individuals carry between 2 and 20 repeats but FTD and ALS patients from 100 to also 1000s of copies of repeat. The minimum repeat length to confer risk of disease is unknown, probably due to the presence of somatic mosaicism. In fact, the length of repeat is different between tissues even in the same individual, and this phenomenon complicates correlative genotype-phenotype studies [62].

C9ORF72 repeat expansion is the most common cause of FTLD (with or without ALS) worldwide, explaining about 25% of familial FTLD and 5% of apparently sporadic FTLD cases. However, there is a particular high frequency in Finland population leading a suggestion of a single founder or predisposing disease haplotype. The frequency in ALS cohort of patients is higher, taking into account that c9orf72 repeat expansion is present in about 22.5% of familial ALS and 21% of sporadic cases. Despite the high presence of this mutation in North America and Europa, studies in Asian cohorts have reported much lower frequencies [63].

The clinical phenotypes are very variable [64] as well as the age at onset and disease duration; in fact, age at onset can range between 27 and 83 years and disease duration from 1 to 22 years [65]. Moreover, few but very important studies about the frequency of *C9ORF72* repeat expansion in older control groups have observed a frequency of 0.17%, suggesting an age-dependent disease penetrance and potential lifelong reduced penetrance associated with this mutation [65].

Regarding the clinical presentation, the most common is FTD, ALS, or both. As mentioned above, in families where FTLD-ALS is the clinical phenotype, *C9ORF72* repeat expansion is very common explaining the disease in more than 50% of families [66]. In FTLD, patients present behavioral disturbances, whereas language impairments (PNFA or SD) are less commonly observed [67]. In addition to classical behavioral presentations, such as apathy, disinhibition, socially inappropriate conducts and loss of empathy, *C9ORF72* expansion carriers present a high frequency of hallucinations, psychosis, and delusions [68], which lead to a primary diagnosis of bipolar disorders and schizophrenia [69]. Sometimes, patients have episodic memory problems at the beginnings of the disease course, receiving a primary diagnosis of AD [70]. It is notable to underlie that 1% of clinically diagnosed AD patients carry a *C9ORF72* expansion with FTLD-TDP pathology [70]. Moreover, *C9ORF72* expansions were reported as the most common cause of Huntington disease photocopies, in which mutations in huntingtin gene were excluded [71]. Early parkinsonism has also been reported in *C9ORF72* expansion carriers similar to *MAPT* and *GRN* mutation carriers, but *C9ORF72* expansion is very rare in patients diagnosed with PD, CBS, and dementia with Lewy bodies.

From a neuropathological point of view, postmortem examination showed that *C9ORF72* expansion carriers present TDP-43-positive inclusions in different brain areas. Most patients present with FTLD-TDP A or B. In addition, they have neuronal inclusions in the cerebellar granule cell layer, hippocampal pyramidal neurons, and other anatomic sites that are positive for ubiquitin and p62 proteins. These inclusions are composed by dipeptide repeat proteins (DPRs), translated from the GGGGCC repeat through unconventional repeat-associated non-ATG translation [72]. Poly-GP, poly-GA, and poly-GR are generated from sense strand and detected in hippocampus and cerebellum of expansion carriers. From the antisense strand, poly-PA, poly-PR, and again poly-GP DPRs are also generated.

However, little is known about the normal function of C9orf72 protein, though homology suggests that it may be part of the DENN (differentially expressed in normal and neoplastic cells) family proteins, which are GDP/GTP exchange factors that activate Rab-GTPases [60]. Regarding the function of the *C9ORF72* product and the mechanisms at the basis of the pathogenesis of the disease in the expansion carriers, quite few information are available. As previously mentioned, the presence of toxic DPRs and/or RNA toxicity is one of the possible pathogenic mechanisms proposed.

Sense and antisense RNA foci are generated from expanded repeat, and they are observed in different brain areas of *C9orf72* expansion carriers [73]. RNA foci, which lead to the sequestration and altered activity of RNA-binding proteins, have been implicated in several neurodegenerative noncoding expansion disorders [73]. Reddy et al. [74] demonstrated that the r(GGGGCC)n RNA forms extremely stable G-quadruplex structures, which are known to theoretically affect promoter activity, genetic instability, RNA splicing, translation, and neurite mRNA localization.

Moreover, several studies, conducted in derived cells and tissue of patients, demonstrated that these foci are able to sequester RNA-binding protein, including hnRNP h, hnRNP A1, and SC35, affecting the mRNA nuclear transport system [75]. Nevertheless, the clear mechanism linking RNA foci and sequestered proteins to neurodegeneration has not been fully understood.

The production of DPRs with unconventional mechanisms of non-ATG-initiated translation called RAN may also contribute to neurodegeneration. In cultured cells and primary neurons, poly-GA overexpression led to the generation of p62-positive inclusions and neurotoxicity attributed to impaired ubiquitin proteasome function [76]. On the other hand, arginine-rich dipeptide (poly-GR and poly-PR) led to the formation of nucleolar inclusions in fly models [77]. Since the clinical utility as well as the significance and the temporal course of DPRs in the pathogenesis of the disease is still unclear, Lehmer et al. established a poly-GP immunoassay from cerebrospinal fluid (CSF) in order to identify and characterized *C9ORF72* patients. They observed the poly-GP CSF levels were already detectable in *C9ORF72* asymptomatic carriers compared to healthy subjects, and these levels are similar in symptomatic expansion carriers demonstrating their possible use as diagnostic biomarker in addition to genetic screening [78].

Recently, in vivo studies observed that toxicity from RNA foci and DPRs can be suppressed by modulation of nuclear transport, suggesting an impairment of nucleo-cytoplasmic transport as common pathogenic mechanism underlying RNA foci and DPRs [77]. Moreover, a defect in the localization of RanGAP1 protein, a component of nuclear pore complex, was observed in the brain of *C9ORF72* expansion carriers supporting the hypothesis that a compromised nucleo-cytoplasmic transport also contributes to the human form of disease [77]. Given that both foci formation and RAN translation in c9FTD/ALS require the synthesis of GGGGCC repeat expansion RNA, therapeutic strategies that target these transcripts and result in their neutralization or degradation could effectively block these two potential pathogenic mechanisms and provide a much needed treatment for c9FTD/ALS. For example, the use of antisense oligonucleotides targeting both strands of *C9ORF72* repeat and the use of small molecules binding the secondary structures formed by *C9ORF72* repeat have been proposed [79]. Another possible therapeutic strategy might be to remove or prevent the DPRs formation by improving the cellular degradation systems. A very recent study demonstrated that while the DPRs are mainly processed via autophagy, this system is unable to fully clear their aggregated forms, and thus they tend to accumulate in basal conditions. Overexpression of the small heat shock protein B8 (HSPB8), which facilitates the autophagy-mediated disposal of a large variety of classical misfolded proteins, significantly decreased the accumulation of most DPR insoluble species. Thus, the induction of HSPB8 might represent a valid approach to decrease DPR-mediated toxicity and maintain neuron viability [80].

Rare Causal Genes

CHMP2B

Few FTLD families display mutations in *CHMP2B*, located on chromosome 3p11.2, which encodes a component of the heteromeric ESCRT III complex, involved in the endosomal trafficking and degradation [81]. In particular, CHMP2B protein is involved in sorting and trafficking surface receptors or proteins into intraluminal vesicles for lysosomal degradation and binding the Vps4 protein responsible for the dissociation of ESCRT components [82]. CHMP2B is a 213-amino acid-long protein that presents a coiled-coil domain at the N-terminus, a microtubule-interacting transport (MIT), and microtubule-interacting region (MIR) at the C-terminus. The first mutation in *CHMP2B* was identified in one large kindred from Denmark, and it occurs in the splice acceptor site for the sixth and final *CHMP2B* exon, leading to a formation of two novel transcripts termed *CHMP2BIntron5* and *CHMP2BDelta10* [84]. To date, 11 different mutations, of which four in five families seem to exert a pathogenic action (http://www.molgen.vib-ua.be/), have been so far described; for this reason, *CHMP2B* is considered an extremely rare genetic cause of FTLD pathology. It is important to note that all mutations described (missense and truncation mutations) show a common mechanism of action: the deletion of the C-terminus of the protein [82]. Probably, the loss of the Vsp-4 binding domain located in C-terminus of the protein causes the accumulation of mutated CHMP2B on the endosomal membrane and prevents the recruitment of other proteins necessary for endosomal fusion with lysosomal. This phenomenon leads to the impairment of the late endosomal trafficking and contributes to neurodegenerative processes in FTD [83]. This can be observed as enlarged and abnormal endosomal structures in postmortem brain tissue from patients [84]. From a histological point of view, patients with *CHMP2B* mutations present FTLD-U with ubiquitin- and p62-positive but TDP-43-negative neuronal cytoplasmic inclusions [85]. Recently, it was observed in transgenic mice expressing either human CHMP2B*intron5* or human wild-type protein, that only CHMP2B*intron5*, but not wild-type or CHMP2B knockout mice developed neuropathology consistent with that seen in FTLD patients carrying CHMP2B mutations [86]. These data support the hypothesis that *CHMP2B* mutations act through a gain-of-function mechanism. Moreover, the use of RNA interference approach against mutant *CHMP2B* in primary patient fibroblasts has shown that this treatment reverses the mutant endosomal phenotype. Importantly, this morphological change is also observed in *CHMP2B* mutation brain tissue, suggesting that RNA interference might be a future therapeutic approach for the treatment of FTLD patients with *CHMP2B* mutations [87].

Behavioral and cognitive impairment associated with extrapyramidal and pyramidal signs are the main clinical manifestations in *CHMP2B*. Myoclonus can occur late in the course of the disease, and motor neuron disorders have been described in only two cases [88]. To assess the earliest neuropsychological changes in *CHMP2B* mutation carriers, a longitudinal prospective study spanning over 8 years and including 17 asymptomatic individuals with *CHMP2B* mutations was carried out. Longitudinal analyses showed a gradual decline in psychomotor speed, working memory capacity, and global executive measures in the mutation carriers group compared with controls. This decline starts several years before they fulfill diagnostic criteria for FTD, but the level of cognitive changes over time varied considerably among different individuals [89].

VCP-1 and SQSTM1

Mutations in *VCP* were firstly described as cause of hereditary inclusion body myopathy (IBM) with Paget's disease of the bone (PDB) and frontotemporal dementia (IBMPFD) [90]. Myopathy is the more frequent clinical symptom, present in about 90% of affected subjects, whereas FTD is seen in about 33%, usually many years after the onset of muscle symptoms. From a histological point of view, brain tissues of patients carrying *VCP* mutations are characterized by FTLD-TDP type D pathology with TDP43- and p62-positive inclusions within neuronal nuclei [91]. In 2015, Taylor et al. introduced the term multisystem proteinopathy (MSP) to describe a multisystem disorder that affects bone, muscle, and nervous system and that is now used to describe VCP mutation carriers [92].

(*VCP*)-*1* is located on chromosome 9p13.3 and encodes a monomeric protein composed by 806 amino acids. The VCP hexamer is a member of the AAA-ATPase superfamily that is composed by six monomers, forming a ring around a central pore with two AAA+ protein domain called D1 and D2 domains [93]. It is known as regulator of many cellular processes, such as ubiquitin-dependent protein quality control, labeling proteins for degradation and coordination of the removal of protein aggregates via multivesicular body formation [94].

More than 30 different mutations in *VCP-1* have been now described [95]. The most relevant mutations are located in the N-terminal that is important for binding substrates and cofactors required for protein processing. R155H (the most frequent) was the first pathogenic missense mutation reported in FTD and is located in the cofactor-binding domain at the N-terminus of the protein; missense mutations associated with IBMFD have been identified in different domains such as N-terminus domain, the linker L1 connecting N-terminus and D1 domain. One missense mutation was identified in linker L2 and one in D2 domain [96]. Moreover, VCP is also present in the inclusions of several diseases including ALS, PD, and HD [93]. Very recently, Komatsu et al. identified a novel mutation (G156S) associated with IBM, Paget disease, and FTD [97].

Another gene involved in the mechanism of protein degradation as well as in FTLD pathogenesis is *SQSTM1*. This gene encodes for p62 protein, a connector between ubiquitinated proteins and autophagy receptor or proteasome degradation pathways [98]. Mutations in *SQSTM1* gene were firstly described in PDB and are responsible for around 30% of familial PDB cases. To date, mutations in this gene are also associated with FTLD and ALS, and *SQSTM1* is now included in the list of causal gene responsible for MPS [99]. Regarding FTLD, *SQSTM1* mutations explain about 3% of cases, but segregation of these mutations with FTLD has been shown in few families. In 2014, Van der Zee et al. published a large-scale resequencing study in FTLD cohort of patients and identified a number of mutations in the C-terminal of the gene that are involved in the binding with ubiquitinated proteins [100]. It is interesting to note that several studies reported cases of FTLD patients with mutations in *SQSTM1* gene but also with *C9ORF72* expansion. The co-presence of *C9ORF72* mutation may influence the phenotype; thus, finding one FTLD-related mutation does not exclude the presence of further influential genetic alterations.

CHCHD10

CHCHD10 gene is located on chromosome 22 and encodes a mitochondrial protein that is enriched at cristae junctions in the intermembrane space. By exome sequencing, it was possible to identify the first pathogenic mutation, p.S59L, in an atypical family with late-onset motor neuron disease, FTD, cerebellar ataxia, and mitochondrial myopathy [100]. Subsequent genetic studies identified additional potential pathogenic mutation in FTLD and ALS patients with 1–3% of frequency [101]. A recent study, conducted on Asian FTLD patients showed that the frequency of *CHCHD10* mutations in this population was 7.7%, whereas mutations in the three major causal genes (*MAPT*, *GRN*, and *C9ORF72*) explained <3% of cases, suggesting that *CHCHD10* is the most common FTLD gene in Asia [102]. In silico prediction programs suggest that *CHCHD10* mutations are pathogenic, but no segregation in families could be performed. However, functional in vitro studies demonstrated that *CHCHD10* mutations disrupt the mitochondrial contact site and cristae organization system leading to respiratory chain deficiency, nucleoid disorganization, and decrease of apoptosis [103]. Very recently, Perrone et al. identified a novel nonsense mutation (p.Gln108*) in a patient with atypical clinical FTD and pathology-confirmed Parkinson's disease (1/459, 0.22%) leading to loss of transcript. They further observed three previously described missense variants (p.Pro34Ser, p.Pro80Leu, and p.Pro96Thr) that were also present in the matched control series [104].

TBK1

In 2015, a large exome sequencing case-control study identified mutations in *TBK1* in sporadic ALS cohort of patients [105]. Subsequent studies showed TBK1 loss-of-function mutations in families with FTLD-ALS but also in clinical FTLD and pathologically confirmed FTLD-TDP even in the absence of motor neuron disease [106]. Most mutation identified are loss-of-function mutations leading in a loss of 50% of TBK1 levels or missense mutations impairing the binding of TBK1 to optineurin (OPTN) or its kinase activity. TBK1 mutations explain 2% of FTLD-TDP and 1% of FTLD-ALS cases. Just as VCP or p62, also TBK1 is involved in protein degradation and autophagy mechanisms. In fact, it phosphorylates p62 and OPTN, another member of autophagy pathway. OPTN gene was first reported as cause of autosomal recessive ALS in Japanese population (deletion in exon5) [106]. In 2015, Potteir et al. discovered, in a pathologically confirmed cohort of patients, one heterozygous mutation and one deletion in OPTN as well as a nonsense mutation in TBK1 suggesting that both genes contribute to FTLD-TDP etiology [106].

The age at onset and also the clinical presentation of TBK1 mutation carriers are very heterogeneous. The onset age ranges from 48 to 80 years, and clinical phenotype included bvFTD, AD, and FTLD-ALS [107].

TARDBP

TARDBP gene is constituted by six exons and located on chromosome 1p36.22. It encodes for TDP-43 protein, whose major form is translated from exons 2 to 6, resulting in a highly conserved 43 kDa protein. TDP-43 is localized in the nucleus of the cell where it is able to form heterogeneous nuclear ribonucleoprotein (hnRNP) complexes with several functions such as RNA regulation, mRNA stability and transport, and splicing control. A link between FTLD and ALS and TDP-43 was supported by the evidence that TDP-43 regulate axon growth in vivo and in vitro suggesting that the capacity of motor neuron to produce and maintain an axon is compromised by TDP-43 dysregulation [107]. Mutations in *TARDBP* are cause of 5% of familial ALS cases and 1% of sporadic patients. However, these mutations are rarely found in FTLD and FTD-MND although TDP-43 is the major component of neuronal ubiquitin-positive inclusion seen in FTLD-TDP patients. To date, 40 missense mutations in TARDBP gene are described, and most of them are localized in the C-terminal glycine-rich domain, encoded by exon 6 and involved in protein–protein interaction. It is interesting to note that more than 6000 RNAs are known to interact with TDP-43, but few of them have been studied. Moreover, it is not clear whether the main pathogenic mechanism is a loss-of-function or a gain-of-function one because the functional consequences of TARDBP mutations are still under investigation. Nevertheless, most evidence for a role of TARBDP mutations in FTLD and not only in ALS pathogenesis comes from Sardinian population. Here, the variant p.A328T is observed in about 21% of familial FTLD cases suggesting a common founder effect [13].

FUS

Similar to TDP-43, fused in sarcoma (*FUS*) is highly conserved ubiquitously expressed protein-coding gene located on chromosome 16. FUS is a component of the hnRNP complex and a member of FET protein family which works to facilitate RNA transport in and out of the nucleus, RNA splicing, and DNA/RNA metabolism [108]. FUS protein contains an amino-terminal glycine-rich domain, an RNA recognition motif and a zinc finger motif. The most important is the prion-like domain which plays a critical role in FUS misfolding. In 2009, FUS mutations were discovered to be the cause of 3% of familial ALS, and they are mostly located in the C-terminal of the protein particularly in nuclear localization sequence resulting in an impairment of transportin (TRN1)-mediated nuclear import of FUS [108]. From a neuropathological point of view, in ALS patients with *FUS* mutations, there are abnormal cytoplasmic neuronal and glial inclusions positive for FUS. However, in FTLD-FUS subset of FTLD patients, no *FUS* mutations have been identified. The pathological inclusions in FTLD-FUS, but not in FUS-ALS, patients are also positive for EWS and TAF15 protein suggesting a different pathogenic mechanism.

Other Rare Causal Genes

A number of other genes have been studied in relation to FTLD. One of these is ubiquilin 2 (*UBQLN2*) gene that is involved in a rare form of chromosome X-linked familial ALS and FTLD-ALS [109]. Mutations are located in proline residues in the highly conserved PXXP repeat domain involved in the degradation of misfolded proteins via ubiquitin proteasome system and autophagy. Moreover, UBQLN2-positive inclusions are observed in the hippocampus of ALS-FTLD patients even in the absence of *UBQLN2* mutations. Another one is tubulin alpha 4a (*TUBA4A*) gene on chromosome 2q35. *TUBA4A* encodes 1 of 8 human a-tubulins (448 amino acids), which polymerize with b-tubulins to form the microtubule cytoskeleton, implicating the neuroskeletal architecture. *TUBA4A* mutations have primarily been associated with ALS although some patients also had cognitive involvement ranging from mild cognitive impairment to FTD.

In TUBA4A, 10 missense, 1 nonsense, and 1 splice donor site mutation have been identified in both sporadic and familial ALS patients, with some also presenting with FTD [104]. Lastly, recessive mutations in the triggering receptor expressed on myeloid cells 2 gene (*TREM2*) have been described in patients with atypical FTLD, very young age at onset and with matter lesions on brain imaging [110].

FTLD Genetic Modifiers

In addition to genes mentioned above and generally involved in families showing autosomal dominant transmission, several genetic risk factors have been studied. The most important and replicated is the transmembrane protein 106b gene (*TMEM106B*). In 2010, Van Deerlin and coworkers published the first GWAS on 515 FTD patients with TDP-43 pathology; they identified a possible susceptibility locus, which encompasses *TMEM106*B gene on chromosome 7p21 [111]. In particular, the study identified three associated single nucleotide polymorphisms (SNPs), rs102004, rs6966915, and rs1990622, which are correlated with an increase of *TMEM106b* expression level [111]. Several subsequently studies showed that the highest association with *TMEM106b* locus was found in FTL-TDP patients with *GRN* mutations [112, 113]. These results increased our knowledge about the genetics of FTLD-TDP and represent a starting point from where researchers can look into a possible new pathogenic pathway. It is also true that these data are specific for a subgroup of FTLD patients, suggesting that the connection between *TMEM106B* and FTLD cannot be extended to the general FTLD population. In *GRN* FTD mutation carriers, the presence of protective C allele of SNP rs1990622 protects these patients from developing disease [113]. TMEM106b is a glycosylated type 2 membrane protein that localized to late endosomes and lysosomes where it seems to have an important function. Overexpression of TMEM106b in cell cultures showed an aberrant vacuole formation and an impairment of endolysosomal pathway [114]. These data suggest a key role for lysosomal biology in FTLD-TDP.

Common SNPs in the major causal genes have been studied to determine their association as FTLD risk factors. For example, rs5848, located in the 3'UTR of GRN gene in a putative miRNA binding site, has been investigated. Unfortunately, its role remains unclear with significant association in initial series of FTD-TDP patients but not in subsequent series of clinical patients [115]. Moreover, the H1/H2 haplotype in *MAPT* gene has been considered as FTLD risk factor. H1 haplotype is predominantly associated with FTLD-tau-related disease such as CBD; however, the genetic association of H1/H2 haplotype with clinical FTLD has been not confirmed [116]. More recently, a two-stage genome-wide association study (GWAS) identified the HLA locus at chromosome 6p21.3 and a locus at chromosome 11q14 encompassing *RAB38* and cathepsin C (*CTSC*). These two genes are especially associated with FTD, and it was observed an association between the top SNP *at RAB8/CTSC* locus and a 50% reduction of RAB8 levels in the blood of patients suggesting that a loss of RAB8 function may play a role in the development of FTLD. RAB8 is a protein involved in the regulation of lysosomal biology and protein trafficking. The HLA locus, instead, suggests a link between FTLD and immune system as well as other neurodegenerative diseases. Ferrari et al., in 2015, identified, in a sub-analysis of Italian FTLD cohort of patients, an association of two additional loci on chromosome 2p16.3 and on chromosome 17q25.3 within a region near *CEP131*, *ENTHD2*, and *C17ORF89* [117].

Conclusions and Future Prospective

The discoveries of the last few years showed that the term "FTLD" actually comprises diseases with a different etiology. It has become clearer and clearer that there are multiple genetic autosomal dominant mutations leading to the development of FTLD. The most frequent are so far *MAPT*, *GRN*, and *C9ORF72* mutations. The description of peculiar clinical phenotypes showed that there is an overlap among neurodegenerative disorders in terms of symptoms and pathogenic events leading to neurodegeneration. From a clinical point of view, the same genetic defect has been observed in patients with different diseases, i.e., bvFTD, MND, or both, raising the question whether there are additional unknown genetic or environmental factors influencing the phenotype. In addition, *GRN* and *C9ORF72* mutations are associated with a wide range of phenotypes and age at disease onset, including memory and psychosis, making difficult to predict the presence of a mutation basing on symptoms and/or familial history. Moreover, the situation is even more complex considering the incomplete penetrance of such mutations. Moreover, emerging data suggest that a significant number of FTLD patients carry potential pathogenic mutations in different genes. This is an oligogenic model disease in which multiple genes contribute to FTLD pathogenesis probably affecting disease penetrance and progression. Initially, several studies reported double mutations considering the second one only a polymorphism, but now the presence of *C9ORF72* expansion in combination with mutations in more than 15 genes significantly challenged this view [118].

Concerning pathogenic mechanisms related to FTLD, a growing number of genes are now discovered, and new disease pathways are emerged and are likely implicated with FTLD. The first pathway clusters around the degradation and clearance of misfolded proteins by proteasome degradation and autophagy and includes *CHMP2B*, *VCP*, *UBQLN2*, *SQSTM1*, *TBK1*, and *OPTN*.

A second pathway includes genes involved in lysosomal/endosomal biology such as *GRN*, *TMEM106B*, and *RAB8*. Moreover, there are mechanisms involved in DNA/RNA metabolism in which *C9ORF72*, *TARDBP*, and *FUS* are implicated.

New findings about genetics and molecular biology of FTLD recently described have some implications for FTLD diagnosis and treatment. First, biomarkers for identifying mutation carriers are needed. So far, given the heterogeneity of age at disease onset and presentation symptoms, it is not possible to predict the presence of a causal mutation basing on the clinical picture only. In this regard, low plasma progranulin levels are very good predictors of the presence of a *GRN* mutation leading to haploinsufficiency. Second, in view of the availability of future tailored therapies aimed to modify the course of the disease by acting on pathogenic mechanisms (i.e., replacing progranulin loss or hampering tau deposition), it would be extremely important to develop tools to predict the ongoing pathology (i.e., tau deposition or TDP-43 altered functioning). In this framework, the study of autosomal dominant families provides an opportunity to learn about the early stage of the disease following asymptomatic subjects in order to discover new disease biomarkers. Thus, in 2012 it was carried out a multicenter study, the genetic and frontotemporal dementia initiative (GENFI), in which British, European, and Canadian research sites recruit symptomatic and asymptomatic known carriers of pathogenic mutations in *MAPT*, *GRN*, or *C9ORF72*. In 2015, Roher et al. analyzed imaging data from 118 GENFI mutation carriers and 102 non-carriers and observed that structural imaging can be identified up to 10 years before the expected onset of symptoms in asymptomatic subjects [119]. Another similar initiative, promoted, for example, by the National Institute of Health, is Longitudinal Evaluation of Familial Frontotemporal Dementia that includes North American and Canadian centers. The aim is to identify robust reliable methods to track disease progression in order to design appropriate clinical trials.

In conclusion, the discovery of several gene mutations provided a high number of information leading to the identification of specific disease pathways that are currently being explored in the search for new therapeutic strategies. In the near future, the use of patient-derived iPS cell lines might contribute to understand the link between genetic and other factors in FTLD development. Nevertheless, next-generation sequencing and innovative biostatistical approach might lead to the identification of new and additional FTLD causal and/or disease modifier genes.

References

1. Coyle-Gilchrist IT, Dick KM, Patterson K, et al. Prevalence, characteristics and survival of frontotemporal lobar degeneration syndromes. Neurology. 2016;86:36–1743.
2. Ioannidis P, Konstantinopoulou E, Maiovis P, Karacostas D. The frontotemporal dementias in a tertiary referral center: classification and demographic characteristics in a series of 232 cases. J Neurol Sci. 2012;318:171–3.
3. Rascovsky K, Hodges JR, Knopman D, et al. Sensitivity of revised diagnostic criteria for the behavioural variant of frontotemporal dementia. Brain. 2011;134:2456–77.
4. Landqvist WM, Gustafson L, Passant U, Englund E. Psychotic symptoms in frontotemporal dementia: a diagnostic dilemma? Int Psychogeriatr. 2015;27:531–9.
5. Neary D, Snowden JS, Gustafson L, et al. Frontotemporal lobar degeneration. A consensus on clinical diagnostic criteria. Neurology. 1998;51:1546–54.
6. Gorno-Tempini ML, Hillis AE, Weintraub S, et al. Classification of primary progressive aphasia and its variants. Neurology. 2011;76:1006–14.
7. Harris JM, Gall C, Thompson JC, et al. Classification and pathology of primary progressive aphasia. Neurology. 2013;81:1832–9.
8. Hodges JR, Patterson K. Semantic dementia: a unique clinicopathological syndrome. Lancet Neurol. 2007;6:1004–14.
9. Snowden JS, Thompson JC, Neary D. Knowledge of famous faces and names in semantic dementia. Brain. 2004;127:860–72.
10. Govaarts R, Beeldman E, Kamelmacher MJ, et al. The frontotemporal syndrome of ALS is associated with poor survival. J Neurol. 2016;263:2476–83.
11. Saxon JA, Harris JM, Thompson JC, et al. Semantic dementia, progressive nonfluent aphasia and their association with amyotrophic lateral sclerosis. J Neurol Neurosurg Psychiatry. 2017;88(8):711–2.
12. Wang Y, Mandelkow E. Tau in physiology and pathology. Nat Rev Neurosci. 2015;17:22–35.
13. Pottier C, Ravenscroft TA, Sanchez-Contreras M, Rademakers R. Genetics of FTLD: overview and what else we can expect from genetic studies. J Neurochem. 2016;138(Suppl 1):32–53.
14. Mann DMA, Snowden JS. Frontotemporal lobar degeneration: pathogenesis, pathology and pathways to phenotype. Brain Pathol. 2017;27:723. https://doi.org/10.1111/bpa.12486.

15. Mackenzie IRA, Neumann M, Baborie A, et al. A harmonized classification system for FTLDTDP pathology. Acta Neuropathol. 2011;122:111–3.
16. Ratnavalli E, Brayne C, Dawson K, Hodges JR. The prevalence of frontotemporal dementia. Neurology. 2002;58:1615–21.
17. Bird T, Knopman D, VanSwieten J, et al. Epidemiology and genetics of frontotemporal dementia/Pick's disease. Ann Neurol. 2003;54:S29–31.
18. Goldman JS, Farmer JS, Wood EM, et al. Comparison of family histories in FTLD subtypes and related tauopathies. Neurology. 2005;65:1817–9.
19. Pickering-Brown SM. The complex aetiology of frontotemporal lobar degeneration. Exp Neurol. 2007;114:39–47.
20. Govaarts R, Beeldman E, Kamelmacher MJ, et al. The frontotemporal syndrome of ALS is associated with poor survival. J Neurol. 2016;3:2476–83.
21. Spillantini MG, Goedert M, Crowther RA, et al. Familial multiple system tauopathy with presenile dementia: a disease with abundant neuronal and glial tau filaments. Proc Natl Acad Sci U S A. 1997;94:4113–8.
22. Hutton M, Lendon CL, Rizzu P, et al. Association of missense and 5′-splice-site mutations in tau with the inherited dementia FTDP-17. Nature. 1998;393:702–5.
23. Neve RL, Harris P, Kosik KS, Kurnit DM, et al. Identification of cDNA clones for the human microtubule associated protein tau and chromosomal localization of the genes for tau and microtubule-associated protein 2. Brain Res. 1986;387:271–80.
24. Ghetti B, Oblak AL, Boeve BF, et al. Invited review: frontotempora. dementia caused by microtubule-associated protein tau gene (MAPT) mutations: a chameleon for neuropathology and neuroimaging. Neuropathol Appl Neurobiol. 2015;41:24–46.
25. Bang J, Spina S, Miller BL. Frontotemporal dementia. Lancet. 2015;386:1672–82.
26. Rademakers R, Cruts M, van Broeckhoven C. The role of tau (MAPT) in frontotemporal dementia and related tauopathies. Hum Mutat. 2004;24:277–95.
27. Malkani R, D'Souza I, Gwinn-Hardy K, et al. A MAPT mutation in a regulatory element upstream of exon 10 causes frontotemporal dementia. Neurobiol Dis. 2006;22:401–3.
28. Hong M, Zhukareva V, Vogelsberg-Ragaglia V, et al. Mutation-specific functional impairments in distinct tau isoforms of hereditary FTDP-17. Science. 1998;282:1914–7.
29. Goedert M, Jakes R, Crowther RA. Effects of frontotemporal dementia FTDP-17 mutations on heparin-induced assembly of tau filaments. FEBS Lett. 1999;450:306–11.
30. Grisart B, Willatt L, Destrée A, et al. 17q21.31 microduplication patients are characterised by behavioural problems and poor social interaction. J Med Genet. 2009;46:524–30.
31. Lladó A, Rodríguez-Santiago B, Antonell A, et al. MAPT gene duplications are not a cause of frontotemporal lobar degeneration. Neurosci Lett. 2007;424:61–5.
32. Rovelet-Lecrux A, Lecourtois M, Thomas-Anterion C, et al. Partial deletion of the MAPT gene: a novel mechanism of FTDP-17. Hum Mutat. 2009;30:591–602.
33. Rovelet-Lecrux A, Hannequin D, Guillin O, et al. Frontotemporal dementia phenotype associated with MAPT gene duplication. J Alzheimers Dis. 2010;21:897–902.
34. Rossi G, Conconi D, Panzeri E, et al. Mutations in MAPT gene cause chromosome instability and introduce copy number variations widely in the genome. J Alzheimers Dis. 2013;33:969–82.
35. Wszolek ZK, Uitti RJ, Hutton MA. Mutation in the microtubule-associated protein tau in pallido-nigro-luysian degeneration. Neurology. 2000;54:2028–30.
36. Rossi G, Tagliavini F. Frontotemporal lobar degeneration: old knowledge and new insight into the pathogenetic mechanisms of tau mutations. Front Aging Neurosci. 2015;7:192.
37. Baker M, Mackenzie IR, Pickering-Brown SM, et al. Mutations in progranulin cause tau-negative frontotemporal dementia linked to chromosome 17. Nature. 2006;442:916–9.
38. Cruts M, Gijselinck I, van der Zee J, et al. Null mutations in progranulin cause ubiquitin-positive frontotemporal dementia linked to chromosome 17q21. Nature. 2006;442:920–4.
39. Rademakers R, Neumann M, Mackenzie IR. Advances in understanding the molecular basis of frontotemporal dementia. Nat Rev Neurol. 2012;8:423–34.

40. Petkau TL, Leavitt BR. Progranulin in neurodegenerative disease. Trends Neurosci. 2014;37:388–98.
41. Gass J, Cannon A, Mackenzie IR, et al. The spectrum of mutations in progranulin: a collaborative study screening 545 cases of neurodegeneration. Arch Neurol. 2010;67:161–70.
42. Shankaran SS, Capell A, Hruscha AT, et al. Missense mutations in the progranulin gene linked to frontotemporal lobar degeneration with ubiquitin-immunoreactive inclusions reduce progranulin production and secretion. J Biol Chem. 2008;283:1744–53.
43. Wang J, Van Damme P, Cruchaga C, et al. Pathogenic cysteine mutations affect progranulin function and production of mature granulins. J Neurochem. 2010;112:1305–15.
44. Neumann M, Sampathu DM, Kwong LK, et al. Ubiquitinated TDP-43 in frontotemporal lobar degeneration and amyotrophic lateral sclerosis. Science. 2006;314:130–3.
45. Zhou X, Sun L, Bastos de Oliveira F, et al. Prosaposin facilitates sortilinin-dependent lysosomal trafficking of progranulin. J Cell Biol. 2015;210:991–1002.
46. Minami SS, Min SW, Krabbe G, et al. Progranulin protects against amyloid beta deposition and toxicity in Alzheimer's disease mouse models. Nat Med. 2014;20:1157–64.
47. Benussi A, Padovani A, Borroni B. Phenotypic heterogeneity of monogenic frontotemporal dementia. Front Aging Neurosci. 2015;7:171.
48. Le Ber I, Camuzat A, Hannequin D, et al. Phenotype variability in progranulin mutation carriers: a clinical, neuropsychological, imaging and genetic study. Brain. 2008;131:732–46.
49. Rainero I, Rubino E, Negro E, et al. Heterosexual pedophilia in a frontotemporal dementia patient with a mutation in the progranulin gene. Biol Psychiatry. 2011;70:43–4.
50. Cerami C, Marcone A, Galimberti D, et al. From genotype to phenotype: two cases of genetic frontotemporal lobar degeneration with premorbid bipolar disorder. J Alzheimers Dis. 2011;27(4):791–7.
51. Pietroboni AM, Fumagalli GG, Ghezzi L, et al. Phenotypic heterogeneity of the GRN Asp22fs mutation in a large Italian kindred. J Alzheimers Dis. 2011;24:253–9.
52. Bruni AC, Momeni P, Bernardi L, et al. Heterogeneity within a large kindred with frontotemporal dementia: a novel progranulin mutation. Neurology. 2007;69:140–7.
53. Carecchio M, Fenoglio C, De Riz M, et al. Progranulin plasma levels as potential biomarker for the identification of GRN deletion carriers. A case with atypical onset as clinical amnestic mild cognitive impairment converted to Alzheimer's disease. J Neurol Sci. 2009;287:291–3.
54. Pickford F, Marcus J, Camargo LM, et al. Progranulin is a chemoattractant for microglia and stimulates their endocytic activity. Am J Pathol. 2011;178(1):284–95.
55. Tang W, Lu Y, Tian QY, et al. The growth factor progranulin binds to TNF receptors and is therapeutic against inflammatory arthritis in mice. Science. 2011;332(6028):478–84.
56. Yin F, Banerjee R, Thomas B, et al. Exaggerated inflammation, impaired host defense, and neuropathology in progranulin-deficient mice. J Exp Med. 2010;207(1):117–28.
57. Filiano AJ, Martens LH, Young AH, et al. Dissociation of frontotemporal dementia-related deficits and neuroinflammation in progranulin haploinsufficient mice. J Neurosci. 2013;33(12):5352–61.
58. Hosler BA, Siddique T, Sapp PC, et al. Linkage of familial amyotrophic lateral sclerosis with frontotemporal dementia to chromosome 9q21–q22. JAMA. 2000;284:1664–9.
59. Morita M, Al-Chalabi A, Andersen PM, et al. A locus on chromosome 9p confers susceptibility to ALS and frontotemporal dementia. Neurology. 2006;66:839–44.
60. DeJesus-Hernandez M, Mackenzie IR, Boeve BF, et al. Expanded GGGGCC hexanucleotide repeat in noncoding region of C9ORF72 causes chromosome 9p-linked FTD and ALS. Neuron. 2011;72:245–56.
61. Renton AE, Majounie E, Waite A, et al. A hexanucleotide repeat expansion in C9ORF72 is the cause of chromosome 9p21-linked ALS-FTD. Neuron. 2011;72:257–68.
62. van Blitterswijk M, DeJesus-Hernandez M, Niemantsverdriet E, et al. Association between repeat sizes and clinical and pathological characteristics in carriers of C9ORF72 repeat expansions (Xpansize-72): a cross-sectional cohort study. Lancet Neurol. 2013;12:978–88.

63. Ishiura H, Tsuji S. Epidemiology and molecular mechanism of frontotemporal lobar degeneration/amyotrophic lateral sclerosis with repeat expansion mutation in C9orf72. J Neurogenet. 2015;29:85–94.
64. Rohrer JD, Isaacs AM, Mizielinska S, et al. C9orf72 expansions in frontotemporal dementia and amyotrophic lateral sclerosis. Lancet Neurol. 2015;14:291–301.
65. Cruts M, Engelborghs S, van der Zee J, Van Broeckhoven C. C9orf72-related amyotrophic lateral sclerosis and frontotemporal dementia. In: Pagon RA, Adam MP, Ardinger HH, et al., editors. GeneReviews (R). Seattle: University of Washington; 2015.
66. Cooper-Knock J, Kirby J, Highley R, Shaw PJ. The spectrum of C9orf72-mediated neurodegeneration and amyotrophic lateral sclerosis. Neurotherapeutics. 2015;12:326–39.
67. Snowden JS, Rollinson S, Thompson JC, et al. Distinct clinical and pathological characteristics of frontotemporal dementia associated with C9ORF72 mutations. Brain. 2012;135:693–708.
68. Galimberti D, Fenoglio C, Serpente M, et al. Autosomal dominant frontotemporal lobar degeneration due to the C9ORF72 hexanucleotide repeat expansion: late-onset psychotic clinical presentation. Biol Psychiatry. 2013;74(5):384–91.
69. Galimberti D, Reif A, Dell'Osso B, et al. C9ORF72 hexanucleotide repeat expansion is a rare cause of schizophrenia. Neurobiol Aging. 2014;35:1214 e7–10.
70. Majounie E, Abramzon Y, Renton AE, et al. Repeat expansion in C9ORF72 in Alzheimer's disease. N Engl J Med. 2012;366:283–4.
71. Hensman Moss DJ, Poulter M, Beck J, et al. C9orf72 expansions are the most common genetic cause of Huntington disease phenocopies. Neurology. 2014;82:292–9.
72. Mori K, Arzberger T, Grasser FA, et al. Bidirectional transcripts of the expanded C9orf72 hexanucleotide repeat are translated into aggregating dipeptide repeat proteins. Acta Neuropathol. 2013;126:881–93.
73. Renoux AJ, Todd PK. Neurodegeneration the RNA way. Prog Neurobiol. 2012;97:173–89.
74. Reddy K, Zamiri B, Stanley SY, et al. The disease-associated r(GGGGCC)n repeat from the C9ORF72 gene forms tract length-dependent uni- and multi-molecular RNA G-quadruplex structures. J Biol Chem. 2013;288(14):9860–6.
75. Mizielinska S, Isaacs AM. C9orf72 amyotrophic lateral sclerosis and frontotemporal dementia: gain or loss of function? Curr Opin Neurol. 2014;27:515–23.
76. May S, Hornburg D, Schludi MH, et al. C9orf72 FTLD/ALS associated Gly-Ala dipeptide repeat proteins cause neuronal toxicity and Unc119 sequestration. Acta Neuropathol. 2014;128:485–503.
77. van Blitterswijk M, Rademakers R. Neurodegenerative disease: C9orf72 repeats compromise nucleocytoplasmic transport. Nat Rev Neurol. 2015;11:670–2.
78. Lehmer C, Oeckl P, Weishaupt JH, et al. Poly-GP in cerebrospinal fluid links C9orf72-associated dipeptide repeat expression to the asymptomatic phase of ALS/FTD. EMBO Mol Med. 2017;9(7):859–68.
79. Su Z, Zhang Y, Gendron TF, et al. Discovery of a biomarker and lead small molecules to target r(GGGGCC)-associated defects in c9FTD/ALS. Neuron. 2014;83:1043–50.
80. Cristofani R, Crippa V, Vezzoli G, et al. The small heat shock protein B8 (HSPB8) efficiently removes aggregating species of dipeptides produced in C9ORF72-related neurodegenerative diseases. Cell Stress Chaperones. 2018;23(1):1–12. https://doi.org/10.1007/s12192-017-0806-9.
81. Skibinski G, Parkinson NJ, Brown JM, et al. Mutations in the endosomal ESCRTIII-complex subunit CHMP2B in frontotemporal dementia. Nat Genet. 2005;37:806–8.
82. Urwin H, Ghazi-Noori S, Collinge J, Isaacs A. The role of CHMP2B in frontotemporal dementia. Biochem Soc Trans. 2009;37:208–12.
83. Lindquist SG, Braedgaard H, Svenstrup K, et al. Frontotemporal dementia linked to chromosome 3 (FTD-3)-current concepts and the detection of a previously unknown branch of the Danish FTD-3 family. Eur J Neurol. 2008;15:667–70.
84. Urwin H, Authier A, Nielsen JE, et al. Disruption of endocytic trafficking in frontotemporal dementia with CHMP2B mutations. Hum Mol Genet. 2010;19:2228–38.

85. Isaacs AM, Johannsen P, Holm I, et al. Frontotemporal dementia caused by CHMP2B mutations. Curr Alzheimer Res. 2011;8:246–51.
86. Holm IE, Englund E, Mackenzie IR, et al. A reassessment of the neuropathology of frontotemporal dementia linked to chromosome 3. J Neuropathol Exp Neurol. 2007;66:884–91.
87. Ghazi-Noori S, Froud KE, Mizielinska S, et al. Progressive neuronal inclusion formation and axonal degeneration in CHMP2B mutant transgenic mice. Brain. 2012;135:819–32.
88. Nielsen TT, Mizielinska S, Hasholt L, et al. Reversal of pathology in CHMP2B-mediated frontotemporal dementia patient cells using RNA interference. J Gene Med. 2012;14:521–9.
89. Parkinson N, Ince PG, Smith MO, et al. ALS phenotypes with mutations in CHMP2B (charged multivesicular body protein 2B). Neurology. 2006;67:1074–7.
90. Watts GDJ, Wymer J, Kovach MJ, et al. Inclusion body myopathy associated with Paget disease of bone and frontotemporal dementia is caused by mutant valosin containing protein. Nat Genet. 2004;36:377–81.
91. Spina S, Van Laar AD, Murrell JR, et al. Phenotypic variability in three families with valosin-containing protein mutation. Eur J Neurol. 2013;20:251–8.
92. Taylor JP. Multisystem proteinopathy: intersecting genetics in muscle, bone, and brain degeneration. Neurology. 2015;85:658–60.
93. Weihl CC, Pestronk A, Kimonis VE. Valosin-containing protein disease: inclusion body myopathy with Paget's disease of the bone and fronto-temporal dementia. Neuromuscul Disord. 2009;19:308–15.
94. J-S J, Weihl CC. Inclusion body myopathy, Paget's disease of the bone and fronto-temporal dementia: a disorder of autophagy. Hum Mol Genet. 2010;19:R38–45.
95. Meyer H, Weihl CC. The VCP/p97 system at a glance: connecting cellular function to disease pathogenesis. J Cell Sci. 2014;127:3877–83.
96. Kimonis VE, Fulchiero E, Vesa J, Watts G. VCP disease associated with myopathy, Paget disease of bone and frontotemporal dementia: review of a unique disorder. Biochim Biophys Acta. 2008;1782:744–8.
97. Komatsu J, Iwasa K, Yanase D, Yamada M. Inclusion body myopathy with Paget disease of the bone and frontotemporal dementia associated with a novel G156S mutation in the VCP gene. Muscle Nerve. 2013;48:995. https://doi.org/10.1002/mus.23960.
98. Ng ASL, Rademakers R, Miller BL. Frontotemporal dementia: a bridge between dementia and neuromuscular disease. Ann N Y Acad Sci. 2015;1338:71–93.
99. van der Zee J, Van Langenhove T, Kovacs GG, et al. Rare mutations in SQSTM1 modify susceptibility to frontotemporal lobar degeneration. Acta Neuropathol. 2014;128:397–410.
100. Bannwarth S, Ait-El-Mkadem S, Chaussenot A, et al. A mitochondrial origin for frontotemporal dementia and amyotrophic lateral sclerosis through CHCHD10 involvement. Brain. 2014;137:2329–45.
101. Zhang M, Xi Z, Zinman L, et al. Mutation analysis of CHCHD10 in different neurodegenerative diseases. Brain. 2015;138:e380.
102. Jiao B, Xiao T, Hou L, et al. High prevalence of CHCHD10 mutation in patients with frontotemporal dementia from China. Brain. 2015;139:1–4.
103. Genin EC, Plutino M, Bannwarth S, et al. CHCHD10 mutations promote loss of mitochondrial cristae junctions with impaired mitochondrial genome maintenance and inhibition of apoptosis. EMBO Mol Med. 2015;8:58–72.
104. Perrone F, Nguyen HP, Van Mossevelde S, et al. Investigating the role of ALS genes CHCHD10 and TUBA4A in Belgian FTD-ALS spectrum patients. Neurobiol Aging. 2017;51:177.e9–177.
105. Cirulli ET, Lasseigne BN, Petrovski S, et al. Exome sequencing in amyotrophic lateral sclerosis identifies risk genes and pathways. Science. 2015;347:1436–41.
106. Pottier C, Bieniek KF, Finch N, et al. Whole-genome sequencing reveals important role for TBK1 and OPTN mutations in frontotemporal lobar degeneration without motor neuron disease. Acta Neuropathol. 2015;130:77–92.
107. Rainero I, Rubino E, Michelerio A, et al. Recent advances in the molecular genetics of frontotemporal lobar degeneration. Funct Neurol. 2017;32(1):7–16.

108. Neumann M, Valori CF, Ansorge O, et al. Transportin 1 accumulates specifically with FET proteins but no other transportin cargos in FTLD-FUS and is absent in FUS inclusions in ALS with FUS mutations. Acta Neuropathol. 2012;124:705–16.
109. Dillen L, Van Langenhove T, Engelborghs S, et al. Explorative genetic study of UBQLN2 and PFN1 in an extended Flanders-Belgian cohort of frontotemporal lobar degeneration patients. Neurobiol Aging. 2013;34:1711.e1–5.
110. Guerreiro RJ, Lohmann E, Bras JM, et al. Using exome sequencing to reveal mutations in TREM2 presenting as a frontotemporal dementia-like syndrome without bone involvement. JAMA Neurol. 2013;70:78–84.
111. Van Deerlin VM, Sleiman PM, Martinez-Lage M, et al. Common variants at 7p21 are associated with frontotemporal lobar degeneration with TDP-43 inclusions. Nat Genet. 2010;42:234–9.
112. Finch N, Carrasquillo MM, Baker M, et al. TMEM106B regulates progranulin levels and the penetrance of FTLD in GRN mutation carriers. Neurology. 2011;76:467–74.
113. Cruchaga C, Graff C, Chiang HH, et al. Association of TMEM106B gene polymorphism with age at onset in granulin mutation carriers and plasma granulin protein levels. Arch Neurol. 2011;68:581–6.
114. Brady OA, Zheng Y, Murphy K, et al. The frontotemporal lobar degeneration risk factor, TMEM106B, regulates lysosomal morphology and function. Hum Mol Genet. 2013;22:685–95.
115. Rollinson S, Rohrer JD, van der Zee J, et al. No association of PGRN 30UTR rs5848 in frontotemporal lobar degeneration. Neurobiol Aging. 2011;32:754–5.
116. Vandrovcova J, Anaya F, Kay V, et al. Disentangling the role of the tau gene locus in sporadic tauopathies. Curr Alzheimer Res. 2010;7:726–34.
117. Ferrari R, Grassi M, Salvi E, et al. A genome-wide screening and SNPs-to-genes approach to identify novel genetic risk factors associated with frontotemporal dementia. Neurobiol Aging. 2015;36(2904):e2913–26.
118. Lattante S, Ciura S, Rouleau GA, Kabashi E. Defining the genetic connection linking amyotrophic lateral sclerosis (ALS) with frontotemporal dementia (FTD). Trends Genet. 2015;31:263–73.
119. Rohrer JD, Nicholas JM, Cash DM, et al. Presymptomatic cognitive and neuroanatomical changes in genetic frontotemporal dementia in the Genetic Frontotemporal dementia Initiative (GENFI) study: a cross-sectional analysis. Lancet Neurol. 2015;14:253–62.

Genetic Risk Factors for Sporadic Frontotemporal Dementia

Raffaele Ferrari, Claudia Manzoni, and Parastoo Momeni

Abstract

Frontotemporal dementia (FTD) is a complex multifactorial disorder characterized by heterogeneous clinical, pathological and genetic features.

FTD is subdivided in familial and sporadic on the basis of the form of inheritance: familial (or Mendelian) cases are those defined by a family history of FTD or closely related neurodegenerative disorders, whilst sporadic cases are those where a family history is not evident. Families are genetically studied to identify genes or genetic markers segregating with (and strongly contributing to) disease through strategies that developed from positional cloning, linkage studies to more recently family-focused whole exome sequencing (WES) approaches. The study of the idiopathic cases is less straightforward: here, besides screening the known candidate (Mendelian) genes (that generally are extremely rare in sporadic cases), the currently most cost-effective strategy is to perform genome-wide association studies (GWAS) to highlight risk-loci. These then need to be further genetically and functionally characterize through, for example, targeted re-sequencing and expression quantitative trait loci (eQTL), to name a few methods.

This chapter focuses on the current status of our genetic understanding of sporadic FTD thanks to the GWAS type of approach. This is followed by conclusive critical remarks on the ways ahead, driven by ever-advancing technologies and integrative strategies, for the dissection of complex disorders, including FTD.

R. Ferrari, Ph.D. (✉)
Department of Molecular Neuroscience, UCL Institute of Neurology, London, UK
e-mail: r.ferrari@ucl.ac.uk

C. Manzoni, Ph.D.
School of Pharmacy, University of Reading, Whiteknights, Reading, UK

P. Momeni, Ph.D.
Rona Holdings, Silicon Valley, CA, USA

© Springer International Publishing AG 2018
D. Galimberti, E. Scarpini (eds.), *Neurodegenerative Diseases*,
https://doi.org/10.1007/978-3-319-72938-1_9

Keywords

Frontotemporal dementia · FTD · FTLD-TDP · Complex disorders · Mendelian · Sporadic · Genome-wide association studies · MAPT · GRN · C9orf72 · TMEM106B · RAB38 · CTSC · CEP131 · ENTHD2 · Genome · Transcriptome · Proteome · Epigenome · Metabolome · Exposome · Personalized medicine

Introduction

The overall improved quality of life in our societies has led to unprecedented population growth over the past seven decades. Although this is in many ways an extraordinary achievement, it cannot be ignored that it has accounted and is accounting for an increased incidence of diseases of the old age, including disorders of the brain such as dementia.

In high-income countries (HIC), as well as low-/middle-income countries (LMIC), ageing represents one of the major risk factor contributing to dementia [1] suggesting that, globally, its incidence and prevalence can only increase in the next decades [2]. In 2010 there were approximately 4.7% of individuals over 60 years of age (~36 million people) affected by dementia and it has been estimated that incidence of dementia will reach ~115 million by 2050, worldwide [1].

Dementia is among the major causes for a wide range of deficits that impact memory, cognition and executive functions and that contribute to a progressive deterioration in the performance of normal daily activities leaving an affected individual in need of assistance of a caregiver [1]. The diagnostic and statistical manual of mental disorders (DSM) indicates that dementias fall within the extended category of neurocognitive disorders (NCDs) [3], where the main feature driving the diagnosis is 'impaired cognition' defined as an 'acquired' rather than 'inborn' condition [3]. Dementias are subdivided in two distinct groups: 'Alzheimer's' and 'non-Alzheimer's' types of dementia that are distinguished by their clinical presentation as well as by their pathological signatures [3]. The former group is exclusively identified by Alzheimer's disease (AD)—defined by its unique (early) clinical signatures that include short-term memory loss, visuo-spatial deficits, along with cognitive decline and language impairment [4]. The latter group encompasses multiple syndromes that, conversely, do not present 'short-term memory loss' and 'visuo-spatial deficits' (at least in the early stages of disease) and include vascular dementia (VD), dementia with Lewy bodies (DLB), Creutzfeldt–Jakob (prion) disease, frontotemporal dementia (FTD) as well as forms of NCDs characterized by a major motor component and occasionally by symptoms of dementia such as in the case of Parkinson's disease (PD), progressive supranuclear palsy (PSP), corticobasal syndrome/degeneration (CBS/D) and amyotrophic lateral sclerosis (ALS) [3].

NCDs are complex and multifactorial disorders meaning that they are the result of a combination of genetic and environmental factors as well as lifestyle [5]. Although this implies the interplay between these elements for a complete disease pathogenesis, it is recognized that there is a genetic component to almost every condition with variable levels of penetrance that span from strong monogenic (Mendelian forms of disease) to small/moderate oligo- or polygenic effects (sporadic forms of disease) [6–8]. Within this scenario, it is fundamental to further the study of the genetic underpinnings of complex disorders to set the basis for a better understanding of their underlying biological and molecular mechanisms as such knowledge will eventually be critical for developing preventive and therapeutic strategies that are currently lacking in NCDs [9–11].

The focus of this chapter is the NCD called frontotemporal dementia (FTD). A broad range of clinical manifestations, pathological signatures and genetic variability characterize FTD (see next sections). Genetics of FTD is currently able to explain a small proportion (10–30%) of cases, called familial or Mendelian, whilst the vast majority of cases, called sporadic (70–90%), are not yet genetically well characterized [12, 13]. Here we will highlight the current status of genetics of sporadic FTD with a particular focus on genome-wide association types of study (GWAS) and their contribution to a better dissection of sporadic FTD. We further explore the major future avenues for improving our understanding of the genetics of sporadic FTD and the molecular mechanisms involved in FTD pathogenesis.

Frontotemporal Dementia (FTD)

Synopsis

FTD is the second most common form of young-onset 'non-Alzheimer's' type of dementia and contributes to ~10–20% of all dementias [14]. It affects approximately 3–15 out of 100,000 individuals that are in their mid- to late 50s or early 60s [15]. FTD has insidious onset, it is categorized in familial (~10–30% of cases) or sporadic (~70–90% of cases) [12, 13], and its incidence is almost equal among men and women [16].

Clinically, FTD is characterized by (1) cognitive decline and behavioural dysfunction (behavioural variant [bvFTD]), which result in changes in personal and social conduct as well as deficits in executive functions such as planning, reasoning and problem-solving [17], and (2) language dysfunctions broadly called primary progressive aphasia (PPA). PPA is subdivided into semantic dementia (SD) (or semantic variant PPA), progressive nonfluent aphasia (PNFA) (or nonfluent/agrammatic variant PPA) and logopenic progressive aphasia (LPA) (or logopenic variant PPA) [17, 18]. SD affects conceptual knowledge through severe word comprehension impairment, whilst speech output remains fluent [18]. PNFA entails deficits in expressive language characterized by effortful nonfluent speech, phonological and grammatical errors and by difficulties in word retrieval [18]. LPA features impairments in word retrieval and repetition deficits [18] yet it has been suggested that LPA may be a subtle and atypical early presentation of AD [19]. However, within this extended and heterogeneous clinical picture, memory and visuo-spatial abnormalities remain initially intact in FTD. The Neary criteria [17] have been the most commonly used diagnostic criteria for FTD since 1998, yet two international consortia recently developed revised guidelines for diagnosing the behavioural and language variants increasing the sensitivity of each syndrome's clinical diagnosis. These guidelines can be reviewed in [18, 20].

From a pathological perspective, the brains of FTD patients show shrinkage in the frontal and temporal areas. Based on the topography of the lesions, frontotemporal dementia is also called FTLD (frontotemporal lobar degeneration). In bvFTD atrophy affects frontal lobes bilaterally, specifically the medial frontal lobes and the anterior temporal lobes [21]. SD shows asymmetric atrophy in the middle, inferior and medial anterior temporal lobe [18, 21], whilst PNFA presents mainly with left posterior frontal and insular regions atrophy [18]. FTLD's molecular pathology is characterized by (1) abnormal accumulation of protein aggregates [22] that cause inclusion bodies (i.e. neuronal cytoplasmic inclusions [NCI] and/or neuronal intranuclear inclusions [NII]) in neurons and/or glial cells and (2) dystrophic neurites (DN) [23]. FTLD pathology is classified on the basis of the type of proteins that constitute the abnormal inclusions in FTD brains: ≤ 40 to 50% of FTLD cases show tau pathology (FTLD-tau), ≤ 40 to 50% ubiquitin/TDP-43 pathology (FTLD-TDP), $\leq 10\%$ FUS pathology (FTLD-FUS) and ≤ 1 to 2% ubiquitin/p62 pathology (FTLD-UPS) [23]. Comprehensive features of FTLD's pathology can be reviewed in [24].

Genetics

In line with the clinical and pathological ones also the genetic features of FTD are heterogeneous (Fig. 9.1). Although there is no clear-cut subdivision between familial (or Mendelian) and non-familial patients (or sporadic), Mendelian and sporadic cases account for up to ~30% and ~70% of all FTD cases, respectively [12, 13] (Fig. 9.1a).

Mendelian FTD, in the vast majority of cases ($\geq 25/30\%$), has been explained by pathogenic mutations in the microtubule-associated protein tau (*MAPT*) and progranulin (*GRN*) genes [12] and by an abnormal repeat expansion in either the promoter region or the first intron of chromosome 9 open reading frame 72 (*C9orf72*) [25, 26]. A remainder of Mendelian cases ($\ll 5/30\%$) has been associated with rare variability in a number of genes that include the charged multivesicular body protein 2B (*CHMP2B*) [27, 28], the valosin-containing protein (*VCP*) [29], sequestosome 1 (*SQSTM1*) [30], ubiquilin 2 (*UBQLN2*) [31], intraflagellar transport 74 (*IFT74*) [32], optineurin (*OPTN*) [33], coiled-coil-helix-coiled-coil-helix domain containing 10 (*CHCHD10*) [34], TANK-binding kinase 1 (*TBK1*) [33, 35, 36] and, most recently, the T-cell-restricted intracellular antigen-1 (*TIA1*) [37]. It is important to note that all such genes have also been associated—with variable prevalence and penetrance—with a number of other syndromes. Although mutations in *MAPT*, *GRN* and *CHMP2B* have been mainly seen in FTD cases, genetic variability in these three genes has also been described (with fairly low prevalence, i.e. $\ll 1\%$) in AD, CBD and PSP pedigrees for *MAPT* [38–47], AD and CBS pedigrees for *GRN* [48–53], and ALS and CBS pedigrees for *CHMP2B* [54, 55]. Conversely, the expansion in *C9orf72* (that is clearly prevalent in FTD as indicated above) has also been

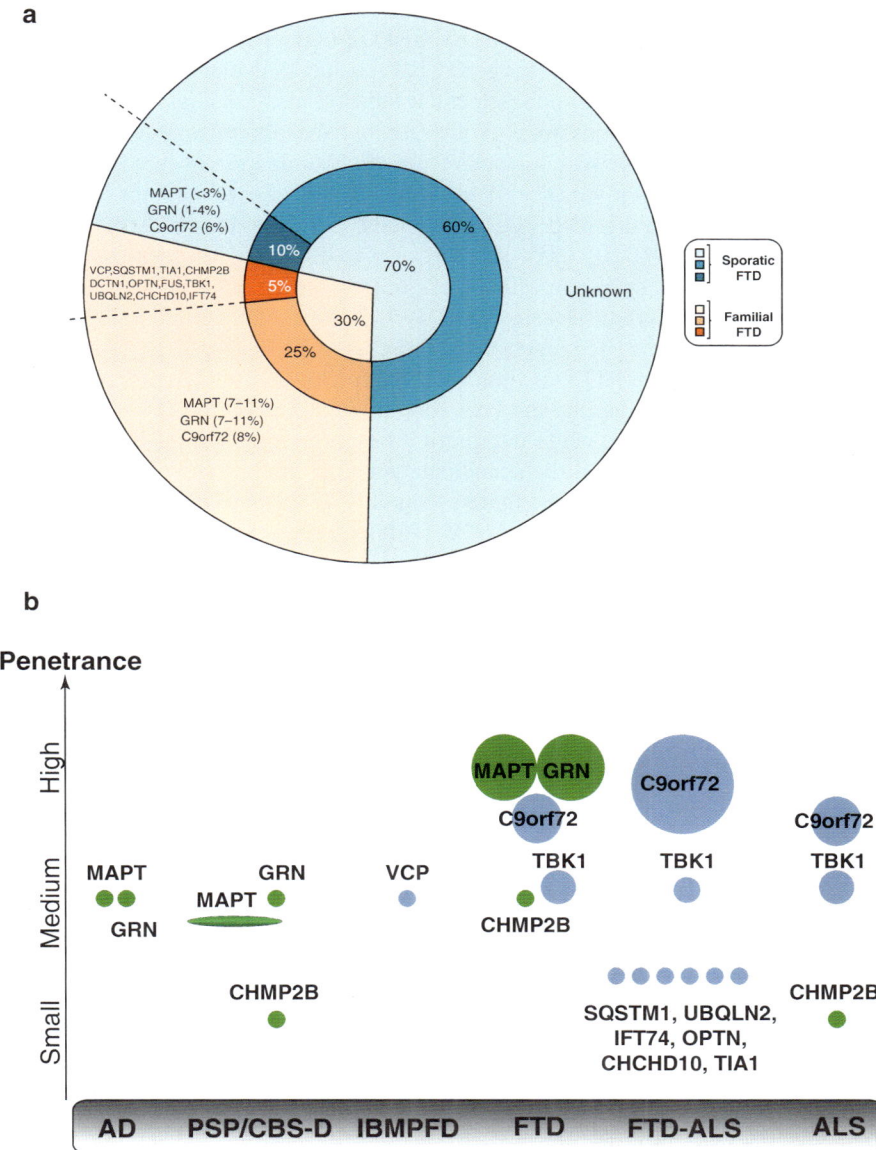

Fig. 9.1 Summary of FTD genetics. (**a**) Genetic prevalence of familial and sporadic FTD; the details of the currently known genetic aetiology of both familial and sporadic FTD is represented in a pie chart. (**b**) Distinction between 'major' (colour green) and 'spectrum' (colour light blue) FTD genes across FTD phenotypes; the size of shapes representing each gene is directly proportional to their level of prevalence (and penetrance)

reported, with variable prevalence (~1–20%) and penetrance, in a wide range of phenotypes that include ALS (and ALS-FTD), AD, Parkinsonian syndromes, Huntington's disease (HD) phenocopies, CBS and ataxia, to name a few (and, not to forget, in a number of normal/non-demented subjects across studies) [6, 26, 56–64]. Interestingly, *VCP* mutations not only are rare, but also appear to underpin a rather more complex phenotype which results from the combination of three conditions, namely, inclusion body myopathy (IBM) with Paget disease of the bone (PDB) and frontotemporal dementia (IBMPFD) [65]. Finally, variability in *SQSTM1*, *UBQLN2*, *IFT74*, *OPTN*, *CHCHD10*, *TBK1* and *TIA1* has mainly been associated with ALS and/or the FTD-ALS spectrum [30–34, 36, 66]. A particular mention is necessary for the TAR DNA-binding protein (*TARDBP*) and fused in sarcoma (*FUS*) genes: although variability has been reported in a handful of FTD cases [67, 68] and TDP-43 and FUS clearly are FTLD pathological hallmarks [23], they unlikely are FTD genes as pathogenic variability has been too rare or equivocal to fully support such claim [68–70]. Therefore, while interpreting Mendelian genetics of FTD, it might be more accurate to label all such candidate genes as 'major' and 'spectrum' FTD genes where the former are those that have mainly or exclusively been identified in FTD cases—*MAPT*, *GRN* and *CHMP2B* (although only two *CHMP2B* mutations appear convincing exclusively in one family, whereas any other has been equivocally associated with disease or replicated [28, 69])—and the latter are those that encompass a rather heterogeneous array of disorders—*C9orf72*, *VCP*, *SQSTM1*, *UBQLN2*, *IFT74*, *OPTN*, *CHCHD10*, *TBK1* and *TIA1* (Fig. 9.1b).

Sporadic FTD includes individuals with no familial history of FTD (or other neurodegenerative conditions) and/or with no clear genetic aetiology [71]. Sporadic cases are routinely screened for the known candidate genes and, to date, pathogenic variants are known to be confined to *MAPT* (0–3%), *GRN* (1–4%) or *C9orf72* (~6%) in about ≤10 of all idiopathic cases [66, 72] (Fig. 9.1a). The genetic architecture underlying idiopathic FTD appears to point to multiple risk markers with small effect size that act in concert and are modulated by variable modifying (including environmental) factors [73]. It is important to note that the best current approaches to identify novel genetic risk factors and candidate genes contributing to sporadic FTD are array-based as well as next-generation sequencing (NGS) techniques. In classical genome-wide association study (GWAS), the design involves comparing two cohorts (diseased vs. control), looking for statistically significant differences in allele frequencies (genome-wide association study [GWAS; see 'Genetic Studies of Complex Disorders' section]) or presence/absence of markers in either cohort (exome-chips array or NGS) [74]. Approaches that use genome-wide arrays (excluding exome-chips) explore differences in allele frequencies for common markers (MAF > 1–50%), whilst approaches that use exome-chip arrays and/or NGS target rare variants (MAF ≪ 1%). Provided availability of well-defined cohorts, large numbers are required in either type of study to identify significant genotype-phenotype associations [75]. It is important to note that, whilst looking at the same trait, a GWAS and a NGS feature same parameters differently: (1) GWAS screens 'common variants' (MAF > 1–50%), whilst NGS tests 'rare variants' (MAF ≪ 1%); (2) in GWAS the association analysis tests each marker individually, thus statistical correction is in the range of millions (Bonferroni correction) leading to a 'significance threshold' of 5×10^{-8}, whilst for NGS tests vary on the basis of the research question (i.e. the focus might be (1) all genes and coding variants or (2) subset of genes and a subset of coding variants or (3) only non-coding variants, thus here one might correct more or less times leading to variable 'thresholds'); and (4) in GWAS common variants are associated with a small-to-moderate effect size defined for the most by odds ratios (OR) of 1.5–2, whilst NGS variants are expected to be very rare and highly penetrant (OR $\gg$ 3). Therefore, the sample numbers of 1000 cases and 1000 controls lead to completely different results in GWAS and NGS; (1) in the case of a GWAS, a 'significance threshold' of 5×10^{-8}, a 'disease prevalence' of 10/100,000, assessing 'common markers' (MAF > 1–50%) and OR of 1.5 lead to a power to detect association of ~82.6% [76]; (2) in the case of NGS, specifically whole exome sequencing (WES), a 'significance threshold' of 5×10^{-5} (considering 20,000 genes), a 'disease prevalence' of 10/100,000, assessing 'rare variants' (MAF < 1%) and OR of 3 lead to a power to detect association of ~64.5% [76]. It follows that different combinations of these parameters influence the power of WES more robustly than in the case of GWAS; thus, the former is more sensitive to the sample size. Additionally, the financial cost of NGS is still about twofold greater than GWAS. Taken all this together, although in the near future, NGS will allow to screen populations' genomes at an incommensurable level of resolution, for many scientific questions, GWAS still remains the platform of choice for genetic investigation due to computational-, time- and cost-effectiveness.

A comprehensive assessment of FTD genetics can be further reviewed in [58, 66]. In the next sections, we review the major features of GWAS and results obtained to date by applying GWAS strategies to the study of sporadic FTD. Overall features of FTD genetics are summarized in Fig. 9.1.

Genetic Studies of Complex Disorders

Synopsis

In the era of modern genetics, there is an array of techniques that can be used to explore the genetics associated with a particular trait. Genetic and, latterly, genomic data have been generated with increasing speed and efficiency, allowing the transition from studies focused on individual genes to comparing genomes of whole populations [77]. Examining genetic background is, therefore, of great importance for identifying individual mutations and/or variants causing or increasing risk for developing diseases [78]. More specifically, these overall variants can be harmful, increasing susceptibility for a condition (i.e. a cluster of variants with low penetrance) or directly causing a disease (i.e. one or few variants with high penetrance) [73]. Clearly, different techniques are used to answer different questions or to support different types of study design. Particularly, it is worth noting that in complex diseases there is a trend to grossly subdivide the patient groups into familial or sporadic. The familial cases generally include a so-called index patient who is affected by a particular condition and a number of close relatives (including parents and siblings) some of whom also present similar or same symptomatology. These individuals are generally screened via next-generation sequencing (NGS) techniques, particularly whole exome sequencing (WES), to identify pathogenic mutations in candidate or novel genes that segregate with the disease within the family. These types of study are likely to identify a gene that causes or strongly associates with the trait or condition under study (Mendelian forms of inheritance).

Nevertheless, for the most part, familial cases represent a small minority of cases affected by a particular trait. Although familial studies are of great importance as they highlight the Mendelian candidate genes (that are also basis of functional/biological studies to understand disease mechanism), at the same time the vast majority of patients fall in the category of the sporadic cases, which are more difficult to characterize genetically. Sporadic cases are generally unique as they present typical clinical symptoms, falling within the clinical description of a particular trait, yet only rarely the known candidate genes appear to be the genetic cause or strongly contribute to their pathogenesis. Sporadic cases are generally investigated via genome-wide association types of study (GWAS). A typical GWAS design involves using microarray to genotype a cohort of interest and to identify variants associating with a particular trait in a hypothesis-free discovery study. A GWAS results in a list of SNPs evaluated for their frequency in relation to the trait under study. Most reported associations in GWAS are intronic or intergenic affecting DNA structure and gene expression rather than protein sequence [73]. Although GWAS identify risk loci, defined by SNPs that might be the actual reason of the signal or just in linkage disequilibrium (LD) with it, the associated variants might be informative implying to causal or susceptibility functional pathways [79]. The exponential growth in the number of GWAS in the past 10 years has led to the discovery of thousands of published associations for a range of traits (over 25,300 unique SNP-trait associations from over 2500 studies in GWAS Catalog [http://www.ebi.ac.uk/gwas]). Although most of the associating SNPs have a small effect size, they provide important clues to the pathobiology of the disease and even may suggest new therapeutic approaches (e.g. in sickle-cell disease, BCL11A was identified as a gene controlling foetal haemoglobin levels [80, 81]; in Crohn's disease, GWAS underlined the pathogenic role of specific processes such as autophagy and the innate immunity [82]). Another opportunity supported by GWAS is the possibility of comparing the genetic architecture between traits (LD score regression, [83]). Conversely, a common criticism is that significant SNPs still do not explain the entire genetic contribution to the trait (i.e. missing heritability [84]). However, models incorporating all SNPs regardless of their statistical significance in GWAS, substantially improve the genetic study of the trait [85] for which, ultimately, the remaining missing heritability is likely explained by rare variants (therefore not captured in GWAS). It follows that in the genetic study of complex disorders, the choice between a microarray or NGS approach (or both) should be based on the pertinent scientific or medical question(s) under consideration [79, 86].

Basics of Genetic Variability

In *Homo sapiens*, the haploid genome consists of three billion DNA base pairs (bps). These make up the coding regions (1–2% of the entire genome) and the remaining 98–99% (non-coding regions) which appear to hold structural and functional relevance [78, 87, 88]. Genetic variability (i.e. difference in genotype) in the genome is among the main driving factors at the basis of differences between individuals and populations, yet the majority of the variants are benign. However, some rare variants ($\ll 1\%$) can be harmful either in a Mendelian fashion (gene/mutation/disease), like in rare or familial forms of disease such as Huntington's disease [89] or cystic fibrosis [90]. As well variants/mutations might contribute to increased risk of developing common/complex disorders through an interplay between genetic and environmental factors with variable effect size as in the case of late onset Alzheimer's disease (LOAD) [91]. Each disorder has a genetic component (with low, intermediate or full penetrance). A better understanding of the variants in the genome and a better genotype-phenotype correlation are critical for identifying the genetic factors that influence health and disease. Currently, two major categories of variants are recognized: the simple nucleotide variations (SNVs) and the structural variations (SVs) (Fig. 9.2). SNVs, comprising single nucleotide polymorphisms (SNPs) and small insertion/deletions (small indels), affect single or few bases. Structural variants (SVs) comprise copy number polymorphisms (CNPs) that include large indels (100 bp–1 kb) and copy number variations (CNVs) (>1 kb) and affect larger genomic regions. Inversions and translocations belong to the category of SVs.

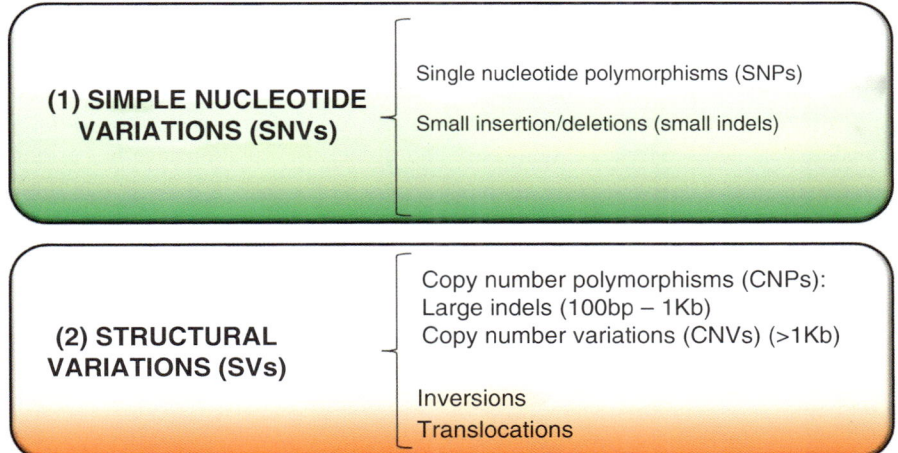

Fig. 9.2 Types of genetic variants. Summary of the main types of variants in the genome involving (*1*) single (or few) base pairs (bp) or (*2*) larger areas of the genome

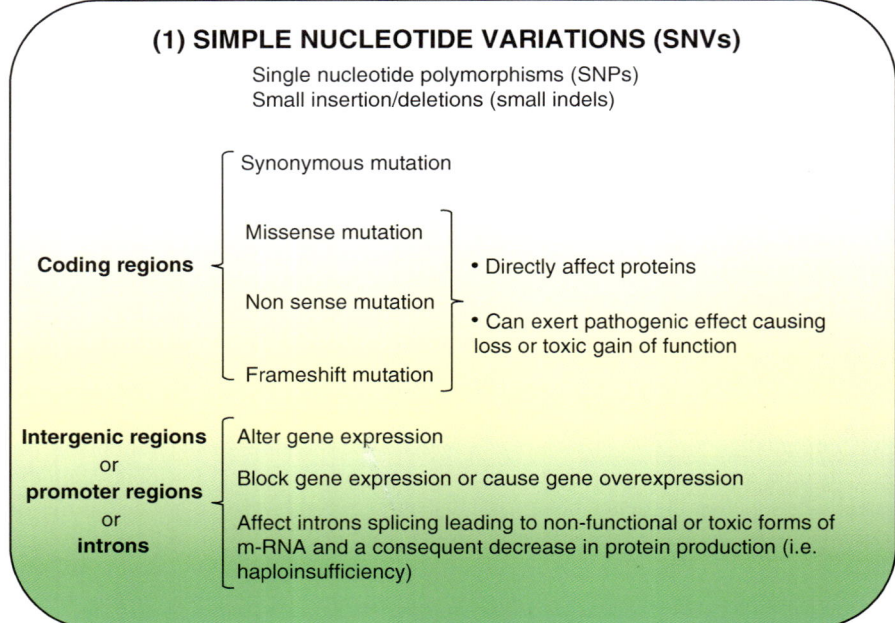

Fig. 9.3 Characteristics of simple nucleotide variations. Schematic view of the main single nucleotide variations (SNVs) and their related effect

In the majority of cases, SNVs (Fig. 9.3) can cause direct changes to proteins (when located in the coding regions), affect *cis* and/or *trans* gene expression, or splicing (when located in intergenic or promoter regions or in introns). Changes within the coding regions can result in synonymous, missense, non-sense and frameshift mutations. Non-sense mutations cause a premature truncation of the protein, and frameshift mutations cause a shift in the reading frame giving rise to novel translated elements. Missense, non-sense and frameshift mutations can be harmful and exert a pathogenic effect through mechanisms such as loss or toxic gain of function. Conversely, changes in promoter regions can affect gene expression by negatively modulating the activity of transcription factors, blocking gene expression or causing aberrant gene expression. Changes in introns can affect splicing leading to non-functional or toxic forms of m-RNA and a decrease in protein production (i.e. haploinsufficiency).

SVs (Fig. 9.4) affect larger parts of the genome. These types of variants can cause the loss of portions of DNA (deletions) that, in turn, may lead to haploinsufficiency or aberrant regulation of gene expression. On the other hand, duplications, which lead to multiple tandem copies of an allele, can cause aberrant phenotypes due to gene over-expression. Duplications can also happen at the level of chromosomes causing over-expression of the set of genes located on that chromosome.

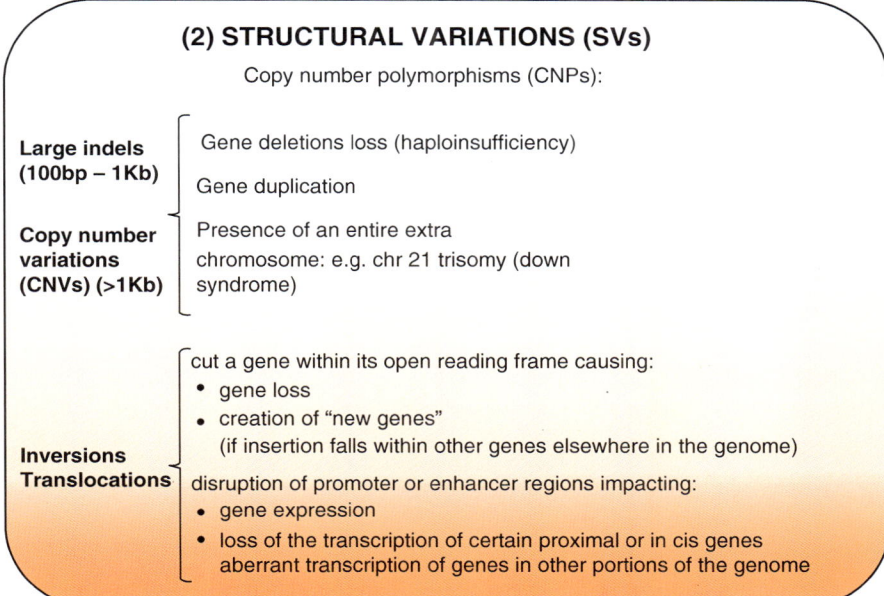

Fig. 9.4 Characteristics of structural variations. Schematic view of the main structural variations (SVs) and their related effect

Current techniques to capture genetic variants such as SNVs and SVs include (1) Sanger sequencing [92], (2) DNA microarrays [93] and (3) next-generation sequencing (NGS) [94]. Both microarrays and NGS approaches allow the identification of SNVs as well as some types of CNVs (Fig. 9.2); nevertheless, microarrays are more limited compared to NGS strategies as they are based on a priori knowledge of sequence and SNVs, whilst NGS allows detection of novel changes. Particularly, NGS allows the sequencing of specifically targeted regions, whole exome (WES) and whole genomes (WGS) of individuals. WES allows the screening of all variants (including rare) in the coding region, including mutations with a direct effect on the protein; WGS allows the identification of all common and rare coding and non-coding variants [94, 95].

The Study of Genetic Variability

The human genome was sequenced and released in the early 2000s by the public Human Genome Project (HGP) [78]. The reference genome is paired with a genome-wide map of common variability, thanks to the International HapMap Project [96]. This project identified common variants (minor allele frequency [MAF] ≥ 5%) across the genome of different populations (African, Asian and European ancestry) leading to the awareness that up to 99.5% of the genome across any two individuals is identical and, in addition, to the mapping of up to 10M SNPs. Importantly, the HapMap project allowed to complement the HGP with additional information such as that of haplotype blocks, based on the concept of linkage disequilibrium (LD) the grounding foundation of GWAS [77]. To increase the resolution achieved by HapMap, the 1000 Genomes Project was concluded in 2015 with 2504 genomes sequenced from 26 populations [97] to produce an extensive public catalog of human genetic variation, including rarer SNPs (MAF $\geq$ 1%) and SVs. This data (reference genome + HapMap + 1000 Genomes projects) is publicly available, greatly fostering high-resolution and population-specific GWAS and filtering of benign common and rare variants for NGS data analysis.

The HGP, HapMap and 1000 Genome projects have laid the cornerstone of today's deep analysis of the human genome and continuing development of platforms and bioinformatics tools available for the study of genetics of disease. By means of evenly distributed known SNPs and based on the ever-developing knowledge on LD blocks, GWAS are able to identify loci associated with disease. The primary outcome of GWAS is the identification of a locus, a genetic region that might be associated with a trait/disease; the association, normally, is further investigated to discover the possible underlying causal variants through fine mapping, dense genotyping and DNA sequencing. The expected outcome of GWAS is not exclusively the identification of one or several coding changes affecting the functions of a protein but the identification of variants affecting transcription and translation or variants that are in LD with the causal variants [98]. GWAS have now reached the level of almost standard technique and have been used to investigate the genetic bases of a large variety of different disorders. For a complete list of GWAS accomplished to date, see http://www.ebi.ac.uk/gwas, whilst for a complete list of GWAS on neurological disorders, see http://www.alzgene.org.

Genome-Wide Association Studies (GWAS)

GWAS: Study Design

GWAS follow the broad hypothesis of 'common disease—common variant(s)' and represent a large-scale example of classical cases vs. control studies to assess differences in the allelic frequencies of genotyped (and imputed) genetic markers between the two study groups. Specifically, differences in the frequencies of the alleles are statistically evaluated for each SNP in order to detect discriminants that may associate with/contribute to disease. Conceptually, GWAS interrogate the genome in an unbiased manner by means of hundreds of thousands of evenly distributed SNPs and allow for the identification of loci that increase susceptibility for disease, i.e. genetic markers within genetic regions with small to moderate effect size. A GWAS consists of two phases: a discovery and a replication phase. The discovery phase (or phase I) is hypothesis free and allows identification of one or more genetic risk-loci. The statistically significant loci and those that are suggestive of association in phase I are selected for replication (phase II) that is to be performed in a novel independent cohort of cases and controls for validation. When and if results of phase I are replicated in phase II, it is sensible to infer that most probably the locus/loci that show association contain or are in LD with the SNP(s) that is/are responsible for the association.

Finally, after completion of phase II, the associated loci are further investigated through fine mapping, i.e. genotyping a smaller number of SNPs (~10K SNPs) within a smaller region (1–5 Mbp) comprising the associated SNP, to identify other associated SNPs and/or, possibly, disease-associated haplotypes, or through direct sequencing of all the neighbouring genes implicated by the associated SNP [98].

GWAS: Good Practice for Success

GWAS is a long, complex and error-prone experimental procedure with confounding elements to contaminate the final outcome of the study. The most common errors include phenotyping, sample quality, genotyping errors/artefacts and population stratification (heterogeneous genetic background within the study cohorts), to name the most relevant. However, there are a number of good practices to implement in order to minimize errors. As such, the requirements and the workflow for a successful GWAS can be summarized as follows:

1. Choice of appropriate genotyping array in order to evenly cover the genome and best target the genetic background of the study population.
2. Detailed characterization of the study cohorts through stringent clinical and/ or pathological inclusion/exclusion criteria and a well-defined disease phenotype.
3. Accurate match of cases and controls and large enough sample and control size in order to increase the power of the study.
4. Use of stringent quality control (QC) steps prior to and after genotyping. Prior to genotyping excluding poor quality samples is fundamental to avoid genotyping errors/artefacts. After genotyping, there are several quality control measures to be implemented in order to target and filter both SNPs and samples included in the association analysis.
5. SNPs: SNPs with call rates ≥ 0.95 should be included. This measure can be more stringent (≥ 0.97–0.99) based, mainly, on study design. To eliminate possible confounding factors, all SNPs with no call, or the outliers, or those that deviate from the Hardy-Weinberg equilibrium law or those that have a MAF < 0.01 need to be excluded. These inclusion/exclusion criteria need to be applied for both cases and control sample sets.
6. Samples: cases and controls will need to be matched based on ancestry in order to exclude possible false positives simply due to differences in the genetic background of the two cohorts (population stratification). Samples with missing data for >5% of SNPs, samples that might be related and samples with discordant gender (gender mismatch) need to be excluded because of their high probability of contaminating the results of the association analysis.

After the preliminary QC steps, the clean dataset (which means all informative cases, controls and SNPs have been identified and filtered) is used for the association analysis. The latter is performed by means of online free open-source whole genome association analysis toolsets such as Plink (http://pngu.mgh.harvard.edu/~purcell/plink/) and R (http://www.r-project.org/). In studies evaluating for dichotomous traits (i.e. cases vs. control), the association needs to be tested for its significance and the effect size. Significance can be assessed through a number of methods including the Chi-squared test with either one degree of freedom (df) (allelic) or two df (genotypic), the Fisher's exact test or logistic regression. The significance is expressed in *p*-values for which, currently, an association is considered genome-wide significant when the *p*-value is $<5 \times 10^{-08}$. Once the association analysis is performed, there is an additional step to evaluate the impact on results of possible confounding factors. This is the assessment of the distribution of the data through the quantile-quantile plots (QQ plots) that allow appreciating inflation/deflation from the expected distribution (Fig. 9.5). The genomic inflation factor (λ) defines the deviation from the expected distribution under the null model (null hypothesis). Inflation ($\lambda > 1$) is generally a signal of possible population stratification, or an issue of relatedness (sample duplicates), or a technical bias or due to DNA poor quality, whilst deflation ($\lambda < 1$) is generally a sign of possible phenotype discordance. However, a value of $\lambda \sim 1.05$ is considered acceptable in GWAS. Effect size is measured in the vast majority of GWAS through odds ratio (OR). An OR

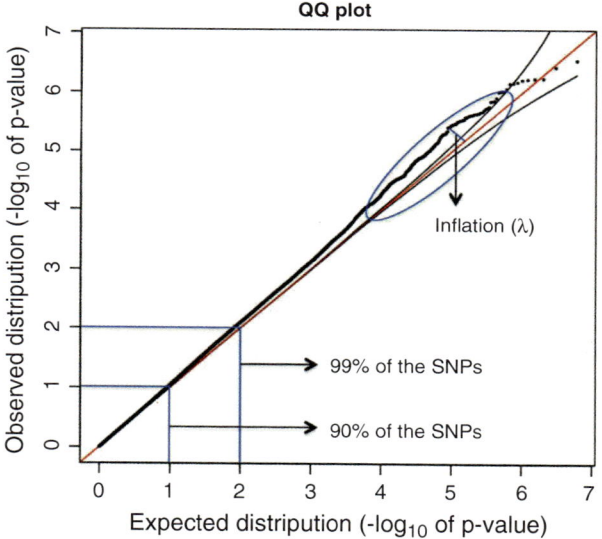

Fig. 9.5 Example of quantile-quantile plot. Example of a QQ plot. The red line identifies the expected distribution in concordance with the null hypothesis. The black line represents the observed distribution. The majority of the observed distribution (up to 99%) overlaps with the expected distribution. Only a minority of SNPs (≤1%) deviates from the expected distribution because of presumable association with the trait (disease). However, the deviation underlying true association is expected to be minimal as a major deviation (inflation: $\lambda > 1$ or deflation $\lambda < 1$) represents confounding issues (e.g. population stratification)

greater than 1 generally suggests increased risk, whilst an odds ratio smaller than 1 indicates protection.

Keywords and Definitions

Null hypothesis: the hypothesis that there is no association between genotype and phenotype (i.e. no association between any allelic frequency and disease).

p-**value**: probability of finding an association exclusively by chance.

Type I error (α): probability of identifying an association when there is actually none (false positive also called spurious association).

Type II error (β): probability of not identifying an association when there is actually one (false negative).

Power: probability of identifying an association when there is actually one ($1 - \beta$). The power is function of (1) sample size, (2) allele frequency, (3) effect size and (4) haplotype structure.

Effect size: magnitude of risk conferred by a certain allele.

Odds ratio (OR): the measure of association by comparing the odds of an event happening in the presence or absence of a specific variable (e.g. an allele). In the specific case of GWAS the OR is:

$$OR = \frac{\text{presence of allele A and presence of phenotype} / \text{presence of allele A and absence of phenotype}}{\text{absence of allele A and presence of phenotype} / \text{absence of allele A and absence of phenotype}}$$

GWAS: Interpretation of the Results

To date, one of the most important lessons in interpretation of GWAS results is that the SNPs with the smallest *p*-values might not necessarily be the real reason for the association. In the majority of cases, the SNPs with highest association are tag-SNPs which act as a surrogate for association indirectly pointing to neighbouring SNPs in LD for the real cause of association at that locus. In addition, the SNPs showing association are rarely coding, rather intronic or intergenic. When a GWAS is concluded (discovery + replication phases), caution is warranted in the interpretation of the outcome, and the following steps are generally recommended (Fig. 9.6): (1) identifying and sequencing all genes in proximity of the top hits to identify possible coding changes, that is, pathogenic coding variants and novel genes associated with the disease under study; (2) selecting all identified known polymorphisms to build haplotypes within and around the associated locus/loci to possibly identify disease-specific haplotypes and/or particular SNPs that are in linkage disequilibrium with the GWAS top hits to be further studied; and (3) evaluating effects of the associated SNPs (and of those in LD with the associated SNPs) on expression and/or splicing. For the latter analysis, all SNPs (intergenic, intronic or even synonymous variants) that show association or are in LD with associate SNPs are informative. In fact, when the associated loci do not affect proteins directly, it is likely that they exert their effect by (1) altering constitutively transcript levels, (2) modulating transcript expression and (3) affecting splicing [98].

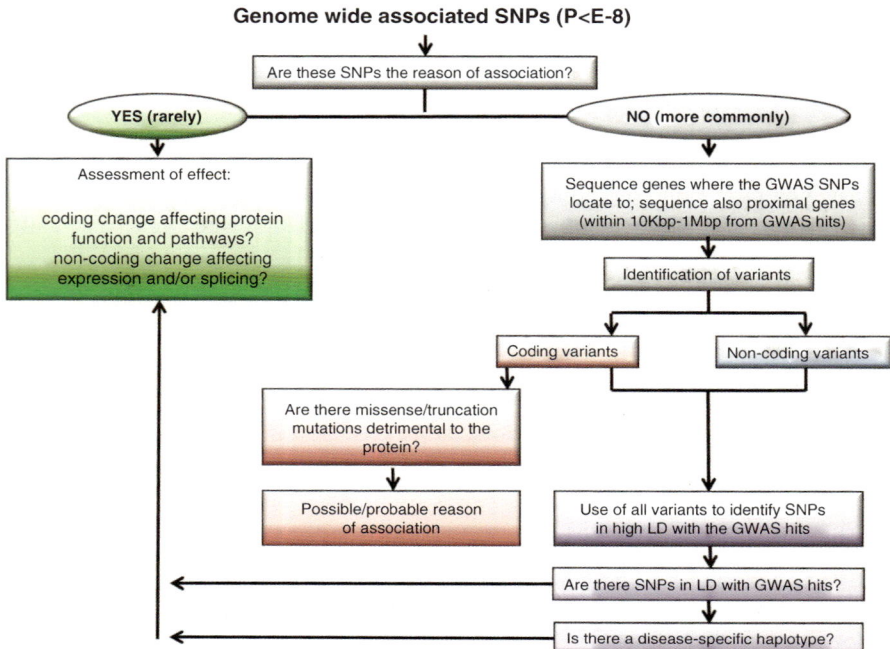

Fig. 9.6 Post-GWAS workflow. Diagram highlighting the recommended workflow subsequent to completion of discovery and replication phases of a GWAS. These are the main (but not exclusive) immediate steps warranted for further interpreting the results of a GWAS

GWAS in FTD

Synopsis

To date, three major GWAS have been performed in the field of frontotemporal dementia. The first study was released in 2010 [99] and focused on a subgroup of the FTD spectrum defined by TDP-43 pathology (FTLD-TDP). The second study was published in 2014 and was performed on clinical FTD including the major syndromes—bvFTD and PPAs—as well as FTD overlapping with motor neuron disease (MND) [100]. The last was a population-specific association study exploring the genetic underpinnings of clinical Italian FTD and was published in 2015 [101]. All such studies were performed to further the understanding of, and to identify novel genetic risk factors associated with, sporadic FTD.

International FTLD-TDP GWAS

The Study and Lessons Learned

This work included FTD cases meeting either pathological or genetic (i.e. presence of pathogenic *GRN* mutations) criteria for FTLD-TDP and was designed as a classical case-control study with a discovery and a replication phase. The discovery phase included 515 FTLD-TDP pathologically confirmed cases (89 of which carried *GRN* mutations) and 2509 disease-free population controls and used standard Illumina array chips such as the 550K and 610K BeadChips to generate genotypes. The replication phase was performed on 89 independent FTLD-TDP cases and 553 Caucasian controls using the TaqMan SNP genotyping technology.

The discovery phase highlighted three significant SNPs—rs6966915, rs1020004 and rs1990622 (Table 9.1A)—mapping to a 68 kb interval on chromosome 7 at 7p21.3 and encompassing the transmembrane protein 106B (*TMEM106B*) gene: the top SNP, rs1990622, locates 6.9 kb downstream of *TMEM106B*, whilst rs1020004 and rs6966915 map to introns 3 and 5, respectively, of *TMEM106B*. When conditioning the analysis upon presence ($n = 89$) or absence ($n = 426$) of *GRN* mutation carriers, the signal at 7p21.3 persisted, yet there was a robust inflation for the *GRN*+ cohort strongly suggesting *TMEM106B* modulating or being modulated by *GRN* mutations. Replication could be only performed for two of the top SNPs (i.e. rs1020004 and rs1990622) indicating significant association and same direction of effect as in discovery phase (Table 9.1A). Of note, however, replication was only achieved in the FTLD-TDP replication-cohort, whilst the results could not be replicated in additional 192 individuals with unspecified FTLD [99]. To evaluate the biological effect of the associated risk alleles, authors firstly interrogated a database for lymphoblastoid cell lines that suggested the risk allele (T) of rs1990622 associating with higher *TMEM106B* mRNA levels. This was supported by additional assessments of *TMEM106B*'s expression in frontal cortex in FTLD-TDP post-mortem ($n = 18$) and neurologically normal control ($n = 7$) brains for rs1020004 and rs1990622 indicating that risk allele carriers had 2.5-fold higher expression rates of *TMEM106B* compared to controls [99]. Finally, testing *TMEM106B* expression levels in FTLD-TDP with and without *GRN* mutations indicated that *GRN*+ carriers displayed highest increase of *TMEM106B* expression comparatively to *GRN* individuals and controls (and this was independent from the rs1990622 risk allele) [99].

Altogether, this study indicated SNPs encompassing the *TMEM106B* gene as risk factors for the FTLD-TDP subtype with the risk alleles exerting their effect by influencing an increase in *TMEM106B* expression levels in brain areas typically affected in FTD (i.e. frontal cortex). Additionally, this study suggested *TMEM106B* risk alleles being particularly enriched in *GRN* mutation carriers strongly suggesting *TMEM106B* as a *GRN* modifier.

Follow-Up Studies

After completion of the original GWAS [99], multiple follow-up studies have been performed in order to further the understanding of TMEM106B's biology.

Such works have both investigated genetic replication of the risk SNPs and tried to shed light on their functional meaning. Genetic replication studies have led, to date, to rather variable results, and this might be partially ascribed to the differences in the study designs that is differences in cohort sizes, in the diagnoses of the analysed cases (i.e. clinical FTD or FTLD-TDP), genetic background (i.e. presence/absence of *GRN* or *C9orf72* variability) or population subtypes. As indicated by the original study, *TMEM106B* appears to be a risk factor confined to the FTLD-TDP

Table 9.1 Summary of associations from current FTD-GWAS

A	Phenotype	Marker	Alleles	Risk allele	Chr	BP	Discovery		Replication			Joint		Study
							p-Value	OR	p-Value	OR		p-Value	OR	
	FTLD-TDP	rs1990622	C/T	T	7	12283787	1.08×10^{-11}	1.64	**0.0002**[a]	1.75		NA	NA	[99]
		rs1020004	A/G	A		12255778	5.00×10^{-11}	1.67	**0.004**[a]	1.89		NA	NA	
		rs6966915	A/C/T	C		12265988	1.63×10^{-11}	1.64	NA	NA		NA	NA	
	bvFTD	rs302652	T/A	T	11	87894831	2.02×10^{-08}	1.37	NA	NA		NA	NA	[100]
		rs302668[b]	T/C	T		87876911	NA	NA	**0.041**	1.14		2.44×10^{-07}	1.23	
		rs74977128	T/C	C		87936874	3.06×10^{-08}	1.81	NA	NA		NA	NA	
		rs16913634[b*]	G/A	G - A		87934068	NA	NA	0.71	1.04		8.15×10^{-04}	1.25	
	Meta clinical FTLD	rs1980493	T/C	T	6	32363215	4.94×10^{-08}	1.39	**0.02**	1.17		$\mathbf{1.57 \times 10^{-08}}$	1.30	
		rs9268877	A/G	A		32431147	1.65×10^{-10}	1.33	0.104	1.08		$\mathbf{1.05 \times 10^{-08}}$	1.20	
		rs9268856	A/C	C		32429719	1.30×10^{-08}	1.33	**0.014**	1.14		$\mathbf{5.51 \times 10^{-09}}$	1.24	

B	Phenotype	Marker	Alleles	Risk allele	Chr	BP	p-Value Discovery	OR	Method Replication	Gene	FDR	Study
	Italian clinical FTD	rs17042852	C/T	C	2	52600067	2.01×10^{-7}	2.82	NA			[101]
		rs1526678	G/A	G		52635727	2.19×10^{-7}	2.83				
		rs17042770	C/G	C		52571393	2.22×10^{-7}	2.83				
		rs12621157	T/G	T		52509876	3.73×10^{-7}	2.75				
		rs12622570	C/G	C		52546301	3.99×10^{-7}	2.76				
		rs12619513	A/G	A		52532874	6.39×10^{-7}	2.53				
		rs12614311	T/C	T		52521716	8.83×10^{-7}	2.55				
		rs906175	T/C	T	17	79173462	1.22×10^{-7}	1.58	GATES	CEP131	0.001469	
		rs2725391	T/C	T		79192430	2.50×10^{-7}	1.52		ENTHD2	0.004264	
		rs969413	A/T	A		79195814	4.26×10^{-7}	1.52		C17orf89	0.004264	
		rs2659030	A/G	A		79177974	4.42×10^{-7}	1.56	sPCA	CEP131	0.000650	
		rs2255166	C/T	C		79213562	6.19×10^{-7}	1.55		ENTHD2	0.001007	
		rs9319617	C/T	T		79192446	6.62×10^{-7}	1.51	SKAT	CEP131	0.000014	
		rs1048775	G/C	C		79202329	8.04×10^{-7}	1.51		ENTHD2	0.000651	
										C17orf89	0.000651	

A. Summary of association results for FTLD-TDP and International clinical FTD GWAS. B. Summary of association results for FTLD-TDP and Italian clinical FTD GWAS

[a]Replication tested in 89 FTLD-TDP cases
[b]Surrogate/proxy SNPs for the bvFTD subtype
*denotes heterogeneity p-value <0.01 in the meta-analysis of the discovery and replication phases combined

endophenotype; in fact, authors failed to replicate their findings in 192 clinical or non-FTLD-TDP cases indicating as a reason the lack of power due to sample size [99]. Nevertheless, when different and larger cohorts were analysed, results were at times equivocal. In a North American study, only nominal significance was reached for the three *TMEM106B* GWAS-SNPs in a cohort of over 600 FTD cases [102]: patients were subdivided in $n = 482$ clinical FTD, $n = 80$ GRN^- FTLD-TDP and $n = 78$ GRN^+ FTLD-TDP. Only rs1990622 resulted significantly associated in the GRN^+ subgroup whilst analysed in a recessive genetic model [102]. Of note, the GRN^+ cases that concomitantly carried the protective (minor) allele (C)—especially in the case of homozygotes—showed a later age of onset [102]. In support of this finding, carriers of the homozygous allele T (rs1990622) showed association with 13 years earlier age of onset in another North American cohort of sporadic FTLD-TDP cases with *GRN* mutations ($n = 50$) [103]. Significant association for the rs1990622 risk alleles was reported in a Belgian, mainly clinical, FTLD cohort ($n \sim 290$) that included *GRN* mutation carriers ($n = 13$), FTLD-ALS ($n = 22$) and FTLD-TDP cases ($n = 14$) [104]. Yet this resulted an isolated replication of the original findings in clinical FTD as association for *TMEM106B* risk alleles failed in a British (Manchester and London) clinical FTLD cohort sized $n > 400$ with no *GRN* mutations carriers [105]. Replication also failed in the largest dataset ($n \sim 2200$; discovery phase) of clinical sporadic North American and European FTD to date (see 'International Clinical FTD GWAS' section and [100]); assessing the three *TMEM106B* GWAS-SNPs, lowest p-values (in the range of 5×10^{-3}) were seen for rs1990622 in the bvFTD subtype and the meta-analysis of the entire cohort, whilst for any other combination of SNPs and phenotype, p-values were negligible (Table 9.2). Additionally, replication failed in an Italian cohort of clinical FTD ($n = 530$; see 'Italian Clinical FTD GWAS' section and [101]) where the three *TMEM106B* GWAS-SNPs reached p-values ranging between 3 and 6×10^{-1} (Table 9.2). Genetic analyses were also carried out in an ALS population ($n = 85$), from North America, excluding association with increased risk of developing ALS for the major risk allele (T) of rs1990622, although there was a hint of association when focusing on those ALS cases with a concomitant cognitive impairment (whilst distribution and severity of TDP-43 pathology seemed independent from *TMEM106B* genotypes) [106]. It follows that, thus far, studies suggested a major implication of *TMEM106B* risk alleles almost exclusively in FTLD-TDP cases carrying *GRN* mutations, with, additionally, a modulating effect on age of onset where risk alleles or protective alleles (especially in the homozygous state) contributed to earlier or later age of onset, respectively. The focus of the next studies became verification of their effects as modifiers in presence of *C9orf72* expansion, all the more considering that cases carrying the expansion, similar to the case of *GRN* mutations, are grossly associated with FTLD-TDP pathology [24]. One study on over 300 FTLD *C9orf72* expansion carriers from North American extraction, subdivided into FTD ($n = 86$), FTD-MND ($n = 78$) and MND ($n = 127$), used a recessive genetic model to verify frequencies of the protective (minor) alleles for rs3173615 and rs1990622. They stratified by disease phenotype, and major protection was yielded in FTD and least in MND cases (although the effect was less penetrant than in the

Table 9.2 Replication across different sporadic FTD GWAS

Phenotype original study	Marker	Alleles	Chr	BP	p-Value					Italian clinical FTD
					International clinical FTD					
					Meta	bvFTD	SD	PNFA	FTD-MND	Meta
FTLD-TDP hits	rs1990622	C/T	7	12283787	7.9×10^{-2}	5.8×10^{-3}	8.3×10^{-1}	8.9×10^{-1}	3.1×10^{-1}	3.1×10^{-1}
	rs1020004	A/G		12255778	4.6×10^{-1}	5.7×10^{-2}	8.5×10^{-1}	5×10^{-1}	1.2×10^{-1}	5.9×10^{-1}
	rs6966915	A/C/T		12265988	1.2×10^{-1}	5.7×10^{-3}	5.2×10^{-1}	7.2×10^{-1}	3.6×10^{-1}	3.7×10^{-1}
International clinical FTD	rs302652	A/T	11	87894831	NA					7.26×10^{-1}
	rs16913634	A/G		87934068						6.23×10^{-1}
	rs9268877	A/G	6	32431147						2.54×10^{-1}
	rs9268856	A/C		32429719						1.82×10^{-1}
	rs1980493	C/T		32363215						1.97×10^{-1}
Italian clinical FTD	rs12621157	G/T	2	52509876	6.5×10^{-1}	9.1×10^{-1}	5×10^{-1}	7.5×10^{-1}	5×10^{-1}	NA
	rs12622570	G/C		52546301	7.1×10^{-1}	8.5×10^{-1}	5.4×10^{-1}	7.5×10^{-1}	5×10^{-1}	
	rs12619513	G/A		52532874	5.4×10^{-1}	9.3×10^{-1}	4×10^{-1}	7.9×10^{-1}	3×10^{-1}	
	rs12614311	C/T		52521716	6.5×10^{-1}	9.4×10^{-1}	5.4×10^{-1}	7.6×10^{-1}	5.1×10^{-1}	
	rs1048775	C/G	17	79202329	3.6×10^{-1}	6.1×10^{-1}	2×10^{-1}	4×10^{-2}	3.3×10^{-1}	

The first two columns indicate the original study and associated phenotype and their relative GWAS-significant SNPs. The last two columns indicate the p-values that those SNPs reached either in the International or Italian clinical FTD cohorts case of *GRN* as previously described [99, 102, 103]). Also, the TDP-43 burden was attenuated in presence of homozygous protective *TMEM106B* alleles [107]. Of note, a reduced association for the MND cases appeared in line with previous reports [106]. However, another study [108], published back-to-back with [107], counterintuitively contradicted findings of the latter: using utterly small discovery ($n = 14$) and replication ($n = 75$) cohorts of North American and British extraction—all defined as FTLD-TDP cases carrying *C9orf72* expansion—Gallagher and colleagues also applied the recessive genetic model and found correlation between homozygous minor (protective) alleles for rs1990622 and earlier disease age of onset as well as shorter disease duration [108]. This was in complete contradiction with [107] and previous literature reports [102, 103]. A third work studied a FTD cohort of southern European extraction (French and Italian) with either no mutations in known genes ($n = 384$) or *C9orf72* expansion ($n = 145$) or *GRN* mutations ($n = 76$) indicated association with increased risk for disease for the major (risk) allele T of rs1990622 exclusively in the *GRN* mutation carriers group, yet no association or effect on age of onset or disease duration was found [109]. A more recent study in the Spanish population of 146 clinical FTD patients illustrated a tendency, yet not fully statistically significant, of increased risk for the risk (major) allele T for rs1990622 [110].

The investigation of biological effects of the *TMEM106B* GWAS-SNPs relied on the assessment of effects on *TMEM106B* expression as well as on plasma levels of TMEM106B and GRN and on functional studies aimed at exploring TMEM106B's biology in relation to lysosomes. In the original study [99], the effect alleles were shown to influence an increased expression of *TMEM106B* although it was also shown that increased *TMEM106B* mRNA levels could be seen in *GRN* mutation carriers, independently of the *TMEM106B* GWAS-SNPs genotypes [99]. Another study indicated that the minor (protective) alleles of the GWAS-SNPs showed association with reduced levels of *TMEM106B* mRNA and concomitant higher *GRN* mRNA and GRN plasma levels [102]. Nevertheless, as in the case of the previously described genetic replication studies, studies evaluating effects on *TMEM106B* expression led to contradicting results, as multiple studies did not detect an increase of *TMEM106B* mRNA in presence of TMEM106B GWAS-risk alleles as evidenced in [103, 104, 111]. It had also been suggested that increased levels of TMEM106B protein may modulate GRN plasma levels and contribute to pathogenic processes in this fashion [112]. Another effect on TMEM106B protein levels appears to be exerted by rs3173615 that leads to the missense mutation T185S. Rs3173615 is the only coding variant in strong LD with the TMEM106B GWAS-SNPs [102]. It was shown that when the T185 isoform was overexpressed there was nearly a twofold increase in TMEM106B protein expression as compared to the S185 isoform, despite equal mRNA expression levels; it was also shown that the higher levels of TMEM106B T185 protein were due to more rapid lysosomal degradation of the TMEM106B S185 protein [113]. Considering these two pieces of information together, one might speculate that a basal higher expression level of TMEM106B is the functional/biological reason contributing to increased risk of developing the disorder.

TMEM106B encodes the 274 amino acid long protein TMEM106B, which is a type 2 integral transmembrane protein located on endosomal and/or lysosomal membranes [114]. TMEM106B protein is found in neurons, glial and endothelial cells [115]. Currently, the suggested disease model indicates the increased level of TMEM106B as detrimental. A number of studies in different (non-neuronal) cell lines indicated that increased levels of TMEM106B protein seemed to interfere with the generation of mature and/or functional lysosomes leading to a delay in endolysosomal-dependent degradation and concomitant cytotoxicity [112, 116, 117] as well as causing translocation of transcription factor EB (TFEB) to the nucleus and to upregulate gene expression from the Coordinated Lysosomal Expression and Regulation (CLEAR) gene network [117], which is a marker of lysosomal stress [118]. These results were only partially replicated in mouse primary neurons: more specifically, in this setting, effects of higher TMEM106B protein levels were compared with effects on lysosomal size, but not on cell viability, TFEB translocation (thus lysosomal stress) or lysosomal acidification [117, 119]. Functional interplay between TMEM106B and GRN, and C9orf72 at the lysosomal cellular level has started to be investigated. A recent animal model indicated that grn and tmem106b have opposite effects on lysosomal enzyme levels (increase following GRN depletion and decrease after TMEM106B depletion) and that tmem106b deletion in grn knockout mice normalized lysosomal protein levels partially rescuing behavioural abnormalities in the mice [120]. Conversely, in multiple cell types (including neurons) it was shown that TMEM106B-induced defects were rescued after *C9orf72* knockdown suggesting that TMEM106B and C9orf72 might interact within lysosomal pathophysiology [116]. Although further functional studies are needed to investigate TMEM106B's interplay with GRN and C9orf72 and its implication in FTLD-TDP, overall, it currently appears that well-regulated TMEM106B levels are important to support correct lysosomal functional.

International Clinical FTD GWAS

The Study and Lessons Learned

This study included FTD cases falling within the four clinical subgroups bvFTD, PPAs (SD and PNFA) and FTD-MND and followed the classical two-phased (discovery + replication) case-control strategy. Cases collected during discovery and replication phases had been diagnosed following the Neary [17] and the Rascovsky & Gorno-Tempini criteria [18, 20], respectively. Metadata accompanying all samples—including diagnosis, pathology and/or imaging data and genetic characterization for the known candidate genes—were collected in order to better stratify analyses (e.g. exclusion of *MAPT* and *GRN* mutation carriers). Up to 2154 cases vs. 4308 neurologically normal controls (discovery phase) and 1372 cases vs. 5092 neurologically normal controls (replication phase)—for a total (discovery + replication phases) of 3526 FTD samples and 9400 controls—were analysed in this study. The samples in discovery phase were genotyped on the Illumina 660K BeadChips, whilst the samples in replication phase were genotyped on a semi-custom exome-chip designed for ad hoc genetic studies in neurodegenerative conditions (as explained in the original study [100]). Of note, during phase I of this study there was no knowledge of the *C9orf72* repeat expansion; thus, cases carrying the expansion were blindly included in the discovery phase; same design was thus kept in replication phase for the *C9orf72*-positive cases to ease post hoc analyses.

The primary association analyses (discovery phase) were performed separately for the four subtypes (bvFTD, SD, PNFA and FTD-MND), followed by a meta-analysis on the entire cohort. For the bvFTD subtype, 1377 cases vs. 2754 controls were analysed leading to the identification of two significant SNPs—rs302652 and rs74977128—mapping to chromosome 11q14 (Table 9.1A). Rs302652 locates to intron 1 of the gene RAB38, member of RAS oncogene family (*RAB38*), whilst rs74977128 maps to the intergenic region between *RAB38* and cathepsin C (*CTSC*), ~25 kb upstream from *RAB38* open reading frame (ORF). For the remainder subtypes, association analysis was performed on 308 SD cases (vs. 616 controls), 269 PNFA cases (vs. 538 controls) and 200 FTD-MND cases (vs. 400 controls), and no SNP reached genome-wide significance likely due to small sample size, thus insufficient power. The meta-analysis on all four subtypes indicated significant SNPs at the 6p21.3 locus: rs1980493, locating to intron 5 of the butyrophilin-like 2 (MHC class II associated) gene (*BTNL2*), and rs9268877 and rs9268856, both mapping ~18.5–20 kb downstream from *HLA-DRA*, between the major histocompatibility complex class II, DR alpha and DR beta 5 genes (*HLA-DRA*; *HLA-DRB5*) (Table 9.1A). Replication was assessed through surrogate/proxy SNPs for the bvFTD subtype in $n = 690$ bvFTD cases (vs. 5094 controls) at chr11q14: rs302668, locating to intron 2 of *RAB38* was significant, whilst rs16913634, locating to the intergenic region between *RAB38* and *CTSC*, wasn't, and, accordingly, joint analysis showed suggestive *p*-values only for rs302668 (Table 9.1A), probably reflecting a decrease in power due to proxy-based replication and smaller sample size (i.e. 690 cases). Replication for the entire cohort at the 6p21.3 locus was significant for rs9268856 and rs1980493, and joint analysis confirmed strong association for the three top SNPs rs9268877, rs9268856 and rs1980493 (Table 9.1A). To then assess the potential biological effects of the risk alleles, expression and methylation quantitative trait loci data (e/mQTL) in brain were evaluated: although no *cis* changes of expression in brain were evident, the top SNP at the *RAB38*/*CTSC* locus (rs302652) was associated with decreased levels of RAB38 expression in blood, suggesting that a loss of RAB38 function might play a role in bvFTD pathogenesis. Also, a significant *cis*-mQTL at 6p21.3 for rs1980493 associating with changes in the methylation levels related to HLA-DRA in the frontal cortex was reported [100].

After the completion of this project, two novel susceptibility loci associated with clinical FTD were identified: one on chromosome 11 encompassing *RAB38*/*CTSC* for the bvFTD subtype and one on chromosome 6 encompassing *BTNL2* and *HLA-DRA*/*DRB5* for the entire cohort, suggesting that FTD pathogenesis might involve lysosomal/phagosomal pathways (link to chromosome 11) and immune system processes (link to chromosome 6) with risk alleles exerting their effect by modulating expression and methylation levels on *cis* genes, respectively.

Follow-Up Studies

To date not many replication studies have been performed for the GWAS-SNPs associated with clinical FTD [100]. One first replication attempt was performed in the Italian clinical FTD GWAS (see 'Italian Clinical FTD GWAS' section and [101]): this resulted in lack of replication of the original findings (Table 9.2) indicating that a population substructure might be at the basis of the lack of association. Particularly, this may be due to the fact that, besides the European ancestry, the Mediterranean population might not exactly share the same risk factors as that of Western/Central/North American-European extraction. Additionally, the only other studies exploring the markers defined by the International clinical FTD GWAS were done in the Chinese population: a first study in the Northern Han Chinese population investigated potential association of theses SNPs in 984 sporadic late-onset Alzheimer's disease (LOAD) cases identifying significant association for rs302668 (*RAB38/CTSC*), rs9268877 (*HLA-DRA/HLA-DRB5*), rs9268856 (*HLA-DRA/HLA-DRB5*) and rs1980493 (*BTNL2*) [121]. Additionally, another Han Chinese population, from southwest China, was assessed for these SNPs (see above) in 400 patients with sporadic ALS, 554 with sporadic PD indicating that the AA genotype for rs9268856 increased risk of ALS and shorter mean survival time [122]. Besides being surprising (but not impossible) that SNPs originally found significant in populations with European ancestry do replicate in a fairly diverse population such as the Chinese one, it is not clear why in these two studies FTD-GWAS SNPs were tested in different neurodegenerative disorders (i.e. AD, ALS and PD) rather than FTD itself in the first place. No other replication studies for these markers saw the light yet.

Italian Clinical FTD GWAS

The Study and Lessons Learned

This study was performed on a population-specific subgroup for which authors had access to raw data of 634 samples from 8 Italian research centres through the International clinical FTD GWAS [101]. The inclusion criteria were in line with those of the original study [100] and the work was executed through a standard cases vs. controls association study in the discovery phase, whilst three different statistical methods such as GATES, supervised PCA (sPCA) and the sequential kernel machine association test (SKAT)—representing three different ways to score and prioritize genes in the same dataset (also called 'SNPs-to-genes' analysis as explained in the original study [101])—were used during replication phase. After quality check (QC) steps, 530 patients diagnosed with bvFTD ($n = 418$), SD ($n = 27$), PNFA ($n = 61$) and FTD-MND ($n = 23$) were included in the study and compared to 926 controls obtained from the European Network for Genetic-Epidemiological Studies Hypergenes [123].

Although no SNP reached genome-wide significant p-values, there were two suggestive loci: one on chromosome 2, at 2p16.3, defined by 7 SNPs with p-values ranging between 2 and 8×10^{-7} with odds ratios (OR) exceeding 2.5 (Table 9.1B) and locating to introns of the uncharacterized gene *LOC730100*, which immediately maps downstream (centromeric) from the neurexin 1 (*NRXN1*) gene. No significant *cis* effects on transcription in brain tissues were seen for the risk alleles. The other locus mapped to chromosome 17, at 17q25.3, and was also defined by seven suggestive SNPs with *p*-values ranging from 1 to 8×10^{-7} and OR barely exceeding 1.5 (Table 9.1B). These markers are located to the introns of two genes, the centrosomal protein 131 (*CEP131*) and the yet uncharacterized *C17orf89*, and to the 3'-UTR of the ENTH domain containing 2 (*ENTHD2*) gene. Interestingly, the risk alleles at 17q25.3 defined a suggestive risk haplotype encompassing *CEP131*, *ENTHD2* and *C17orf89* and causing decreased expression of the *cis* genes *RFNG*, *AATK* and *MIR1250* suggesting their cumulative effect on transcription processes as the biological mechanism underlying the association at this locus. During replication, the GATES analysis indicated 13 FDR significant genes, and *CEP131*, *ENTHD2* and *C17orf89* had lowest *p*-values; the sPCA analysis revealed 30 FDR significant genes, and *CEP131* and *ENTHD2* had lowest *p*-values; and the SKAT analysis indicated four FDR significant genes, and *CEP131*, *ENTHD2* and *C17orf89* had lowest *p*-values again (Table 9.1B and [101]). All this taken together, replication phase revealed that the two genes (*CEP131* and *ENTHD2*) were consistently identified across the three analysis methods (GATES, sPCA or SKAT); these genes map exactly to the strongly suggestive locus 17q25.3 as per association analysis (Table 9.1B and [101]). Of note, there was no replication for the suggestive locus 2p16.3 because of power issues as explained in [101].

In summary, this study, the first of this size in the Italian FTD population, identified two novel potential loci for FTD. Particularly, one of the two new loci (17q25.3) showed a haplotype substructure significantly associating with disease and affecting expression (decrease) of nearby *cis* genes. From a functional perspective, this study directly (*CEP131* and *ENTHD2*) or indirectly (*NRXN1*, *RFNG* and *AATK*) pointed to genes involved in variable processes spanning from control of genome stability and neuronal apoptosis to trans-Golgi vesicular network, neuronal development, differentiation and maturation, and axonogenesis.

Follow-Up Studies

To date there have not been replication studies for these markers in the Italian population yet. The only evaluation made to date was a retrospective assessment in the extended International FTD GWAS [100] dataset showing a lack of association given the non-significant *p*-values ranging from 4 to 7×10^{-1}. As previously indicated whilst assessing replication of the GWAS-SNPs of the original study [100] in the Italian cohort [101], the lack of replication in a bidirectional manner as seen here may underlie the fact that differences in populations play an important role in determining genetic association, even more than sample size and statistical power.

Summary and Significance of These Studies

An overlap of all loci associated with FTD, as per the GWAS performed to date, is summarized in Fig. 9.7.

9 Genetic Risk Factors for Sporadic Frontotemporal Dementia

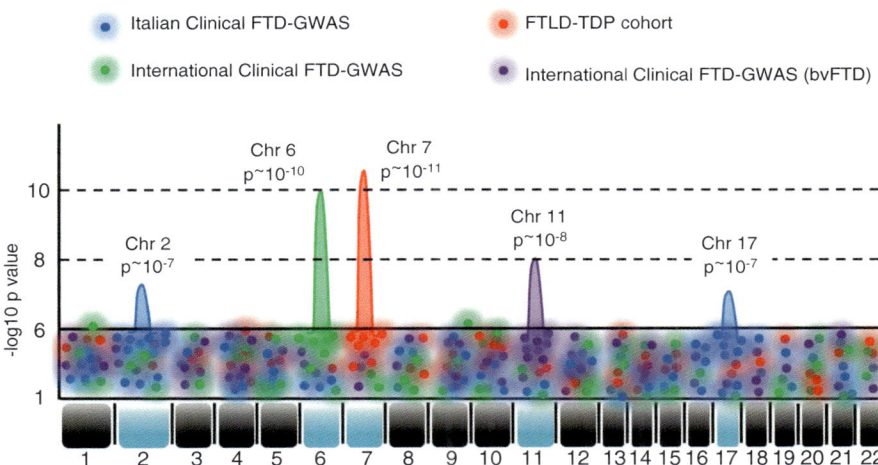

Fig. 9.7 Visual summary of major GWAS loci published to date in FTD. This cartoon indicates that the genetic architecture underlying FTD might differ based on the different FTD phenotypes or populations analysed

The FTLD-TDP GWAS screened a rather homogeneous cohort defined by TDP-43 pathology; this was a sensible choice in that homogeneity in the tested cohort is an important feature in GWAS type of studies (see 'Genetic Studies of Complex Disorders' section). This clearly balanced a potential issue represented by sample size as $n = 515$ as discovery cohort is a remarkably small number and led to the identification of a locus associated with this pathologically defined subtype with good confidence. The SNPs defining the identified locus were three markers encompassing the *TMEM106B* gene; this study indicated three major outcomes: (1) the association was the strongest in cases carrying *GRN* mutations; (2) the risk alleles influence TMEM106B expression levels being associated with an increase of *TMEM106B* mRNA; (3) *TMEM106B* expression levels were increased also just in presence of *GRN* mutations and independently from the TMEM106B allele genotypes [99]; and (4) the increase of expression might exert a functional detrimental effect by influencing lysosomal biology, that is, lysosomal size and function. This is important as it immediately indicates that *TMEM106B* is a disease modifier mainly restricted to FTLD-TDP pathology and presence of *GRN* mutations. Replication studies in fact supported this view in that, for the most, clinical FTD cohorts and cohorts negative for *GRN* mutations did not replicate the original genetic findings [99–102, 105]. Nevertheless, in presence of *GRN* mutations multiple studies tended to confirm the original findings and, all the more, indicated a likely effect on disease age of onset [102, 103] that is an earlier age of onset and shorter disease duration. Similar results were observed in the case of presence of *C9orf72* expansion although strength of association was less than in the case of *GRN* mutation carriers [107] yet results were rather equivocal [108]. It follows that, as both GRN^+ and $C9orf72^+$ cases are associated with TDP-43 pathology, it is sensible to consider *TMEM106B*, a marker of FTLD-TDP, as a modifier in presence of both genetic mutations. Not least, an important message of the body of work characterizing the biology of TMEM106B is the putative effect on lysosomal biology; the interplay between TMEM106B and GRN in the lysosomal biology needs to be further studied to identify their exact implications, whilst that between TMEM106B and C9orf72 and lysosomal biology needs to be clarified as to date no convincing results have been published.

The International clinical FTD GWAS by far exceeded cohort size of the FTLD-TDP GWAS study in both discovery and replication phases. This was critical to overcome an associated caveat with the International clinical project deriving from the inclusion of the two major FTD syndromes, behavioural and language variants, and FTD overlapping with MND, thus cohort heterogeneity and power issues when focusing on single subtypes. This study is to date the largest on sporadic FTD and resulted also in the foundation of the International FTD-Genomics Consortium (IFGC; https://ifgcsite.wordpress.com/), a group of International experts in FTD with the goal of expanding on the genetics and functional biology of sporadic FTD. The study by itself indicated association with a locus pointing to lysosomal biology in the case of bvFTD, that is, the *RAB38/CTSC* locus, and indicated that a risk factor for FTD globally might reside in an aberrant behaviour of the adaptive immune system through the *HLA* locus on chromosome 6. Additionally, although no statistical significance was reached in the language (SD and PNFA) as well as the FTD-MND subtypes, results in those cohorts pointed to a number of unique suggestive loci (p-values ~ $10^{-6/7}$) that will need to be repeated and further explored when bigger cohorts become available for novel larger discovery and meta-analyses. Nevertheless, this study provided a number of informative points on sporadic FTD. First, based on the study design, there was no association at the expected or known loci such as those encompassing *MAPT* and *GRN* genes on chromosome 17, or *C9orf72* on chromosome 9, the most likely reasons being (1) the fact that all known chromosome 17 mutation carriers were excluded from analysis and (2) the frequency of *C9orf72* expansion carriers within the whole discovery cohort was ~8% (n = 194/2412) as revealed by post hoc analysis (which reflects currently known or expected prevalence of the expansion in the FTD sporadic population [see 'Frontotemporal Dementia (FTD)' section, 'Genetics' section]), therefore probably insufficient for contributing to a genome-wide significant signal. Considering the FTD-MND subgroup, the expansion frequency was much higher (=52/221, 23.5%) compared to the other subtypes (Table 9.3). As such, if any signal was absent in the bvFTD, SD and PNFA subtypes (as well as in the whole cohort), a signal at the *C9orf72* locus was barely detectable in the FTD-MND subtype with an associated p-value = 2.12×10^{-06}.

The locus associating with bvFTD includes the two genes *RAB38* and *CTSC*. *RAB38*, an oncogene that was reported being mutated in melanoma [124], encodes the transmembrane protein RAB38 which is ubiquitously expressed across tissues, including the brain. From a functional perspective, RAB38 has been suggested to be involved in mediation of protein trafficking to lysosomal-related organelles within the trans-Golgi network (TGN) [125, 126], maturation of phagosomes that envelop pathogens [127] and neurite outgrowth [128]. *CTSC* is a lysosomal cysteine-proteinase that participates in the activation of serine proteinases in immune/

Table 9.3 Frequency of *C9orf72*-positive cases within the International clinical FTD discovery cohort

	Discovery phase								
	bvFTD		SD		PNFA		FTD-MND		Total
	n+/nS	%	n+/nS	%	n+/nS	%	n+/nS	%	n+/nS	%
C9orf72+	121/1537	7.9	12/350	3.4	9/304	3.0	52/221	23.5	194/2412	8.0

Summary of the frequency of *C9orf72*-positive cases within the discovery cohort shown for each subtype separately and for the totality of samples. *n+* number of cases positive for the repeat expansion, *nS* number of cases screened inflammatory cells that are involved in immune and inflammatory processes including phagocytosis of pathogens and local activation/deactivation of inflammatory factors (e.g. cytokines) (OMIM: #602365). Both RAB38 and CTSC correlate with lysosomal as well as phagosomal biology suggesting that autophagosomal/lysosomal dysfunctions might play a critical role in the development and progression of bvFTD. Conversely, the locus encompasses the *HLA-DRA/B* and *BTNL2* genes. HLA-DRA/B encode monomorphic/polymorphic class II HLA-DR transmembrane chains, which are expressed on the surface of antigen-presenting immune cells. The HLA-DR molecules are known to be expressed on the surface of microglia, and it has been suggested that increased expression of HLA-DR molecules on microglia may reflect pathological activity, as previously reported, for example, in AD and PD [129]. *BTNL2*, which encodes a membrane protein that is ubiquitously expressed across different tissues including the brain, is involved in repressing T-cells proliferation [130]. The immune system is highly important in modulating several processes in the central nervous system (CNS): for example, in normal conditions, microglial cells play an important role during brain development by pruning neurons and maintain CNS homeostasis through removal of either debris and apoptotic cells or pathogens via phagocytosis [131, 132]. All together, these notions offer insight into a possible role of aberrant/detrimental immune responses in the brain affecting neurodegeneration and a potential role for the adaptive immune system in FTD. Undoubtedly, these loci will need to be replicated in other FTD cohorts and further investigated in order to establish a link between the genetic association and biological processes underlying disease. These findings however hold promise for a better understanding of the pathogenesis of FTD and for the development of tools to be implemented for preventive and therapeutic measures.

The Italian clinical FTD GWAS was the first (and to date still unique) study of this size in the Italian clinical FTD population. This study highlighted two suggestive loci. The one at the 2p16.3 locus—showing high OR (>2.5) for the risk alleles—encompasses the *LOC730100* gene, a long non-coding RNA (lncRNA): lncRNAs are known to be implicated in a number of complex processes that include chromatin stabilization, histone methylation as well as pre-transcriptional and post-transcriptional (cis- and trans-) regulation [133]. The other locus at 17q25.3 showed suggestive association with OR > 1.5 for each risk alleles. These SNPs map to three genes: *CEP131*, *ENTHD2* and *C17orf89*. If *C17orf89* is still uncharacterized, *CEP131* encodes a centrosomal protein of 131 kDa weight, which is part of the centrosomal complex and seems involved in cilia formation and genome stability processes [134]. *ENTHD2* encodes a protein that localizes to the cytoplasm and seems to be involved in trans-Golgi network vesicular processes [135]. Interestingly, the seven risk alleles of these suggestive SNPs define a haplotype substructure that is significantly associated with disease status (OR = 1.45), and second, each of the risk alleles had significant or suggestive effects on transcription, specifically, causing a decrease in expression of *cis* genes such as *RFNG*, *AATK* and *MIR1250*. *RFNG* encodes an N-acetylglucosaminyltransferase for which involvement in neurogenesis and a role in modulating Notch signalling have been previously suggested [136]. *AATK* was shown to have a potential role in apoptotic processes in mature neurons and neuronal differentiation [137] or axon outgrowth [138]. Conversely, a general implication in regulation of transcription and/or gene expression applies to MIR1250. Taken together, these results suggest that neuronal development, maturation and axonogenesis, as well as regulation of gene expression, might be impacted in the Italian FTD population. Additionally, and finally, in this population-specific GWAS other expected candidate loci such as those including *MAPT* or *C9orf72* resulted non-significant. Similarly, the risk alleles at the *C9orf72* locus (p-value = 3×10^{-2}, OR = 1.2) and *MAPT* (p-value = $7.57 \times 10^{-1} - 4.77 \times 10^{-2}$, OR = 1.03–1.2) were non-significant suggesting that these genetic risk factors seem not to associate with the Italian clinical FTD population. All this might indicate that for some loci population specificity is an important factor for discriminating genetic variants and their contribution to disease.

Future Studies to Untangle Sporadic FTD

Synopsis

To date few GWAS have been performed in FTD. This is due to a couple of major reasons: on the one hand, FTD is a rather rare neurodegenerative condition (e.g. compared to Alzheimer's or Parkinson's diseases), and on the other it represents a spectrum of heterogeneous syndromes. This means that it is not straightforward to gather large and well-defined cohorts when studying sporadic FTD. The first and second published GWAS benefitted from the collaboration of multiple research centres worldwide [99, 100]. Yet, because of the reasons above, even a homogeneous, thus well-defined, subgroup characterized by TDP-43 pathology only reached a sample size of 515 [99]. Conversely, the clinical International study reached higher numbers (in the order of $n \sim 22,000$ and ~1400 in discovery [phase-I] and replication [phase-II] phases, respectively); however, due to heterogeneity across the various syndromes (i.e. bvFTD, PPAs [SD and PNFA] and FTD-MND), stratifying by subtypes reduced the number of cases available to study homogeneous cohorts, negatively impacting the power of subtype-specific studies (see 'GWAS in FTD' section—'International Clinical FTD GWAS' section and [100]).

For these reasons, the IFGC (https://ifgcsite.wordpress.com/) is currently expanding the original study [100] by generating data for the phase-III of this extended International clinical FTD-GWAS (and genomics) project. The study design is the same as in the original study [100], and it is being performed for over 2500 new (since the completion of the original study [100]) sporadic cases that fall within the four major FTD syndromes (i.e. bvFTD, PPAs [SD and PNFA] and FTD-MND). This study—expected to be completed in 2018—will allow to robustly increase sample sizes for the different FTD syndromes, tremendously helping coping with power issues that affected the original study as indicated above and support a number of critical study designs that include (1) replicating previous results obtained during the original study [100] in a powerful replication cohort; (2) pooling together all samples from phases-I, -II and -III to perform a large discovery study on close to 6000 sporadic FTD samples; (3) increasing sample sizes for each subtype (i.e. bvFTD, PPAs [SD and PNFA] and FTD-MND) to increase the power of dissecting syndrome-specific genetic underpinnings; and (4) set the basis for large-scale meta-analyses with other closely related neurodegenerative conditions, such as ALS, exploring the FTD-ALS spectrum—with and without *C9orf72* expansion—at highest resolution.

Prospective Approaches

Strategies to look at the genetics of sporadic FTD (and any other form of complex disorder) need to take into account the opportunities that are offered by ever-advancing technologies and the ever-shifting questions that biomedical research is striving to answer. The following considerations will be aimed at the more general dissection of complex disorders.

A combination of GWAS, exome-chips and NGS is highly promising in covering better common and genetic variability of complex disorders. However, the focus of biomedical research is facing a paradigm shift in that basic research, to be fully and comprehensively supportive to applied research, needs to grow beyond genetics, particularly aiming at characterizing molecular mechanisms at the basis of disease in order to highlight biomarkers and drug targets for developing measures for disease prediction, prevention, monitoring and therapy. Ways to tackle these issues can be developed, and these depend on the available technologies and on approaches based on data integration. All such concepts are discussed in a structured format here below:

1. The study of genetics of complex disorders can currently be assessed at much higher resolution than ever before keeping in mind that different technical approaches allow to address different types of genetic questions such as (1) GWAS approaches mainly allow to study contribution to disease exerted by common markers that affect a phenotype with small to moderate effect size and that rather constitute polygenic risk factors whose cumulative effect globally represents the genetic architecture (i.e. risk-architecture) that predisposes to disease; and (2) exome-chip and NGS approaches allow to investigate the contribution of rarer and more penetrant variants to disease. This is equally important as these markers represent an additional layer on top of the risk-architecture that may robustly impact the disease phenotype, explain familial or cases defined by private mutations and help in characterizing missing heritability. Particularly, the use of ad hoc developed exome-chips is becoming a standard approach to rapidly assess known mutations or known genes for specific traits in large cohorts as well as discovering novel genes for a particular trait or re-evaluate the prevalence of certain variants across multiple (closely related or divergent) phenotypes. As well exome-chips, WES and WGS techniques impact the study of missing heritability as they aid fine-mapping classical GWAS loci, provide support in exploring the (likely) oligogenic nature of complex disorders as well as allow the

identification of novel genes and improve genotype-phenotype correlation for complex disorders (including FTD and its subtypes).
2. As stated above, new needs in the field urge moving beyond 'just' identifying causative or risk variants and genes. For example, there is the need for developing strategies to better characterize GWAS loci, i.e. methods to confidently identify the real reason for association and their functional effect. If on the one hand this can be done by evaluating effects on expression (see 'Genetics of Complex Disorders' section, 'GWAS: Interpretation of Results' section), on the other, the rise of the so-called burden tests allows to collapse multiple markers around an open reading frame to score and prioritize genes in loci that would have been ignored because of not reaching the (strict) Bonferroni correction significance. Also, there is a need for a better interpreting GWAS signals that are just below genome-wide significance, considering the global contribution of markers below a certain threshold. Clearly, methods need to be developed not only for understanding the effects of the SNPs at the associated loci, but also for prioritizing genes within GWAS loci. Multiple methods to interpret GWAS data have recently emerged including burden scoring at gene or pathway level (e.g. Pascal [139] or MAGMA [140]) as well as GWAS data integration with *cis*-eQTL signals [141] or epigenetic markers (e.g. methylation profiling) using tools such as summary data-based Mendelian randomization (SMR) [142]; thus the need to design additional and complementary pipelines to further and better characterize GWAS loci as well as the impacted biological processes, risk pathways and therein key functional players for potential future targeting is real [143, 144].
3. Next, it is fundamental to find strategies to translate the genetic into functional molecular understanding of molecular mechanisms of disease. Functional and biological analysis of molecular genetics of human diseases, has to date, relied heavily on Mendelian genetics (accounting for the minority of cases for given trait) and applied high-resolution but low-throughput approaches to investigate one gene at a time. This is not only time-consuming and underpowered, but also has ignored the genetic risk variants that drive phenotypes in sporadic cases (accounting for the majority of cases for given trait) by not taking into consideration the global genetic architecture contributing to the trait. One gathers that there is a clear and urgent requirement for a more holistic strategy across genetic and functional investigations to better reflect the contribution of genetic variability to human disease. In particular, there is a need to improve systemic approaches to identify causative genes within associated loci resulting from GWAS and to characterize the impacted biological processes, risk pathways and therein key functional players [145]. For example, using in silico methods, that consider genes in a functional annotation analysis format, allows to better put into perspective the biological processes and pathways that are impacted by genetic variability. Specifically in the case of FTD, there are a few examples of novel systems biology approaches that have started aiding in this respect [146, 147]. Weighted gene co-expression network and weighted protein–protein interaction network analysis are among the methods that can be used for highlighting

biological processes and pathways impacted in complex disorders (including FTD) on the basis of their Mendelian genetics. In turn, this will aid functional biologists prioritizing and designing more focused and coherent functional assessment to not only validate risk markers and/or genes but risk pathways.
4. Also, and finally, the future of the study of complex disorders (or any disorder) is harmonized access to data obtained from the same sample source including clinical, pathological, imaging, blood, serum, CSF markers, genome, methylome, transcriptome, proteome and metabolome (even microbiome and exposome [i.e. exposure to environmental factors]) in large numbers for a specific phenotype. This will help the global understanding of disease as well as the specific personal/private cases, making personalized medicine possible.

Harmonization of all such strategies will not be immediate and straightforward, yet it is the way to provide support and solutions for both basic and applied research in that these will aid in furthering our dissection and understanding of complex disorders and their molecular underpinnings setting the basis for providing personalized solutions in terms of preventive, monitoring and therapeutic measures.

References

1. Sosa-Ortiz AL, Acosta-Castillo I, Prince MJ. Epidemiology of dementias and Alzheimer's disease. Arch Med Res. 2012;43(8):600–8.
2. Prince M, et al. The global prevalence of dementia: a systematic review and metaanalysis. Alzheimers Dement. 2013;9(1):63–75 e2.
3. American Psychiatric Association. Diagnostic and statistical manual of mental disorders, 5th Edition: DSM-5. 5th ed. Arlington: American Psychiatric Publishing; 2013.
4. Sabbagh MN, et al. Increasing precision of clinical diagnosis of Alzheimer's disease using a combined algorithm incorporating clinical and novel biomarker data. Neurol Ther. 2017;6(Suppl 1):83–95.
5. Olsson T, Barcellos LF, Alfredsson L. Interactions between genetic, lifestyle and environmental risk factors for multiple sclerosis. Nat Rev Neurol. 2017;13(1):25–36.
6. Al-Chalabi A, van den Berg LH, Veldink J. Gene discovery in amyotrophic lateral sclerosis: implications for clinical management. Nat Rev Neurol. 2017;13(2):96–104.
7. Chatterjee N, Shi J, Garcia-Closas M. Developing and evaluating polygenic risk prediction models for stratified disease prevention. Nat Rev Genet. 2016;17(7):392–406.
8. Eilbeck K, Quinlan A, Yandell M. Settling the score: variant prioritization and Mendelian disease. Nat Rev Genet. 2017;18(10):599–612.
9. Organization, W.H. Dementia: a public health priority. Manila: WHO Regional Office for the Western Pacific; 2012.
10. Riedl L, et al. Frontotemporal lobar degeneration: current perspectives. Neuropsychiatr Dis Treat. 2014;10:297–310.
11. Wimo A, et al. The worldwide economic impact of dementia 2010. Alzheimers Dement. 2013;9(1):1–11 e3.
12. Seelaar H, et al. Clinical, genetic and pathological heterogeneity of frontotemporal dementia: a review. J Neurol Neurosurg Psychiatry. 2011;82(5):476–86.
13. Snowden JS, Neary D, Mann DM. Frontotemporal dementia. Br J Psychiatry. 2002;180:140–3.
14. Ratnavalli E, et al. The prevalence of frontotemporal dementia. Neurology. 2002;58(11):1615–21.

15. Rabinovici GD, Miller BL. Frontotemporal lobar degeneration: epidemiology, pathophysiology, diagnosis and management. CNS Drugs. 2010;24(5):375–98.
16. Degeneration, T.A.f.F. 2013. http://www.theaftd.org/frontotemporal-degeneration/ftd-overview.
17. Neary D, et al. Frontotemporal lobar degeneration: a consensus on clinical diagnostic criteria. Neurology. 1998;51(6):1546–54.
18. Gorno-Tempini ML, et al. Classification of primary progressive aphasia and its variants. Neurology. 2011;76(11):1006–14.
19. Rohrer JD, Warren JD. Phenotypic signatures of genetic frontotemporal dementia. Curr Opin Neurol. 2011;24(6):542–9.
20. Rascovsky K, et al. Sensitivity of revised diagnostic criteria for the behavioural variant of frontotemporal dementia. Brain. 2011;134(Pt 9):2456–77.
21. Josephs KA. Frontotemporal dementia and related disorders: deciphering the enigma. Ann Neurol. 2008;64(1):4–14.
22. Kurz A, Perneczky R. Neurobiology of cognitive disorders. Curr Opin Psychiatry. 2009;22(6):546–51.
23. Halliday G, et al. Mechanisms of disease in frontotemporal lobar degeneration: gain of function versus loss of function effects. Acta Neuropathol. 2012;124(3):373–82.
24. Mackenzie IR, Neumann M. Molecular neuropathology of frontotemporal dementia: insights into disease mechanisms from postmortem studies. J Neurochem. 2016;138(Suppl 1):54–70.
25. DeJesus-Hernandez M, et al. Expanded GGGGCC hexanucleotide repeat in noncoding region of C9ORF72 causes chromosome 9p-linked FTD and ALS. Neuron. 2011;72(2):245–56.
26. van der Zee J, et al. A pan-European study of the C9orf72 repeat associated with FTLD: geographic prevalence, genomic instability, and intermediate repeats. Hum Mutat. 2013;34(2):363–73.
27. Brown J, et al. Familial non-specific dementia maps to chromosome 3. Hum Mol Genet. 1995;4(9):1625–8.
28. Skibinski G, et al. Mutations in the endosomal ESCRTIII-complex subunit CHMP2B in frontotemporal dementia. Nat Genet. 2005;37(8):806–8.
29. Weihl CC, Pestronk A, Kimonis VE. Valosin-containing protein disease: inclusion body myopathy with Paget's disease of the bone and fronto-temporal dementia. Neuromuscul Disord. 2009;19(5):308–15.
30. Le Ber I, et al. SQSTM1 mutations in French patients with frontotemporal dementia or frontotemporal dementia with amyotrophic lateral sclerosis. JAMA Neurol. 2013;70(11):1403–10.
31. Synofzik M, et al. Screening in ALS and FTD patients reveals 3 novel UBQLN2 mutations outside the PXX domain and a pure FTD phenotype. Neurobiol Aging. 2012;33(12):2949 e13–7.
32. Momeni P, et al. Analysis of IFT74 as a candidate gene for chromosome 9p-linked ALS-FTD. BMC Neurol. 2006;6:44.
33. Pottier C, et al. Whole-genome sequencing reveals important role for TBK1 and OPTN mutations in frontotemporal lobar degeneration without motor neuron disease. Acta Neuropathol. 2015;130(1):77–92.
34. Bannwarth S, et al. A mitochondrial origin for frontotemporal dementia and amyotrophic lateral sclerosis through CHCHD10 involvement. Brain. 2014;137(Pt 8):2329–45.
35. Freischmidt A, et al. Haploinsufficiency of TBK1 causes familial ALS and fronto-temporal dementia. Nat Neurosci. 2015;18(5):631–6.
36. Gijselinck I, et al. Loss of TBK1 is a frequent cause of frontotemporal dementia in a Belgian cohort. Neurology. 2015;85(24):2116–25.
37. Mackenzie IR, et al. TIA1 mutations in amyotrophic lateral sclerosis and frontotemporal dementia promote phase separation and alter stress granule dynamics. Neuron. 2017;95(4):808–816 e9.
38. Coppola G, et al. Evidence for a role of the rare p.A152T variant in MAPT in increasing the risk for FTD-spectrum and Alzheimer's diseases. Hum Mol Genet. 2012;21(15):3500–12.

39. Jin SC, et al. Pooled-DNA sequencing identifies novel causative variants in PSEN1, GRN and MAPT in a clinical early-onset and familial Alzheimer's disease Ibero-American cohort. Alzheimers Res Ther. 2012;4(4):34.
40. Kouri N, et al. Novel mutation in MAPT exon 13 (p.N410H) causes corticobasal degeneration. Acta Neuropathol. 2014;127(2):271–82.
41. Momeni P, et al. Clinical and pathological features of an Alzheimer's disease patient with the MAPT Delta K280 mutation. Neurobiol Aging. 2009;30(3):388–93.
42. Pastor P, et al. Familial atypical progressive supranuclear palsy associated with homozygosity for the delN296 mutation in the tau gene. Ann Neurol. 2001;49(2):263–7.
43. Poorkaj P, et al. An R5L tau mutation in a subject with a progressive supranuclear palsy phenotype. Ann Neurol. 2002;52(4):511–6.
44. Rohrer JD, et al. Novel L284R MAPT mutation in a family with an autosomal dominant progressive supranuclear palsy syndrome. Neurodegener Dis. 2011;8(3):149–52.
45. Ros R, et al. A new mutation of the tau gene, G303V, in early-onset familial progressive supranuclear palsy. Arch Neurol. 2005;62(9):1444–50.
46. Sala Frigerio C, et al. On the identification of low allele frequency mosaic mutations in the brains of Alzheimer's disease patients. Alzheimers Dement. 2015;11(11):1265–76.
47. Van Cauwenberghe C, Van Broeckhoven C, Sleegers K. The genetic landscape of Alzheimer disease: clinical implications and perspectives. Genet Med. 2016;18(5):421–30.
48. Brouwers N, et al. Alzheimer and Parkinson diagnoses in progranulin null mutation carriers in an extended founder family. Arch Neurol. 2007;64(10):1436–46.
49. Brouwers N, et al. Genetic variability in progranulin contributes to risk for clinically diagnosed Alzheimer disease. Neurology. 2008;71(9):656–64.
50. Coppola C, et al. A progranulin mutation associated with cortico-basal syndrome in an Italian family expressing different phenotypes of fronto-temporal lobar degeneration. Neurol Sci. 2012;33(1):93–7.
51. Perry DC, et al. Progranulin mutations as risk factors for Alzheimer disease. JAMA Neurol. 2013;70(6):774–8.
52. Redaelli V, et al. Alzheimer neuropathology without frontotemporal lobar degeneration hallmarks (TAR DNA-binding protein 43 inclusions) in missense progranulin mutation Cys139Arg. Brain Pathol. 2018;28(1):72–6.
53. Spina S, et al. Corticobasal syndrome associated with the A9D Progranulin mutation. J Neuropathol Exp Neurol. 2007;66(10):892–900.
54. Parkinson N, et al. ALS phenotypes with mutations in CHMP2B (charged multivesicular body protein 2B). Neurology. 2006;67(6):1074–7.
55. van der Zee J, et al. CHMP2B C-truncating mutations in frontotemporal lobar degeneration are associated with an aberrant endosomal phenotype in vitro. Hum Mol Genet. 2008;17(2):313–22.
56. Cooper-Knock J, Shaw PJ, Kirby J. The widening spectrum of C9ORF72-related disease; genotype/phenotype correlations and potential modifiers of clinical phenotype. Acta Neuropathol. 2014;127(3):333–45.
57. Ferrari R, et al. Screening for C9ORF72 repeat expansion in FTLD. Neurobiol Aging. 2012;33(8):1850 e1–11.
58. Ferrari R, Thumma A, Momeni P. Molecular genetics of frontotemporal dementia. In: eLS. Chichester: Wiley; 2013.
59. Galimberti D, et al. Incomplete penetrance of the C9ORF72 hexanucleotide repeat expansions: frequency in a cohort of geriatric non-demented subjects. J Alzheimers Dis. 2014;39(1):19–22.
60. Hensman Moss DJ, et al. C9orf72 expansions are the most common genetic cause of Huntington disease phenocopies. Neurology. 2014;82(4):292–9.
61. Lindquist SG, et al. Corticobasal and ataxia syndromes widen the spectrum of C9ORF72 hexanucleotide expansion disease. Clin Genet. 2013;83(3):279–83.
62. Majounie E, et al. Frequency of the C9orf72 hexanucleotide repeat expansion in patients with amyotrophic lateral sclerosis and frontotemporal dementia: a cross-sectional study. Lancet Neurol. 2012;11(4):323–30.

63. Simon-Sanchez J, et al. The clinical and pathological phenotype of C9ORF72 hexanucleotide repeat expansions. Brain. 2012;135(Pt 3):723–35.
64. Smith BN, et al. The C9ORF72 expansion mutation is a common cause of ALS+/-FTD in Europe and has a single founder. Eur J Hum Genet. 2013;21(1):102–8.
65. Watts GD, et al. Inclusion body myopathy associated with Paget disease of bone and frontotemporal dementia is caused by mutant valosin-containing protein. Nat Genet. 2004; 36(4):377–81.
66. Pottier C, et al. Genetics of FTLD: overview and what else we can expect from genetic studies. J Neurochem. 2016;138(Suppl 1):32–53.
67. Borroni B, et al. TARDBP mutations in frontotemporal lobar degeneration: frequency, clinical features, and disease course. Rejuvenation Res. 2010;13(5):509–17.
68. Huey ED, et al. FUS and TDP43 genetic variability in FTD and CBS. Neurobiol Aging. 2012;33(5):1016 e9–17.
69. Ferrari R, Hardy J, Momeni P. Frontotemporal dementia: from Mendelian genetics towards genome wide association studies. J Mol Neurosci. 2011;45(3):500–15.
70. Hardy J, Rogaeva E. Motor neuron disease and frontotemporal dementia: sometimes related, sometimes not. Exp Neurol. 2014;262(Pt B):75–83.
71. Turner MR, et al. Genetic screening in sporadic ALS and FTD. J Neurol Neurosurg Psychiatry. 2017;88(12):1042–4.
72. Takada LT. The genetics of monogenic frontotemporal dementia. Dement Neuropsychol. 2015;9(3):219–29.
73. Manolio TA, et al. Finding the missing heritability of complex diseases. Nature. 2009;461(7265):747–53.
74. Alonso N, Lucas G, Hysi P. Big data challenges in bone research: genome-wide association studies and next-generation sequencing. Bonekey Rep. 2015;4:635.
75. Sham PC, Purcell SM. Statistical power and significance testing in large-scale genetic studies. Nat Rev Genet. 2014;15(5):335–46.
76. Johnson JL, Abecasis GR. GAS power calculator: web-based power calculator for genetic association studies. BioRxiv; 2017.
77. Manolio TA, Brooks LD, Collins FS. A HapMap harvest of insights into the genetics of common disease. J Clin Invest. 2008;118(5):1590–605.
78. International Human Genome Sequencing Consortium. Finishing the euchromatic sequence of the human genome. Nature. 2004;431(7011):931–45.
79. Pearson TA, Manolio TA. How to interpret a genome-wide association study. JAMA. 2008; 299(11):1335–44.
80. Menzel S, et al. A QTL influencing F cell production maps to a gene encoding a zinc-finger protein on chromosome 2p15. Nat Genet. 2007;39(10):1197–9.
81. Uda M, et al. Genome-wide association study shows BCL11A associated with persistent fetal hemoglobin and amelioration of the phenotype of beta-thalassemia. Proc Natl Acad Sci U S A. 2008;105(5):1620–5.
82. Jostins L, et al. Host-microbe interactions have shaped the genetic architecture of inflammatory bowel disease. Nature. 2012;491(7422):119–24.
83. Bulik-Sullivan B, et al. An atlas of genetic correlations across human diseases and traits. Nat Genet. 2015;47(11):1236–41.
84. Eichler EE, et al. Missing heritability and strategies for finding the underlying causes of complex disease. Nat Rev Genet. 2010;11(6):446–50.
85. Yang J, et al. Common SNPs explain a large proportion of the heritability for human height. Nat Genet. 2010;42(7):565–9.
86. Londin E, et al. Use of linkage analysis, genome-wide association studies, and next-generation sequencing in the identification of disease-causing mutations. Methods Mol Biol. 2013; 1015:127–46.
87. Harrow J, et al. GENCODE: the reference human genome annotation for The ENCODE Project. Genome Res. 2012;22(9):1760–74.
88. Venter JC, Smith HO, Adams MD. The sequence of the human genome. Clin Chem. 2015;61(9):1207–8.

89. Gusella JF, et al. A polymorphic DNA marker genetically linked to Huntington's disease. Nature. 1983;306(5940):234–8.
90. Riordan JR, et al. Identification of the cystic fibrosis gene: cloning and characterization of complementary DNA. Science. 1989;245(4922):1066–73.
91. Rao AT, Degnan AJ, Levy LM. Genetics of Alzheimer Disease. AJNR Am J Neuroradiol. 2014;35:457–8.
92. Sanger F, Nicklen S, Coulson AR. DNA sequencing with chain-terminating inhibitors. Proc Natl Acad Sci U S A. 1977;74(12):5463–7.
93. Bumgarner R. Overview of DNA microarrays: types, applications, and their future. Curr Protoc Mol Biol. 2013;Chapter 22:Unit 22.1.
94. van Dijk EL, et al. Ten years of next-generation sequencing technology. Trends Genet. 2014;30(9):418–26.
95. Metzker ML. Sequencing technologies—the next generation. Nat Rev Genet. 2010;11(1):31–46.
96. International HapMap Consortium. The International HapMap Project. Nature. 2003;426(6968):789–96.
97. http://www.internationalgenome.org/.
98. Hardy J, Singleton A. Genomewide association studies and human disease. N Engl J Med. 2009;360(17):1759–68.
99. Van Deerlin VM, et al. Common variants at 7p21 are associated with frontotemporal lobar degeneration with TDP-43 inclusions. Nat Genet. 2010;42(3):234–9.
100. Ferrari R, et al. Frontotemporal dementia and its subtypes: a genome-wide association study. Lancet Neurol. 2014;13(7):686–99.
101. Ferrari R, et al. A genome-wide screening and SNPs-to-genes approach to identify novel genetic risk factors associated with frontotemporal dementia. Neurobiol Aging. 2015;36(10):2904 e13–26.
102. Finch N, et al. TMEM106B regulates progranulin levels and the penetrance of FTLD in GRN mutation carriers. Neurology. 2011;76(5):467–74.
103. Cruchaga C, et al. Association of TMEM106B gene polymorphism with age at onset in granulin mutation carriers and plasma granulin protein levels. Arch Neurol. 2011;68(5):581–6.
104. van der Zee J, et al. TMEM106B is associated with frontotemporal lobar degeneration in a clinically diagnosed patient cohort. Brain. 2011;134(Pt 3):808–15.
105. Rollinson S, et al. Frontotemporal lobar degeneration genome wide association study replication confirms a risk locus shared with amyotrophic lateral sclerosis. Neurobiol Aging. 2011;32(4):758 e1–7.
106. Vass R, et al. Risk genotypes at TMEM106B are associated with cognitive impairment in amyotrophic lateral sclerosis. Acta Neuropathol. 2011;121(3):373–80.
107. van Blitterswijk M, et al. TMEM106B protects C9ORF72 expansion carriers against frontotemporal dementia. Acta Neuropathol. 2014;127(3):397–406.
108. Gallagher MD, et al. TMEM106B is a genetic modifier of frontotemporal lobar degeneration with C9orf72 hexanucleotide repeat expansions. Acta Neuropathol. 2014;127(3):407–18.
109. Lattante S, et al. Defining the association of TMEM106B variants among frontotemporal lobar degeneration patients with GRN mutations and C9orf72 repeat expansions. Neurobiol Aging. 2014;35(11):2658 e1–5.
110. Hernandez I, et al. Association of TMEM106B rs1990622 marker and frontotemporal dementia: evidence for a recessive effect and meta-analysis. J Alzheimers Dis. 2015;43(1):325–34.
111. Yu L, et al. The TMEM106B locus and TDP-43 pathology in older persons without FTLD. Neurology. 2015;84(9):927–34.
112. Brady OA, et al. The frontotemporal lobar degeneration risk factor, TMEM106B, regulates lysosomal morphology and function. Hum Mol Genet. 2013;22(4):685–95.
113. Nicholson AM, et al. TMEM106B p.T185S regulates TMEM106B protein levels: implications for frontotemporal dementia. J Neurochem. 2013;126(6):781–91.
114. Lang CM, et al. Membrane orientation and subcellular localization of transmembrane protein 106B (TMEM106B), a major risk factor for frontotemporal lobar degeneration. J Biol Chem. 2012;287(23):19355–65.

115. Busch JI, et al. Expression of TMEM106B, the frontotemporal lobar degeneration-associated protein, in normal and diseased human brain. Acta Neuropathol Commun. 2013;1:36.
116. Busch JI, et al. Increased expression of the frontotemporal dementia risk factor TMEM106B causes C9orf72-dependent alterations in lysosomes. Hum Mol Genet. 2016;25(13):2681–97.
117. Stagi M, et al. Lysosome size, motility and stress response regulated by fronto-temporal dementia modifier TMEM106B. Mol Cell Neurosci. 2014;61:226–40.
118. Sardiello M, et al. A gene network regulating lysosomal biogenesis and function. Science. 2009;325(5939):473–7.
119. Schwenk BM, et al. The FTLD risk factor TMEM106B and MAP6 control dendritic trafficking of lysosomes. EMBO J. 2014;33(5):450–67.
120. Klein ZA, et al. Loss of TMEM106B ameliorates lysosomal and frontotemporal dementia-related phenotypes in progranulin-deficient mice. Neuron. 2017;95(2):281–296 e6.
121. Tan CC, et al. Association of frontotemporal dementia GWAS loci with late-onset Alzheimer's disease in a northern Han Chinese population. J Alzheimers Dis. 2016;52(1):43–50.
122. Yang X, et al. HLA-DRA/HLA-DRB5 polymorphism affects risk of sporadic ALS and survival in a southwest Chinese cohort. J Neurol Sci. 2017;373:124–8.
123. Salvi E, et al. Genomewide association study using a high-density single nucleotide polymorphism array and case-control design identifies a novel essential hypertension susceptibility locus in the promoter region of endothelial NO synthase. Hypertension. 2012;59(2):248–55.
124. Jager D, et al. Serological cloning of a melanocyte rab guanosine 5′-triphosphate-binding protein and a chromosome condensation protein from a melanoma complementary DNA library. Cancer Res. 2000;60(13):3584–91.
125. Bultema JJ, et al. BLOC-2, AP-3, and AP-1 proteins function in concert with Rab38 and Rab32 proteins to mediate protein trafficking to lysosome-related organelles. J Biol Chem. 2012;287(23):19550–63.
126. Wasmeier C, et al. Rab38 and Rab32 control post-Golgi trafficking of melanogenic enzymes. J Cell Biol. 2006;175(2):271–81.
127. Seto S, Tsujimura K, Koide Y. Rab GTPases regulating phagosome maturation are differentially recruited to mycobacterial phagosomes. Traffic. 2011;12(4):407–20.
128. Fukuda M. Multiple roles of VARP in endosomal trafficking: rabs, retromer components and R-SNARE VAMP7 meet on VARP. Traffic. 2016;17(7):709–19.
129. McGeer PL, et al. Reactive microglia are positive for HLA-DR in the substantia nigra of Parkinson's and Alzheimer's disease brains. Neurology. 1988;38(8):1285–91.
130. Valentonyte R, et al. Sarcoidosis is associated with a truncating splice site mutation in BTNL2. Nat Genet. 2005;37(4):357–64.
131. Amor S, Woodroofe N. Review series on immune responses in neurodegenerative diseases: innate and adaptive immune responses in neurodegeneration and repair. Immunology. 2014;141(3):287–91.
132. Safieh-Garabedian B, Mayasi Y, Saade NE. Targeting neuroinflammation for therapeutic intervention in neurodegenerative pathologies: a role for the peptide analogue of thymulin (PAT). Expert Opin Ther Targets. 2012;16(11):1065–73.
133. Mercer TR, Dinger ME, Mattick JS. Long non-coding RNAs: insights into functions. Nat Rev Genet. 2009;10(3):155–9.
134. Staples CJ, et al. The centriolar satellite protein Cep131 is important for genome stability. J Cell Sci. 2012;125(Pt 20):4770–9.
135. Borner GH, et al. Multivariate proteomic profiling identifies novel accessory proteins of coated vesicles. J Cell Biol. 2012;197(1):141–60.
136. Mikami T, et al. Radical fringe negatively modulates Notch signaling in postmitotic neurons of the rat brain. Brain Res Mol Brain Res. 2001;86(1–2):138–44.
137. Baker SJ, et al. Characterization of an alternatively spliced AATYK mRNA: expression pattern of AATYK in the brain and neuronal cells. Oncogene. 2001;20(9):1015–21.
138. Takano T, et al. LMTK1/AATYK1 is a novel regulator of axonal outgrowth that acts via Rab11 in a Cdk5-dependent manner. J Neurosci. 2012;32(19):6587–99.
139. Lamparter D, et al. Fast and rigorous computation of gene and pathway scores from SNP-based summary statistics. PLoS Comput Biol. 2016;12(1):e1004714.

140. de Leeuw CA, et al. MAGMA: generalized gene-set analysis of GWAS data. PLoS Comput Biol. 2015;11(4):e1004219.
141. Marigorta UM, et al. Transcriptional risk scores link GWAS to eQTLs and predict complications in Crohn's disease. Nat Genet. 2017;49:1517–21.
142. Zhu Z, et al. Integration of summary data from GWAS and eQTL studies predicts complex trait gene targets. Nat Genet. 2016;48(5):481–7.
143. Hasin Y, Seldin M, Lusis A. Multi-omics approaches to disease. Genome Biol. 2017;18(1):83.
144. Manzoni C, et al. Genome, transcriptome and proteome: the rise of omics data and their integration in biomedical sciences. Brief Bioinform. 2016.
145. Furlong LI. Human diseases through the lens of network biology. Trends Genet. 2013;29(3):150–9.
146. Ferrari R, et al. Frontotemporal dementia: insights into the biological underpinnings of disease through gene co-expression network analysis. Mol Neurodegener. 2016;11:21.
147. Ferrari R, et al. Weighted protein interaction network analysis of frontotemporal dementia. J Proteome Res. 2017;16(2):999–1013.

Alzheimer's Disease and Frontotemporal Lobar Degeneration: Mouse Models

10

Lars M. Ittner, Wei S. Lee, Kristie Stefanoska, Prita R. Asih, and Yazi D. Ke

Abstract

Genetically modified mouse models have been instrumental in deciphering pathomechanisms in a large variety of human conditions. Accordingly, transgenic and knockout mice have contributed to understanding neurodegenerative processes in Alzheimer's disease (AD) and frontotemporal lobar degeneration (FTLD). While initial models for AD and FTLD based on mutations in APP and tau have been generated more than a decade ago, identification of novel genes involved in disease has markedly increased the spectrum of available FTLD mouse models. This chapter provides an overview of APP and tau-based mouse models of AD and FTLD and how these models have advanced our understanding

L.M. Ittner (✉)
Dementia Research Unit, School of Medical Sciences, The University of New South Wales, Sydney, NSW, Australia

Transgenic Animal Unit, School of Medical Sciences, The University of New South Wales, Sydney, NSW, Australia

Neuroscience Research Australia, Sydney, NSW, Australia
e-mail: l.ittner@unsw.edu.au

W.S. Lee
Motor Neuron Disease Unit, School of Medical Sciences, The University of New South Wales, Sydney, NSW, Australia

Dementia Research Unit, School of Medical Sciences, The University of New South Wales, Sydney, NSW, Australia

K. Stefanoska • P.R. Asih
Dementia Research Unit, School of Medical Sciences, The University of New South Wales, Sydney, NSW, Australia

Y.D. Ke
Motor Neuron Disease Unit, School of Medical Sciences, The University of New South Wales, Sydney, NSW, Australia

© Springer International Publishing AG 2018
D. Galimberti, E. Scarpini (eds.), *Neurodegenerative Diseases*,
https://doi.org/10.1007/978-3-319-72938-1_10

of disease mechanisms as well as discusses more recent FTLD models of novel disease genes.

Keywords

Mouse model · APP · Tau · TDP-43 · FUS

Different Methods to Genetically Modify Mice

Transgenesis techniques to generate mouse models of disease rely on both gene transfer methods and methods to manipulate the early mouse embryo [1]. To date, the most commonly used technique involves microinjection of DNA constructs into the pronucleus of a developing zygote, leading to random integration of a transgene into the endogenous DNA [2]. The resulting "transgenic animals" have the foreign gene(s) stably incorporated into their genome through human intervention. This integrated recombinant double-stranded DNA is called a "transgene" and commonly drives overexpression of the integrated gene, using either ubiquitous or cell-specific promoters. An example of a frequently used promoter for transgene expression limited to neurons is the murine Thy1.2 promoter.

Over time, the development of more sophisticated models has allowed for better control of transgene expression, both temporally and spatially. This includes both inducible and conditional mouse models. Inducible mouse models enable the study of transgene expression in a strictly regulated and timely manner, whereby transgene expression can be induced by either the presence or absence of a drug, in a dose-dependent manner. This allows researchers to overcome some of the problems associated with constitutive transgene expression, such as embryonic lethality. The most frequently used inducible promoter for transgene expression in animals is still the tetracycline-responsive element system that allows gene expression to be switched on or off, depending on the genetic variant of the transactivator expressed and the delivery of doxycycline to the animals [3]. Conditional models involve the generation of mice with altered gene expression in a cell-specific manner, through the expression of recombinase enzymes, which are under the control of a selected promoter that can remove, invert, or translocate DNA segments to regulate gene expression. In contrast to inducible systems, the conditional gene recombination is an absolute event and cannot be reversed, thereby allowing the induction of gene expression, which cannot be switched off again.

Site-specific manipulation of the genome (gene targeting) allows for the disruption of a specific gene (knockout approach) or the insertion of a transgene in a defined locus (knockin approach). For a long time, gene targeting has relied on the use of homologous recombination and embryonic stem cells, which made the process of generating knockout/knockin animals a costly and time-consuming process. Furthermore, the limited availability of embryonic stem cells further hampers the use of this technology in mice. In recent years, gene targeting for the generation of knockout/knockin mice (and other species) has seen a major revolution. While the introduction of engineered nucleases, such as zinc finger nucleases (ZfN) [4] or transcription activator-like effector nucleases (TALEN) [5], provided a first glimpse at the possibilities of direct genome editing, it was the introduction of the clustered regularly interspaced short palindromic repeats (CRISPR)/CRISPR-associated protein 9 (Cas9) system that transformed the generation of knockout/knockin mice [6–8]. The CRISPR/Cas9 technology enables investigators to manipulate virtually any gene in a diverse range of cell types and organisms with extreme precision (single base pair) within a very short time. The first CRISPR/Cas9 generated models of AD/FTLD have recently been introduced [9], and many more are expected to emerge in years to come. Targeted transgenesis, used either for stable overexpression of a transgene, or for disruption of endogenous genes, ultimately remains the most powerful tool to understand the mechanisms underlying physiological processes, and their pathological counterparts.

Mouse Models of Alzheimer's Disease (AD)

The past two decades have seen the generation of a large number of transgenic mouse models of AD, with a focus on amyloid-β (Aβ)-forming models. These have assisted in a large number of studies investigating mechanisms underlying neuronal dysfunction and neurodegeneration in AD, as well as in developing and testing novel treatments. Aβ-forming transgenic mouse models have been extensively reviewed before [e.g., [10]]. Therefore, this part of the chapter will provide a general overview and highlight only some discoveries made using AD mouse models.

Amyloid-β Precursor Protein (APP) Models

Intensive efforts have been made to develop transgenic mouse models that recapitulate the pathology and symptoms of AD over the past decades. Overexpression of human non-mutant APP did not result in plaque formation and memory deficits. It was the identification of pathogenic mutations in APP, in familial cases of AD, that paved the way for generating the first disease models [11]. Since then, expression of human mutant APP reproduced Aβ plaque pathology in a large number of transgenic mouse models [10]. In most models, expression of mutant APP results in the production of Aβ throughout the brain with plaque formation, affecting memory performance of mice in different test paradigms, such as the Morris water maze. APP transgenic models have also been the basis for showing a prion-like transfer of Aβ pathology between APP transgenic mice in a strain-dependent manner [12]. A feature of APP transgenic mice that receives more attention recently is the occurrence of neuronal network aberrations including non-overt (=silent) seizures recorded by electroencephalography (EEG) [13, 14]. Similar EEG abnormalities have been reported in AD [15]. Their early occurrence in APP transgenic mice provided further evidence for Aβ exerting toxicity prior to its deposition.

While initial studies did not report an overt neuronal loss, a limited number of subsequent studies of established lines reported a decrease in numbers of neurons in certain brain areas [16, 17]. However, the absence of pronounced neuronal loss remains a limitation of Aβ-forming APP transgenic mice.

To determine if loss of APP function contributes to the development of AD, APP knockout mice have been generated. However, their phenotypes are rather mild and possibly due to developmental anomalies [18]. Interestingly, early postnatal death of double knockout mice with deletion of APP and APLP2, the latter belonging to the same protein family, suggests a functional overlap between the family members during development [19]. APP-deficient mice have contributed to the understanding of the possible physiological functions of APP, some of which have implications for the disease [20–22].

A new generation of APP transgenic mice was generated by humanization of Aβ (=changing the murine to the human sequence). However, humanization of Aβ alone was not sufficient to produce pathology or deficits in APP knockin mice [23]. It required the inclusion of multiple pathogenic mutations into the humanized Aβ sequence to achieve pathology in APP knockin mice. Accordingly, homozygous APP^{NL-G-F} mice develop Aβ plaques in the absence of neuronal loss already at 2 months of age [23]. While some features previously identified in conventional mutant APP transgenic mice were not reproduced in this mutant APP knockin mouse [24], others were confirmed [25, 26], providing evidence of the value that conventional APP transgenic mice still hold in AD research. Interestingly, mutant APP knockin mice also have reduced survival [25], highlighting the significance of premature mortality established in conventional mutant APP transgenic lines [27, 28]. Newer studies using these mutant APP knockin mice will further elucidate the contribution of these models to the understanding of AD pathogenesis.

In summary, APP transgenic mice have been instrumental in reproducing aspects of AD pathology in vivo and in deciphering the underlying mechanisms in disease. Furthermore, APP transgenic mice are a valuable tool for the development and testing of treatments for AD.

Combinatorial AD models

In an attempt to accelerate Aβ pathology onset and progression and to more closely model the human pathology, mutant APP transgenic mice have been crossed with other gene mutation-harboring mice. For instance, mutations in the presenilin-encoding (*PSEN*) genes altered the activity of the γ (gamma)-secretase complex in which presenilins are part of. Expression of mutant PSEN1 in mice crossed with Aβ-forming APP transgenic mice resulted in accelerated Aβ formation and early onset of behavioral deficits as well as neuronal loss [29, 30]. Interestingly, the effects of mutant PSEN were even more pronounced in the absence of the murine PSEN, achieved by a mutant human *PSEN1* knockin approach [31]. Conversely, reduced β-secretase activity in beta-secretase 1 (BACE)-deficient mice reduced Aβ formation and ameliorated behavioral deficits when crossed on an Aβ-forming APP transgenic strain [32–34], while overexpression of BACE on an APP background increased pathology [35].

Carriers of the *APOE* epsilon 4 (ε 4) allele have a 20-fold increase risk of developing AD, making it the number one risk gene for developing sporadic late-onset AD [36]. In support of a role for apolipoprotein E (ApoE) in Aβ pathology, crossing APP transgenic mice on a ApoE$^{-/-}$ background reduced both Aβ levels and its deposition [37]. Conversely, expressing human APOE4 in APP transgenic mice by viral gene delivery increased pathology [38].

Aβ-forming APP mice were used to provide the first direct in vivo evidence for the amyloid cascade hypothesis that places Aβ upstream of tau pathology and neurodegeneration in the sequence of pathogenic events. Accordingly, crossing of APP transgenic with human mutant tau-expressing mice resulted in increased neurofibrillary tangle (NFT) formation [39]. A similar result has been achieved by injecting synthetic aggregated Aβ1-42 into brains of P301L mutant tau transgenic pR5 mice [40].

The central role of tau in AD development, particularly in mediating neuronal deficits induced by Aβ, has been shown when APP transgenic mice were crossed on a tau-deficient background [41]. This approach prevented premature mortality and behavioral deficits associated with Aβ formation, in the absence of any change in Aβ levels or plaque numbers. In this context, we showed mechanistically that tau mediates Aβ-induced excitotoxicity by controlling Fyn levels at the post-synapse and sensitizing NMDA receptors to hyper-excitation [28]. Importantly, this work provided the first evidence for a non-axonal function of tau in the dendritic compartment of neurons [42], which has since been supported by several other studies [43, 44]. Very recently, we significantly advanced our understanding of the role of post-synaptic tau in Aβ toxicity. Specifically, we showed that the sensitizing function of tau/Fyn is regulated by the p38γ (gamma) mitogen-activated protein kinase (MAPK) [26]. Depletion of p38γ (gamma) in Aβ-forming APP23 mice by crossing them on a p38γ$^{-/-}$ background, exacerbated memory deficits, neuronal network hypersynchronicity, and premature mortality. Conversely, crossing APP23 mice with a transgenic line that overexpressed a constitutive active variant of p38γ (p38γ^{CA}) in neurons, or delivering p38γ^{CA} via adeno-associated viruses, prevented deficits. Interestingly, the effects of p38γ on limiting Aβ toxicity at the post-synapse of neurons were found to be mediated by phosphorylation of tau, specifically at threonine 205. This was the first report of protective tau phosphorylation, challenging the paradigm that tau phosphorylation in AD is purely a disease-promoting mechanism [45]. Interestingly, the reciprocal approach to APPtg/tau$^{-/-}$ mice, the crossing of mutant tau transgenic mice on an *App*-deficient background, exacerbated tau expression-dependent neuropathology and functional deficits [46]; however, the underlying mechanisms remain unclear and further confirmation is required in independent strains.

In another line of research, combinatorial mouse models begin to shed light on a possible role of TAR DNA-binding protein 43 (TDP-43) in AD. The TDP-43 pathology has been previously reported as an age-related comorbidity in late-onset AD brains [47–49]. Furthermore, neuronal deletion of *Tardbp* (encoding *Tdp*-43 in mice) in Aβ-forming APP/PS1 transgenic mice accelerated neurodegeneration and increased toxic Aβ oligomers formation but reduced plaque deposition, suggesting a possible loss of function of TDP-43 in AD [50]. Moreover, overexpression of human TDP-43 in APP/PS1 transgenic mice induced hyperphosphorylated tau pathology and changes to APP trafficking [51]. In addition, depletion of TDP-43 in microglia promoted Aβ clearance but resulted in enhanced synapse loss [52]. Together, this data challenges the theory of TDP-43 pathology being an "innocent" bystander in AD, consistent with an increased likelihood of cognitive deficits in AD with TDP-43 pathology [53].

Taken together, combinatorial approaches using APP transgenic mice together with additional mutant strains have provided exciting new insights into the pathogenesis of AD. Although only a selected number of studies have been presented here, it is reasonable to expect that combinatorial approaches using APP-based AD mouse models will continue to extend our understanding of AD.

Mouse Models of Frontotemporal Lobar Degeneration

Frontotemporal lobar degeneration (FTLD; also referred to as frontotemporal dementia (FTD)) umbrellas a large number of related neurodegenerative conditions with overlapping clinical symptoms. This is paralleled by an increasing number of proteins that have been found to be present in deposits in FTLD brains, as well as the identification of increasing numbers of genes carrying pathogenic mutations, further distinguishing subforms of FTLD [54]. Furthermore, FTLD is part of a disease continuum with amyotrophic lateral sclerosis (ALS), sharing clinical, neuropathological, and genetic features [reviewed in [55]]. This chapter will discuss transgenic mouse models generated by expressing or deleting different genes, with an emphasis on more recent models and mechanistic discoveries. Tau models, some of which have been generated nearly two decades ago, will only be addressed generally with emphasis given to some of the more recent findings in these mice.

Tau Models

Tau deposits in neurons together with the formation of extracellular Aβ plaques are the neuropathologic and diagnostic hallmarks of AD. In contrast, tau forms inclusions in the absence of overt Aβ pathology in human FTLD brains [56]. To model the tau pathology of AD and FTLD in mice, the first transgenic strain was generated to express the longest human isoform of tau without mutations in neurons [57]. These mice presented with accumulation of hyperphosphorylated forms of tau, resembling a pre-tangle state, but they failed to reproduce NFT formation. It took close to five more years, until transgenic expression of human tau carrying a pathogenic FTDP-17 mutation, P301L, achieved NFT formation in vivo [58]. These mice are characterized by severe motor and behavioral deficits, axonal degeneration, and early death, resembling aspects of the human disease. Notably, motor deficits have since been recognized as a major feature of many mutant tau transgenic strains. Since the generation of this first mutant tau-expressing mouse model, many additional lines have been generated that recapitulate different aspects of the human condition [10]. Interestingly, neuronal loss that characterizes the human disease has not been reproduced in the earlier mutant tau transgenic mice, but only, when expressing distinct mutations (N279K [59] or P301S [60, 61]) with conventional neuronal promoters, or particularly high levels of P301L mutant or aggregation-prone truncation variants of human tau using an inducible modified minimal CMV promoter showed pronounced neuronal loss [62, 63]. The latter line in combination with a complementary model that expresses the same truncated tau variant but with inclusion of two aggregation-preventing point mutations (I277P and I308P) forms an excellent in vivo tool to study tau fibril formation and test anti-aggregation drugs [63].

Since tau pathology in human FTLD is not limited to neurons, transgenic mouse model with non-neuronal mutant tau expression has been generated [64, 65]. Interestingly, both expression in astrocytes and in oligodendrocytes resulted in neuronal dysfunction and axonal degeneration, likely due to impairment of neuronal support by glial cells.

More recently, we introduced a novel P301S mutant tau transgenic strain with rapid NFT development and pronounced motor deficits [66], as well as behavioral changes with disinhibition reminiscent of symptoms presented in behavioral variant FTD [67, 68]. Interestingly, these mice revealed neuropathological changes with lesions that stained positive for the neuronal structure protein neurofilament, but negative for tau [66]. Similar lesions were subsequently found in a number of other tau transgenic lines and, more importantly, FTLD with tau (but not TDP-43) pathology [66], suggesting neurofilament lesion formation is a secondary process induced by pathological tau.

Mutant tau transgenic mice have become a highly valuable tool for studying pathomechanisms underlying tau pathology and neurodegeneration in FTLD, but also in AD. Accordingly, transgenic mice were extensively used to investigate the prion-like disease progression hypothesis for tau, which includes release of distinct tau species from diseased neurons that are then taken up by healthy neurons to form seeds for disease propagation [69]. So far, it has been shown that tau pathology can be transferred from a mutant tau transgenic line with NFT formation to a transgenic strain expressing non-mutant human tau with no NFT formation unless inoculated with brain extracts from NFT-forming mice [70] or human patient brains with tau pathology [71] by stereotaxic injection. Notably, different tau strains have been identified that cause strain-specific tau pathology when inoculated into the brains of P301S mutant tau transgenic mice [72, 73]. Furthermore, inducible mutant tau expression limited to the entorhinal cortex led to NFT formation in connected areas of the hippocampus as mice age [74]. Seeding of tau aggregation was reported to be mediated by small tau fibrils, but not by oligomeric tau [75], while another study suggested that disulfide cross-linked tau dimers were responsible for inducing tau pathology [76]. Exosomes harboring misfolded tau may furthermore contribute to the spreading of tau pathology between neurons [77, 78], a process that requires the presence of microglia in the mouse brain [77]. Release of tau from neurons increases with activity [79], as elegantly shown by optogenetic stimulation of primary neurons in culture and hippocampal neurons in the brains of P301L mutant tau transgenic mice, resulting in enhanced tau pathology and neurodegeneration in the latter [80]. Interestingly, the absence of endogenous tau in *Mapt* knockout mice did not prevent propagation of tau pathology, indicating that the presence of tau is not required as a template for prion-like propagation in this model [81]. Nevertheless, decreasing transgenic human tau mRNA with antisense oligonucleotides prevented seeding of tau pathology in reporter cells and P301S mutant tau transgenic mice [82].

Recently, neuronal network aberrations have been reported in mutant tau transgenic mice. Cortical surface EEG recordings in P301S mutant tau transgenic mice revealed altered sleep patterns and a progressive reduction of EEG power that was associated with cortical brain atrophy [83]. Furthermore, surface recordings showed network hypersynchronicity with epileptiform spike activity in inducible A152T mutant tau transgenic mice [84] and loss of EEG power in P301L/R406W tau transgenic mice [85]. In contrast to detailed EEG analysis of APP transgenic mice [14], further studies in tau transgenic lines including recordings from specific brain areas may be required to consolidate these findings.

Transgenic mice expressing non-mutant but truncated variants of tau have provided further insight into the role of tau in AD and FTLD. Accordingly, we have reported mice that express the N-terminal half of tau (aa 1-255), lacking microtubule binding motifs in neurons [28]. While these Δtau mice were phenotypically normal and did not present with tau hyperphosphorylation, aggregation, or NFT formation, they prevent Aβ-induced memory deficits and death mediated by dominant-negative action on endogenous postsynaptic tau [42]. On the other hand, expression of a C-terminal truncation variant of tau, lacking the last 20 aa in neurons and mimicking a caspase 3 cleavage product of tau found in disease, resulted in severe memory deficits and synaptic loss, together with tau aggregation [86]. Caspase 2 cleaves tau at Asp314 to produce a truncated tau species in AD, and reduction of caspase 2 activity or mutation of the cleavage site in tau prevented deficits in P301L tau transgenic mice [87]. Interestingly, low-level expression of a disease-relevant N-terminal truncation of tau comprising aa 187-441, under the control of the human tau promoter induced aggregation and hyperphosphorylation, as well as functional deficits that resembled features of the human disease, despite the transgene not harboring a pathogenic FTLD mutation [88].

While aberrant and increased phosphorylation of tau remains the focus of the majority of studies into tau pathology and disease mechanisms in mutant tau transgenic mice, it is important to note that other secondary modifications of tau may contribute to disease and, therefore, present as possible drug targets. For example, acetylated tau has been identified in AD brains, and mimicking acetylation of tau in transgenic mice induced cognitive deficits [89]. Furthermore, this study showed that reducing acetylation of tau with the compound salsalate improved memory deficits and prevented neuronal loss in P301S mutant tau transgenic mice.

Mutant tau transgenic mice are also regularly used for preclinical drug development and testing. For instance, more recently, several groups have developed vaccination strategies targeting pathological tau, either by active or passive immunization [90–94]. Each of these studies used different mutant tau transgenic mouse lines to show efficacy and safety of this approach, providing the preclinical evidence needed for translation, with first clinical trials using tau-specific antibodies well on the way. With minimal success of Aβ-targeting immunization strategies, it awaits to be shown whether tau-targeted vaccination is similarly efficient in humans as it is in mice. While the mechanisms of anti-tau immunotherapy remain to be completely understood, a recent study suggested that virally expressed single-chain variable fragment antibodies that lack the F_c domain is sufficient to reduce tau pathology in P301S mutant tau transgenic mice [95]. The deletion of the F_c domain from therapeutic antibodies may prevent unwanted brain inflammation. The effector function of the antibodies F_c domain, which mediates microglial uptake of bound tau and subsequent proinflammatory cytokine release, was not required for neuropathology in P301L mutant tau transgenic mice [96]. Similarly, a 2N tau isoform-specific single-chain antibody fragment with assisted brain delivery using scanning ultrasound reduced behavioral deficits and tau phosphorylation [97]. Reduction of tau phosphorylation in tau transgenic mice was furthermore achieved by expressing a DNA vaccine of a B cell epitope of the 18 N-terminal amino acids of tau, resulting in high anti-tau antibody titers [98], illustrating the significant advances made with antibodies, vaccines, and delivery methods for tau-targeted immunotherapy using transgenic mouse models of FTLD and AD.

Apart from antibodies to tau, mutant tau transgenic mice have been used to determine the effects of a variety of small molecules on different aspects of tau pathology [61, 99, 100]. Some recent examples include the beneficial effects of the antioxidants lycopene and vitamin E on memory deficits and tau pathology in P301L mutant tau transgenic mice [101], the testing of the novel tau anti-aggregation active compound altenusin in P301S tau transgenic mice [102], the prevention of neurodegeneration with the new microtubule stabilizer dictyostatin in P301S mutant tau transgenic mice [103], and restoring memory function and normalization of synaptic transmission in ΔK280 mutant tau transgenic mice with the adenosine A_1 receptor antagonist rolofylline [104]. The tau anti-aggregation compound anle138b improved neuropathology, survival, and cognition of P301S mutant tau transgenic mice [105]. Similarly, mild chronic neuroinflammation induced by systemic delivery of lipopolysaccharide enhanced autophagy reduced tau phosphorylation and cognitive deficits in P301S tau transgenic mice [106]. The glucagon-like peptide 1 receptor agonist liraglutide ameliorated neurological deficits of P301L mutant tau transgenic mice and significantly reduced levels of tau phosphorylation [107]. And lastly, activating cAMP-protein kinase A with rolipram ameliorates tau pathology and improves cognitive deficits in P301L mutant tau transgenic mice [108].

Taken together, the generation of mutant tau transgenic mice provides in vivo evidence that pathogenic FTLD mutations accelerate tau aggregate formation and deposition and drive neuronal dysfunction and loss. Furthermore, mutant tau transgenic mice are important tools for studying pathomechanisms in vivo and to develop and test new therapeutic approaches. Notably, while pathogenic mutations expressed in these lines originate from FTLD patients, tau transgenic mice are also valuable for studying tau-related aspects of AD, given the overlapping features of tau pathology in AD and FTLD.

TDP-43 Models

In 2006, Neumann and colleagues identified TDP-43 as the major component of ubiquitin-positive deposits in FTLD [109]. Moreover, they showed that similar deposits in ALS (also referred to as Lou Gehrig's disease or motor neuron disease (MND)) are also made up of TDP-43. Interestingly, TDP-43-positive lesions are also found in approximately half of AD brains [53], possibly extending its pathomechanistic role beyond FTLD/ALS. TDP-43 is a nuclear protein with two RNA/DNA binding motifs. Consistent with these domains, TDP-43 is involved in RNA/DNA-related processes in cells, including RNA trafficking, alternative splicing, and promoter binding [110]. In disease, TDP-43 accumulates in the cytoplasm and undergoes secondary modifications, such as truncation, phosphorylation, and ubiquitination, eventually leading to the formation of aggregates [111].

Similar to tau transgenic mice, the identification of mutations in the TDP-43-encoding *TARDBP* gene has paved the way for the generation of a number of transgenic mouse models with TDP-43 expression. Furthermore, non-disease mutants of TDP-43 with deletion of specific functional domains from the protein have been expressed in mice.

The first TDP-43 mouse model published in 2009 expressed human TDP-43 carrying the A315T mutation under the murine prion protein promoter to generate the Prp-TDP-43^{A315T} mice [112]. These mice have an approximate threefold expression over endogenous TDP-43 with highest expression present in the brain and spinal cord. Ubiquitination of proteins in layer V neurons of the cortex concomitantly occurred with loss of nuclear staining of TDP-43 in selective neurons in these mice. Reactive gliosis was also present in this region of degenerating neurons. It was later shown that reduced survival and wasting, initially attributed to an ALS-like phenotype, were indeed caused by gastrointestinal complications with gut paralysis due to aberrant TDP-43^{A315T} expression in the mesenteric plexus [113–115], highlighting a potential problem originating from aberrant activity of transgenic promoters. Treating the gastrointestinal problems of TDP-43 transgenic mice prolonged their survival, which allowed sufficient time for TDP-43 pathology to develop in the central nervous system [116].

This initial TDP-43 transgenic line [112] was followed by several new models generated over the last few years [117–124]. Wils and colleagues expressed nonmutant human TDP-43 under the neuronal murine Thy1 promoter to generate the TDP-43^{WT} lines TAR4 and TAR6 [117]. Hemizygous TAR4 and TAR6 have 2.8- and 1.9-fold and homozygous TAR4/4 and TAR6/6 have 5.1- and 3.8-fold expression over endogenous TDP-43. These mice have nuclear and cytoplasmic inclusions in cortical layer V neurons that are ubiquitinated and phosphorylated as well as marked astrogliosis. The limited neuronal loss observed in these mice correlated with the expression levels of TDP-43. In addition, homozygous TAR4 have an accumulation of cytoplasmic full-length TDP-43 as well as the 25 and 35 kDa C-terminal fragments. Phenotypically, these mice exhibit complex motor impairments, with hind limb clasping, reduced footstep length, reduced motor performance on the Rota Rod as well as reduced survival rate with disease onset and severity dependent on TDP-43 expression levels.

Xu and colleagues expressed non-mutant human TDP-43 under the murine prion protein promoter to generate the TDP-43$_{PrP}$ with a 1.9–2.5-fold expression over endogenous TDP-43 [118]. Increased human TDP-43 mRNA levels were observed with a concomitant decrease in mouse TDP-43 mRNA levels. These mice produce ~25kDa C-terminal TDP-43 fragments, which are urea insoluble, as well as phosphorylated and ubiquitinated cytoplasmic inclusions, reactive gliosis, and argyrophilic degenerating neurites and neurons in the spinal cord. Interestingly, these mice also have abnormal clustering and degeneration of mitochondria in their spinal cord neurons. TDP-43$_{PrP}$ mice displayed lower body weights compared to wild-type littermates at 14 days, together with hind limb clasping, body tremors, and a "swimming" gait at 21 days. Their survival was limited as they die between 1 and 2 months of age.

Swarup and colleagues generated three TDP-43 transgenic mice (non-mutant human TDP-43, TDP-43^{A315T}, and TDP-43^{G348C}) from DNA subcloned from *TARDBP* bacterial artificial chromosomes containing the endogenous ~4 kb promoter [121]. These mice present with an approximately threefold overexpression of transgenic TDP-43 over the endogenous protein. Significantly more ~25 and 35 kDa C-terminal fragments were observed in TDP-43^{A315T} and TDP-43^{G348C} compared to non-mutant TDP-43-expressing mice. Ubiquitination of cytoplasmic TDP-43 was observed only in the mutant TDP-43 lines. Abnormal aggregates containing peripherin and neurofilament proteins were also present in TDP-43^{G348C} mice. In addition, gliosis and neuroinflammation were observed in all lines. Furthermore, all lines presented with cognitive and motor deficits in the passive avoidance test, Barnes maze test, and Rota Rod at 7–10 months with these impairments being most severe in the TDP-43^{G348C} line. Interestingly, they revealed that there is a significant increase of GFAP promoter activity or astrogliosis before the onset of behavioral impairments.

Igaz and colleagues generated transgenic mice with inducible overexpression of either non-mutant human TDP-43 (hTDP-43 WT) or human TDP-43 with mutated nuclear localization signal (hTDP-43-ΔNLS) [119]. Mutation of the NLS prevents TDP-43 from entering the nucleus, and hence it accumulates in the cytoplasm [125]. Neuronal expression was achieved by using a CaMK2α promoter to drive tet-off rTA and a tetracycline responsive promoter to drive hTDP-43 expression. hTDP-43 WT mice had an eight- to ninefold expression over endogenous TDP-43 and hTDP-43-ΔNLS mice 0.4–1.7-fold, respectively. Both models present with UREA-soluble TDP-43 with no concomitant presence of C-terminal fragments. In addition, ubiquitinated and phosphorylated TDP-43 aggregates were found to be present in hTDP-43-ΔNLS mice. Significant neuronal loss was observed in the dentate gyrus of both lines with the hTDP-43-ΔNLS mice having more acute and severe dentate gyrus degeneration. The presence of axonal loss and gliosis of the corticospinal tract of hTDP-43-ΔNLS mice occurs in a time-dependent manner relative to the development of motor deficits. Interestingly, motor and memory, but not social deficits, which all developed rapidly after induction of transgenic hTDP-43-ΔNLS expression at weaning were reversible before overt neurodegeneration prevented the improvements [126]. Conversely, the inducible hTDP-43 WT line developed social and memory deficits in the absence of motor problems [127]. Inducible expression of non-mutant hTDP-43 in an independent line produced limited TDP-43 pathology, neurodegeneration, and survival when expressed throughout development, with FTLD-like TDP-43 neuropathology without mortality when transgene expression was initiated later in life [128]. A similar difference in phenotypic presentation was observed between two inducible TDP-43^{M337V} models that differed in expression levels [129]; high transgene levels were associated with shorter survival and neurodegeneration in the absence of significant TDP-43 pathology, while lower expression levels did not affect survival, but showed accumulation and fragmentation of TDP-43 in the absence of neurodegeneration. This suggests that both the choice of mutations and levels contribute to the phenotypes of TDP-43 mice. Using a neurofilament promoter-driven inducer line to drive neuronal hTDP-43-ΔNLS expression resulted in a more severe phenotype [123]; cytoplasmic TDP-43 pathology was accompanied by rapid brain atrophy, progressive motor neuron loss with muscle wasting, and eventually fatal motor deficits, resembling clinical features of ALS. Again, suppression of transgenic hTDP-43-ΔNLS expression reverted neuropathological changes and functional deficits and prolonged survival. In parallel, we introduced an iTDP-43^{A315T} inducible model with A315T mutant TDP-43 expression driven by a Thy1.2 inducer line [122] that presented with ALS-like motor deficits as well as memory impairments and behavioral changes reminiscent of FTLD. Remarkably, suppression of transgenic TDP-43^{A315T} expression for only one week recovered most of the functional impairments, despite overt degeneration, suggesting both a prominent role of pathological soluble TDP-43 species and significant compensatory capacity of neurons once TDP-43 is removed. This reversal of behavioral deficits in several inducible TDP-43 models holds promise for efficacy of future TDP-43-reducing therapies. Interestingly, we found a selective and progressive loss of cortical layer V neurons, with layer II/II neurons spared despite pronounced transgenic TDP-43 expression [122]. This selective vulnerability may be a result of neuronal disinhibition, supported by the discovery of hyperactive somatostatin-positive interneurons that disinhibited layer V neurons in TDP-43^{A315T} transgenic mice [130]. Handley and colleagues recently showed in an elegant study using neuronal YFP transgenic mice crossed with TDP-43^{A315T} model that synaptic dysfunction proceeds degeneration of layer V neurons due to pathological TDP-43 [131]. Selective vulnerability was recently also reported in inducible TDP-43-ΔNLS mice, where hypoglossal and fast fatigable spinal cord motor neurons were rapidly lost, while slow spinal cord motor neurons and those of other cranial nerve nuclei were spared despite transgene expression [124]. Taken together, mouse models provided first insight into mechanisms underlying selective vulnerability of distinct neuronal population to pathological TDP-43. Further studies utilizing these mouse models will likely contribute to understanding the processes mediating selective vulnerability/resistance, thereby revealing new therapeutic targets.

TDP-43 accumulation in mitochondria may contribute to neuronal dysfunction and degeneration in FTLD, and pathogenic TDP-43 mutations favor its import into mitochondria [132]. Accordingly, a brain permeable peptide that blocks import of TDP-43 into mitochondria ameliorated their impaired function, motor deficits, and muscle atrophy in TDP-43^{A315T} transgenic mice [132]. This was recently confirmed in a TDP-43^{M337V} transgenic line by the same group [133]. Furthermore, the tyrosine kinase inhibitor nilotinib that reduced cell death in wild-type TDP-43 transgenic mice [134] reversed mitochondrial impairment in these animals [135]. However, a recent study did not find bioenergetics defects of mitochondria in TDP-43^{A315T} transgenic mice [136]. This warrants for further detailed studies into the role of mitochondria in neuronal dysfunction and degeneration associated with TDP-43 pathology.

Transgenic mice expressing a 25 kDa truncation product of full-length TDP-43 that is found in FTLD and ALS brains, which were earlier described to have cognitive deficits due to accumulation of soluble TDP-43 fragments [137], showed moderate and more severe memory and motor deficits in hetero- and homozygous aged mice, respectively [138]. These defects were associated with reduced proteasome and autophagy activity. Inducible expression of a C-terminal truncation fragment of TDP-43 (aa 208-414) caused progressive hippocampal loss, astrogliosis, and TDP-43 phosphorylation, which were mitigated upon suppression of transgene expression [139]. Together, these models suggest that TDP-43 fragments play a pathogenic role in disease, rather than being surrogate events.

Combination of non-mutant and Q331K mutant transgenic mice that develop no overt or slowly progressive (nonlethal) motor deficits, respectively, resulted in a rapid and fatal neurodegenerative phenotype with FTLD/ALS-like neuropathology, including nuclear clearance of endogenous TDP-43 from spinal cord motor neurons [140]. This model suggests that mutant, aggregation-prone TDP-43 recruits non-mutant TDP-43 into insolubility. TDP-43 transgenic models commonly presented with microgliosis and astrogliosis, which may contribute to neuronal pathology in the mice. Supporting a role of inflammation in neuronal pathology, systemic lipopolysaccharide administration exacerbated TDP-43 deposition and mislocalization including in TDP-43^{A315T} transgenic mice [141].

While exosomes may contribute to the propagation of pathological TDP-43 between neurons (similar to tau pathology as outlined above), inhibition of exosome formation induced TDP-43 aggregation in cultured cells and exacerbated the phenotype of TDP-43^{A315T} mice, suggesting exosome as a way for neurons to get rid of pathological TDP-43 [142]. There may be further overlaps between tau and TDP-43, as suggested by the finding of cytoplasmic accumulation of phosphorylated TDP-43 in two different tau transgenic mouse models (rTg4510 and JNPL3), while there was no such pathology in non-tau neurodegenerative models (Aβ, α-synuclein, or huntingtin) [143]. In addition, TDP-43 may directly control tau expression by mediating instability of its mRNA, with increased tau levels in the brains of TDP-43^{M337V} transgenic mice [144].

Since the abnormal localization of TDP-43 in disease means that the protein is depleted from the nucleus, TDP-43 might not be able to execute its normal functions (=loss of function). To test this in vivo, Kraemer and colleagues employed a gene trap insertion strategy to generate mice lacking TDP-43 [145]. Heterozygous mice are viable in contrast to homozygous mice, which are embryonically lethal. Heterozygous (Tardbp$^{+/-}$) mice have reduced grip strength with no reportable differences in pathology observed.

TDP-43 transgenic mice have more recently been used to test novel therapeutic approaches and compounds. For example, reducing the ALS risk gene ataxin-2 in mutant TDP-43 transgenic mice either by crossing them on an ataxin-2-deficient background, or by using ASOs improved survival and reduced TDP-43 pathology, providing a novel therapeutic approach [146]. Similarly, overexpression of survival motor neuron (SMN) in neurons prolonged survival and delayed symptom onset in TDP-43^{A315T} transgenic mice [147]. Treating transgenic mice that expressed a fragment of TDP-43 together with the pathogenic A315T mutation with a herbal extract of *Withania somnifera* improved motor and cognitive functions, potentially by targeting NF-κB signaling [148].

Taken together, TDP-43 transgenic mice have recapitulated both neuropathological and clinical features of FTLD and ALS, provided insight into underlying pathomechanisms, and are instrumental in the development of novel therapeutic approaches.

Progranulin (PGRN) Models

Mutations in PGRN have been shown to cause tau-negative, ubiquitin- and TDP-43-positive FTLD [149, 150]. The majority of these mutations are known to cause mRNA instability (resulting in degradation), while other mutations can cause loss of the entire mutant allele [150]; prematurely truncated protein [150]; or result in the generation of mutant PGRN protein that cannot be secreted efficiently [151] or appropriately cleaved [152]. Therefore, through a variety of mechanisms these mutations all result in either reduced PGRN levels or loss of PGRN function. It is for this reason that *Pgrn* knockout mice have been used to study this particular disorder.

A variety of *PGRN* knockout mice have been generated [153–157], and with the exception of one report [158], all of these strains produce knockout offsprings at an expected Mendelian ratio, suggesting that loss of PGRN does not impair embryonic development and/or survival. One common feature of all strains is that aged, homozygote mice develop severe astrogliosis and microgliosis that increases with age (generally first detected around 12 months of age). Hence, neuroinflammation may play a role in the disease process. Interestingly, homozygous *PGRN* knockout mice react less efficiently and with more severe inflammation to bacterial listeria infections [154]; and both *PGRN*-deficient microglia and macrophages are more cytotoxic to cultured neurons [154, 156]. In addition to this, hippocampal slices from homozygous *PGRN* knockout mice show greater neuronal sensitivity to glucose and oxygen starvation [154]. This suggests that FTLD may arise from a combination of deregulated inflammation and increased neuronal vulnerability to certain stressors.

In all but one strain [155], homozygous *PGRN* knockout mice have been found to display significantly more ubiquitinated structures in various brain regions by as early as 7 months (ranging from 7 to 18 months), which increase with age. In support of a compromised ubiquitin-proteasome system, increased p62 and cathepsin D (markers of autophagy and lysosomes) were found in addition to increased neuronal ubiquitin in *PGRN* knockout mice [157]. These pathological changes are common features of FTLD-TDP but are also associated with aging. Furthermore, in three of the *PGRN* knockout strains, levels of lipofuscin, a marker of cellular aging, were significantly increased (throughout the brain and also in the liver in one strain) by as early as 8 months. Hence, *PGRN* knockout mice may undergo accelerated aging, thereby potentially contributing to the disease process. Interestingly, levels of PGRN progressively increased in the brains of aging wild-type animals, suggesting a role for PGRN in aging [158]. Interestingly, however, no neuronal loss or markers of apoptosis have been observed in any of the strains though some lines have shorter life spans [157, 159].

Although PGRN mutations are associated with TDP-43 neuropathology in humans, it is not clear whether this is also the case in *PGRN* knockout mice. To date, only some pathologically phosphorylated TDP-43 have been identified in brains of two strains [154, 157, 160]. Therefore it remains unclear what role PGRN mutations play in the development of TDP-43 pathology.

The behavioral assessment of different *PGRN* knockout lines produced variable results. This could be the result of variation in genetic background or differences in protocols and equipment used. *PGRN* knockout mice do not have any significant motor impairments (although reduced muscle strength has been reported by Ghoshal and colleagues); however, there have been multiple reports of reduced social engagement and aggression [155, 159, 160] and depression-like behavior and disinhibition [160], which mimics several major behavioral hallmarks of FTLD. In addition, aged *PGRN* knockout mice show reduced performance in the Morris water maze [157, 159, 160] and novel object testing [155], suggesting late-onset learning and memory impairments. Although the mechanism by which PGRN deficiency causes these behavioral phenotypes is unclear, Petkau and colleagues [155] utilized electrophysiological recordings to demonstrate that hippocampal slices from homozygous *PGRN* knockout mice display reduced postsynaptic responsiveness and occasional LTP dysfunction. Furthermore, CA1 pyramidal neurons showed reduced dendritic length and reduced spine density. Therefore, synaptic dysfunction may play a role in the disease process underlying FTLD. Alternatively, increased lysosomal activity markers and cytoplasmic TDP-43 aggregates in neurons were found in the brains of aged *PGRN* knockout mice, suggesting lysosomal dysfunction may contribute to pathology [161].

Although *PGRN* mutations have initially been identified in tau-negative FTLD-TDP-43, *PGRN* mutations have since been found in a range of neurodegenerative conditions, including AD with tau pathology [162, 163]. This is supported by increased tau pathology in P301L tau transgenic mice that were crossed on a

$Pgrn^{+/-}$ background, suggesting a direct connection between granulins and tau pathology [164].

It should be noted that the majority of studies discussed above utilized homozygous PGRN knockout mice, despite the fact that *PGRN* mutations cause haploinsufficiency in humans. For this reason, it is important to highlight some results obtained from heterozygous *PGRN* knockout mice [165]. These mice express approximately 50% less *PGRN* mRNA and protein (and were maintained on two different genetic backgrounds), but unlike homozygous *PGRN* knockout mice, they do not develop any significant astrogliosis, microgliosis, and lipofuscinosis or show any electrophysiological changes, nor do they have any motor impairment or memory and learning impairments. Nevertheless, these animals (regardless of the genetic background) still show social and emotional dysfunction.

In summary, *PGRN* knockout mice recapitulate a number of hallmark features of FTLD-TDP-43, including neuroinflammation, ubiquitinated aggregates, and behavioral impairments. However, the exact role of TDP-43 in this disease and the exact effects of *PGRN* haploinsufficiency versus homozygous deficiency remain to be determined.

Valosin-Containing Protein (VCP) Models

Mutations in the valosin-containing protein (VCP) gene are known to cause the multisystem degenerative disorder called inclusion body myopathy associated with Paget's disease of the bone and frontotemporal dementia (IBMPFD) [166]. Although muscle weakness and myopathy are the most common clinical features of this disorder, approximately 30% of patients also develop language and behavioral impairments typical of FTLD [167]. Furthermore, TDP-43- and ubiquitin-positive inclusions are found in both the brain and muscle of IBMPFD patients. Interestingly, some reports also link VCP mutations to ALS [168, 169]. Over 20 mutations have been identified in *VCP*, all of which are thought to alter the 3D structure of VCP and thereby perturb the interactions between VCP and its various substrates [170]. Substitution of arginine 155 to histidine (R155H) is the mutation most commonly associated with IBMPFD. It is for this reason that the majority of mouse models utilize this particular mutation. Another mutation, A232E, is associated with a particularly severe clinical presentation in humans [166].

To develop an animal model of IBMPFD, a number of groups have generated transgenic mice that express mutant VCP [171–175]. Although these strains all express a similar mutant protein, there are a number of inherent differences amongst the strains. For example, because mouse VCP differs from the human protein by only one amino acid, some groups chose to express human mutant VCP in the mouse model, whereas other models express mutant mouse VCP. Various promoters have been used to generate mice that overexpress the mutant protein exclusively in muscle [172], the brain [171], or ubiquitous expression in all tissues [173], while other groups have generated knockin mice that express mutant VCP at levels similar to that of the endogenous protein [174, 175].

Despite these inherent differences, all mutant VCP mouse strains have been reported to develop VCP-negative, TDP-43-positive, and ubiquitin-positive aggregates. These aggregates develop in regions where the mutant protein is expressed, i.e., the muscle, brain, and spinal cord. In heterozygote animals, these aggregates appear at around 10-15 months in the muscle and the spinal cord and at 14-20 months in the brain, while in homozygous mice [175], TDP-43 aggregates were observed as early as 15 days in the muscle, brain, and spinal cord. In some strains, cytoplasmic and nuclear clearance of TDP-43 was observed, as well as insoluble and high molecular weight TDP-43 species [171, 173, 176]. In one particular strain, TDP-43 aggregates were observed to co-localize with the stress granule marker TIA-1, and overall levels of TIA-1 were increased, suggesting an increased stress response, which could potentially alter mRNA transport and translation. Altered stress granule dynamics and/or altered mRNA metabolism may therefore play a role in the disease processes associated with TDP-43 proteinopathies. Despite the presence of TDP-43 aggregates, none of the strains show any sign of neurodegeneration in the brain [171, 173, 174], although loss of motor neurons in the spinal cord has been reported [176].

Other pathological features commonly observed in these mice include a significant increase in the levels of general protein ubiquitination [171, 172, 175, 176] and upregulation of markers of autophagy [174-176] in the muscle, brain, and spinal cord. Combined with the knowledge that VCP is known to play a role in regulating ubiquitin degradation of a number of proteins, this data suggests that dysfunctional protein degradation and accumulation of ubiquitinated proteins may play a role in the development of this disorder. In addition to this, high molecular weight species of TDP-43 were found to pull down with VCP, suggesting a direct interaction between VCP and high molecular weight TDP-43 isoforms in these mice [171]. One possible explanation for this interaction is that VCP may be trying to direct TDP-43 to the proteasome for degradation and that disruptions to this interaction may cause TDP-43 to accumulate in the cytoplasm and eventually aggregate.

IBMPFD is most commonly characterized by myopathy. In accordance with this, in all the mutant VCP mice strains that express the transgene in muscle tissue, significant pathology was observed. This includes vacuoles, disordered architecture, variation in muscle fiber size, and swollen mitochondria [172-176]. On average, these features were observed at around 6-15 months of age; however, in mice that were bred to homozygosity, muscle abnormalities were already observed after 15 days. Radiographic and biochemical bone deformities consistent with Paget's disease are also commonly observed in IBMPFD. Similar characteristics have been reproduced in the mutant VCP mice, including loss of bone structure, decreased bone density, hypomineralization, and sclerotic lesions at around 13-16 months of age [173-175]. Therefore, these mice recapitulate the wide range of pathological features associated with IBMPFD within the muscle, brain, and bone.

In general, all mutant VCP mouse strains show signs of muscle weakness and reduced Rota Rod performance, which is in accordance with the clinical presentation in human patients [172-175]. Although some reports show weight loss and reduced survival in certain strains [173, 176], particularly in the homozygote mice which only survive 14–21 days [175], this has not been observed in all strains. Interestingly, short survival of homozygous VCP$^{R155H/R155H}$ mice was significantly improved, as were motor deficits when mice were fed a lipid-rich diet [177]. Custer and colleagues reported increased anxiety in these mice in the elevated zero maze and reduced performance in the novel object test, while other strains did not show any memory deficits [173–175]. Rodriguez-Ortiz and colleagues used a neuron-specific promoter to overexpress mutant VCP specifically in the forebrain [171]. These mice showed no difference in swim speed and distance in the Morris water maze, but showed significant impairment in the probe trial, as well as impairment in object recognition testing, indicating learning and memory deficits. Furthermore, higher mutant VCP-expressing mice were shown in these studies to have greater cognitive deficits than lower expressing mice, with both lines showing greater impairment with age, suggesting that neuronal mutant VCP expression impairs cognition in an age- and dose-dependent manner in these mice.

In summary, mutant VCP mice develop muscle and brain pathology as well as bone abnormalities that closely match with what is observed in human IBMPFD patients. In addition, the spinal cord pathology closely matches that observed in human ALS patients. This therefore raises the question whether inclusion body myopathy, Paget's disease, ALS, and FTLD share a common underlying mechanism. Because these mice developed ubiquitin-positive, TDP-43 aggregates and showed re-localization of TDP-43, they can be used not only to study IBMPFD but also the mechanisms underlying the development of TDP-43 pathology in general, particularly the neuron-specific expressing mice.

Charged Multivesicular Body Protein 2B (CMBP2B) Models

Although rare, mutations in the charged multivesicular body protein 2B (*CHMP2B*) gene are associated with familial forms of FTLD that display ubiquitin- and p62-positive inclusions that are negative for tau, FUS, and TDP-43 [178]. All mutations identified have been shown to cause a loss of the C-terminus of CHMP2B; therefore, the disease pathogenesis could be caused by either loss of normal CHMP2B function, or more specifically, loss of the CHMP2B C-terminus. To investigate this in greater depth, Ghazi-Noori and colleagues generated both wild-type (CHMP2B^{wt}) and C-terminally truncated (CHMP2B^{Intron5}) CHMP2B transgenic mice, as well as CHMP2B knockout mice [179]. Initially, both the CHMP2B transgenic and knockout mice showed normal survival curves; however, after 500 days the CHMP2B^{Intron5} mice showed increased mortality. Interestingly, the CHMP2B^{Intron5} mice were shown to develop p62- and ubiquitin-positive inclusions (but TDP-43- and FUS-negative) that were absent in the CHMP2B^{wt} and knockout mice, suggesting that the formation of these inclusions was dependent on the expression of mutant CHMP2B. Since these inclusions were absent in the knockout mice, this suggests that the pathology is not caused by a loss of function but rather a gain of toxic function. These inclusions were found in a number of brain regions and motor neurons in the spinal cord, as early as 6 months and were found abundantly by 18 months of age. In addition to the formation of inclusions, the CHMP2B^{Intron5} were also shown to develop astrogliosis and microgliosis, which were absent in the CHMP2B^{wt} and knockout mice. Interestingly, there were no signs of astrogliosis in the CHMP2B^{Intron5} mice until 12 months of age, and thus occurred only after the formation of inclusions, whereas reactive microglia was already present at 6 months of age and therefore coincided with the formation of inclusions. Another feature that was found to develop exclusively in the CHMP2B^{Intron5} mice were axonal swellings. These swellings were apparent at 6 months and increased with age and were found to contain mitochondria as well as vesicles from the lysosomal and autophagy degradation pathways. This suggests that axonal dysfunction and impairment, and possibly even axonal transport, may play a role in the disease process underlying FTLD caused by *CMHP2B* mutations. A second CHMP2B^{Intron5} line using a Thy1.2 instead of a PrP promoter to achieve neuronal expression presented with decreased survival (higher mortality in homo-, than in heterozygous mice) due to paralysis and muscle atrophy with denervation reminiscent of ALS, but also FTLD-like behavioral changes including disinhibition and social deficits [180]. P62-positive neuronal inclusions in these mice were negative for TDP-43 and FUS. A recent behavioral characterization of the original CHMP2B^{Intron5} mouse line showed slowly progressing motor and social deficits as mice reached 18 months of age, which was in contrast to early-onset neuroinflammation already detectable at 3 months of age [181]. Therefore, neuroinflammation may significantly contribute to the neurodegeneration in FTLD with *CHMP2B* mutations.

Fused in Sarcoma (FUS) Models

Mutations in the fused in sarcoma (FUS) gene have been identified not only in rare cases of FTLD [182], but also in a number of familial ALS cases [183, 184]. In contrast to the pathology in ALS however, FUS-positive inclusions identified in cases of FTLD co-localize with the RNA binding proteins TAF15 and EWS and are also ubiquitinated. The majority of FUS mutations cluster within the extreme C-terminus of the protein and interfere with the nuclear localization sequence residing in the C-terminus [185]. However, it has been demonstrated that overexpression of non-mutant FUS is sufficient to cause an aggressive phenotype and neuropathology in mice [186] as well as in rats [187].

Mitchell and colleagues generated both heterozygote (FUS$^{tg/+}$) and homozygote (FUS$^{tg/tg}$) mice overexpressing human non-mutant FUS in the brain, spinal cord, and testis [186]. Although the FUS$^{tg/tg}$ mice expressed higher levels of transgenic human FUS, this was found to decrease endogenous levels of murine FUS. FUS$^{tg/tg}$ mice were found to have a significantly shorter life span that only averaged 82 days, whereas FUS$^{tg/+}$ mice showed normal survival. In addition to nuclear localization of transgenic FUS, FUS$^{tg/tg}$ mice harbored perinuclear inclusions throughout the brain and spinal cord and cytoplasmic FUS within cortical neurons of end-stage FUS$^{tg/tg}$ mice, whereas only some perinuclear inclusions were found in the brains of FUS$^{tg/+}$ mice. However, there was no obvious co-localization between FUS and increased ubiquitin. Furthermore, these FUS aggregates did not co-localize with EWS and TAF15, as is observed in FTLD. Neuronal loss and gliosis were limited to the spinal cord in FUS$^{tg/tg}$ mice, resulting in muscle atrophy, early-onset motor deficits, and eventually limb paralysis. Adeno-associated virus (AAV)-mediated neuronal expression of R521C mutant or C-terminally truncated (Δ14) FUS in neonatal mice resulted in cytoplasmic accumulation and aggregation of FUS with co-aggregation of p62, but not TDP-43, similar to FTLD [188]. For comparison, transgenic expression of nuclear localization-deficient (ΔNLS) FUS or a variant depleted of its RNA binding motif and harboring a pathogenic R522G mutation produced ALS-like neuropathology and death due to severe motor dysfunction [189–191]. Similarly, transgenic neuronal expression of R521C mutant FUS resulted in severe motor deficits and ALS-like neuropathology [192]. Dendritic and synaptic atrophy in these mice was associated with DNA damage and partially reversed by BDNF treatment. Systemic overexpression of both mutant and wild-type FUS resulted in short survival, severe muscle atrophy, and neurodegeneration [193]. However, pan-neuronal or motor neuron-specific expression of mutant FUS from the endogenous *Mapt* promoter resulted in motor neuron dysfunction and loss, supporting a gain of toxic function in ALS [194], and cell autonomous pathogenic processes, as shown also in *Fus* ΔNLS knockin mice [195]. In contrast, aged homozygous FUS knockout mice lacked ALS-like symptoms, but rather presented with hippocampal degeneration and behavioral deficits [196]. If these loss-of-function phenotypes relate to mechanisms relevant for FTLD remains to be shown.

In summary, these mice recapitulate various pathological and behavioral features of both ALS and FTLD patients, making them good models to study these disorders. Exactly how overexpression of FUS causes these features and whether a similar process occurs in the presence of mutant FUS and whether the same process occurs in both ALS and FTLD remains to be determined [197].

C9orf72 Models

Hexanucleotide GGGGCC (G_4C_2) expansion in intron 1 of the *C9orf72* locus has been identified as a major genetic cause of FTLD and ALS [198, 199]. In Europe, up to 70% of patients with familial and up to 20% with apparent sporadic FTLD/ALS carry the *C9orf72* repeat expansion [55, 200].

Models of C9orf72 are the most recent addition to the spectrum of mouse models of FTLD. AAV-mediated neuronal expression of up to 66 G_4C_2 repeats in mice at birth resulted in typical RNA foci and c9RAN pathology, with inclusion of phosphorylated TDP-43 and neuronal loss, accompanied by FTLD-like behavioral and motor deficits 6 months after inoculation [201]. This model recently facilitated insight into sequestration with loss of function of nucleocytoplasmic transport proteins as possible disease mechanism [202]. Transgenic C9orf72 mice were generated carrying bacterial artificial chromosomes (BACs) from human *C9orf72* poly-G_4C_2 carriers [203, 204]. Different from the AAV-induced model, these BAC transgenic mice presented with neuronal RNA foci and c9RAN pathology, but did not develop functional deficits, suggesting further drivers of disease are required. However, more recent *C9orf72* poly-G_4C_2 BAC transgenic lines presented with neuronal loss, behavioral deficits, and c9RAN and TDP-43 pathology, resembling key features of FTLD and ALS [205, 206]. Single doses of antisense oligonucleotides (ASOs) to *C9orf72* repeat-containing RNAs were sufficient to ameliorate neuropathology and behavioral deficits [206]. Interestingly, a recent study reported similar epigenetic changes in *C9orf72* poly-G_4C_2 BAC transgenic mice and *C9orf72* poly-G_4C_2 disease carrier, highlighting the value of these models [207]. To study the effect of poly-GA polypeptides translated from aberrant G_4G_2 hexanucleotide repeats in *C9orf72* independent of possible other related mechanisms, novel transgenic mice expressing codon-modified poly-GA-CFP were developed, showing co-aggregation of proteins found in human disease (e.g., p62 Mlf2), phosphorylation of TDP-43 and some motor, but no memory deficits [208].

In contrast to transgenic models, neuronal and glial knockout of C9orf72 did not develop neurological deficits [209]. Systemic *C9orf72* knockout resulted in an immune phenotype, neoplastic lesions, and decreased survival, without neurological deficits, suggesting that loss of C9orf72 function is unlikely to contribute to FTD [206, 210–212]. Interestingly, loss of *C9orf72* enhanced autophagic activity, suggesting a regulative role in cell metabolism [213].

In summary, the newly developed *C9orf72* mouse models have significantly extended the spectrum of current FTLD mouse models, allowing novel insight into disease pathogenesis and providing platforms for new therapeutic strategies.

Concluding Remarks

Genetically modified mouse models are central to in vivo studies in AD and FTLD. Such models have provided insight and in some aspects a detailed understanding of pathological processes. With the identification of new proteins that form intracellular inclusion and novel pathogenic mutations in genes in FTLD, the number of different mouse models has dramatically increased. However, keeping in mind that each of the models reproduces and addresses only certain aspects of the human condition, and with the ease to rapidly generate knockout/knockin mice by CRISPR/Cas9 genome editing, it is likely that we see a plethora of transgenic models of even long-known candidates such as APP and tau. In addition, many of the new models of FTLD are awaiting to be used in combination with other genetically modified strains to address complex pathological processes in vivo.

References

1. Brinster RL, Cross PC. Effect of copper on the preimplantation mouse embryo. Nature. 1972;238(5364):398–9.
2. Ittner LM, Gotz J. Pronuclear injection for the production of transgenic mice. Nat Protoc. 2007;2(5):1206–15.

3. Delerue F, White M, Ittner LM. Inducible, tightly regulated and non-leaky neuronal gene expression in mice. Transgenic Res. 2014;23(2):225–33.
4. Geurts AM, Cost GJ, Freyvert Y, Zeitler B, Miller JC, Choi VM, et al. Knockout rats via embryo microinjection of zinc-finger nucleases. Science. 2009;325(5939):433.
5. Sung YH, Baek IJ, Kim DH, Jeon J, Lee J, Lee K, et al. Knockout mice created by TALEN-mediated gene targeting. Nat Biotechnol. 2013;31(1):23–4.
6. Delerue F, Ittner LM. Genome editing in mice using CRISPR/Cas9: achievements and prospects. Clon Transgen. 2015;4:2.
7. Wang H, Yang H, Shivalila CS, Dawlaty MM, Cheng AW, Zhang F, et al. One-step generation of mice carrying mutations in multiple genes by CRISPR/Cas-mediated genome engineering. Cell. 2013;153(4):910–8.
8. Yang H, Wang H, Shivalila CS, Cheng AW, Shi L, Jaenisch R. One-step generation of mice carrying reporter and conditional alleles by CRISPR/Cas-mediated genome engineering. Cell. 2013;154(6):1370–9.
9. Kleinberger G, Brendel M, Mracsko E, Wefers B, Groeneweg L, Xiang X, et al. The FTD-like syndrome causing TREM2 T66M mutation impairs microglia function, brain perfusion, and glucose metabolism. EMBO J. 2017;36(13):1837–53.
10. Gotz J, Ittner LM. Animal models of Alzheimer's disease and frontotemporal dementia. Nat Rev Neurosci. 2008;9(7):532–44.
11. Games D, Adams D, Alessandrini R, Barbour R, Berthelette P, Blackwell C, et al. Alzheimer-type neuropathology in transgenic mice overexpressing V717F beta-amyloid precursor protein. Nature. 1995;373(6514):523–7.
12. Meyer-Luehmann M, Coomaraswamy J, Bolmont T, Kaeser S, Schaefer C, Kilger E, et al. Exogenous induction of cerebral beta-amyloidogenesis is governed by agent and host. Science. 2006;313(5794):1781–4.
13. Verret L, Mann EO, Hang GB, Barth AM, Cobos I, Ho K, et al. Inhibitory interneuron deficit links altered network activity and cognitive dysfunction in Alzheimer model. Cell. 2012;149(3):708–21.
14. Ittner AA, Gladbach A, Bertz J, Suh LS, Ittner LM. p38 MAP kinase-mediated NMDA receptor-dependent suppression of hippocampal hypersynchronicity in a mouse model of Alzheimer inverted question marks disease. Acta Neuropathol Commun. 2014;2(1):149.
15. Lam AD, Deck G, Goldman A, Eskandar EN, Noebels J, Cole AJ. Silent hippocampal seizures and spikes identified by foramen ovale electrodes in Alzheimer's disease. Nat Med. 2017;23(6):678–80.
16. Calhoun ME, Wiederhold KH, Abramowski D, Phinney AL, Probst A, Sturchler-Pierrat C, et al. Neuron loss in APP transgenic mice. Nature. 1998;395(6704):755–6.
17. Wright AL, Zinn R, Hohensinn B, Konen LM, Beynon SB, Tan RP, et al. Neuroinflammation and neuronal loss precede Abeta plaque deposition in the hAPP-J20 mouse model of Alzheimer's disease. PLoS One. 2013;8(4):e59586.
18. Guo Q, Wang Z, Li H, Wiese M, Zheng H. APP physiological and pathophysiological functions: insights from animal models. Cell Res. 2012;22(1):78–89.
19. Wang P, Yang G, Mosier DR, Chang P, Zaidi T, Gong YD, et al. Defective neuromuscular synapses in mice lacking amyloid precursor protein (APP) and APP-Like protein 2. J Neurosci. 2005;25(5):1219–25.
20. Zheng H, Jiang M, Trumbauer ME, Sirinathsinghji DJ, Hopkins R, Smith DW, et al. beta-Amyloid precursor protein-deficient mice show reactive gliosis and decreased locomotor activity. Cell. 1995;81(4):525–31.
21. Li ZW, Stark G, Gotz J, Rulicke T, Gschwind M, Huber G, et al. Generation of mice with a 200-kb amyloid precursor protein gene deletion by Cre recombinase-mediated site-specific recombination in embryonic stem cells. Proc Natl Acad Sci U S A. 1996;93(12):6158–62.
22. Duce JA, Tsatsanis A, Cater MA, James SA, Robb E, Wikhe K, et al. Iron-export ferroxidase activity of beta-amyloid precursor protein is inhibited by zinc in Alzheimer's disease. Cell. 2010;142(6):857–67.

23. Saito T, Matsuba Y, Mihira N, Takano J, Nilsson P, Itohara S, et al. Single App knock-in mouse models of Alzheimer's disease. Nat Neurosci. 2014;17(5):661–3.
24. Saito T, Matsuba Y, Yamazaki N, Hashimoto S, Saido TC. Calpain activation in Alzheimer's model mice is an artifact of APP and presenilin overexpression. J Neurosci. 2016;36(38):9933–6.
25. Masuda A, Kobayashi Y, Kogo N, Saito T, Saido TC, Itohara S. Cognitive deficits in single App knock-in mouse models. Neurobiol Learn Mem. 2016;135:73–82.
26. Ittner A, Chua SW, Bertz J, Volkerling A, van der Hoven J, Gladbach A, et al. Site-specific phosphorylation of tau inhibits amyloid-beta toxicity in Alzheimer's mice. Science. 2016;354(6314):904–8.
27. Chin J, Palop JJ, GQ Y, Kojima N, Masliah E, Mucke L. Fyn kinase modulates synaptotoxicity, but not aberrant sprouting, in human amyloid precursor protein transgenic mice. J Neurosci. 2004;24(19):4692–7.
28. Ittner LM, Ke YD, Delerue F, Bi M, Gladbach A, van Eersel J, et al. Dendritic function of tau mediates amyloid-beta toxicity in Alzheimer's disease mouse models. Cell. 2010;142(3):387–97.
29. Holcomb L, Gordon MN, McGowan E, Yu X, Benkovic S, Jantzen P, et al. Accelerated Alzheimer-type phenotype in transgenic mice carrying both mutant amyloid precursor protein and presenilin 1 transgenes. Nat Med. 1998;4(1):97–100.
30. Schmitz C, Rutten BP, Pielen A, Schafer S, Wirths O, Tremp G, et al. Hippocampal neuron loss exceeds amyloid plaque load in a transgenic mouse model of Alzheimer's disease. Am J Pathol. 2004;164(4):1495–502.
31. Wang R, Wang B, He W, Zheng H. Wild-type presenilin 1 protects against Alzheimer disease mutation-induced amyloid pathology. J Biol Chem. 2006;281(22):15330–6.
32. Ohno M, Sametsky EA, Younkin LH, Oakley H, Younkin SG, Citron M, et al. BACE1 deficiency rescues memory deficits and cholinergic dysfunction in a mouse model of Alzheimer's disease. Neuron. 2004;41(1):27–33.
33. McConlogue L, Buttini M, Anderson JP, Brigham EF, Chen KS, Freedman SB, et al. Partial reduction of BACE1 has dramatic effects on Alzheimer plaque and synaptic pathology in APP transgenic mice. J Biol Chem. 2007;282(36):26326–34.
34. Ma H, Lesne S, Kotilinek L, Steidl-Nichols JV, Sherman M, Younkin L, et al. Involvement of beta-site APP cleaving enzyme 1 (BACE1) in amyloid precursor protein-mediated enhancement of memory and activity-dependent synaptic plasticity. Proc Natl Acad Sci U S A. 2007;104(19):8167–72.
35. Willem M, Dewachter I, Smyth N, Van Dooren T, Borghgraef P, Haass C, et al. beta-site amyloid precursor protein cleaving enzyme 1 increases amyloid deposition in brain parenchyma but reduces cerebrovascular amyloid angiopathy in aging BACE x APP[V717I] double-transgenic mice. Am J Pathol. 2004;165(5):1621–31.
36. Bertram L, Tanzi RE. The genetic epidemiology of neurodegenerative disease. J Clin Invest. 2005;115(6):1449–57.
37. Bales KR, Verina T, Dodel RC, Du Y, Altstiel L, Bender M, et al. Lack of apolipoprotein E dramatically reduces amyloid beta-peptide deposition. Nat Genet. 1997;17(3):263–4.
38. Dodart JC, Marr RA, Koistinaho M, Gregersen BM, Malkani S, Verma IM, et al. Gene delivery of human apolipoprotein E alters brain Abeta burden in a mouse model of Alzheimer's disease. Proc Natl Acad Sci U S A. 2005;102(4):1211–6.
39. Lewis J, Dickson DW, Lin WL, Chisholm L, Corral A, Jones G, et al. Enhanced neurofibrillary degeneration in transgenic mice expressing mutant tau and APP. Science. 2001;293(5534):1487–91.
40. Gotz J, Chen F, van Dorpe J, Nitsch RM. Formation of neurofibrillary tangles in P301l tau transgenic mice induced by Abeta 42 fibrils. Science. 2001;293(5534):1491–5.
41. Roberson ED, Scearce-Levie K, Palop JJ, Yan F, Cheng IH, Wu T, et al. Reducing endogenous tau ameliorates amyloid beta-induced deficits in an Alzheimer's disease mouse model. Science. 2007;316(5825):750–4.

42. Ittner LM, Gotz J. Amyloid-beta and tau—a toxic pas de deux in Alzheimer's disease. Nat Rev Neurosci. 2011;12(2):67–72.
43. Mondragon-Rodriguez S, Trillaud-Doppia E, Dudilot A, Bourgeois C, Lauzon M, Leclerc N, et al. Interaction of endogenous tau protein with synaptic proteins is regulated by N-methyl-D-aspartate receptor-dependent tau phosphorylation. J Biol Chem. 2012;287(38):32040–53.
44. Nakanishi N, Ryan SD, Zhang X, Khan A, Holland T, Cho EG, et al. Synaptic protein alpha1-Takusan mitigates amyloid-beta-induced synaptic loss via interaction with tau and postsynaptic density-95 at postsynaptic sites. J Neurosci. 2013;33(35):14170–83.
45. DeVos SL, Hyman BT. Tau at the crossroads between neurotoxicity and neuroprotection. Neuron. 2017;94(4):703–4.
46. Vanden Dries V, Stygelbout V, Pierrot N, Yilmaz Z, Suain V, De Decker R, et al. Amyloid precursor protein reduction enhances the formation of neurofibrillary tangles in a mutant tau transgenic mouse model. Neurobiol Aging. 2017;55:202–12.
47. Amador-Ortiz C, Lin WL, Ahmed Z, Personett D, Davies P, Duara R, et al. TDP-43 immunoreactivity in hippocampal sclerosis and Alzheimer's disease. Ann Neurol. 2007;61(5):435–45.
48. Uryu K, Nakashima-Yasuda H, Forman MS, Kwong LK, Clark CM, Grossman M, et al. Concomitant TAR-DNA-binding protein 43 pathology is present in Alzheimer disease and corticobasal degeneration but not in other tauopathies. J Neuropathol Exp Neurol. 2008;67(6):555–64.
49. Arai T, Mackenzie IR, Hasegawa M, Nonoka T, Niizato K, Tsuchiya K, et al. Phosphorylated TDP-43 in Alzheimer's disease and dementia with Lewy bodies. Acta Neuropathol. 2009;117(2):125–36.
50. LaClair KD, Donde A, Ling JP, Jeong YH, Chhabra R, Martin LJ, et al. Depletion of TDP-43 decreases fibril and plaque beta-amyloid and exacerbates neurodegeneration in an Alzheimer's mouse model. Acta Neuropathol. 2016;132(6):859–73.
51. Davis SA, Gan KA, Dowell JA, Cairns NJ, Gitcho MA. TDP-43 expression influences amyloidbeta plaque deposition and tau aggregation. Neurobiol Dis. 2017;103:154–62.
52. Paolicelli RC, Jawaid A, Henstridge CM, Valeri A, Merlini M, Robinson JL, et al. TDP-43 depletion in microglia promotes amyloid clearance but also induces synapse loss. Neuron. 2017;95(2):297–308 e6.
53. Josephs KA, Whitwell JL, Weigand SD, Murray ME, Tosakulwong N, Liesinger AM, et al. TDP-43 is a key player in the clinical features associated with Alzheimer's disease. Acta Neuropathol. 2014;127(6):811–24.
54. Mackenzie IR, Munoz DG, Kusaka H, Yokota O, Ishihara K, Roeber S, et al. Distinct pathological subtypes of FTLD-FUS. Acta Neuropathol. 2011;121(2):207–18.
55. Burrell JR, Halliday GM, Kril JJ, Ittner LM, Gotz J, Kiernan MC, et al. The frontotemporal dementia-motor neuron disease continuum. Lancet. 2016;388(10047):919–31.
56. Ittner LM, Halliday GM, Kril JJ, Gotz J, Hodges JR, Kiernan MCFTD. ALS-translating mouse studies into clinical trials. Nat Rev Neurol. 2015;11(6):360–6.
57. Gotz J, Probst A, Spillantini MG, Schafer T, Jakes R, Burki K, et al. Somatodendritic localization and hyperphosphorylation of tau protein in transgenic mice expressing the longest human brain tau isoform. EMBO J. 1995;14(7):1304–13.
58. Lewis J, McGowan E, Rockwood J, Melrose H, Nacharaju P, Van Slegtenhorst M, et al. Neurofibrillary tangles, amyotrophy and progressive motor disturbance in mice expressing mutant (P301L) tau protein. Nat Genet. 2000;25(4):402–5.
59. Dawson HN, Cantillana V, Chen L, Vitek MP. The tau N279K exon 10 splicing mutation recapitulates frontotemporal dementia and parkinsonism linked to chromosome 17 tauopathy in a mouse model. J Neurosci. 2007;27(34):9155–68.
60. Allen B, Ingram E, Takao M, Smith MJ, Jakes R, Virdee K, et al. Abundant tau filaments and nonapoptotic neurodegeneration in transgenic mice expressing human P301S tau protein. J Neurosci. 2002;22(21):9340–51.
61. Yoshiyama Y, Higuchi M, Zhang B, Huang SM, Iwata N, Saido TC, et al. Synapse loss and microglial activation precede tangles in a P301S tauopathy mouse model. Neuron. 2007;53(3):337–51.

62. Santacruz K, Lewis J, Spires T, Paulson J, Kotilinek L, Ingelsson M, et al. Tau suppression in a neurodegenerative mouse model improves memory function. Science. 2005;309(5733):476–81.
63. Mocanu MM, Nissen A, Eckermann K, Khlistunova I, Biernat J, Drexler D, et al. The potential for beta-structure in the repeat domain of tau protein determines aggregation, synaptic decay, neuronal loss, and coassembly with endogenous Tau in inducible mouse models of tauopathy. J Neurosci. 2008;28(3):737–48.
64. Forman MS, Lal D, Zhang B, Dabir DV, Swanson E, Lee VM, et al. Transgenic mouse model of tau pathology in astrocytes leading to nervous system degeneration. J Neurosci. 2005;25(14):3539–50.
65. Higuchi M, Zhang B, Forman MS, Yoshiyama Y, Trojanowski JQ, Lee VM. Axonal degeneration induced by targeted expression of mutant human tau in oligodendrocytes of transgenic mice that model glial tauopathies. J Neurosci. 2005;25(41):9434–43.
66. van Eersel J, Stevens CH, Przybyla M, Gladbach A, Stefanoska K, Chan CK, et al. Early-onset axonal pathology in a novel P301S-Tau transgenic mouse model of frontotemporal lobar degeneration. Neuropathol Appl Neurobiol. 2015;41(7):906–25.
67. Przybyla M, Stevens CH, van der Hoven J, Harasta A, Bi M, Ittner A, et al. Disinhibition-like behavior in a P301S mutant tau transgenic mouse model of frontotemporal dementia. Neurosci Lett. 2016;631:24–9.
68. Van der Jeugd A, Vermaercke B, Halliday GM, Staufenbiel M, Gotz J. Impulsivity, decreased social exploration, and executive dysfunction in a mouse model of frontotemporal dementia. Neurobiol Learn Mem. 2016;130:34–43.
69. Aguzzi A, Rajendran L. The transcellular spread of cytosolic amyloids, prions, and prionoids. Neuron. 2009;64(6):783–90.
70. Clavaguera F, Bolmont T, Crowther RA, Abramowski D, Frank S, Probst A, et al. Transmission and spreading of tauopathy in transgenic mouse brain. Nat Cell Biol. 2009;11(7):909–13.
71. Clavaguera F, Akatsu H, Fraser G, Crowther RA, Frank S, Hench J, et al. Brain homogenates from human tauopathies induce tau inclusions in mouse brain. Proc Natl Acad Sci U S A. 2013;110(23):9535–40.
72. Sanders DW, Kaufman SK, DeVos SL, Sharma AM, Mirbaha H, Li A, et al. Distinct tau prion strains propagate in cells and mice and define different tauopathies. Neuron. 2014;82(6):1271–88.
73. Kaufman SK, Sanders DW, Thomas TL, Ruchinskas AJ, Vaquer-Alicea J, Sharma AM, et al. Tau prion strains dictate patterns of cell pathology, progression rate, and regional vulnerability in vivo. Neuron. 2016;92(4):796–812.
74. Liu L, Drouet V, JW W, Witter MP, Small SA, Clelland C, et al. Trans-synaptic spread of tau pathology in vivo. PLoS One. 2012;7(2):e31302.
75. Jackson SJ, Kerridge C, Cooper J, Cavallini A, Falcon B, Cella CV, et al. Short fibrils constitute the major species of seed-competent tau in the brains of mice transgenic for human P301S tau. J Neurosci. 2016;36(3):762–72.
76. Kim D, Lim S, Haque MM, Ryoo N, Hong HS, Rhim H, et al. Identification of disulfide cross-linked tau dimer responsible for tau propagation. Sci Rep. 2015;5:15231.
77. Asai H, Ikezu S, Tsunoda S, Medalla M, Luebke J, Haydar T, et al. Depletion of microglia and inhibition of exosome synthesis halt tau propagation. Nat Neurosci. 2015;18(11):1584–93.
78. Baker S, Polanco JC, Gotz J. Extracellular vesicles containing P301L mutant tau accelerate pathological tau phosphorylation and oligomer formation but do not seed mature neurofibrillary tangles in ALZ17 mice. J Alzheimers Dis. 2016;54(3):1207–17.
79. Yamada K, Holth JK, Liao F, Stewart FR, Mahan TE, Jiang H, et al. Neuronal activity regulates extracellular tau in vivo. J Exp Med. 2014;211(3):387–93.
80. JW W, Hussaini SA, Bastille IM, Rodriguez GA, Mrejeru A, Rilett K, et al. Neuronal activity enhances tau propagation and tau pathology in vivo. Nat Neurosci. 2016;19(8):1085–92.
81. Wegmann S, Maury EA, Kirk MJ, Saqran L, Roe A, DeVos SL, et al. Removing endogenous tau does not prevent tau propagation yet reduces its neurotoxicity. EMBO J. 2015;34(24):3028–41.

82. DeVos SL, Miller RL, Schoch KM, Holmes BB, Kebodeaux CS, Wegener AJ, et al. Tau reduction prevents neuronal loss and reverses pathological tau deposition and seeding in mice with tauopathy. Sci Transl Med. 2017;9(374):eaag0481.
83. Holth JK, Mahan TE, Robinson GO, Rocha A, Holtzman DM. Altered sleep and EEG power in the P301S Tau transgenic mouse model. Ann Clin Transl Neurol. 2017;4(3):180–90.
84. Maeda S, Djukic B, Taneja P, GQ Y, Lo I, Davis A, et al. Expression of A152T human tau causes age-dependent neuronal dysfunction and loss in transgenic mice. EMBO Rep. 2016;17(4):530–51.
85. Koss DJ, Robinson L, Drever BD, Plucinska K, Stoppelkamp S, Veselcic P, et al. Mutant Tau knock-in mice display frontotemporal dementia relevant behaviour and histopathology. Neurobiol Dis. 2016;91:105–23.
86. Kim Y, Choi H, Lee W, Park H, Kam TI, Hong SH, et al. Caspase-cleaved tau exhibits rapid memory impairment associated with tau oligomers in a transgenic mouse model. Neurobiol Dis. 2016;87:19–28.
87. Zhao X, Kotilinek LA, Smith B, Hlynialuk C, Zahs K, Ramsden M, et al. Caspase-2 cleavage of tau reversibly impairs memory. Nat Med. 2016;22(11):1268–76.
88. Bondulich MK, Guo T, Meehan C, Manion J, Rodriguez Martin T, Mitchell JC, et al. Tauopathy induced by low level expression of a human brain-derived tau fragment in mice is rescued by phenylbutyrate. Brain. 2016;139(Pt 8):2290–306.
89. Min SW, Chen X, Tracy TE, Li Y, Zhou Y, Wang C, et al. Critical role of acetylation in tau-mediated neurodegeneration and cognitive deficits. Nat Med. 2015;21(10):1154–62.
90. Asuni AA, Boutajangout A, Quartermain D, Sigurdsson EM. Immunotherapy targeting pathological tau conformers in a tangle mouse model reduces brain pathology with associated functional improvements. J Neurosci. 2007;27(34):9115–29.
91. Bi M, Ittner A, Ke YD, Gotz J, Ittner LM. Tau-targeted immunization impedes progression of neurofibrillary histopathology in aged P301L tau transgenic mice. PLoS One. 2011;6(12):e26860.
92. Boimel M, Grigoriadis N, Lourbopoulos A, Haber E, Abramsky O, Rosenmann H. Efficacy and safety of immunization with phosphorylated tau against neurofibrillary tangles in mice. Exp Neurol. 2010;224(2):472–85.
93. Chai X, Wu S, Murray TK, Kinley R, Cella CV, Sims H, et al. Passive immunization with anti-Tau antibodies in two transgenic models: reduction of Tau pathology and delay of disease progression. J Biol Chem. 2011;286(39):34457–67.
94. Yanamandra K, Patel TK, Jiang H, Schindler S, Ulrich JD, Boxer AL, et al. Anti-tau antibody administration increases plasma tau in transgenic mice and patients with tauopathy. Sci Transl Med. 2017;9(386):eaal2029.
95. Ising C, Gallardo G, Leyns CEG, Wong CH, Stewart F, Koscal LJ, et al. AAV-mediated expression of anti-tau scFvs decreases tau accumulation in a mouse model of tauopathy. J Exp Med. 2017;214(5):1227–38.
96. Lee SH, Le Pichon CE, Adolfsson O, Gafner V, Pihlgren M, Lin H, et al. Antibody-mediated targeting of Tau in vivo does not require effector function and microglial engagement. Cell Rep. 2016;16(6):1690–700.
97. Nisbet RM, Van der Jeugd A, Leinenga G, Evans HT, Janowicz PW, Gotz J. Combined effects of scanning ultrasound and a tau-specific single chain antibody in a tau transgenic mouse model. Brain. 2017;140(5):1220–30.
98. Davtyan H, Chen WW, Zagorski K, Davis J, Petrushina I, Kazarian K, et al. MultiTEP platform-based DNA epitope vaccine targeting N-terminus of tau induces strong immune responses and reduces tau pathology in THY-Tau22 mice. Vaccine. 2017;35(16):2015–24.
99. van Eersel J, Ke YD, Liu X, Delerue F, Kril JJ, Gotz J, et al. Sodium selenate mitigates tau pathology, neurodegeneration, and functional deficits in Alzheimer's disease models. Proc Natl Acad Sci U S A. 2010;107(31):13888–93.
100. Brunden KR, Zhang B, Carroll J, Yao Y, Potuzak JS, Hogan AM, et al. Epothilone D improves microtubule density, axonal integrity, and cognition in a transgenic mouse model of tauopathy. J Neurosci. 2010;30(41):13861–6.

101. Yu L, Wang W, Pang W, Xiao Z, Jiang Y, Hong Y. Dietary lycopene supplementation improves cognitive performances in tau transgenic mice expressing P301L mutation via inhibiting oxidative stress and tau hyperphosphorylation. J Alzheimers Dis. 2017;57(2):475–82.
102. Chua SW, Cornejo A, van Eersel J, Stevens CH, Vaca I, Cueto M, et al. The polyphenol altenusin inhibits in vitro fibrillization of tau and reduces induced tau pathology in primary neurons. ACS Chem Nerosci. 2017;8(4):743–51.
103. Makani V, Zhang B, Han H, Yao Y, Lassalas P, Lou K, et al. Evaluation of the brain-penetrant microtubule-stabilizing agent, dictyostatin, in the PS19 tau transgenic mouse model of tauopathy. Acta Neuropathol Commun. 2016;4(1):106.
104. Dennissen FJ, Anglada-Huguet M, Sydow A, Mandelkow E, Mandelkow EM. Adenosine A1 receptor antagonist rolofylline alleviates axonopathy caused by human Tau DeltaK280. Proc Natl Acad Sci U S A. 2016;113(41):11597–602.
105. Wagner J, Krauss S, Shi S, Ryazanov S, Steffen J, Miklitz C, et al. Reducing tau aggregates with anle138b delays disease progression in a mouse model of tauopathies. Acta Neuropathol. 2015;130(5):619–31.
106. Qin Y, Liu Y, Hao W, Decker Y, Tomic I, Menger MD, et al. Stimulation of TLR4 attenuates Alzheimer's disease-related symptoms and pathology in tau-transgenic mice. J Immunol. 2016;197(8):3281–92.
107. Hansen HH, Barkholt P, Fabricius K, Jelsing J, Terwel D, Pyke C, et al. The GLP-1 receptor agonist liraglutide reduces pathology-specific tau phosphorylation and improves motor function in a transgenic hTauP301L mouse model of tauopathy. Brain Res. 2016;1634:158–70.
108. Myeku N, Clelland CL, Emrani S, Kukushkin NV, Yu WH, Goldberg AL, et al. Tau-driven 26S proteasome impairment and cognitive dysfunction can be prevented early in disease by activating cAMP-PKA signaling. Nat Med. 2016;22(1):46–53.
109. Neumann M, Sampathu DM, Kwong LK, Truax AC, Micsenyi MC, Chou TT, et al. Ubiquitinated TDP-43 in frontotemporal lobar degeneration and amyotrophic lateral sclerosis. Science. 2006;314(5796):130–3.
110. Buratti E, Baralle FE. Multiple roles of TDP-43 in gene expression, splicing regulation, and human disease. Front Biosci. 2008;13:867–78.
111. Liscic RM, Grinberg LT, Zidar J, Gitcho MA, Cairns NJ. ALS and FTLD: two faces of TDP-43 proteinopathy. Eur J Neurol. 2008;15(8):772–80.
112. Wegorzewska I, Bell S, Cairns NJ, Miller TM, Baloh RH. TDP-43 mutant transgenic mice develop features of ALS and frontotemporal lobar degeneration. Proc Natl Acad Sci U S A. 2009;106(44):18809–14.
113. Guo Y, Wang Q, Zhang K, An T, Shi P, Li Z, et al. HO-1 induction in motor cortex and intestinal dysfunction in TDP-43 A315T transgenic mice. Brain Res. 2012;1460:88–95.
114. Esmaeili MA, Panahi M, Yadav S, Hennings L, Kiaei M. Premature death of TDP-43 (A315T) transgenic mice due to gastrointestinal complications prior to development of full neurological symptoms of amyotrophic lateral sclerosis. Int J Exp Pathol. 2013;94(1):56–64.
115. Hatzipetros T, Bogdanik LP, Tassinari VR, Kidd JD, Moreno AJ, Davis C, et al. C57BL/6J congenic Prp-TDP43A315T mice develop progressive neurodegeneration in the myenteric plexus of the colon without exhibiting key features of ALS. Brain Res. 2014;1584:59–72.
116. Herdewyn S, Cirillo C, Van Den Bosch L, Robberecht W, Vanden Berghe P, Van Damme P. Prevention of intestinal obstruction reveals progressive neurodegeneration in mutant TDP-43 (A315T) mice. Mol Neurodegener. 2014;9:24.
117. Wils H, Kleinberger G, Janssens J, Pereson S, Joris G, Cuijt I, et al. TDP-43 transgenic mice develop spastic paralysis and neuronal inclusions characteristic of ALS and frontotemporal lobar degeneration. Proc Natl Acad Sci U S A. 2010;107(8):3858–63.
118. YF X, Gendron TF, Zhang YJ, Lin WL, D'Alton S, Sheng H, et al. Wild-type human TDP-43 expression causes TDP-43 phosphorylation, mitochondrial aggregation, motor deficits, and early mortality in transgenic mice. J Neurosci. 2010;30(32):10851–9.
119. Igaz LM, Kwong LK, Lee EB, Chen-Plotkin A, Swanson E, Unger T, et al. Dysregulation of the ALS-associated gene TDP-43 leads to neuronal death and degeneration in mice. J Clin Invest. 2011;121(2):726–38.

120. Stallings NR, Puttaparthi K, Luther CM, Burns DK, Elliott JL. Progressive motor weakness in transgenic mice expressing human TDP-43. Neurobiol Dis. 2010;40(2):404–14.
121. Swarup V, Phaneuf D, Bareil C, Robertson J, Rouleau GA, Kriz J, et al. Pathological hallmarks of amyotrophic lateral sclerosis/frontotemporal lobar degeneration in transgenic mice produced with TDP-43 genomic fragments. Brain. 2011;134(Pt 9):2610–26.
122. Ke YD, van Hummel A, Stevens CH, Gladbach A, Ippati S, Bi M, et al. Short-term suppression of A315T mutant human TDP-43 expression improves functional deficits in a novel inducible transgenic mouse model of FTLD-TDP and ALS. Acta Neuropathol. 2015;130(5):661–78.
123. Walker AK, Spiller KJ, Ge G, Zheng A, Xu Y, Zhou M, et al. Functional recovery in new mouse models of ALS/FTLD after clearance of pathological cytoplasmic TDP-43. Acta Neuropathol. 2015;130(5):643–60.
124. Spiller KJ, Cheung CJ, Restrepo CR, Kwong LK, Stieber AM, Trojanowski JQ, et al. Selective motor neuron resistance and recovery in a new inducible mouse model of TDP-43 proteinopathy. J Neurosci. 2016;36(29):7707–17.
125. Winton MJ, Igaz LM, Wong MM, Kwong LK, Trojanowski JQ, Lee VM. Disturbance of nuclear and cytoplasmic TAR DNA-binding protein (TDP-43) induces disease-like redistribution, sequestration, and aggregate formation. J Biol Chem. 2008;283(19):13302–9.
126. Alfieri JA, Pino NS, Igaz LM. Reversible behavioral phenotypes in a conditional mouse model of TDP-43 proteinopathies. J Neurosci. 2014;34(46):15244–59.
127. Alfieri JA, Silva PR, Igaz LM. Early cognitive/social deficits and late motor phenotype in conditional wild-type TDP-43 transgenic mice. Front Aging Neurosci. 2016;8:310.
128. Cannon A, Yang B, Knight J, Farnham IM, Zhang Y, Wuertzer CA, et al. Neuronal sensitivity to TDP-43 overexpression is dependent on timing of induction. Acta Neuropathol. 2012;123(6):807–23.
129. D'Alton S, Altshuler M, Cannon A, Dickson DW, Petrucelli L, Lewis J. Divergent phenotypes in mutant TDP-43 transgenic mice highlight potential confounds in TDP-43 transgenic modeling. PLoS One. 2014;9(1):e86513.
130. Zhang W, Zhang L, Liang B, Schroeder D, Zhang ZW, Cox GA, et al. Hyperactive somatostatin interneurons contribute to excitotoxicity in neurodegenerative disorders. Nat Neurosci. 2016;19(4):557–9.
131. Handley EE, Pitman KA, Dawkins E, Young KM, Clark RM, Jiang TC, et al. Synapse dysfunction of layer V pyramidal neurons precedes neurodegeneration in a mouse model of TDP-43 proteinopathies. Cereb Cortex. 2017;27(7):3630–47.
132. Wang W, Wang L, Lu J, Siedlak SL, Fujioka H, Liang J, et al. The inhibition of TDP-43 mitochondrial localization blocks its neuronal toxicity. Nat Med. 2016;22(8):869–78.
133. Wang W, Arakawa H, Wang L, Okolo O, Siedlak SL, Jiang Y, et al. Motor-coordinative and cognitive dysfunction caused by mutant TDP-43 could be reversed by inhibiting its mitochondrial localization. Mol Ther. 2017;25(1):127–39.
134. Wenqiang C, Lonskaya I, Hebron ML, Ibrahim Z, Olszewski RT, Neale JH, et al. Parkin-mediated reduction of nuclear and soluble TDP-43 reverses behavioral decline in symptomatic mice. Hum Mol Genet. 2014;23(18):4960–9.
135. Heyburn L, Hebron ML, Smith J, Winston C, Bechara J, Li Z, et al. Tyrosine kinase inhibition reverses TDP-43 effects on synaptic protein expression, astrocytic function and amino acid dis-homeostasis. J Neurochem. 2016;139(4):610–23.
136. Kawamata H, Peixoto P, Konrad C, Palomo G, Bredvik K, Gerges M, et al. Mutant TDP-43 does not impair mitochondrial bioenergetics in vitro and in vivo. Mol Neurodegener. 2017;12(1):37.
137. Caccamo A, Majumder S, Oddo S. Cognitive decline typical of frontotemporal lobar degeneration in transgenic mice expressing the 25-kDa C-terminal fragment of TDP-43. Am J Pathol. 2012;180(1):293–302.
138. Caccamo A, Shaw DM, Guarino F, Messina A, Walker AW, Oddo S. Reduced protein turnover mediates functional deficits in transgenic mice expressing the 25 kDa C-terminal fragment of TDP-43. Hum Mol Genet. 2015;24(16):4625–35.

139. Walker AK, Tripathy K, Restrepo CR, Ge G, Xu Y, Kwong LK, et al. An insoluble frontotemporal lobar degeneration-associated TDP-43 C-terminal fragment causes neurodegeneration and hippocampus pathology in transgenic mice. Hum Mol Genet. 2015;24(25):7241–54.
140. Mitchell JC, Constable R, So E, Vance C, Scotter E, Glover L, et al. Wild type human TDP-43 potentiates ALS-linked mutant TDP-43 driven progressive motor and cortical neuron degeneration with pathological features of ALS. Acta Neuropathol Commun. 2015;3:36.
141. Correia AS, Patel P, Dutta K, Julien JP. Inflammation induces TDP-43 mislocalization and aggregation. PLoS One. 2015;10(10):e0140248.
142. Iguchi Y, Eid L, Parent M, Soucy G, Bareil C, Riku Y, et al. Exosome secretion is a key pathway for clearance of pathological TDP-43. Brain. 2016;139(Pt 12):3187–201.
143. Clippinger AK, D'Alton S, Lin WL, Gendron TF, Howard J, Borchelt DR, et al. Robust cytoplasmic accumulation of phosphorylated TDP-43 in transgenic models of tauopathy. Acta Neuropathol. 2013;126(1):39–50.
144. Gu J, Wu F, Xu W, Shi J, Hu W, Jin N, et al. TDP-43 suppresses tau expression via promoting its mRNA instability. Nucleic Acids Res. 2017;45(10):6177–93.
145. Kraemer BC, Schuck T, Wheeler JM, Robinson LC, Trojanowski JQ, Lee VM, et al. Loss of murine TDP-43 disrupts motor function and plays an essential role in embryogenesis. Acta Neuropathol. 2010;119(4):409–19.
146. Becker LA, Huang B, Bieri G, Ma R, Knowles DA, Jafar-Nejad P, et al. Therapeutic reduction of ataxin-2 extends lifespan and reduces pathology in TDP-43 mice. Nature. 2017;544(7650):367–71.
147. Perera ND, Sheean RK, Crouch PJ, White AR, Horne MK, Turner BJ. Enhancing survival motor neuron expression extends lifespan and attenuates neurodegeneration in mutant TDP-43 mice. Hum Mol Genet. 2016;25(18):4080–93.
148. Dutta K, Patel P, Rahimian R, Phaneuf D, Julien JP. Withania somnifera reverses transactive response DNA binding protein 43 proteinopathy in a mouse model of amyotrophic lateral sclerosis/frontotemporal lobar degeneration. Neurotherapeutics. 2017;14(2):447–62.
149. Baker M, Mackenzie IR, Pickering-Brown SM, Gass J, Rademakers R, Lindholm C, et al. Mutations in progranulin cause tau-negative frontotemporal dementia linked to chromosome 17. Nature. 2006;442(7105):916–9.
150. Cruts M, Gijselinck I, van der Zee J, Engelborghs S, Wils H, Pirici D, et al. Null mutations in progranulin cause ubiquitin-positive frontotemporal dementia linked to chromosome 17q21. Nature. 2006;442(7105):920–4.
151. Shankaran SS, Capell A, Hruscha AT, Fellerer K, Neumann M, Schmid B, et al. Missense mutations in the progranulin gene linked to frontotemporal lobar degeneration with ubiquitin-immunoreactive inclusions reduce progranulin production and secretion. J Biol Chem. 2008;283(3):1744–53.
152. Wang J, Van Damme P, Cruchaga C, Gitcho MA, Vidal JM, Seijo-Martinez M, et al. Pathogenic cysteine mutations affect progranulin function and production of mature granulins. J Neurochem. 2010;112(5):1305–15.
153. Kayasuga Y, Chiba S, Suzuki M, Kikusui T, Matsuwaki T, Yamanouchi K, et al. Alteration of behavioural phenotype in mice by targeted disruption of the progranulin gene. Behav Brain Res. 2007;185(2):110–8.
154. Yin F, Banerjee R, Thomas B, Zhou P, Qian L, Jia T, et al. Exaggerated inflammation, impaired host defense, and neuropathology in progranulin-deficient mice. J Exp Med. 2010;207(1):117–28.
155. Petkau TL, Neal SJ, Milnerwood A, Mew A, Hill AM, Orban P, et al. Synaptic dysfunction in progranulin-deficient mice. Neurobiol Dis. 2012;45(2):711–22.
156. Martens LH, Zhang J, Barmada SJ, Zhou P, Kamiya S, Sun B, et al. Progranulin deficiency promotes neuroinflammation and neuron loss following toxin-induced injury. J Clin Invest. 2012;122(11):3955–9.
157. Wils H, Kleinberger G, Pereson S, Janssens J, Capell A, Van Dam D, et al. Cellular ageing, increased mortality and FTLD-TDP-associated neuropathology in progranulin knockout mice. J Pathol. 2012;228(1):67–76.

158. Ahmed Z, Sheng H, Xu YF, Lin WL, Innes AE, Gass J, et al. Accelerated lipofuscinosis and ubiquitination in granulin knockout mice suggest a role for progranulin in successful aging. Am J Pathol. 2010;177(1):311–24.
159. Ghoshal N, Dearborn JT, Wozniak DF, Cairns NJ. Core features of frontotemporal dementia recapitulated in progranulin knockout mice. Neurobiol Dis. 2012;45(1):395–408.
160. Yin F, Dumont M, Banerjee R, Ma Y, Li H, Lin MT, et al. Behavioral deficits and progressive neuropathology in progranulin-deficient mice: a mouse model of frontotemporal dementia. FASEB J. 2010;24(12):4639–47.
161. Tanaka Y, Chambers JK, Matsuwaki T, Yamanouchi K, Nishihara M. Possible involvement of lysosomal dysfunction in pathological changes of the brain in aged progranulin-deficient mice. Acta Neuropathol Commun. 2014;2:78.
162. Brouwers N, Nuytemans K, van der Zee J, Gijselinck I, Engelborghs S, Theuns J, et al. Alzheimer and Parkinson diagnoses in progranulin null mutation carriers in an extended founder family. Arch Neurol. 2007;64(10):1436–46.
163. Cortini F, Fenoglio C, Guidi I, Venturelli E, Pomati S, Marcone A, et al. Novel exon 1 progranulin gene variant in Alzheimer's disease. Eur J Neurol. 2008;15(10):1111–7.
164. Hosokawa M, Arai T, Masuda-Suzukake M, Kondo H, Matsuwaki T, Nishihara M, et al. Progranulin reduction is associated with increased tau phosphorylation in P301L tau transgenic mice. J Neuropathol Exp Neurol. 2015;74(2):158–65.
165. Filiano AJ, Martens LH, Young AH, Warmus BA, Zhou P, Diaz-Ramirez G, et al. Dissociation of frontotemporal dementia-related deficits and neuroinflammation in progranulin haploinsufficient mice. J Neurosci. 2013;33(12):5352–61.
166. Watts GD, Wymer J, Kovach MJ, Mehta SG, Mumm S, Darvish D, et al. Inclusion body myopathy associated with Paget disease of bone and frontotemporal dementia is caused by mutant valosin-containing protein. Nat Genet. 2004;36(4):377–81.
167. Kimonis VE, Mehta SG, Fulchiero EC, Thomasova D, Pasquali M, Boycott K, et al. Clinical studies in familial VCP myopathy associated with Paget disease of bone and frontotemporal dementia. Am J Med Genet A. 2008;146A(6):745–57.
168. Johnson JO, Mandrioli J, Benatar M, Abramzon Y, Van Deerlin VM, Trojanowski JQ, et al. Exome sequencing reveals VCP mutations as a cause of familial ALS. Neuron. 2010;68(5):857–64.
169. Abramzon Y, Johnson JO, Scholz SW, Taylor JP, Brunetti M, Calvo A, et al. Valosin-containing protein (VCP) mutations in sporadic amyotrophic lateral sclerosis. Neurobiol Aging. 2012;33(9):2231 e1–6.
170. Tang WK, Li D, Li CC, Esser L, Dai R, Guo L, et al. A novel ATP-dependent conformation in p97 N-D1 fragment revealed by crystal structures of disease-related mutants. EMBO J. 2010;29(13):2217–29.
171. Rodriguez-Ortiz CJ, Hoshino H, Cheng D, Liu-Yescevitz L, Blurton-Jones M, Wolozin B, et al. Neuronal-specific overexpression of a mutant valosin-containing protein associated with IBMPFD promotes aberrant ubiquitin and TDP-43 accumulation and cognitive dysfunction in transgenic mice. Am J Pathol. 2013;183(2):504–15.
172. Weihl CC, Miller SE, Hanson PI, Pestronk A. Transgenic expression of inclusion body myopathy associated mutant p97/VCP causes weakness and ubiquitinated protein inclusions in mice. Hum Mol Genet. 2007;16(8):919–28.
173. Custer SK, Neumann M, Lu H, Wright AC, Taylor JP. Transgenic mice expressing mutant forms VCP/p97 recapitulate the full spectrum of IBMPFD including degeneration in muscle, brain and bone. Hum Mol Genet. 2010;19(9):1741–55.
174. Badadani M, Nalbandian A, Watts GD, Vesa J, Kitazawa M, Su H, et al. VCP associated inclusion body myopathy and paget disease of bone knock-in mouse model exhibits tissue pathology typical of human disease. PLoS One. 2010;5(10):e13183.
175. Nalbandian A, Llewellyn KJ, Badadani M, Yin HZ, Nguyen C, Katheria V, et al. A progressive translational mouse model of human valosin-containing protein disease: the VCP(R155H/+) mouse. Muscle Nerve. 2013;47(2):260–70.

176. Yin HZ, Nalbandian A, Hsu CI, Li S, Llewellyn KJ, Mozaffar T, et al. Slow development of ALS-like spinal cord pathology in mutant valosin-containing protein gene knock-in mice. Cell Death Dis. 2012;3:e374.
177. Llewellyn KJ, Nalbandian A, Jung KM, Nguyen C, Avanesian A, Mozaffar T, et al. Lipid-enriched diet rescues lethality and slows down progression in a murine model of VCP-associated disease. Hum Mol Genet. 2014;23(5):1333–44.
178. Skibinski G, Parkinson NJ, Brown JM, Chakrabarti L, Lloyd SL, Hummerich H, et al. Mutations in the endosomal ESCRTIII-complex subunit CHMP2B in frontotemporal dementia. Nat Genet. 2005;37(8):806–8.
179. Ghazi-Noori S, Froud KE, Mizielinska S, Powell C, Smidak M, Fernandez de Marco M, et al. Progressive neuronal inclusion formation and axonal degeneration in CHMP2B mutant transgenic mice. Brain. 2012;135(Pt 3):819–32.
180. Vernay A, Therreau L, Blot B, Risson V, Dirrig-Grosch S, Waegaert R, et al. A transgenic mouse expressing CHMP2Bintron5 mutant in neurons develops histological and behavioural features of amyotrophic lateral sclerosis and frontotemporal dementia. Hum Mol Genet. 2016;25(15):3341–60.
181. Clayton EL, Mancuso R, Nielsen TT, Mizielinska S, Holmes H, Powell N, et al. Early microgliosis precedes neuronal loss and behavioural impairment in mice with a frontotemporal dementia-causing CHMP2B mutation. Hum Mol Genet. 2017;26(5):873–87.
182. Van Langenhove T, van der Zee J, Sleegers K, Engelborghs S, Vandenberghe R, Gijselinck I, et al. Genetic contribution of FUS to frontotemporal lobar degeneration. Neurology. 2010;74(5):366–71.
183. Kwiatkowski TJ Jr, Bosco DA, Leclerc AL, Tamrazian E, Vanderburg CR, Russ C, et al. Mutations in the FUS/TLS gene on chromosome 16 cause familial amyotrophic lateral sclerosis. Science. 2009;323(5918):1205–8.
184. Vance C, Rogelj B, Hortobagyi T, De Vos KJ, Nishimura AL, Sreedharan J, et al. Mutations in FUS, an RNA processing protein, cause familial amyotrophic lateral sclerosis type 6. Science. 2009;323(5918):1208–11.
185. Dormann D, Rodde R, Edbauer D, Bentmann E, Fischer I, Hruscha A, et al. ALS-associated fused in sarcoma (FUS) mutations disrupt transportin-mediated nuclear import. EMBO J. 2010;29(16):2841–57.
186. Mitchell JC, McGoldrick P, Vance C, Hortobagyi T, Sreedharan J, Rogelj B, et al. Overexpression of human wild-type FUS causes progressive motor neuron degeneration in an age- and dose-dependent fashion. Acta Neuropathol. 2013;125(2):273–88.
187. Huang C, Zhou H, Tong J, Chen H, Liu YJ, Wang D, et al. FUS transgenic rats develop the phenotypes of amyotrophic lateral sclerosis and frontotemporal lobar degeneration. PLoS Genet. 2011;7(3):e1002011.
188. Verbeeck C, Deng Q, Dejesus-Hernandez M, Taylor G, Ceballos-Diaz C, Kocerha J, et al. Expression of fused in sarcoma mutations in mice recapitulates the neuropathology of FUS proteinopathies and provides insight into disease pathogenesis. Mol Neurodegener. 2012;7:53.
189. Shelkovnikova TA, Peters OM, Deykin AV, Connor-Robson N, Robinson H, Ustyugov AA, et al. Fused in sarcoma (FUS) protein lacking nuclear localization signal (NLS) and major RNA binding motifs triggers proteinopathy and severe motor phenotype in transgenic mice. J Biol Chem. 2013;288(35):25266–74.
190. Robinson HK, Deykin AV, Bronovitsky EV, Ovchinnikov RK, Ustyugov AA, Shelkovnikova TA, et al. Early lethality and neuronal proteinopathy in mice expressing cytoplasm-targeted FUS that lacks the RNA recognition motif. Amyotroph Lateral Scler Frontotemporal Degener. 2015;16(5-6):402–9.
191. Shiihashi G, Ito D, Yagi T, Nihei Y, Ebine T, Suzuki N. Mislocated FUS is sufficient for gain-of-toxic-function amyotrophic lateral sclerosis phenotypes in mice. Brain. 2016;139(Pt 9):2380–94.
192. Qiu H, Lee S, Shang Y, Wang WY, Au KF, Kamiya S, et al. ALS-associated mutation FUS-R521C causes DNA damage and RNA splicing defects. J Clin Invest. 2014;124(3):981–99.

193. Sephton CF, Tang AA, Kulkarni A, West J, Brooks M, Stubblefield JJ, et al. Activity-dependent FUS dysregulation disrupts synaptic homeostasis. Proc Natl Acad Sci U S A. 2014;111(44):E4769–78.
194. Sharma A, Lyashchenko AK, Lu L, Nasrabady SE, Elmaleh M, Mendelsohn M, et al. ALS-associated mutant FUS induces selective motor neuron degeneration through toxic gain of function. Nat Commun. 2016;7:10465.
195. Scekic-Zahirovic J, Sendscheid O, El Oussini H, Jambeau M, Sun Y, Mersmann S, et al. Toxic gain of function from mutant FUS protein is crucial to trigger cell autonomous motor neuron loss. EMBO J. 2016;35(10):1077–97.
196. Kino Y, Washizu C, Kurosawa M, Yamada M, Miyazaki H, Akagi T, et al. FUS/TLS deficiency causes behavioral and pathological abnormalities distinct from amyotrophic lateral sclerosis. Acta Neuropathol Commun. 2015;3:24.
197. Neumann M, Bentmann E, Dormann D, Jawaid A, DeJesus-Hernandez M, Ansorge O, et al. FET proteins TAF15 and EWS are selective markers that distinguish FTLD with FUS pathology from amyotrophic lateral sclerosis with FUS mutations. Brain. 2011;134(Pt 9):2595–609.
198. DeJesus-Hernandez M, Mackenzie IR, Boeve BF, Boxer AL, Baker M, Rutherford NJ, et al. Expanded GGGGCC hexanucleotide repeat in noncoding region of C9ORF72 causes chromosome 9p-linked FTD and ALS. Neuron. 2011;72(2):245–56.
199. Renton AE, Majounie E, Waite A, Simon-Sanchez J, Rollinson S, Gibbs JR, et al. A hexanucleotide repeat expansion in C9ORF72 is the cause of chromosome 9p21-linked ALS-FTD. Neuron. 2011;72(2):257–68.
200. van der Zee J, Gijselinck I, Dillen L, Van Langenhove T, Theuns J, Engelborghs S, et al. A pan-European study of the C9orf72 repeat associated with FTLD: geographic prevalence, genomic instability, and intermediate repeats. Hum Mutat. 2013;34(2):363–73.
201. Chew J, Gendron TF, Prudencio M, Sasaguri H, Zhang YJ, Castanedes-Casey M, et al. Neurodegeneration. C9ORF72 repeat expansions in mice cause TDP-43 pathology, neuronal loss, and behavioral deficits. Science. 2015;348(6239):1151–4.
202. Zhang YJ, Gendron TF, Grima JC, Sasaguri H, Jansen-West K, Xu YF, et al. C9ORF72 poly(GA) aggregates sequester and impair HR23 and nucleocytoplasmic transport proteins. Nat Neurosci. 2016;19(5):668–77.
203. O'Rourke JG, Bogdanik L, Muhammad AK, Gendron TF, Kim KJ, Austin A, et al. C9orf72 BAC transgenic mice display typical pathologic features of ALS/FTD. Neuron. 2015;88(5):892–901.
204. Peters OM, Cabrera GT, Tran H, Gendron TF, McKeon JE, Metterville J, et al. Human C9ORF72 hexanucleotide expansion reproduces RNA foci and dipeptide repeat proteins but not neurodegeneration in BAC transgenic mice. Neuron. 2015;88(5):902–9.
205. Liu Y, Pattamatta A, Zu T, Reid T, Bardhi O, Borchelt DR, et al. C9orf72 BAC mouse model with motor deficits and neurodegenerative features of ALS/FTD. Neuron. 2016;90(3):521–34.
206. Jiang J, Zhu Q, Gendron TF, Saberi S, McAlonis-Downes M, Seelman A, et al. Gain of toxicity from ALS/FTD-linked repeat expansions in C9ORF72 is alleviated by antisense oligonucleotides targeting GGGGCC-containing RNAs. Neuron. 2016;90(3):535–50.
207. Esanov R, Cabrera GT, Andrade NS, Gendron TF, Brown RH Jr, Benatar M, et al. A C9ORF72 BAC mouse model recapitulates key epigenetic perturbations of ALS/FTD. Mol Neurodegener. 2017;12(1):46.
208. Schludi MH, Becker L, Garrett L, Gendron TF, Zhou Q, Schreiber F, et al. Spinal poly-GA inclusions in a C9orf72 mouse model trigger motor deficits and inflammation without neuron loss. Acta Neuropathol. 2017;134(2):241–54.
209. Koppers M, Blokhuis AM, Westeneng HJ, Terpstra ML, Zundel CA, Vieira de Sa R, et al. C9orf72 ablation in mice does not cause motor neuron degeneration or motor deficits. Ann Neurol. 2015;78(3):426–38.
210. Atanasio A, Decman V, White D, Ramos M, Ikiz B, Lee HC, et al. C9orf72 ablation causes immune dysregulation characterized by leukocyte expansion, autoantibody production, and glomerulonephropathy in mice. Sci Rep. 2016;6:23204.

211. Sudria-Lopez E, Koppers M, de Wit M, van der Meer C, Westeneng HJ, Zundel CA, et al. Full ablation of C9orf72 in mice causes immune system-related pathology and neoplastic events but no motor neuron defects. Acta Neuropathol. 2016;132(1):145–7.
212. Burberry A, Suzuki N, Wang JY, Moccia R, Mordes DA, Stewart MH, et al. Loss-of-function mutations in the C9ORF72 mouse ortholog cause fatal autoimmune disease. Sci Transl Med. 2016;8(347):347ra93.
213. Ugolino J, Ji YJ, Conchina K, Chu J, Nirujogi RS, Pandey A, et al. Loss of C9orf72 enhances autophagic activity via deregulated mTOR and TFEB signaling. PLoS Genet. 2016;12(11):e1006443.

Fluid Biomarkers in Alzheimer's Disease and Frontotemporal Dementia

11

Niklas Mattsson, Sotirios Grigoriou, and Henrik Zetterberg

Abstract

Fluid biomarkers, including cerebrospinal fluid (CSF) biomarkers and blood-based biomarkers, may reflect different pathological processes in Alzheimer's disease (AD) and frontotemporal dementia (FTD). The most used biomarkers are CSF β-amyloid42, total tau, phosphorylated tau, and neurofilament light, which have been studied for diagnosis, prognosis, and treatment follow-up and in relation to genetics and neuroimaging. These biomarkers are now increasingly used in research, drug development, and clinical settings to increase our understanding of AD and FTD and to improve patient management. Recent progress in stable, automated assays for CSF biomarkers and ultrasensitive assays for blood-based biomarkers and the incorporation of fluid biomarkers in clinical practice and in clinical trials have accelerated the field. Key issues for further research include more extensive studies of the earliest stages of neurodegenerative

---

N. Mattsson, M.D., Ph.D. (✉)
Clinical Memory Research Unit, Faculty of Medicine, Lund University, Lund, Sweden

Memory Clinic, Skåne University Hospital, Malmö, Sweden

Department of Neurology, Skåne University Hospital, Lund, Sweden
e-mail: niklas.mattsson@med.lu.se

S. Grigoriou, M.D.
Department of Neurology, Skåne University Hospital, Lund, Sweden

H. Zetterberg, M.D., Ph.D.
Department of Psychiatry and Neurochemistry, Institute of Neuroscience and Physiology, The Sahlgrenska Academy at the University of Gothenburg, Mölndal, Sweden

Clinical Neurochemistry Laboratory, Sahlgrenska University Hospital, Mölndal, Sweden

Department of Molecular Neuroscience, UCL Institute of Neurology, London, UK

UK Dementia Research Institute, London, UK

© Springer International Publishing AG 2018
D. Galimberti, E. Scarpini (eds.), *Neurodegenerative Diseases*,
https://doi.org/10.1007/978-3-319-72938-1_11

diseases, better biomarkers for distinct proteinopathies, and the creation of universally accepted guidelines specifying the role of fluid biomarkers in relation to clinical measures and neuroimaging findings.

Keywords

Biomarker · Alzheimer · Frontotemporal lobe dementia · Tau · Amyloid · Neurofilament

Introduction

Fluid biomarkers are used ubiquitously by physicians and life scientists to measure normal physiology, pathological processes, and effects of therapeutic interventions [1]. In neurology, cerebrospinal fluid (CSF) biomarkers have been used to map brain diseases since the beginning of the twentieth century [2]. Modern technology makes it possible to measure CSF biomarkers related to accumulation of proteins, neuronal injury, inflammation, and other pathological processes. Several brain-derived biomarkers can also be measured in blood. This has been used in neurodegenerative diseases, including Alzheimer's disease (AD) and frontotemporal dementia (FTD), to provide early and accurate diagnosis, to elucidate disease mechanisms, and to facilitate development of new therapies.

We here provide an up-to-date review of CSF and blood-based biomarkers in AD and FTD. This is an updated version of a review written in 2013 (chapter by Mattsson and Zetterberg in [3]). We therefore focus mainly on papers published in 2013–2017. Priority was given to original papers and meta-analyses, but we occasionally direct the reader to specialized reviews for subtopics.

The Brain, Cerebrospinal Fluid, and Blood

Due to its proximity to the brain parenchyma, CSF may be the most useful fluid for measuring biomarkers related to brain physiology. CSF is a clear liquid which occupies the ventricles and the subarachnoid space around the brain and the spinal cord [4]. CSF is essentially a highly diluted filtrate of plasma (about 99% water), which is mainly produced by the choroid plexus in the ventricles, but also released from other structures, including blood vessels and the remaining ventricular ependyma. CSF circulates from the ventricles deep inside the brain to the subarachnoid space and is reabsorbed to the venous blood stream through arachnoid granulations, as well as through meningeal lymphatic vessels [5]. The discovery of the so-called glymphatic system has helped explaining how subarachnoid CSF may enter and exit the brain along paravascular spaces and clear the brain parenchyma from extracellular metabolites and other breakdown products [6]. Histologically, CSF is in close contact with the cells of the CNS and is not separated from the brain tissue by the blood–brain barrier.

CSF has several normal functions, including the creation of neutral buoyancy for the brain, to reduce its net weight and protect the blood supply and the integrity of neurons, especially in the lower sections of the brain; supplying nutrients, peptides, and hormones to widespread neuronal networks; clearing waste products from the normal metabolism into blood stream; providing mechanical protection for the brain, by distributing the impact of an incoming force; and helping to maintain a constant intracranial pressure.

The total CSF volume in an adult human is about 150 mL, with a formation rate of about 0.4 mL per minute, and an overall turnover rate of about 3–4 volumes per day [4]. CSF in the caudal lumbar sac is available for sampling by lumbar puncture. The normally acquired volume is about 10–20 mL, which is quickly replenished. Lumbar puncture is a relatively easy procedure that can be performed on outpatients [7]. The only significant complication is headache, which has an incidence that is most often reported to be between 2 and 10%, depending primarily on age (older people have low incidence) [7–10]. The headache is often mild, can be symptomatically treated, and resolves by itself within a day or two.

The other major biofluid for biomarker analysis is blood (serum or plasma), which communicates with the brain and CSF compartments through the glymphatic system [6]. There are several issues, both biological and technical, with the measurement of CNS-related biomarkers in blood, however. First, a biomarker that has its origin in the CNS has to cross the blood–brain barrier in order to be detected in the periphery and, if the concentration is low in CSF, it will be even lower in the blood due to the blood:CSF volume ratio causing a substantial dilution. Second, if the biomarker is not specific for the CNS but also expressed in peripheral tissues that may be injured in trauma, the contribution from CNS will potentially drown in the high biological background caused by non-CNS sources (a good tool to assess the risk for this is the publicly available web-based Human Protein Atlas, http://www.proteinatlas.org/, which presents mRNA and protein expression in 44 different human tissues of close to 20,000 proteins). Third, the huge amount of other proteins in blood (e.g., albumin and immunoglobulins) introduces analytical challenges due to possible interference. Fourth, heterophilic antibodies may be present in blood, which may interfere in immunoassays. Fifth, the analyte of interest may undergo proteolytic degradation by various proteases in plasma. Sixth, clearance of the biomarker in the liver or by the kidneys may introduce variability. Nevertheless, recent advances in ultrasensitive measurement techniques have overcome some of these hurdles, and the field of blood biomarkers for CNS diseases now looks much more promising and several biomarker candidates, reviewed below, exist.

Alzheimer's Disease

AD is characterized by the presence of extracellular amyloid-β (Aβ) pathology and intracellular tau pathology. AD is believed to have a preclinical stage, when Aβ pathology appears. This is followed by clinical stages, when tau pathology spreads throughout the brain, in a process that is paralleled by hypometabolism, atrophy, and cognitive decline. The cognitive impairment is dominated by memory loss but may also include language, visuospatial, and executive dysfunction and in rare cases even motor problems. The overall duration from preclinical debut to advanced dementia stages may be several decades.

The literature on fluid biomarkers in AD is huge, with thousands of papers on diagnosis, prognosis, and associations with neuroimaging and treatment effects. But most of these papers deal only with three biomarkers: CSF β-amyloid1-42 (Aβ42), total tau (T-tau), and phosphorylated tau (P-tau). These are sometimes referred to as the core AD biomarkers and have been incorporated into research diagnostic criteria for AD, as presented by both the international working group (IWG) [11–13] and the National Institute on Aging-Alzheimer's Association (NIA-AA) workgroup [14–16]. These criteria emphasize both that CSF Aβ42, T-tau and P-tau may be used to identify clinical AD and that CSF Aβ42 may be altered already in preclinical stages of AD, prior to any symptoms. Several other CSF biomarkers have also been studied in AD (Table 11.1). The last few years have also seen a rising interest in blood-based biomarkers for AD.

Table 11.1 Biomarkers in AD and FTD

Biomarker	Main pathological feature	AD	FTD
Aβ-related biomarkers			
CSF Aβ42	Aβ pathology	↓↓	↓/−
CSF Aβ42:Aβ40 ratio	Aβ pathology	↓↓	−
CSF BACE1	Altered Aβ metabolism	↑	?
CSF α-sAPP and β-sAPP	Altered Aβ metabolism	↑/−	↓/−
CSF Aβ oligomers	Altered Aβ metabolism	?	?
Plasma/serum Aβ40 and Aβ42	Aβ pathology and/or extra-cerebral APP metabolism	↑/−/↓	?
Tau-related biomarkers			
CSF T-tau	Neurodegeneration	↑↑	↑
CSF P-tau	Neurodegeneration/tau pathology	↑↑	−
Plasma/serum tau	Neurodegeneration	↑/−	?
Other biomarkers of degeneration and proteinopathy			
CSF neurogranin	Synaptic degeneration	↑↑	↓
CSF NFL	Neurodegeneration	↑	↑↑
Plasma/serum NFL	Neurodegeneration	↑	↑↑
Plasma/CSF TDP-43	TDP-43 pathology and/or extra-cerebral TDP-43 metabolism	?	↑/−
Plasma/CSF progranulin	*GNR* mutation status	−	↓
CSF ubiquitin	Neurodegeneration	↑	−
Inflammation-related markers			
CSF YKL-40	Astrocytosis/microglial activity	↑	↑
CSF sTREM2	Microglial activity	↑	?
CSF IL-8	Neuroinflammation	↓	?

The table summarizes the key biomarkers discussed in this review. The main biomarker changes in AD and FTD are presented as increases (marked ↑↑ or mild ↑), decreases (marked ↓↓ or mild ↓), or no significant changes (−). We list the main features associated with each biomarker, but other processes may also contribute to altered biomarker levels, as explained in the main text. Note that this selection does not constitute a comprehensive list of all studied fluid biomarkers in AD and FTD

Frontotemporal Dementia

FTD refers to a spectrum of heterogeneous neurodegenerative disorders [17], which collectively represents one of the most common causes of early-onset dementia. The diagnosis of FTD versus other dementias and the differentiation of different FTD variants are important both for clinical practice and research but can be very challenging. FTD biomarkers are actively being explored by many researchers, but disease-specific biomarkers are still lacking.

Clinically, FTD is characterized by changes in behavior, executive dysfunction, and/or language impairment. The most common type of FTD is the behavioral variant (bvFTD), with behavioral change, inappropriate social conduct, and executive dysfunction. Another type is the semantic variant of primary progressive aphasia (svPPA, also called semantic dementia, SD), with progressive language decline and speech difficulties leading to fluent speech with anomia, impaired single word comprehension, and surface dyslexia due to loss of semantic memory. A third type is the nonfluent variant of PPA (nfvPPA, also called progressive nonfluent aphasia, PNFA) [17], with effortful speech production with agrammatism, apraxia of speech, and impaired sentence comprehension. The logopenic variant PPA (lvPPA), which leads to word-finding pauses and impaired sentence repetition, is mostly associated with AD pathology [18]. Finally, there is a significant clinical, pathological, and genetic overlap between FTD and amyotrophic lateral sclerosis (ALS), progressive supranuclear palsy (PSP), and corticobasal syndrome (CBS).

The clinical FTD variants may be caused by several different underlying frontotemporal lobar degeneration (FTLD) pathologies, which partly overlap between the clinical variants. FTLD pathologies are classified based upon their predominant neuropathological protein. Most patients have a dominance of either (1) tau (FTLD-Tau), seen in about 35–50% of patients; (2) TAR DNA-binding protein-43 (FTLD-TDP), seen in about 50% of patients; or (3) fused in sarcoma protein (FTLD-FUS), seen in most remaining patients [17].

Biomarkers for Pathological Processes in AD and FTD

We now turn to the most studied biomarkers in AD and FTD. We focus on biomarkers related to Aβ, tau, axonal degeneration, synapses, inflammation, and some FTLD-related inclusions. In general, we suggest that each biomarker should be regarded as an indicator of a specific pathological process. Since these may partly occur in several different diseases, most individual biomarkers are not disease-specific. It is therefore crucial to interpret biomarker results in the context of other biomarkers, clinical presentation, and neuroimaging.

Amyloid-β-Related Biomarkers

Several biomarkers related to Aβ metabolism have been explored in dementing diseases, especially in AD.

Aβ Peptides

Aβ peptides are derived from the type-I transmembrane amyloid precursor protein (APP), which is ubiquitously expressed by neurons in the brain. APP can be processed by different enzymes, including α-secretase, β-secretase (BACE1), and γ-secretase. The combined activity of BACE1 and γ-secretase leads to production of Aβ peptides of different lengths, where the most studied species is Aβ1-42 (Aβ42). CSF Aβ42 is reduced in AD patients [19] and correlates inversely with brain Aβ accumulation, both in neuropathology [20, 21] and PET Aβ imaging studies [22, 23]. The main theory for this is that Aβ42 is sequestered in plaques and thus has limited access to CSF in the presence of Aβ pathology. Hypothetically, CSF Aβ42 may also be reduced by other processes, including altered release of Aβ, formation of Aβ oligomers that are not detected by common assays [24], binding of Aβ to other proteins that block antibody epitopes [25], or intracellular Aβ accumulation [26]. Infection and inflammation may also affect APP metabolism and lower CSF Aβ peptide levels without formation of plaques [27–31].

Reduced CSF Aβ42 has been reported not only in AD, but also in some patients with vascular dementia [32], Lewy body dementia [33], FTD [34], Creutzfeldt-Jakob's disease [35], ALS [36], and multiple system atrophy [37]. In some cases, this could represent actual brain Aβ aggregation, for example, in cases of AD comorbidity [38]. But it is also possible that CSF Aβ42 could be reduced by other mechanisms than brain Aβ aggregation, for example, as a consequence of white matter pathology [39].

Besides Aβ42, APP processing may also give rise to many other Aβ isoforms, which to varying degrees are present in plasma and CSF (Fig. 11.1). One of these is

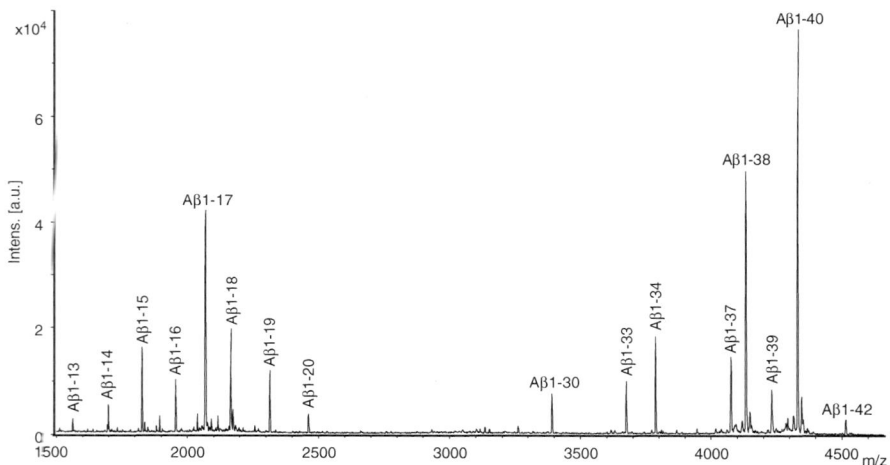

Fig. 11.1 Aβ peptides present in CSF. A large number of different Aβ variants are present in CSF, besides the commonly studied Aβ1-40 and Aβ1-42. The figure shows different Aβ isoforms present in normal human CSF, as detected by immunoprecipitation and matrix-assisted laser desorption/ionization time-of-flight mass spectrometry (IP-MALDI-TOF-MS) using the anti-Aβ antibodies 6E10 and 4G8. Courtesy of Erik Portelius, University of Gothenburg

Aβ1-40 (Aβ40). Although some patients with Aβ pathology have increased CSF Aβ40 levels [40], Aβ40 is generally unaltered in AD. The ratio between Aβ42 and Aβ40 appears to be a better indicator of Aβ pathology and AD than Aβ42 alone [41]. Other Aβ peptides include C-terminally truncated peptides formed by γ-secretase cleavage (e.g., Aβ1-37 and Aβ1-38 [42]), C-terminally truncated short isoforms formed by combined β-secretase and α-secretase activity (e.g., Aβ1-16 [43, 44]), and N-terminally truncated isoforms formed by other enzymatic activities (e.g., Aβ5-40 [45]). These other isoforms of Aβ are explored in different settings but have not yet become established as disease biomarkers.

Aβ Generating Enzymes

BACE1, which has a rate-limiting function in the formation of Aβ peptides, exists in a soluble form that is measurable in CSF [46]. CSF BACE1 activity may be increased in AD or MCI [47–49], but this has not been replicated in all studies [50]. One possibility is that BACE1 levels are increased only early in the disease [50, 51].

The other enzyme that participates in formation of Aβ, γ-secretase, is a general proteolytic enzyme residing in the cellular membrane. It has more than 100 known substrates, and several of these are present in CSF. One of these is alcadein, which is processed by γ-secretase into several smaller peptides, like APP. Some alcadein peptides are present in CSF, and this may be useful to explore γ-secretase function in humans [52].

sAPP Peptides

APP processing also gives rise to the N-terminal soluble fragments sAPP-α (formed after α-secretase cleavage) and sAPP-β (formed after BACE1 cleavage). A few studies have found increased CSF sAPP-α or sAPP-β in MCI or AD (especially in subjects with pathological CSF Aβ42 or T-tau) [53, 54], but not all studies have replicated this [55]. Furthermore, some studies (but not all [56]) have found reduced CSF sAPP-α or sAPP-β in FTD [57–59].

Aβ Oligomers

Several studies have measured CSF levels of Aβ oligomers, although these are difficult to quantify and characterize, and results have varied [60–64]. There has been little progress in this field during recent years, and it is clear that CSF Aβ oligomers represent a difficult biomarker category.

Blood-Based Measures of Aβ

In plasma, current assays allow for the measurement of Aβ40 and Aβ42 (although additional species most likely exist), and several studies have examined their association with dementia, AD, and/or cerebral β-amyloidosis. However, the results are less clear than those derived from CSF or PET studies and the significant associations, if any, are weak and go in either direction [65]. As the correlation of CSF with plasma Aβ concentrations is low [66], it is possible that most of the Aβ peptides measured in plasma are derived from extra-cerebral sources such as platelets in which APP expression is high. If ultrasensitive assays are used, samples can be diluted which may mitigate matrix effects that may disturb the measurement of Aβ. Using such an assay, there were weak positive correlations between plasma and CSF concentrations for both Aβ42 and Aβ40 and negative correlations between plasma Aβ42 and neocortical amyloid deposition (measured with PET) [66]. These disease-related changes were not clear enough to be diagnostically useful, but the data still represent a step forward towards a blood test for cerebral β-amyloidosis.

Tau-Related Biomarkers

Tau is a neuronal protein, which is mainly found in thin, unmyelinated, cortical axons, where it stabilizes microtubule and facilitates axonal transport mechanisms. Alternative splicing of exon 10 leads to tau isoforms with three (3R-tau) or four (4R-tau) microtubule-binding repeat domains with only 3R-tau in embryonic brain and comparable levels of 3R- and 4R-tau in normal adult brain [67]. In general, released tau is thought to reflect at least two different processes, namely, neuronal injury and accumulation of tau aggregates.

Total Tau

T-tau denotes tau proteins measured by unspecific tau assays. CSF T-tau is increased in many diseases with significant neuronal loss [68–70]. The highest CSF T-tau concentrations are seen in conditions with the most severe injury, including stroke and Creutzfeldt-Jakob's disease.

Phospho-Tau

P-tau denotes tau proteins phosphorylated at specific threonine or serine residues. Phosphorylation leads to altered properties of tau and may cause it to aggregate into paired helical filaments and neurofibrillary tangles. Some studies have found that CSF P-tau correlates to the amount of neurofibrillary tangles and phosphorylated tau in the brain [23, 71, 72]. However, the correlations are modest, and sometimes seen also for T-tau, and not replicated in all studies [73]. Furthermore, it is not clear why CSF P-tau is increased in AD, but not in other dementias with neurofibrillary tangles. CSF P-tau may also be elevated in the absence of tangles, for example, in some cerebral infections [27], and during normal brain development [74].

Tau Isoforms

It is possible that specific isoforms or modified variants of tau may give additional information in neurodegenerative diseases. For example, a study using 3R/4R-tau-specific assays revealed selective decreases of 4R-tau in CSF of PSP and AD patients compared with controls and lower 4R-tau levels in AD compared with Parkinson's disease with dementia [75]. It has also been suggested that neurons with *MAPT* mutations have reduced release of extracellular tau [76], which could partly explain why FTD patients with *MAPT* mutations lack prominently increased CSF tau levels.

Blood-Based Measurements of Tau

Tau has recently been measured also in serum and plasma using ultrasensitive technologies. AD patients have slightly increased plasma tau concentration, but not sufficient for clinical use [77].

Synaptic Biomarkers

Synaptic loss is a hallmark of AD [78]. Several presynaptic and postsynaptic proteins, including rab3A, synaptotagmin [79], growth-associated protein 43 (GAP43), synaptosomal-associated protein 25 (SNAP25) [80], and neurogranin, have been identified in CSF using protein purification and mass spectrometric techniques [81].

One of the most promising synaptic markers is neurogranin, which is increased in CSF in AD patients, where it correlates with tau proteins [82–84]. A striking and presently unexplained result regarding neurogranin is its AD specificity [85, 86]. Whereas AD patients have robust increases in CSF neurogranin concentration, FTD patients show low concentrations.

Neurofilament Light

NFL is one of three neurofilament proteins (the others are the heavy [NFH] and intermediate [NFM] chains), which are important cytoskeletal proteins, predominantly found in large diameter myelinated axons. CSF NFL is markedly increased in several conditions with neuronal injury, including acute cerebral infarctions and vascular dementia [87], white matter disease [88], FTD [89], CBD [90], and ALS [91]. Recently, CSF NFL has been shown to be slightly increased also in AD [92]. Overall in dementia, CSF NFL correlates with more severe cognitive impairment and shorter survival [93]. CSF NFL likely represents another pathway of neuronal injury than tau-related biomarkers since elevated CSF NFL predicts neurodegeneration independently of T-tau and P-tau [83].

Blood-Based Measurements of NFL

NFL can also be measured also in serum and plasma. Slightly increased plasma NFL is seen in AD [94]. Blood-based NFL is also increased in PSP and CBS [95], as well as in FTD [96].

TDP-43

Several studies have explored fluid biomarkers linked to TDP-43, which is a major FTLD proteinopathy. Plasma and CSF levels of TDP-43 may be increased in FTD and ASL [97–99], although the results have varied for different FTLD mutations [100] and for different variants of TDP-43 [101]. One criticism against using CSF TDP-43 measurements as a biomarker of brain pathology is that the origin of blood-based TDP-43 is unclear, and blood-based TDP-43 may contaminate the CSF measurements [102].

Progranulin

Up to 10% of FTD cases are caused by mutations in the *GRN* gene, which encodes the secreted protein progranulin [103]. *GNR* mutations have no specificity to any clinical entity in the FTD spectrum, and different diseases may even appear among mutation carriers within the same families. But individuals carrying *GNR* mutations have reduced plasma (and CSF) levels of progranulin, enabling screening tests with high sensitivity and specificity for mutation carriers versus controls or patients with other dementias [103, 104].

Ubiquitin

Ubiquitin is a small protein which can be attached to proteins as labeling for subsequent degradation. The ubiquitin-proteasome system may be impaired in neurodegenerative diseases [105], and CSF ubiquitin has been reported to be increased in AD compared to controls and FTLD [106].

Inflammation and Microglial and Astrocytic Activation

Many markers of inflammatory activity are altered in neurodegenerative diseases [107]. Several of these are believed to be related to microglia activity, including chitotriosidase activity [108] and concentrations of YKL-40 [109], which are upregulated in CSF from AD and FTD patients [85, 110]. Recent reports suggest that the CSF concentration of the secreted ectodomain of triggering receptor expressed on myeloid cells 2 (sTREM2), a molecule that is selectively expressed on microglia in the CNS and genetically linked to AD, is increased in AD in a disease-specific manner and correlates with CSF T-tau and P-tau [111–113]. Other inflammatory markers, including IL-8, may be reduced in both CSF and serum of AD patients [114].

Biomarkers in AD

While the previous sections focused on individual biomarkers, we will now change perspective and focus on AD.

The Dynamic Biomarker Model

Many studies have found that AD dementia patients have about 50% reduced levels of CSF Aβ42, and several times increased CSF T-tau and P-tau levels, compared to cognitively healthy controls [115], with 80–85% sensitivity and specificity [116]. These biomarker changes are thought to start decades before patients become demented and may develop in a specific sequence, as summarized in "the dynamic biomarker model" [117]. The presumed ordering of biomarkers in this model follows the amyloid cascade hypothesis, which states that the initial pathological event in AD is abnormal aggregation of Aβ peptides and that this secondarily leads to neuroinflammation, synaptic dysfunction, tau pathology, and neuronal degeneration [118]. In line with this, CSF Aβ42 is thought to change before CSF T-tau and P-tau, both in autosomal-dominant AD [119] and sporadic AD [120, 121]. However, novel results using longitudinal data suggests that subtle effects on brain metabolism and cognition may appear several years before the conventional threshold for Aβ biomarker positivity is reached [122]. The second proposition of the dynamic biomarker model is that the biomarkers' trajectories are sigmoid. This is based on several sources, including the finding that CSF Aβ42, T-tau, and P-tau are stable in clinical stages of AD [123], suggesting that they have reached a plateau phase, as well as autopsy studies showing that amyloid accumulation plateaus with increasing disease duration [124]. Ultimately proving that the trajectories are sigmoid requires longitudinal studies with multiple time-points per subject, but most data published so far have had short follow-up or been cross-sectional with derived longitudinal measurements based on cognitive scales [125].

Biomarkers in Preclinical AD

AD can be identified by fluid biomarkers prior to clinical symptoms [16]. The earliest definitive biochemical alteration in preclinical AD is thought to be reduced CSF Aβ42. However, CSF biomarkers related to inflammation predict future decline of CSF Aβ42 already in healthy controls, suggesting that CSF biomarkers may be used to detect inflammatory activities that are important for development of the first stages of AD pathology [126]. Also, on average, cognitively healthy people with low CSF Aβ42 have increased CSF T-tau and P-tau (and other markers of neuronal injury), suggesting that deleterious effects on axons and tau metabolism are partly present in the preclinical stage of AD [127]. Baseline CSF Aβ42, and sometimes CSF tau, predict future impairment in people who are cognitively normal [128–131] or who have subjective cognitive impairment [132, 133]. Combinations of pathological CSF Aβ42 and tau may be more likely to result in cognitive impairment than individual biomarker positivity [134, 135]. Besides future cognitive impairment, reduced CSF Aβ42 is also linked to increased brain atrophy rates in cognitively healthy controls [136].

The predictive accuracy of biomarkers to determine future cognitive decline may be increased by also adjusting for factors that are related to cognitive reserve, such as age, education, and brain volume [137]. For example, among cognitively normal people with high levels of T-tau or P-tau, long education and large brain volumes are related to slower development of cognitive impairment, suggesting that the preclinical disease indicated by elevated tau levels results in symptoms later in subjects who also have protective factors [138].

Biomarkers in Clinical AD

The earliest clinical stage of AD may be referred to as prodromal AD, or mild cognitive impairment (MCI) due to AD. This is a stage with objective cognitive dysfunction that does not interfere significantly with daily functioning. In general, MCI patients have increased risk of progression to dementia, but from the clinical symptoms alone it is difficult to predict when or if an individual patient will progress and to determine which underlying pathology causes the symptoms. The current status and challenges for CSF Aβ42, T-tau, and P-tau in prodromal AD have recently been reviewed [139]. In sum, CSF Aβ42, T-tau, and P-tau are altered already in MCI patients that later progress to AD dementia, with sensitivities and specificities 70–90% compared to patients who develop other dementias or remain cognitively stable [140–143]. The negative predictive values are around 90%, while the positive predictive values vary from 60 to 90%. Although the diagnostic accuracies decrease with age (mainly due to increased accumulation of Aβ in non-demented subjects, i.e., more prevalence of preclinical AD), CSF Aβ42, T-tau, and P-tau still have stable positive and negative predictive values for AD dementia in older age groups [144].

Different definitions of MCI exist, and some studies only include "amnestic" MCI patients, while others include unselected MCI with or without dominating amnestic symptoms. Such differences may contribute to a variability in results between studies. Carefully controlled mono-center studies may achieve very high diagnostic accuracies. For example, in one longitudinal MCI study, the combination of CSF Aβ42 and T-tau had a sensitivity of 95% and a specificity of 83% for conversion to AD dementia at a median follow-up of 5 years [140]. When the same study population was evaluated at 9 years follow-up, the ratio of Aβ42 to P-tau at baseline had sensitivities and specificities 85–90% for future AD dementia [120]. Furthermore, among MCI patients with biomarker evidence of Aβ pathology, high T-tau and P-tau are associated with shorter time to dementia [120, 145].

CSF Aβ42 reaches a plateau already in the preclinical or early clinical stage, when Aβ pathology is widespread throughout the brain. Once AD patients reach the dementia stage, CSF T-tau and P-tau are also essentially stable [123, 146, 147]. The fact that CSF T-tau and P-tau are relatively stable throughout the clinical stages of the disease suggests that their concentrations are proportional to the rate of neuronal loss rather than to the accumulated loss [123, 148].

Biomarkers in Autosomal-Dominant AD

In autosomal-dominant familial AD, the known deterministic relationship between mutations and future clinical disease provides a unique opportunity to investigate preclinical biomarker changes. A cross-sectional study on presymptomatic mutation carriers in a Colombian kindred found that CSF Aβ42 levels were increased in mutation carriers more than two decades prior to expected age of symptom onset [149]. This was in line with experimental data showing that similar mutations resulted in increased Aβ42 production. In a study by the Dominantly Inherited Alzheimer Network (DIAN), CSF Aβ42 levels started to decline about 25 years before expected symptom onset, and this was about 10 years earlier than any other biomarker alteration, including increased CSF tau [119]. When testing mutation carriers closer to onset of dementia, other studies have found reduced CSF Aβ42 and increased CSF T-tau and P-tau [150–152].

Biomarkers in Atypical Variants of AD

Besides the typical amnestic form of AD, some patients may present with a predominance of language deficits, visuospatial deficits, or behavioral/executive deficits. In general, CSF Aβ42, T-tau, and P-tau do not differ between these clinical presentations of AD [153].

Biomarkers in FTD

We now turn towards biomarkers for FTD. The combination of high CSF Aβ42 and low CSF P-tau has high discrimination for FTD versus AD [154, 155]. The elevated CSF NFL in many patients with FTD may also help to discriminate against AD, but not against many other dementias [93, 156].

One important factor that may influence CSF biomarkers diagnostic performance is the clinical and pathophysiological heterogeneity of the FTLD spectrum. The different FTD and FTLD variants could potentially have different profiles of CSF biomarkers. One common finding is that FTLD-TDP have lower CSF P-tau/T-tau ratio compared to FTLD-Tau [157–159]. FTLD-TDP patients may also have higher CSF NFL [160]. Patients with PSP have been reported to have reduced CSF T-tau and P-tau levels compared to controls [161], which may hypothetically be a consequence of altered processing of tau in PSP.

CSF biomarkers may be associated with disease severity in FTD. For example, CSF (and serum) NFL is elevated in symptomatic but not presymptomatic FTLD mutation carriers [162]. High CSF NFL also correlates with neuropsychological measures and atrophy in FTD [156]. Both high CSF T-tau [163] and CSF NFL [93, 159] are associated with shorter survival in FTD. Low CSF Aβ42 levels have been associated with worse general cognitive function and worse executive function in patients with bvFTD [164].

Biomarkers in Clinical Trials

Fluid biomarkers may facilitate drug development in neurodegenerative diseases by (1) enrichment of participants with underlying specific pathologies, (2) measurement of pharmacodynamic effects, and (3) monitoring of toxicity and side effects. Most of this work has been done in AD.

Biomarkers to Enrich Study Populations

Most early AD drug trials included patients based only on clinical characteristics. This may have resulted in inclusion of some participants who did not have underlying AD pathology. This has now changed, since most novel AD trials use biomarkers, especially for Aβ, to select study participants [165]. This is believed to increase the power and lower the costs of trials, although savings are partly offset by prolonged trial duration, since biomarker-based enrichment means that more study subjects must undergo screening. One alternative may be to use basic demographic information together with *APOE* genotype information in a prescreening, to select people for further phenotyping with biomarkers [166]. The European Medicines Agency (EMA) supports the use of CSF Aβ42 and T-tau to enrich clinical populations with prodromal AD [167]. At the point of writing, the US counterpart agency Food and Drug Administration has still not released a corresponding statement. There is also an interest in using Aβ biomarkers to identify participants for preclinical prevention studies [168]. However, Aβ information alone may be insufficient to reach adequate power in preclinical trials, and combinations of biomarkers [169], or incorporation of measures of cognition may also be needed to select suitable trial participants [170].

Biomarkers of Toxicity and Side Effects

Fluid biomarkers may detect signs of drug-induced side effects, including meningoencephalitis, which was a side effect of active Aβ immunotherapy in early trials [171]. CSF profiling at baseline may also identify immunoactivities that are present already before treatment (e.g., chronic infection or inflammation) to avoid the risk of misinterpreting inflammatory reactions as adversary effects [172].

Biomarkers of Treatment Effects

Biomarkers of drug effects may be classified as primary, secondary, or exploratory pharmacodynamic biomarkers.

Primary biomarkers reflect the intended drug target, for example, CSF measurements of Aβ metabolism for anti-Aβ therapies. Proof-of-concept studies have shown that several classes of therapies directed against Aβ, including aggregation inhibitors [173], BACE1-inhibitors [45, 174, 175], and γ-secretase inhibitors and modulators [176], result in altered CSF (and plasma) levels of different Aβ-related biomarkers. Many different Aβ peptides are potentially useful to measure treatment response. For example, γ-secretase inhibition resulted in increased CSF levels of short Aβ isoforms, such as Aβ1-14, Aβ1-15, and Aβ1-16, and increased levels of long isoforms, from Aβ1-17 and up [177]. Other Aβ peptides (Aβ5-40 and Aβ5-42) are upregulated by BACE1 inhibition ([45], p. 1). Measurement of these peptides may be a useful complement to the core biomarkers for specific drug classes.

Secondary pharmacodynamic biomarkers reflect effects on pathological processes downstream of the intended drug target. This includes CSF tau for anti-Aβ drugs since reduced CSF tau levels may indicate reduced axonal degeneration after successful blockage of pathological Aβ metabolism. Some Aβ immunotherapy trials have reported reduced CSF tau levels in patients receiving active treatment, suggesting beneficial drug effects on axonal degeneration [178, 179].

CSF biomarkers may also be used as exploratory pharmacodynamic biomarkers, to identify novel drug effects. For example, in presymptomatic carriers of PSEN1 mutations, HMG-CoA reductase inhibitors lowered CSF sAPP-α and sAPP-β levels, without changing CSF Aβ42, P-tau, or T-tau, suggesting that the treatment interfered with APP processing, but not with Aβ plaque pathology or axonal degeneration [180].

Surrogate Biomarkers

The term surrogate biomarker is a regulatory term, indicating a measurement that may serve as a surrogate for a clinical outcome in a specific treatment [181]. The regulatory framework for surrogate markers is stringent and requires extensive studies of drug effects on both clinical outcome and biomarker response. The extensive studies necessary to qualify a surrogate marker are essentially the same studies that the surrogate was intended to avoid, making the number of surrogate biomarkers in all of medicine very small. Fluid biomarkers are unlikely to have broad use as surrogate markers in the regulatory meaning anytime soon. However, if multiple AD drugs show clinical effects coupled to a specific biomarker response, it may result in the qualification of a surrogate biomarker, facilitating the development of coming generations of AD drugs.

Biomarkers and Genetics

The concept of using fluid biomarkers to enrich genetic studies with patients with AD pathology and to exclude preclinical AD from the controls is supported by a study showing that the odds ratio of *APOE* increased from 4 to around 10 when combining clinical with fluid biomarker data [182]. However, another study failed to show any association between the AD risk genes *BIN1*, *CLU*, *CR1*, and *PICALM* and CSF Aβ42 and P-tau, despite being powered to detect very small effects, suggesting that some AD risk genes mediate risk through Aβ- and tau-independent mechanisms [183]. CSF biomarkers have also been used as quantitative traits for genetic analysis, to find new risk loci for AD [184].

Biomarkers and Imaging

Several studies have compared fluid biomarkers with neuroimaging in AD and FTD, primarily using PET imaging of Aβ and tau, and structural and functional MRI. A main finding is that CSF Aβ42 and PET Aβ overall have similar diagnostic accuracy for AD [185]. However, studies comparing CSF Aβ42 and PET Aβ imaging typically identify a proportion of subjects with reduced CSF Aβ42 levels despite normal PET Aβ signal [186]. Direct comparisons of CSF Aβ42 and PET Aβ, using several different assays for CSF Aβ42 and several different PET tracers, suggest that reductions in CSF Aβ42 may occur slightly prior to increases in PET Aβ, and be especially common in cognitively healthy controls [187, 188].

The classification of controls, MCI and AD dementia, and the prediction of conversion from MCI to AD dementia may be improved by combining CSF and imaging markers (structural MRI [189–191] and functional imaging with FDG PET [192]). One study in 250 MCI patients found that the addition of CSF tests to standard clinical and imaging tests improved the predictive accuracy of future dementia in 56% of the participants [193].

In early clinical stages of AD, CSF Aβ42 and T-tau at baseline are correlated with longitudinal hippocampal atrophy rates [194]. In cognitively healthy elderly, reduced CSF Aβ42 and increased CSF P-tau have been correlated with increased brain atrophy rates [136, 195]. However, there is much heterogeneity in biomarker patterns among healthy controls and MCI subjects. For example, healthy controls with MRI gray matter loss indicative of AD are at risk of developing cognitive impairment, but only 60% of those with an AD-like pattern have reduced CSF Aβ42, compared to 19% of those without [196]. Considering the dynamic biomarker model, it may be surprising that 40% of healthy controls with AD-like brain atrophy have nonreduced CSF Aβ42 levels. This suggests that AD-like brain atrophy may develop without concomitant brain Aβ pathology as measurable by current available methods [197].

Biomarker Technologies

Biomarker research may be done either by targeted methods, where a pre-hoc identified molecule is tested for a certain performance, or general methods, where many different molecules are screened and tested simultaneously. Furthermore, identification of novel biomarkers may be done using either clinical information (e.g., comparing biomarker levels between controls, MCI or AD), or a biological trait, for example, Aβ42 pathology or tau pathology, as measured by CSF biomarkers.

The most commonly used assays for Aβ42 include an enzyme-linked immunosorbent assay (ELISA) [198], a bead-based multiplex assay for the xMAP platform [199] (both the ELISA and the xMAP assay measure peptides containing the N-terminal 1st amino acid and the C-terminal 42nd amino acid of the Aβ sequence, Aβ1-42), and a plate-based multiplex assay for the Meso Scale Discovery platform [200] (which also detects N-terminal truncated isoforms, AβX-42, although these have minor concentrations relative to Aβ1-42). These assays are believed to measure monomeric Aβ42, rather than aggregated or oligomeric peptides, but concentrations correlate well with the total Aβ42 amount, as measured by a selected reaction monitoring mass-spectrometry method [201]. The most commonly used assays for T-tau and P-tau are also immunoassays, where T-tau assays are constructed to be independent of tau phosphorylation state [202] and P-tau assays are constructed to be specific to phosphorylated tau (typically at amino acid residues 181 or 231 [199, 203, 204]). A common multiplex xMAP assay simultaneously measures CSF Aβ42, T-tau, and P-tau [199]. Different technologies report different absolute quantifications and may also differ in terms of specific molecules that they actually measure. However, comparisons between ELISA, xMAP, and Meso Scale Discovery show good agreement between the different technologies, especially for T-tau and P-tau [205], and conversion factors may be used to transfer data between technologies.

Mass spectrometry-based methods have been used to identify and quantify a large number of different Aβ isoforms (Fig. 11.1), which may be used both for clinical applications and basic research [43, 206]. In combination with mass spectrometry, Stable Isotope Labeling Kinetics (SILK) may be used to measure production and clearance rates of proteins. For this, subjects are administered a stable isotope-labeled amino acid (e.g., $^{13}C_6$ leucine), which becomes incorporated into proteins during normal protein synthesis. Body fluid samples, including CSF, may then be analyzed to compare fractions of labeled versus unlabeled proteins. This technique has been used to determine production and clearance rates of Aβ peptides [207–209].

There is an ongoing rapid development of large-scale fully automated systems, which will facilitate measurements of CSF Aβ42, T-tau, and P-tau outside expert centers [210]. There are currently two fully automated platforms for the measurement of CSF Aβ42 (Cobas from Roche and Lumipulse from Fujirebio).

Finally, there has been a rapid development in regard to ultrasensitive measurement techniques [211]. Most of these rely on antibody-based detection of the target molecule; but in single molecule array (Simoa), the detection reaction is compartmentalized into a small volume (50 femtoliters), so that the reporter molecule accumulates at a very high concentration; in single molecule counting (Singulex), the labeled detection antibodies, specifically captured by the target molecule/capture antibody complex, are released and counted one by one in a small detection cell, which allows for a single molecule readout; and in proximity extension assay (PEA), partly overlapping complementary DNA strands are attached to the different antibodies allowing the strands to form a polymerase chain reaction-amplifiable template if immobilized close to each other on the same molecule. These variations in signal generation/detection may result in assays that can be 10- to a 1000-fold as sensitive as the corresponding regular ELISA using the same antibody pair.

Standardizing Biomarker Measurements

Biomarker measurements vary within and across centers [212], due to many different pre-analytical and analytical confounding factors that affect the biomarker results [213]. This type of variability is not unique to dementia biomarkers, but a general concern in laboratory medicine, and external quality control programs have been initiated to monitor it [214]. The largest of these programs is the Alzheimer's Association Quality Control program, which runs with several rounds every year and which has reported biomarker variability around 25–30% across centers [215, 216]. The variability has been reduced with the use of mass spectrometry-based methods for CSF Aβ42 [201], and work is ongoing for tau. Reference materials are being constructed in collaboration with the Institute for Reference Materials and Measurements [210]. These materials will be made available at self-cost for assay vendors to harmonize calibration systems for the same analyte.

Conclusions and Future Challenges

Ideas about what constitutes an optimal biomarker differ. One definition is that an ideal dementia biomarker should (1) be linked to fundamental features of the underlying pathology, (2) be validated in neuropathologically confirmed cases, (3) detect the disease early, (4) distinguish the disease from other dementias, (5) be noninvasive, (6) be simple to use, and (7) be inexpensive [217]. These requirements may now be fulfilled for CSF Aβ42 in AD, and partly also for CSF T-tau and P-tau. CSF NFL is the strongest candidate for FTD, but it is a nonspecific marker of neurodegeneration that is not linked to any fundamental proteinopathy. One difficulty for fluid biomarkers in FTD is the rapidly evolving terminology in this field, which makes it difficult to compare studies over time. As the research community reaches consensus on definitions for different stages and variants of AD and FTD, biomarker studies may become more precise and definitive. One striking feature when reviewing the CSF biomarker literature on AD and FTD is that these two disorders stand out as extreme opposites. Whereas AD patients have abnormal CSF Aβ42 and tau biomarkers, FTD patients typically have very normal concentrations. CSF neurogranin is increased in AD but decreased in FTD. CSF NFL is clearly increased in FTD but comparably normal or only slightly increased in AD. Differentiating AD from FTD using CSF biomarkers is therefore not that hard.

One challenge for future studies in AD and FTD is to do truly longitudinal studies in the earliest stages of the diseases. For example, most studies in preclinical AD have been cross-sectional or have only had a few years follow-up, which should be compared to the two or three decades that it likely takes from the first biomarker signs of pathological Aβ metabolism to dementia. Studies with longer follow-up are needed to clarify exactly how biomarkers develop over time.

Another challenge is to decide on validated, standardized cutoffs for biomarkers for different purposes. For example, although many studies on CSF Aβ42 use a cutoff defined in AD dementia versus controls [142], it is not clear that this is the best cutoff for prodromal AD, since acceleration of atrophy and cognitive decline may be detected already at CSF Aβ42 levels above the traditional cutoff [170, 218].

Finally, more work is needed to identify specific biomarkers for several of the proteinopathies present in AD, FTD, and other neurodegenerative diseases.

After decades of research, fluid biomarkers are now increasingly gaining grounds in clinical practice and in clinical trials. As patients and doctors move towards molecular-based diagnostics for neurodegenerative diseases, we expect the use of fluid biomarkers to increase even further. It will therefore be necessary to construct universally accepted clinical guidelines for the use of fluid biomarkers together with clinical data and neuroimaging, for management of contradictory biomarker results, and for disclosure of biomarker information in early disease stages [219]. We anticipate that the development of such guidelines for AD and FTD will be major topics in research during the next coming years.

References

1. Biomarkers Definitions Working Group. Biomarkers and surrogate endpoints: preferred definitions and conceptual framework. Clin Pharmacol Ther. 2001;69:89–95.
2. Quincke H. Die Technik der Lumbalpunction. Berlin: Urban & Schwarzenberg; 1902.
3. Scarpini E. Neurodegenerative diseases: clinical aspects, molecular genetics and biomarkers. London: Springer; 2014.
4. Johanson CE, Duncan JA, Klinge PM, Brinker T, Stopa EG, Silverberg GD. Multiplicity of cerebrospinal fluid functions: new challenges in health and disease. Cerebrospinal Fluid Res. 2008;5:10.
5. Louveau A, Smirnov I, Keyes TJ, Eccles JD, Rouhani SJ, Peske JD, Derecki NC, Castle D, Mandell JW, Lee KS, Harris TH, Kipnis J. Structural and functional features of central nervous system lymphatic vessels. Nature. 2015;523:337–41.
6. Iliff JJ, Wang M, Liao Y, Plogg BA, Peng W, Gundersen GA, Benveniste H, Vates GE, Deane R, Goldman SA, Nagelhus EA, Nedergaard M. A paravascular pathway facilitates CSF flow through the brain parenchyma and the clearance of interstitial solutes, including amyloid β. Sci Transl Med. 2012;4:147ra111.
7. Zetterberg H, Tullhog K, Hansson O, Minthon L, Londos E, Blennow K. Low incidence of post-lumbar puncture headache in 1,089 consecutive memory clinic patients. Eur Neurol. 2010;63:326–30.
8. Evans RW, Armon C, Frohman EM, Goodin DS. Assessment: prevention of post-lumbar puncture headaches: report of the therapeutics and technology assessment subcommittee of the American Academy of Neurology. Neurology. 2000;55:909–14.
9. Peskind ER, Riekse R, Quinn JF, Kaye J, Clark CM, Farlow MR, Decarli C, Chabal C, Vavrek D, Raskind MA, Galasko D. Safety and acceptability of the research lumbar puncture. Alzheimer Dis Assoc Disord. 2005;19:220–5.
10. Vilming ST, Kloster R. Post-lumbar puncture headache: clinical features and suggestions for diagnostic criteria. Cephalalgia. 1997;17:778–84.
11. Dubois B, Feldman HH, Jacova C, Cummings JL, Dekosky ST, Barberger-Gateau P, Delacourte A, Frisoni G, Fox NC, Galasko D, Gauthier S, Hampel H, Jicha GA, Meguro K, O'Brien J, Pasquier F, Robert P, Rossor M, Salloway S, Sarazin M, et al. Revising the definition of Alzheimer's disease: a new lexicon. Lancet Neurol. 2010;9:1118–27.
12. Dubois B, Feldman HH, Jacova C, Dekosky ST, Barberger-Gateau P, Cummings J, Delacourte A, Galasko D, Gauthier S, Jicha G, Meguro K, O'brien J, Pasquier F, Robert P, Rossor M, Salloway S, Stern Y, Visser PJ, Scheltens P. Research criteria for the diagnosis of Alzheimer's disease: revising the NINCDS-ADRDA criteria. Lancet Neurol. 2007;6:734–46.
13. Dubois B, Feldman HH, Jacova C, Hampel H, Molinuevo JL, Blennow K, DeKosky ST, Gauthier S, Selkoe D, Bateman R, Cappa S, Crutch S, Engelborghs S, Frisoni GB, Fox NC, Galasko D, Habert M-O, Jicha GA, Nordberg A, Pasquier F, et al. Advancing research diagnostic criteria for Alzheimer's disease: the IWG-2 criteria. Lancet Neurol. 2014;13:614–29.
14. Albert MS, DeKosky ST, Dickson D, Dubois B, Feldman HH, Fox NC, Gamst A, Holtzman DM, Jagust WJ, Petersen RC, Snyder PJ, Carrillo MC, Thies B, Phelps CH. The diagnosis of mild cognitive impairment due to Alzheimer's disease: recommendations from the National Institute on Aging-Alzheimer's Association workgroups on diagnostic guidelines for Alzheimer's disease. Alzheimers Dement. 2011;7:270–9.

15. McKhann GM, Knopman DS, Chertkow H, Hyman BT, Jack CR, Kawas CH, Klunk WE, Koroshetz WJ, Manly JJ, Mayeux R, Mohs RC, Morris JC, Rossor MN, Scheltens P, Carrillo MC, Thies B, Weintraub S, Phelps CH. The diagnosis of dementia due to Alzheimer's disease: recommendations from the National Institute on Aging-Alzheimer's Association workgroups on diagnostic guidelines for Alzheimer's disease. Alzheimers Dement. 2011;7:263–9.
16. Sperling RA, Aisen PS, Beckett LA, Bennett DA, Craft S, Fagan AM, Iwatsubo T, Jack CRJ, Kaye J, Montine TJ, Park DC, Reiman EM, Rowe CC, Siemers E, Stern Y, Yaffe K, Carrillo MC, Thies B, Morrison-Bogorad M, Wagster MV, et al. Toward defining the preclinical stages of Alzheimer's disease: recommendations from the National Institute on Aging-Alzheimer's Association workgroups on diagnostic guidelines for Alzheimer's disease. Alzheimers Dement. 2011;7:280–92.
17. Bang J, Spina S, Miller BL. Frontotemporal dementia. Lancet. 2015;386:1672–82.
18. Woollacott IOC, Rohrer JD. The clinical spectrum of sporadic and familial forms of frontotemporal dementia. J Neurochem. 2016;138(Suppl 1):6–31.
19. Motter R, Vigo-Pelfrey C, Kholodenko D, Barbour R, Johnson-Wood K, Galasko D, Chang L, Miller B, Clark C, Green R, et al. Reduction of beta-amyloid peptide42 in the cerebrospinal fluid of patients with Alzheimer's disease. Ann Neurol. 1995;38:643–8.
20. Strozyk D, Blennow K, White LR, Launer LJ. CSF Abeta 42 levels correlate with amyloid-neuropathology in a population-based autopsy study. Neurology. 2003;60:652–6.
21. Tapiola T, Alafuzoff I, Herukka SK, Parkkinen L, Hartikainen P, Soininen H, Pirttila T. Cerebrospinal fluid {beta}-amyloid 42 and tau proteins as biomarkers of Alzheimer-type pathologic changes in the brain. Arch Neurol. 2009;66:382–9.
22. Fagan AM, Mintun MA, Mach RH, Lee SY, Dence CS, Shah AR, LaRossa GN, Spinner ML, Klunk WE, Mathis CA, DeKosky ST, Morris JC, Holtzman DM. Inverse relation between in vivo amyloid imaging load and cerebrospinal fluid Abeta42 in humans. Ann Neurol. 2006;59:512–9.
23. Grimmer T, Riemenschneider M, Förstl H, Henriksen G, Klunk WE, Mathis CA, Shiga T, Wester H-J, Kurz A, Drzezga A. Beta amyloid in Alzheimer's disease: increased deposition in brain is reflected in reduced concentration in cerebrospinal fluid. Biol Psychiatry. 2009;65:927–34.
24. Stenh C, Englund H, Lord A, Johansson AS, Almeida CG, Gellerfors P, Greengard P, Gouras GK, Lannfelt L, Nilsson LN. Amyloid-beta oligomers are inefficiently measured by enzyme-linked immunosorbent assay. Ann Neurol. 2005;58:147–50.
25. Kanekiyo T, Ban T, Aritake K, Huang ZL, Qu WM, Okazaki I, Mohri I, Murayama S, Ozono K, Taniike M, Goto Y, Urade Y. Lipocalin-type prostaglandin D synthase/beta-trace is a major amyloid beta-chaperone in human cerebrospinal fluid. Proc Natl Acad Sci U S A. 2007;104:6412–7.
26. LaFerla FM, Green KN, Oddo S. Intracellular amyloid-beta in Alzheimer's disease. Nat Rev Neurosci. 2007;8:499–509.
27. Krut JJ, Zetterberg H, Blennow K, Cinque P, Hagberg L, Price RW, Studahl M, Gisslén M. Cerebrospinal fluid Alzheimer's biomarker profiles in CNS infections. J Neurol. 2013;260:620–6.
28. Augutis K, Axelsson M, Portelius E, Brinkmalm G, Andreasson U, Gustavsson MK, Malmeström C, Lycke J, Blennow K, Zetterberg H, Mattsson N. Cerebrospinal fluid biomarkers of β-amyloid metabolism in multiple sclerosis. Mult Scler. 2013;19:543–52.
29. Gisslen M, Krut J, Andreasson U, Blennow K, Cinque P, Brew BJ, Spudich S, Hagberg L, Rosengren L, Price RW, Zetterberg H. Amyloid and tau cerebrospinal fluid biomarkers in HIV infection. BMC Neurol. 2009;9:63.
30. Sjogren M, Gisslen M, Vanmechelen E, Blennow K. Low cerebrospinal fluid beta-amyloid 42 in patients with acute bacterial meningitis and normalization after treatment. Neurosci Lett. 2001;314:33–6.
31. Mattsson N, Bremell D, Anckarsäter R, Blennow K, Anckarsäter H, Zetterberg H, Hagberg L. Neuroinflammation in Lyme neuroborreliosis affects amyloid metabolism. BMC Neurol. 2010;10:51.

32. Bibl M, Mollenhauer B, Esselmann H, Schneider M, Lewczuk P, Welge V, Gross M, Falkai P, Kornhuber J, Wiltfang J. Cerebrospinal fluid neurochemical phenotypes in vascular dementias: original data and mini-review. Dement Geriatr Cogn Disord. 2008;25:256–65.
33. Ewers M, Mattsson N, Minthon L, Molinuevo JL, Antonell A, Popp J, Jessen F, Herukka SK, Soininen H, Maetzler W, Leyhe T, Bürger K, Taniguchi M, Urakami K, Lista S, Dubois B, Blennow K, Hampel H. CSF biomarkers for the differential diagnosis of Alzheimer's disease: a large-scale international multicenter study. Alzheimers Dement. 2015;11:1306–15.
34. Bibl M, Gallus M, Welge V, Esselmann H, Wolf S, Rüther E, Wiltfang J. Cerebrospinal fluid amyloid-β 2-42 is decreased in Alzheimer's, but not in frontotemporal dementia. J Neural Transm. 2012;119:805–13.
35. Van Everbroeck B, Green AJE, Pals P, Martin JJ, Cras P. Decreased levels of amyloid-beta 1-42 in cerebrospinal fluid of creutzfeldt-jakob disease patients. J Alzheimers Dis. 1999;1:419–24.
36. Sjogren M, Davidsson P, Wallin A, Granerus AK, Grundstrom E, Askmark H, Vanmechelen E, Blennow K. Decreased CSF-beta-amyloid 42 in Alzheimer's disease and amyotrophic lateral sclerosis may reflect mismetabolism of beta-amyloid induced by disparate mechanisms. Dement Geriatr Cogn Disord. 2002;13:112–8.
37. Holmberg B, Johnels B, Blennow K, Rosengren L. Cerebrospinal fluid Abeta42 is reduced in multiple system atrophy but normal in Parkinson's disease and progressive supranuclear palsy. Mov Disord. 2003;18:186–90.
38. Toledo JB, Brettschneider J, Grossman M, Arnold SE, Hu WT, Xie SX, Lee VM, Shaw LM, Trojanowski JQ. CSF biomarkers cutoffs: the importance of coincident neuropathological diseases. Acta Neuropathol. 2012;124:23–35.
39. van Westen D, Lindqvist D, Blennow K, Minthon L, Nägga K, Stomrud E, Zetterberg H, Hansson O. Cerebral white matter lesions—associations with Aβ isoforms and amyloid PET. Sci Rep. 2016;6:20709.
40. Mattsson N, Insel PS, Palmqvist S, Stomrud E, van Westen D, Minthon L, Zetterberg H, Blennow K, Hansson O. Increased amyloidogenic APP processing in APOE ε4-negative individuals with cerebral β-amyloidosis. Nat Commun. 2016;7:10918.
41. Janelidze S, Zetterberg H, Mattsson N, Palmqvist S, Vanderstichele H, Lindberg O, van Westen D, Stomrud E, Minthon L, Blennow K, Swedish BioFINDER study group & Hansson O. CSF Aβ42/Aβ40 and Aβ42/Aβ38 ratios: better diagnostic markers of Alzheimer disease. Ann Clin Transl Neurol. 2016;3:154–65.
42. Wiltfang J, Esselmann H, Bibl M, Smirnov A, Otto M, Paul S, Schmidt B, Klafki HW, Maler M, Dyrks T, Bienert M, Beyermann M, Ruther E, Kornhuber J. Highly conserved and disease-specific patterns of carboxyterminally truncated Abeta peptides 1-37/38/39 in addition to 1-40/42 in Alzheimer's disease and in patients with chronic neuroinflammation. J Neurochem. 2002;81:481–96.
43. Portelius E, Westman-Brinkmalm A, Zetterberg H, Blennow K. Determination of beta-amyloid peptide signatures in cerebrospinal fluid using immunoprecipitation-mass spectrometry. J Proteome Res. 2006;5:1010–6.
44. Portelius E, Zhang B, Gustavsson MK, Brinkmalm G, Westman-Brinkmalm A, Zetterberg H, Lee VM, Trojanowski JQ, Blennow K. Effects of gamma-secretase inhibition on the amyloid beta isoform pattern in a mouse model of Alzheimer's disease. Neurodegener Dis. 2009;6:258–62.
45. Mattsson N, Rajendran L, Zetterberg H, Gustavsson M, Andreasson U, Olsson M, Brinkmalm G, Lundkvist J, Jacobson LH, Perrot L, Neumann U, Borghys H, Mercken M, Dhuyvetter D, Jeppsson F, Blennow K, Portelius E. BACE1 inhibition induces a specific cerebrospinal fluid beta-amyloid pattern that identifies drug effects in the central nervous system. PLoS One. 2012;7:e31084.
46. Timmers M, Barão S, Van Broeck B, Tesseur I, Slemmon J, De Waepenaert K, Bogert J, Shaw LM, Engelborghs S, Moechars D, Mercken M, Van Nueten L, Tritsmans L, de Strooper B, Streffer JR. BACE1 dynamics upon inhibition with a BACE inhibitor and correlation to downstream Alzheimer's disease markers in elderly healthy participants. J Alzheimers Dis. 2017;56(4):1437–49.

47. Holsinger RM, Lee JS, Boyd A, Masters CL, Collins SJ. CSF BACE1 activity is increased in CJD and Alzheimer disease versus [corrected] other dementias. Neurology. 2006;67:710–2.
48. Holsinger RM, McLean CA, Collins SJ, Masters CL, Evin G. Increased beta-Secretase activity in cerebrospinal fluid of Alzheimer's disease subjects. Ann Neurol. 2004;55:898–9.
49. Verheijen JH, Huisman LG, van Lent N, Neumann U, Paganetti P, Hack CE, Bouwman F, Lindeman J, Bollen EL, Hanemaaijer R. Detection of a soluble form of BACE-1 in human cerebrospinal fluid by a sensitive activity assay. Clin Chem. 2006;52:1168–74.
50. Rosen C, Andreasson U, Mattsson N, Marcusson J, Minthon L, Andreasen N, Blennow K, Zetterberg H. Cerebrospinal fluid profiles of amyloid beta-related biomarkers in Alzheimer's disease. NeuroMolecular Med. 2012;14:65–73.
51. Zhong Z, Ewers M, Teipel S, Burger K, Wallin A, Blennow K, He P, McAllister C, Hampel H, Shen Y. Levels of beta-secretase (BACE1) in cerebrospinal fluid as a predictor of risk in mild cognitive impairment. Arch Gen Psychiatry. 2007;64:718–26.
52. Hata S, Fujishige S, Araki Y, Taniguchi M, Urakami K, Peskind E, Akatsu H, Araseki M, Yamamoto K, Martins RN, Maeda M, Nishimura M, Levey A, Chung KA, Montine T, Leverenz J, Fagan A, Goate A, Bateman R, Holtzman DM, et al. Alternative processing of γ-secretase substrates in common forms of mild cognitive impairment and Alzheimer's disease: evidence for γ-secretase dysfunction. Ann Neurol. 2011;69:1026–31.
53. Gabelle A, Roche S, Gény C, Bennys K, Labauge P, Tholance Y, Quadrio I, Tiers L, Gor B, Chaulet C, Vighetto A, Croisile B, Krolak-Salmon P, Touchon J, Perret-Liaudet A, Lehmann S. Correlations between soluble α/β forms of amyloid precursor protein and Aβ38, 40, and 42 in human cerebrospinal fluid. Brain Res. 2010;1357:175–83.
54. Olsson A, Hoglund K, Sjogren M, Andreasen N, Minthon L, Lannfelt L, Buerger K, Moller HJ, Hampel H, Davidsson P, Blennow K. Measurement of alpha- and beta-secretase cleaved amyloid precursor protein in cerebrospinal fluid from Alzheimer patients. Exp Neurol. 2003;183:74–80.
55. Hertze J, Minthon L, Zetterberg H, Vanmechelen E, Blennow K, Hansson O. Evaluation of CSF biomarkers as predictors of Alzheimer's disease: a clinical follow-up study of 4.7 years. J Alzheimers Dis. 2010;21:1119–28.
56. Magdalinou NK, Paterson RW, Schott JM, Fox NC, Mummery C, Blennow K, Bhatia K, Morris HR, Giunti P, Warner TT, de Silva R, Lees AJ, Zetterberg H. A panel of nine cerebrospinal fluid biomarkers may identify patients with atypical parkinsonian syndromes. J Neurol Neurosurg Psychiatry. 2015;86:1240–7.
57. Alcolea D, Carmona-Iragui M, Suárez-Calvet M, Sánchez-Saudinós MB, Sala I, Antón-Aguirre S, Blesa R, Clarimón J, Fortea J, Lleó A. Relationship between β-secretase, inflammation and core cerebrospinal fluid biomarkers for Alzheimer's disease. J Alzheimers Dis. 2014;42:157–67.
58. Gabelle A, Roche S, Gény C, Bennys K, Labauge P, Tholance Y, Quadrio I, Tiers L, Gor B, Boulanghien J, Chaulet C, Vighetto A, Croisile B, Krolak-Salmon P, Perret-Liaudet A, Touchon J, Lehmann S. Decreased sAβPPβ, Aβ38, and Aβ40 cerebrospinal fluid levels in frontotemporal dementia. J Alzheimers Dis. 2011;26:553–63.
59. Perneczky R, Tsolakidou A, Arnold A, Diehl-Schmid J, Grimmer T, Förstl H, Kurz A, Alexopoulos P. CSF soluble amyloid precursor proteins in the diagnosis of incipient Alzheimer disease. Neurology. 2011;77:35–8.
60. Fukumoto H, Tokuda T, Kasai T, Ishigami N, Hidaka H, Kondo M, Allsop D, Nakagawa M. High-molecular-weight beta-amyloid oligomers are elevated in cerebrospinal fluid of Alzheimer patients. FASEB J. 2010;24:2716–26.
61. Gao CM, Yam AY, Wang X, Magdangal E, Salisbury C, Peretz D, Zuckermann RN, Connolly MD, Hansson O, Minthon L, Zetterberg H, Blennow K, Fedynyshyn JP, Allauzen S. Aβ40 oligomers identified as a potential biomarker for the diagnosis of Alzheimer's disease. PLoS One. 2010;5:e15725.
62. Hölttä M, Hansson O, Andreasson U, Hertze J, Minthon L, Nägga K, Andreasen N, Zetterberg H, Blennow K. Evaluating amyloid-β oligomers in cerebrospinal fluid as a biomarker for Alzheimer's disease. PLoS One. 2013;8(6):e66381. https://doi.org/10.1371/journal.pone.0066381.

63. Pitschke M, Prior R, Haupt M, Riesner D. Detection of single amyloid beta-protein aggregates in the cerebrospinal fluid of Alzheimer's patients by fluorescence correlation spectroscopy. Nat Med. 1998;4:832–4.
64. Santos AN, Torkler S, Nowak D, Schlittig C, Goerdes M, Lauber T, Trischmann L, Schaupp M, Penz M, Tiller FW, Bohm G. Detection of amyloid-beta oligomers in human cerebrospinal fluid by flow cytometry and fluorescence resonance energy transfer. J Alzheimers Dis. 2007;11:117–25.
65. Zetterberg H. Plasma amyloid β-quo vadis? Neurobiol Aging. 2015;36:2671–3.
66. Janelidze S, Stomrud E, Palmqvist S, Zetterberg H, van Westen D, Jeromin A, Song L, Hanlon D, Tan Hehir CA, Baker D, Blennow K, Hansson O. Plasma β-amyloid in Alzheimer's disease and vascular disease. Sci Rep. 2016;6:26801. https://doi.org/10.1038/srep26801.
67. Goedert M, Spillantini MG, Jakes R, Rutherford D, Crowther RA. Multiple isoforms of human microtubule-associated protein tau: sequences and localization in neurofibrillary tangles of Alzheimer's disease. Neuron. 1989;3:519–26.
68. Hesse C, Rosengren L, Andreasen N, Davidsson P, Vanderstichele H, Vanmechelen E, Blennow K. Transient increase in total tau but not phospho-tau in human cerebrospinal fluid after acute stroke. Neurosci Lett. 2001;297:187–90.
69. Otto M, Wiltfang J, Tumani H, Zerr I, Lantsch M, Kornhuber J, Weber T, Kretzschmar HA, Poser S. Elevated levels of tau-protein in cerebrospinal fluid of patients with Creutzfeldt-Jakob disease. Neurosci Lett. 1997;225:210–2.
70. Zetterberg H, Hietala MA, Jonsson M, Andreasen N, Styrud E, Karlsson I, Edman A, Popa C, Rasulzada A, Wahlund LO, Mehta PD, Rosengren L, Blennow K, Wallin A. Neurochemical aftermath of amateur boxing. Arch Neurol. 2006;63:1277–80.
71. Buerger K, Ewers M, Pirttila T, Zinkowski R, Alafuzoff I, Teipel SJ, DeBernardis J, Kerkman D, McCulloch C, Soininen H, Hampel H. CSF phosphorylated tau protein correlates with neocortical neurofibrillary pathology in Alzheimer's disease. Brain. 2006;129:3035–41.
72. Seppala TT, Nerg O, Koivisto AM, Rummukainen J, Puli L, Zetterberg H, Pyykko OT, Helisalmi S, Alafuzoff I, Hiltunen M, Jaaskelainen JE, Rinne J, Soininen H, Leinonen V, Herukka SK. CSF biomarkers for Alzheimer disease correlate with cortical brain biopsy findings. Neurology. 2012;78:1568–75.
73. Engelborghs S, Sleegers K, Cras P, Brouwers N, Serneels S, De Leenheir E, Martin J-J, Vanmechelen E, Van Broeckhoven C, De Deyn PP. No association of CSF biomarkers with APOEepsilon4, plaque and tangle burden in definite Alzheimer's disease. Brain. 2007;130:2320–6.
74. Mattsson N, Savman K, Osterlundh G, Blennow K, Zetterberg H. Converging molecular pathways in human neural development and degeneration. Neurosci Res. 2010;66:330–2.
75. Luk C, Compta Y, Magdalinou N, Martí MJ, Hondhamuni G, Zetterberg H, Blennow K, Constantinescu R, Pijnenburg Y, Mollenhauer B, Trenkwalder C, Van Swieten J, Chiu WZ, Borroni B, Cámara A, Cheshire P, Williams DR, Lees AJ, de Silva R. Development and assessment of sensitive immuno-PCR assays for the quantification of cerebrospinal fluid three- and four-repeat tau isoforms in tauopathies. J Neurochem. 2012;123:396–405.
76. Cruchaga C, Kauwe JSK, Harari O, Jin SC, Cai Y, Karch CM, Benitez BA, Jeng AT, Skorupa T, Carrell D, Bertelsen S, Bailey M, McKean D, Shulman JM, De Jager PL, Chibnik L, Bennett DA, Arnold SE, Harold D, Sims R, et al. GWAS of cerebrospinal fluid tau levels identifies risk variants for Alzheimer's disease. Neuron. 2013;78:256–68.
77. Mattsson N, Zetterberg H, Janelidze S, Insel PS, Andreasson U, Stomrud E, Palmqvist S, Baker D, Tan Hehir CA, Jeromin A, Hanlon D, Song L, Shaw LM, Trojanowski JQ, Weiner MW, Hansson O, Blennow K, ADNI Investigators. Plasma tau in Alzheimer disease. Neurology. 2016;87(17):1827–35.
78. Selkoe DJ. Alzheimer's disease is a synaptic failure. Science. 2002;298:789–91.
79. Öhrfelt A, Brinkmalm A, Dumurgier J, Brinkmalm G, Hansson O, Zetterberg H, Bouaziz-Amar E, Hugon J, Paquet C, Blennow K. The pre-synaptic vesicle protein synaptotagmin is a novel biomarker for Alzheimer's disease. Alzheimers Res Ther. 2016;8:41.
80. Brinkmalm A, Brinkmalm G, Honer WG, Frölich L, Hausner L, Minthon L, Hansson O, Wallin A, Zetterberg H, Blennow K, Öhrfelt A. SNAP-25 is a promising novel cerebrospinal fluid biomarker for synapse degeneration in Alzheimer's disease. Mol Neurodegener. 2014;9:53. https://doi.org/10.1186/1750-1326-9-53.

81. Davidsson P, Puchades M, Blennow K. Identification of synaptic vesicle, pre- and post-synaptic proteins in human cerebrospinal fluid using liquid-phase isoelectric focusing. Electrophoresis. 1999;20:431–7.
82. Kester MI, Teunissen CE, Crimmins DL, Herries EM, Ladenson JH, Scheltens P, van der Flier WM, Morris JC, Holtzman DM, Fagan AM. Neurogranin as a cerebrospinal fluid biomarker for synaptic loss in symptomatic Alzheimer disease. JAMA Neurol. 2015;72:1275–80.
83. Mattsson N, Insel PS, Palmqvist S, Portelius E, Zetterberg H, Weiner M, Blennow K, Hansson O, Alzheimer's Disease Neuroimaging Initiative. Cerebrospinal fluid tau, neurogranin, and neurofilament light in Alzheimer's disease. EMBO Mol Med. 2016;8:1184–96.
84. Portelius E, Zetterberg H, Skillbäck T, Törnqvist U, Andreasson U, Trojanowski JQ, Weiner MW, Shaw LM, Mattsson N, Blennow K, Alzheimer's Disease Neuroimaging Initiative. Cerebrospinal fluid neurogranin: relation to cognition and neurodegeneration in Alzheimer's disease. Brain. 2015;138:3373–85.
85. Janelidze S, Hertze J, Zetterberg H, Landqvist Waldö M, Santillo A, Blennow K, Hansson O. Cerebrospinal fluid neurogranin and YKL-40 as biomarkers of Alzheimer's disease. Ann Clin Transl Neurol. 2016;3:12–20.
86. Wellington H, Paterson RW, Portelius E, Törnqvist U, Magdalinou N, Fox NC, Blennow K, Schott JM, Zetterberg H. Increased CSF neurogranin concentration is specific to Alzheimer disease. Neurology. 2016;86:829–35.
87. Norgren N, Rosengren L, Stigbrand T. Elevated neurofilament levels in neurological diseases. Brain Res. 2003;987:25–31.
88. Sjögren M, Blomberg M, Jonsson M, Wahlund LO, Edman A, Lind K, Rosengren L, Blennow K, Wallin A. Neurofilament protein in cerebrospinal fluid: a marker of white matter changes. J Neurosci Res. 2001;66:510–6.
89. Petzold A, Keir G, Warren J, Fox N, Rossor MN. A systematic review and meta-analysis of CSF neurofilament protein levels as biomarkers in dementia. Neurodegener Dis. 2007;4:185–94.
90. Bech S, Hjermind LE, Salvesen L, Nielsen JE, Heegaard NH, Jorgensen HL, Rosengren L, Blennow K, Zetterberg H, Winge K. Amyloid-related biomarkers and axonal damage proteins in parkinsonian syndromes. Parkinsonism Relat Disord. 2012;18:69–72.
91. Zetterberg H, Jacobsson J, Rosengren L, Blennow K, Andersen PM. Cerebrospinal fluid neurofilament light levels in amyotrophic lateral sclerosis: impact of SOD1 genotype. Eur J Neurol. 2007;14:1329–33.
92. Zetterberg H, Skillbäck T, Mattsson N, Trojanowski JQ, Portelius E, Shaw LM, Weiner MW, Blennow K, Alzheimer's Disease Neuroimaging Initiative. Association of cerebrospinal fluid neurofilament light concentration with Alzheimer disease progression. JAMA Neurol. 2015;73(1):60–7.
93. Skillbäck T, Farahmand B, Bartlett JW, Rosén C, Mattsson N, Nägga K, Kilander L, Religa D, Wimo A, Winblad B, Rosengren L, Schott JM, Blennow K, Eriksdotter M, Zetterberg H. CSF neurofilament light differs in neurodegenerative diseases and predicts severity and survival. Neurology. 2014;83:1945–53.
94. Mattsson N, Andreasson U, Zetterberg H, Blennow K, Alzheimer's Disease Neuroimaging Initiative. Association of plasma neurofilament light with neurodegeneration in patients with Alzheimer disease. JAMA Neurol. 2017;74(5):557–66.
95. Hansson O, Janelidze S, Hall S, Magdalinou N, Lees AJ, Andreasson U, Norgren N, Linder J, Forsgren L, Constantinescu R, Zetterberg H, Blennow K, Swedish BioFINDER Study. Blood-based NfL: a biomarker for differential diagnosis of parkinsonian disorder. Neurology. 2017;88:930–7.
96. Rohrer JD, Woollacott IOC, Dick KM, Brotherhood E, Gordon E, Fellows A, Toombs J, Druyeh R, Cardoso MJ, Ourselin S, Nicholas JM, Norgren N, Mead S, Andreasson U, Blennow K, Schott JM, Fox NC, Warren JD, Zetterberg H. Serum neurofilament light chain protein is a measure of disease intensity in frontotemporal dementia. Neurology. 2016;87:1329–36.

97. Foulds P, McAuley E, Gibbons L, Davidson Y, Pickering-Brown SM, Neary D, Snowden JS, Allsop D, Mann DMA. TDP-43 protein in plasma may index TDP-43 brain pathology in Alzheimer's disease and frontotemporal lobar degeneration. Acta Neuropathol. 2008;116:141–6.
98. Kasai T, Tokuda T, Ishigami N, Sasayama H, Foulds P, Mitchell DJ, Mann DMA, Allsop D, Nakagawa M. Increased TDP-43 protein in cerebrospinal fluid of patients with amyotrophic lateral sclerosis. Acta Neuropathol. 2009;117:55–62.
99. Steinacker P, Hendrich C, Sperfeld AD, Jesse S, von Arnim CAF, Lehnert S, Pabst A, Uttner I, Tumani H, Lee VM-Y, Trojanowski JQ, Kretzschmar HA, Ludolph A, Neumann M, Otto M. TDP-43 in cerebrospinal fluid of patients with frontotemporal lobar degeneration and amyotrophic lateral sclerosis. Arch Neurol. 2008;65:1481–7.
100. Junttila A, Kuvaja M, Hartikainen P, Siloaho M, Helisalmi S, Moilanen V, Kiviharju A, Jansson L, Tienari PJ, Remes AM, Herukka S-K. Cerebrospinal fluid TDP-43 in frontotemporal lobar degeneration and amyotrophic lateral sclerosis patients with and without the C9ORF72 hexanucleotide expansion. Dement Geriatr Cogn Dis Extra. 2016;6:142–9.
101. Suárez-Calvet M, Dols-Icardo O, Lladó A, Sánchez-Valle R, Hernández I, Amer G, Antón-Aguirre S, Alcolea D, Fortea J, Ferrer I, van der Zee J, Dillen L, Van Broeckhoven C, Molinuevo JL, Blesa R, Clarimón J, Lleó A. Plasma phosphorylated TDP-43 levels are elevated in patients with frontotemporal dementia carrying a C9orf72 repeat expansion or a GRN mutation. J Neurol Neurosurg Psychiatry. 2014;85:684–91.
102. Feneberg E, Steinacker P, Lehnert S, Schneider A, Walther P, Thal DR, Linsenmeier M, Ludolph AC, Otto M. Limited role of free TDP-43 as a diagnostic tool in neurodegenerative diseases. Amyotroph Lateral Scler Frontotemporal Degener. 2014;15:351–6.
103. Ghidoni R, Paterlini A, Benussi L. Circulating progranulin as a biomarker for neurodegenerative diseases. Am J Neurodegener Dis. 2012;1:180–90.
104. Meeter LHH, Patzke H, Loewen G, Dopper EGP, Pijnenburg YAL, van Minkelen R, van Swieten JC. Progranulin levels in plasma and cerebrospinal fluid in granulin mutation carriers. Dement Geriatr Cogn Dis Extra. 2016;6:330–40.
105. Oeckl P, Steinacker P, Feneberg E, Otto M. Cerebrospinal fluid proteomics and protein biomarkers in frontotemporal lobar degeneration: current status and future perspectives. Biochim Biophys Acta. 2015;1854:757–68.
106. Oeckl P, Steinacker P, von Arnim CAF, Straub S, Nagl M, Feneberg E, Weishaupt JH, Ludolph AC, Otto M. Intact protein analysis of ubiquitin in cerebrospinal fluid by multiple reaction monitoring reveals differences in Alzheimer's disease and frontotemporal lobar degeneration. J Proteome Res. 2014;13:4518–25.
107. Fagan AM, Perrin RJ. Upcoming candidate cerebrospinal fluid biomarkers of Alzheimer's disease. Biomark Med. 2012;6:455–76.
108. Mattsson N, Tabatabaei S, Johansson P, Hansson O, Andreasson U, Mansson JE, Johansson JO, Olsson B, Wallin A, Svensson J, Blennow K, Zetterberg H. Cerebrospinal fluid microglial markers in Alzheimer's disease: elevated chitotriosidase activity but lack of diagnostic utility. NeuroMolecular Med. 2011;13:151–9.
109. Craig-Schapiro R, Perrin RJ, Roe CM, Xiong C, Carter D, Cairns NJ, Mintun MA, Peskind ER, Li G, Galasko DR, Clark CM, Quinn JF, D'Angelo G, Malone JP, Townsend RR, Morris JC, Fagan AM, Holtzman DM. YKL-40: a novel prognostic fluid biomarker for preclinical Alzheimer's disease. Biol Psychiatry. 2010;68:903–12.
110. Teunissen CE, Elias N, Koel-Simmelink MJA, Durieux-Lu S, Malekzadeh A, Pham TV, Piersma SR, Beccari T, Meeter LHH, Dopper EGP, van Swieten JC, Jimenez CR, Pijnenburg YAL. Novel diagnostic cerebrospinal fluid biomarkers for pathologic subtypes of frontotemporal dementia identified by proteomics. Alzheimers Dement (Amst). 2016;2:86–94.
111. Heslegrave A, Heywood W, Paterson R, Magdalinou N, Svensson J, Johansson P, Öhrfelt A, Blennow K, Hardy J, Schott J, Mills K, Zetterberg H. Increased cerebrospinal fluid soluble TREM2 concentration in Alzheimer's disease. Mol Neurodegener. 2016;11:3. https://doi.org/10.1186/s13024-016-0071-x.

112. Piccio L, Deming Y, Del-Águila JL, Ghezzi L, Holtzman DM, Fagan AM, Fenoglio C, Galimberti D, Borroni B, Cruchaga C. Cerebrospinal fluid soluble TREM2 is higher in Alzheimer disease and associated with mutation status. Acta Neuropathol. 2016;131:925–33.
113. Suárez-Calvet M, Kleinberger G, Araque Caballero MÁ, Brendel M, Rominger A, Alcolea D, Fortea J, Lleó A, Blesa R, Gispert JD, Sánchez-Valle R, Antonell A, Rami L, Molinuevo JL, Brosseron F, Traschütz A, Heneka MT, Struyfs H, Engelborghs S, Sleegers K, et al. sTREM2 cerebrospinal fluid levels are a potential biomarker for microglia activity in early-stage Alzheimer's disease and associate with neuronal injury markers. EMBO Mol Med. 2016;8:466–76.
114. Hesse R, Wahler A, Gummert P, Kirschmer S, Otto M, Tumani H, Lewerenz J, Schnack C, von Arnim CAF. Decreased IL-8 levels in CSF and serum of AD patients and negative correlation of MMSE and IL-1β. BMC Neurol. 2016;16:185.
115. Bloudek LM, Spackman DE, Blankenburg M, Sullivan SD. Review and meta-analysis of biomarkers and diagnostic imaging in Alzheimer's disease. J Alzheimers Dis. 2011;26:627–45.
116. Blennow K, Hampel H, Weiner M, Zetterberg H. Cerebrospinal fluid and plasma biomarkers in Alzheimer disease. Nat Rev Neurol. 2010;6:131–44.
117. Jack CRJ, Knopman DS, Jagust WJ, Petersen RC, Weiner MW, Aisen PS, Shaw LM, Vemuri P, Wiste HJ, Weigand SD, Lesnick TG, Pankratz VS, Donohue MC, Trojanowski JQ. Tracking pathophysiological processes in Alzheimer's disease: an updated hypothetical model of dynamic biomarkers. Lancet Neurol. 2013;12:207–16.
118. Hardy J, Selkoe DJ. The amyloid hypothesis of Alzheimer's disease: progress and problems on the road to therapeutics. Science. 2002;297:353–6.
119. Bateman RJ, Xiong C, Benzinger TL, Fagan AM, Goate A, Fox NC, Marcus DS, Cairns NJ, Xie X, Blazey TM, Holtzman DM, Santacruz A, Buckles V, Oliver A, Moulder K, Aisen PS, Ghetti B, Klunk WE, McDade E, Martins RN, et al. Clinical and biomarker changes in dominantly inherited Alzheimer's disease. N Engl J Med. 2012;367:795–804.
120. Buchhave P, Minthon L, Zetterberg H, Wallin AK, Blennow K, Hansson O. Cerebrospinal fluid levels of beta-amyloid 1-42, but not of tau, are fully changed already 5 to 10 years before the onset of Alzheimer dementia. Arch Gen Psychiatry. 2012;69:98–106.
121. Jack CRJ, Vemuri P, Wiste HJ, Weigand SD, Aisen PS, Trojanowski JQ, Shaw LM, Bernstein MA, Petersen RC, Weiner MW, Knopman DS, Alzheimer's Disease Neuroimaging Initiative. Evidence for ordering of Alzheimer disease biomarkers. Arch Neurol. 2011;68:1526–35.
122. Insel PS, Ossenkoppele R, Gessert D, Jagust W, Landau S, Hansson O, Weiner MW, Mattsson N, Alzheimer's Disease Neuroimaging Initiative. Time to amyloid positivity and preclinical changes in brain metabolism, atrophy, and cognition: evidence for emerging amyloid pathology in Alzheimer's disease. Front Neurosci. 2017;11:281.
123. Zetterberg H, Pedersen M, Lind K, Svensson M, Rolstad S, Eckerstrom C, Syversen S, Mattsson UB, Ysander C, Mattsson N, Nordlund A, Vanderstichele H, Vanmechelen E, Jonsson M, Edman A, Blennow K, Wallin A. Intra-individual stability of CSF biomarkers for Alzheimer's disease over two years. J Alzheimers Dis. 2007;12:255–60.
124. Ingelsson M, Fukumoto H, Newell KL, Growdon JH, Hedley-Whyte ET, Frosch MP, Albert MS, Hyman BT, Irizarry MC. Early Abeta accumulation and progressive synaptic loss, gliosis, and tangle formation in AD brain. Neurology. 2004;62:925–31.
125. Jack CRJ, Vemuri P, Wiste HJ, Weigand SD, Lesnick TG, Lowe V, Kantarci K, Bernstein MA, Senjem ML, Gunter JL, Boeve BF, Trojanowski JQ, Shaw LM, Aisen PS, Weiner MW, Petersen RC, Knopman DS, Alzheimer's Disease Neuroimaging Initiative. Shapes of the trajectories of 5 major biomarkers of Alzheimer disease. Arch Neurol. 2012;69:856–67.
126. Mattsson N, Insel P, Nosheny R, Zetterberg H, Trojanowski JQ, Shaw LM, Tosun D, Weiner M, Alzheimer's Disease Neuroimaging Initiative. CSF protein biomarkers predicting longitudinal reduction of CSF β-amyloid42 in cognitively healthy elders. Transl Psychiatry. 2013;3:e293.
127. Höglund K, Kern S, Zettergren A, Börjesson-Hansson A, Zetterberg H, Skoog I, Blennow K. Preclinical amyloid pathology biomarker positivity: effects on tau pathology and neurodegeneration. Transl Psychiatry. 2017;7:e995.

128. Fagan AM, Roe CM, Xiong C, Mintun MA, Morris JC, Holtzman DM. Cerebrospinal fluid tau/beta-amyloid(42) ratio as a prediction of cognitive decline in nondemented older adults. Arch Neurol. 2007;64:343–9.
129. Gustafson DR, Skoog I, Rosengren L, Zetterberg H, Blennow K. Cerebrospinal fluid beta-amyloid 1-42 concentration may predict cognitive decline in older women. J Neurol Neurosurg Psychiatry. 2007;78:461–4.
130. Li G, Sokal I, Quinn JF, Leverenz JB, Brodey M, Schellenberg GD, Kaye JA, Raskind MA, Zhang J, Peskind ER, Montine TJ. CSF tau/Abeta42 ratio for increased risk of mild cognitive impairment: a follow-up study. Neurology. 2007;69:631–9.
131. Skoog I, Davidsson P, Aevarsson O, Vanderstichele H, Vanmechelen E, Blennow K. Cerebrospinal fluid beta-amyloid 42 is reduced before the onset of sporadic dementia: a population-based study in 85-year-olds. Dement Geriatr Cogn Disord. 2003;15:169–76.
132. Stomrud E, Hansson O, Blennow K, Minthon L, Londos E. Cerebrospinal fluid biomarkers predict decline in subjective cognitive function over 3 years in healthy elderly. Dement Geriatr Cogn Disord. 2007;24:118–24.
133. van Harten AC, Visser PJ, Pijnenburg YAL, Teunissen CE, Blankenstein MA, Scheltens P, van der Flier WM. Cerebrospinal fluid Aβ42 is the best predictor of clinical progression in patients with subjective complaints. Alzheimers Dement. 2012;9(5):481–7.
134. Desikan RS, McEvoy LK, Thompson WK, Holland D, Brewer JB, Aisen PS, Sperling RA, Dale AM, Alzheimer's Disease Neuroimaging Initiative. Amyloid-β—associated clinical decline occurs only in the presence of elevated P-tau. Arch Neurol. 2012;69:709–13.
135. Roe CM, Fagan AM, Grant EA, Hassenstab J, Moulder KL, Maue Dreyfus D, Sutphen CL, Benzinger TLS, Mintun MA, Holtzman DM, Morris JC. Amyloid imaging and CSF biomarkers in predicting cognitive impairment up to 7.5 years later. Neurology. 2013;80:1784–91.
136. Schott JM, Bartlett JW, Fox NC, Barnes J, Alzheimer's Disease Neuroimaging Initiative Investigators. Increased brain atrophy rates in cognitively normal older adults with low cerebrospinal fluid Aβ1-42. Ann Neurol. 2010;68:825–34.
137. Roe CM, Fagan AM, Williams MM, Ghoshal N, Aeschleman M, Grant EA, Marcus DS, Mintun MA, Holtzman DM, Morris JC. Improving CSF biomarker accuracy in predicting prevalent and incident Alzheimer disease. Neurology. 2011;76:501–10.
138. Roe CM, Fagan AM, Grant EA, Marcus DS, Benzinger TLS, Mintun MA, Holtzman DM, Morris JC. Cerebrospinal fluid biomarkers, education, brain volume, and future cognition. Arch Neurol. 2011;68:1145–51.
139. Mattsson N, Lönneborg A, Boccardi M, Blennow K, Hansson O, Geneva Task Force for the Roadmap of Alzheimer's Biomarkers. Clinical validity of cerebrospinal fluid Aβ42, tau, and phospho-tau as biomarkers for Alzheimer's disease in the context of a structured 5-phase development framework. Neurobiol Aging. 2017;52:196–213.
140. Hansson O, Zetterberg H, Buchhave P, Londos E, Blennow K, Minthon L. Association between CSF biomarkers and incipient Alzheimer's disease in patients with mild cognitive impairment: a follow-up study. Lancet Neurol. 2006;5:228–34.
141. Mattsson N, Zetterberg H, Hansson O, Andreasen N, Parnetti L, Jonsson M, Herukka SK, van der Flier WM, Blankenstein MA, Ewers M, Rich K, Kaiser E, Verbeek M, Tsolaki M, Mulugeta E, Rosen E, Aarsland D, Visser PJ, Schroder J, Marcusson J, et al. CSF biomarkers and incipient Alzheimer disease in patients with mild cognitive impairment. JAMA. 2009;302:385–93.
142. Shaw LM, Vanderstichele H, Knapik-Czajka M, Clark CM, Aisen PS, Petersen RC, Blennow K, Soares H, Simon A, Lewczuk P, Dean R, Siemers E, Potter W, Lee VM, Trojanowski JQ. Cerebrospinal fluid biomarker signature in Alzheimer's disease neuroimaging initiative subjects. Ann Neurol. 2009;65:403–13.
143. Visser PJ, Verhey F, Knol DL, Scheltens P, Wahlund LO, Freund-Levi Y, Tsolaki M, Minthon L, Wallin AK, Hampel H, Burger K, Pirttila T, Soininen H, Rikkert MO, Verbeek MM, Spiru L, Blennow K. Prevalence and prognostic value of CSF markers of Alzheimer's disease pathology in patients with subjective cognitive impairment or mild cognitive impairment in the DESCRIPA study: a prospective cohort study. Lancet Neurol. 2009;8:619–27.

144. Mattsson N, Rosen E, Hansson O, Andreasen N, Parnetti L, Jonsson M, Herukka SK, van der Flier WM, Blankenstein MA, Ewers M, Rich K, Kaiser E, Verbeek MM, Olde Rikkert M, Tsolaki M, Mulugeta E, Aarsland D, Visser PJ, Schroder J, Marcusson J, et al. Age and diagnostic performance of Alzheimer disease CSF biomarkers. Neurology. 2012;78(7):468–76. https://doi.org/10.1212/WNL.0b013e3182477eed.
145. van Rossum IA, Vos SJB, Burns L, Knol DL, Scheltens P, Soininen H, Wahlund L-O, Hampel H, Tsolaki M, Minthon L, L'italien G, van der Flier WM, Teunissen CE, Blennow K, Barkhof F, Rueckert D, Wolz R, Verhey F, Visser PJ. Injury markers predict time to dementia in subjects with MCI and amyloid pathology. Neurology. 2012;79:1809–16.
146. Blennow K, Zetterberg H, Minthon L, Lannfelt L, Strid S, Annas P, Basun H, Andreasen N. Longitudinal stability of CSF biomarkers in Alzheimer's disease. Neurosci Lett. 2007;419:18–22.
147. Mattsson N, Portelius E, Rolstad S, Gustavsson M, Andreasson U, Stridsberg M, Wallin A, Blennow K, Zetterberg H. Longitudinal cerebrospinal fluid biomarkers over four years in mild cognitive impairment. J Alzheimers Dis. 2012;30:767–78.
148. Buchhave P, Blennow K, Zetterberg H, Stomrud E, Londos E, Andreasen N, Minthon L, Hansson O. Longitudinal study of CSF biomarkers in patients with Alzheimer's disease. PLoS One. 2009;4:e6294.
149. Reiman EM, Quiroz YT, Fleisher AS, Chen K, Velez-Pardo C, Jimenez-Del-Rio M, Fagan AM, Shah AR, Alvarez S, Arbelaez A, Giraldo M, Acosta-Baena N, Sperling RA, Dickerson B, Stern CE, Tirado V, Munoz C, Reiman RA, Huentelman MJ, Alexander GE, et al. Brain imaging and fluid biomarker analysis in young adults at genetic risk for autosomal dominant Alzheimer's disease in the presenilin 1 E280A kindred: a case-control study. Lancet Neurol. 2012;11:1048–56.
150. Moonis M, Swearer JM, Dayaw MPE, St George-Hyslop P, Rogaeva E, Kawarai T, Pollen DA. Familial Alzheimer disease: decreases in CSF Abeta42 levels precede cognitive decline. Neurology. 2005;65:323–5.
151. Ringman JM, Coppola G, Elashoff D, Rodriguez-Agudelo Y, Medina LD, Gylys K, Cummings JL, Cole GM. Cerebrospinal fluid biomarkers and proximity to diagnosis in preclinical familial Alzheimer's disease. Dement Geriatr Cogn Disord. 2012;33:1–5.
152. Ringman JM, Younkin SG, Pratico D, Seltzer W, Cole GM, Geschwind DH, Rodriguez-Agudelo Y, Schaffer B, Fein J, Sokolow S, Rosario ER, Gylys KH, Varpetian A, Medina LD, Cummings JL. Biochemical markers in persons with preclinical familial Alzheimer disease. Neurology. 2008;71:85–92.
153. Ossenkoppele R, Mattsson N, Teunissen CE, Barkhof F, Pijnenburg Y, Scheltens P, van der Flier WM, Rabinovici GD. Cerebrospinal fluid biomarkers and cerebral atrophy in distinct clinical variants of probable Alzheimer's disease. Neurobiol Aging. 2015;36:2340–7.
154. Rivero-Santana A, Ferreira D, Perestelo-Pérez L, Westman E, Wahlund L-O, Sarría A, Serrano-Aguilar P. Cerebrospinal fluid biomarkers for the differential diagnosis between Alzheimer's disease and frontotemporal lobar degeneration: systematic review, HSROC analysis, and confounding factors. J Alzheimers Dis. 2017;55:625–44.
155. de Souza LC, Lamari F, Belliard S, Jardel C, Houillier C, De Paz R, Dubois B, Sarazin M. Cerebrospinal fluid biomarkers in the differential diagnosis of Alzheimer's disease from other cortical dementias. J Neurol Neurosurg Psychiatry. 2011;82:240–6.
156. Scherling CS, Hall T, Berisha F, Klepac K, Karydas A, Coppola G, Kramer JH, Rabinovici G, Ahlijanian M, Miller BL, Seeley W, Grinberg LT, Rosen H, Meredith J, Boxer AL. Cerebrospinal fluid neurofilament concentration reflects disease severity in frontotemporal degeneration. Ann Neurol. 2014;75:116–26.
157. Borroni B, Benussi A, Archetti S, Galimberti D, Parnetti L, Nacmias B, Sorbi S, Scarpini E, Padovani A. Csf p-tau181/tau ratio as biomarker for TDP pathology in frontotemporal dementia. Amyotroph Lateral Scler Frontotemporal Degener. 2015;16:86–91.
158. Hu WT, Watts K, Grossman M, Glass J, Lah JJ, Hales C, Shelnutt M, Van Deerlin V, Trojanowski JQ, Levey AI. Reduced CSF p-Tau181 to Tau ratio is a biomarker for FTLD-TDP. Neurology. 2013;81:1945–52.

159. Pijnenburg YAL, Verwey NA, van der Flier WM, Scheltens P, Teunissen CE. Discriminative and prognostic potential of cerebrospinal fluid phosphoTau/tau ratio and neurofilaments for frontotemporal dementia subtypes. Alzheimers Dement (Amst). 2015;1:505–12.
160. Landqvist Waldö M, Frizell Santillo A, Passant U, Zetterberg H, Rosengren L, Nilsson C, Englund E. Cerebrospinal fluid neurofilament light chain protein levels in subtypes of frontotemporal dementia. BMC Neurol. 2013;13:54.
161. Wagshal D, Sankaranarayanan S, Guss V, Hall T, Berisha F, Lobach I, Karydas A, Voltarelli L, Scherling C, Heuer H, Tartaglia MC, Miller Z, Coppola G, Ahlijanian M, Soares H, Kramer JH, Rabinovici GD, Rosen HJ, Miller BL, Meredith J, et al. Divergent CSF τ alterations in two common tauopathies: Alzheimer's disease and progressive supranuclear palsy. J Neurol Neurosurg Psychiatry. 2015;86:244–50.
162. Meeter LH, Dopper EG, Jiskoot LC, Sanchez-Valle R, Graff C, Benussi L, Ghidoni R, Pijnenburg YA, Borroni B, Galimberti D, Laforce RJ, Masellis M, Vandenberghe R, Ber IL, Otto M, van Minkelen R, Papma JM, Rombouts SA, Balasa M, Öijerstedt L, et al. Neurofilament light chain: a biomarker for genetic frontotemporal dementia. Ann Clin Transl Neurol. 2016;3:623–36.
163. Borroni B, Benussi A, Cosseddu M, Archetti S, Padovani A. Cerebrospinal fluid tau levels predict prognosis in non-inherited frontotemporal dementia. Neurodegener Dis. 2014;13:224–9.
164. Koedam ELGE, van der Vlies AE, van der Flier WM, Verwey NA, Koene T, Scheltens P, Blankenstein MA, Pijnenburg YAL. Cognitive correlates of cerebrospinal fluid biomarkers in frontotemporal dementia. Alzheimers Dement. 2013;9:269–75.
165. Mattsson N, Carrillo MC, Dean RA, Devous MD Sr, Nikolcheva T, Pesini P, Salter H, Potter WZ, Sperling RS, Bateman RJ, Bain LJ, Liu E. Revolutionizing Alzheimer's disease and clinical trials through biomarkers. Alzheimers Dement (Amst). 2015;1:412–9.
166. Insel PS, Palmqvist S, Mackin RS, Nosheny RL, Hansson O, Weiner MW, Mattsson N. Assessing risk for preclinical β-amyloid pathology with APOE, cognitive, and demographic information. Alzheimers Dement (Amst). 2016;4:76–84.
167. Isaac M, Vamvakas S, Abadie E, Jonsson B, Gispen C, Pani L. Qualification opinion of novel methodologies in the predementia stage of Alzheimer's disease: cerebro-spinal-fluid related biomarkers for drugs affecting amyloid burden—regulatory considerations by European Medicines Agency focusing in improving benefit/risk in regulatory trials. Eur Neuropsychopharmacol. 2011;21:781–8.
168. Donohue MC, Sperling RA, Salmon DP, Rentz DM, Raman R, Thomas RG, Weiner M, Aisen PS, Australian Imaging, Biomarkers, and Lifestyle Flagship Study of Ageing, Alzheimer's Disease Neuroimaging Initiative & Alzheimer's Disease Cooperative Study. The preclinical Alzheimer cognitive composite: measuring amyloid-related decline. JAMA Neurol. 2014;71:961–70.
169. Insel PS, Mattsson N, Mackin RS, Kornak J, Nosheny R, Tosun-Turgut D, Donohue MC, Aisen PS, Weiner MW, Alzheimer's Disease Neuroimaging Initiative. Biomarkers and cognitive endpoints to optimize trials in Alzheimer's disease. Ann Clin Transl Neurol. 2015;2:534–47.
170. Insel PS, Donohue MC, Mackin RS, Aisen PS, Hansson O, Weiner MW, Mattsson N, Alzheimer's Disease Neuroimaging Initiative. Cognitive and functional changes associated with Aβ pathology and the progression to mild cognitive impairment. Neurobiol Aging. 2016;48:172–81.
171. Orgogozo JM, Gilman S, Dartigues JF, Laurent B, Puel M, Kirby LC, Jouanny P, Dubois B, Eisner L, Flitman S, Michel BF, Boada M, Frank A, Hock C. Subacute meningoencephalitis in a subset of patients with AD after Abeta42 immunization. Neurology. 2003;61:46–54.
172. Andreasen N, Blennow K, Zetterberg H. Neuroinflammation screening in immunotherapy trials against Alzheimer's disease. Int J Alzheimers Dis. 2010;2010:638379.
173. Lannfelt L, Blennow K, Zetterberg H, Batsman S, Ames D, Harrison J, Masters CL, Targum S, Bush AI, Murdoch R, Wilson J, Ritchie CW. Safety, efficacy, and biomarker findings of PBT2 in targeting Abeta as a modifying therapy for Alzheimer's disease: a phase IIa, doubleblind, randomised, placebo-controlled trial. Lancet Neurol. 2008;7:779–86.

174. Kennedy ME, Stamford AW, Chen X, Cox K, Cumming JN, Dockendorf MF, Egan M, Ereshefsky L, Hodgson RA, Hyde LA, Jhee S, Kleijn HJ, Kuvelkar R, Li W, Mattson BA, Mei H, Palcza J, Scott JD, Tanen M, Troyer MD, et al. The BACE1 inhibitor verubecestat (MK-8931) reduces CNS β-amyloid in animal models and in Alzheimer's disease patients. Sci Transl Med. 2016;8:363ra150.
175. May PC, Dean RA, Lowe SL, Martenyi F, Sheehan SM, Boggs LN, Monk SA, Mathes BM, Mergott DJ, Watson BM, Stout SL, Timm DE, Smith Labell E, Gonzales CR, Nakano M, Jhee SS, Yen M, Ereshefsky L, Lindstrom TD, Calligaro DO, et al. Robust central reduction of amyloid-beta in humans with an orally available, non-peptidic beta-secretase inhibitor. J Neurosci. 2011;31:16507–16.
176. Portelius E, Dean RA, Gustavsson MK, Andreasson U, Zetterberg H, Siemers E, Blennow K. A novel Abeta isoform pattern in CSF reflects gamma-secretase inhibition in Alzheimer disease. Alzheimers Res Ther. 2010;2:7.
177. Portelius E, Price E, Brinkmalm G, Stiteler M, Olsson M, Persson R, Westman-Brinkmalm A, Zetterberg H, Simon AJ, Blennow K. A novel pathway for amyloid precursor protein processing. Neurobiol Aging. 2011;32:1090–8.
178. Blennow K, Zetterberg H, Rinne JO, Salloway S, Wei J, Black R, Grundman M, Liu E, AAB-001 201/202 Investigators. Effect of immunotherapy with bapineuzumab on cerebrospinal fluid biomarker levels in patients with mild to moderate Alzheimer disease. Arch Neurol. 2012;69:1002–10.
179. Gilman S, Koller M, Black RS, Jenkins L, Griffith SG, Fox NC, Eisner L, Kirby L, Rovira MB, Forette F, Orgogozo JM. Clinical effects of Abeta immunization (AN1792) in patients with AD in an interrupted trial. Neurology. 2005;64:1553–62.
180. Hinerfeld DA, Moonis M, Swearer JM, Baker SP, Caselli RJ, Rogaeva E, St George-Hyslop P, Pollen DA. Statins differentially affect amyloid precursor protein metabolism in presymptomatic PS1 and non-PS1 subjects. Arch Neurol. 2007;64:1672–3.
181. Katz R. Biomarkers and surrogate markers: an FDA perspective. NeuroRx. 2004;1:189–95.
182. Andreasson U, Lautner R, Schott JM, Mattsson N, Hansson O, Herukka S-K, Helisalmi S, Ewers M, Hampel H, Wallin A, Minthon L, Hardy J, Blennow K, Zetterberg H. CSF biomarkers for Alzheimer's pathology and the effect size of APOE ε4. Mol Psychiatry. 2013;19(2):148–9.
183. Kauwe JSK, Cruchaga C, Karch CM, Sadler B, Lee M, Mayo K, Latu W, Su'a M, Fagan AM, Holtzman DM, Morris JC, Alzheimer's Disease Neuroimaging Initiative, Goate AM. Fine mapping of genetic variants in BIN1, CLU, CR1 and PICALM for association with cerebrospinal fluid biomarkers for Alzheimer's disease. PLoS One. 2011;6:e15918.
184. Cruchaga C, Kauwe JSK, Mayo K, Spiegel N, Bertelsen S, Nowotny P, Shah AR, Abraham R, Hollingworth P, Harold D, Owen MM, Williams J, Lovestone S, Peskind ER, Li G, Leverenz JB, Galasko D, Alzheimer's Disease Neuroimaging Initiative, Morris JC, Fagan AM, et al. SNPs associated with cerebrospinal fluid phospho-tau levels influence rate of decline in Alzheimer's disease. PLoS Genet. 2010;6:e1001101.
185. Mattsson N, Insel PS, Landau S, Jagust W, Donohue M, Shaw LM, Trojanowski JQ, Zetterberg H, Blennow K, Weiner M, the Alzheimer's Disease Neuroimaging Initiative. Diagnostic accuracy of CSF Ab42 and florbetapir PET for Alzheimer's disease. Ann Clin Transl Neurol. 2014;1(8):534–43.
186. Cairns NJ, Ikonomovic MD, Benzinger T, Storandt M, Fagan AM, Shah AR, Reinwald LT, Carter D, Felton A, Holtzman DM, Mintun MA, Klunk WE, Morris JC. Absence of Pittsburgh compound B detection of cerebral amyloid beta in a patient with clinical, cognitive, and cerebrospinal fluid markers of Alzheimer disease: a case report. Arch Neurol. 2009;66:1557–62.
187. Mattsson N, Insel PS, Donohue M, Landau S, Jagust WJ, Shaw LM, Trojanowski JQ, Zetterberg H, Blennow K, Weiner MW, Alzheimer's Disease Neuroimaging Initiative. Independent information from cerebrospinal fluid amyloid-β and florbetapir imaging in Alzheimer's disease. Brain. 2015;138:772–83.
188. Palmqvist S, Mattsson N, Hansson O, Alzheimer's Disease Neuroimaging Initiative. Cerebrospinal fluid analysis detects cerebral amyloid-β accumulation earlier than positron emission tomography. Brain. 2016;139:1226–36.

189. Brys M, Glodzik L, Mosconi L, Switalski R, De Santi S, Pirraglia E, et al. Magnetic resonance imaging improves cerebrospinal fluid biomarkers in the early detection of Alzheimer's disease. J Alzheimers Dis. 2009;16:351–62.
190. Schoonenboom NS, van der Flier WM, Blankenstein MA, Bouwman FH, Van Kamp GJ, Barkhof F, et al. CSF and MRI markers independently contribute to the diagnosis of Alzheimer's disease. Neurobiol Aging. 2008;29:669–75.
191. Vos S, van Rossum I, Burns L, Knol D, Scheltens P, Soininen H, et al. Test sequence of CSF and MRI biomarkers for prediction of AD in subjects with MCI. Neurobiol Aging. 2012;33:2272–81.
192. Fellgiebel A, Scheurich A, Bartenstein P, Müller MJ. FDG-PET and CSF phospho-tau for prediction of cognitive decline in mild cognitive impairment. Psychiatry Res. 2007;155:167–71.
193. Handels RLH, Vos SJB, Kramberger MG, Jelic V, Blennow K, van Buchem M, van der Flier W, Freund-Levi Y, Hampel H, Olde Rikkert M, Oleksik A, Pirtosek Z, Scheltens P, Soininen H, Teunissen C, Tsolaki M, Wallin AK, Winblad B, Verhey FRJ, Visser PJ. Predicting progression to dementia in persons with mild cognitive impairment using cerebrospinal fluid markers. Alzheimers Dement. 2017;13(8):903–12.
194. Wang L, Fagan AM, Shah AR, Beg MF, Csernansky JG, Morris JC, Holtzman DM. Cerebrospinal fluid proteins predict longitudinal hippocampal degeneration in early-stage dementia of the Alzheimer type. Alzheimer Dis Assoc Disord. 2012;26:314–21.
195. Mattsson N, Insel P, Nosheny R, Trojanowski JQ, Shaw LM, Jack CR, Tosun D, Weiner M, Alzheimer's Disease Neuroimaging Initiative. Effects of cerebrospinal fluid proteins on brain atrophy rates in cognitively healthy older adults. Neurobiol Aging. 2014;35:614–22.
196. Dickerson BC, Wolk DA, Alzheimer's Disease Neuroimaging Initiative. MRI cortical thickness biomarker predicts AD-like CSF and cognitive decline in normal adults. Neurology. 2012;78:84–90.
197. Fjell AM, Walhovd KB. Neuroimaging results impose new views on Alzheimer's disease—the role of amyloid revised. Mol Neurobiol. 2012;45:153–72.
198. Andreasen N, Hesse C, Davidsson P, Minthon L, Wallin A, Winblad B, Vanderstichele H, Vanmechelen E, Blennow K. Cerebrospinal fluid beta-amyloid(1-42) in Alzheimer disease: differences between early- and late-onset Alzheimer disease and stability during the course of disease. Arch Neurol. 1999;56:673–80.
199. Olsson A, Vanderstichele H, Andreasen N, De Meyer G, Wallin A, Holmberg B, Rosengren L, Vanmechelen E, Blennow K. Simultaneous measurement of beta-amyloid(1-42), total tau, and phosphorylated tau (Thr181) in cerebrospinal fluid by the xMAP technology. Clin Chem. 2005;51:336–45.
200. Zetterberg H, Andreasson U, Hansson O, Wu G, Sankaranarayanan S, Andersson ME, Buchhave P, Londos E, Umek RM, Minthon L, Simon AJ, Blennow K. Elevated cerebrospinal fluid BACE1 activity in incipient Alzheimer disease. Arch Neurol. 2008;65:1102–7.
201. Pannee J, Portelius E, Oppermann M, Atkins A, Hornshaw M, Zegers I, Höjrup P, Minthon L, Hansson O, Zetterberg H, Blennow K, Gobom J. A selected reaction monitoring (SRM)-based method for absolute quantification of Aβ38, Aβ40, and Aβ42 in cerebrospinal fluid of Alzheimer's disease patients and healthy controls. J Alzheimers Dis. 2013;33:1021–32.
202. Blennow K, Wallin A, Agren H, Spenger C, Siegfried J, Vanmechelen E. Tau protein in cerebrospinal fluid: a biochemical marker for axonal degeneration in Alzheimer disease? Mol Chem Neuropathol. 1995;26:231–45.
203. Kohnken R, Buerger K, Zinkowski R, Miller C, Kerkman D, DeBernardis J, Shen J, Moller HJ, Davies P, Hampel H. Detection of tau phosphorylated at threonine 231 in cerebrospinal fluid of Alzheimer's disease patients. Neurosci Lett. 2000;287:187–90.
204. Vanmechelen E, Vanderstichele H, Davidsson P, Van Kerschaver E, Van Der Perre B, Sjogren M, Andreasen N, Blennow K. Quantification of tau phosphorylated at threonine 181 in human cerebrospinal fluid: a sandwich ELISA with a synthetic phosphopeptide for standardization. Neurosci Lett. 2000;285:49–52.
205. Wang L-S, Leung YY, Chang S-K, Leight S, Knapik-Czajka M, Baek Y, Shaw LM, Lee VM-Y, Trojanowski JQ, Clark CM. Comparison of xMAP and ELISA assays for detecting cerebrospinal fluid biomarkers of Alzheimer's disease. J Alzheimers Dis. 2012;31:439–45.

206. Portelius E, Zetterberg H, Andreasson U, Brinkmalm G, Andreasen N, Wallin A, Westman-Brinkmalm A, Blennow K. An Alzheimer's disease-specific beta-amyloid fragment signature in cerebrospinal fluid. Neurosci Lett. 2006;409:215–9.
207. Bateman RJ, Siemers ER, Mawuenyega KG, Wen G, Browning KR, Sigurdson WC, Yarasheski KE, Friedrich SW, Demattos RB, May PC, Paul SM, Holtzman DM. A gamma-secretase inhibitor decreases amyloid-beta production in the central nervous system. Ann Neurol. 2009;66(1):48–54. https://doi.org/10.1002/ana.21623.
208. Bateman RJ, Munsell LY, Morris JC, Swarm R, Yarasheski KE, Holtzman DM. Human amyloid-beta synthesis and clearance rates as measured in cerebrospinal fluid in vivo. Nat Med. 2006;12:856–61.
209. Mawuenyega KG, Sigurdson W, Ovod V, Munsell L, Kasten T, Morris JC, Yarasheski KE, Bateman RJ. Decreased clearance of CNS beta-amyloid in Alzheimer's disease. Science. 2010;330:1774.
210. Carrillo MC, Blennow K, Soares H, Lewczuk P, Mattsson N, Oberoi P, Umek R, Vandijck M, Salamone S, Bittner T, Shaw LM, Stephenson D, Bain L, Zetterberg H. Global standardization measurement of cerebral spinal fluid for Alzheimer's disease: an update from the Alzheimer's Association Global Biomarkers Consortium. Alzheimers Dement. 2013;9:137–40.
211. Andreasson U, Blennow K, Zetterberg H. Update on ultrasensitive technologies to facilitate research on blood biomarkers for central nervous system disorders. Alzheimers Dement (Amst). 2016;3:98–102.
212. Mattsson N, Zetterberg H, Blennow K. Lessons from multicenter studies on CSF biomarkers for Alzheimer's disease. Int J Alzheimers Dis. 2010;2010:610613.
213. Bjerke M, Portelius E, Minthon L, Wallin A, Anckarsater H, Anckarsater R, Andreasen N, Zetterberg H, Andreasson U, Blennow K. Confounding factors influencing amyloid Beta concentration in cerebrospinal fluid. Int J Alzheimers Dis. 2010;21:221–8.
214. Mattsson N, Andreasson U, Carrillo MC, Persson S, Shaw LM, Zegers I, Zetterberg H, Blennow K. Proficiency testing programs for Alzheimer's disease cerebrospinal fluid biomarkers. Biomark Med. 2012;6:401–7.
215. Mattsson N, Andreasson U, Persson S, Arai H, Batish SD, Bernardini S, Bocchio-Chiavetto L, Blankenstein MA, Carrillo MC, Chalbot S, Coart E, Chiasserini D, Cutler N, Dahlfors G, Duller S, Fagan AM, Forlenza O, Frisoni GB, Galasko D, Galimberti D, et al. The Alzheimer's Association external quality control program for cerebrospinal fluid biomarkers. Alzheimers Dement. 2011;7:386–395.e6.
216. Mattsson N, Andreasson U, Persson S, Carrillo MC, Collins S, Chalbot S, Cutler N, Dufour-Rainfray D, Fagan AM, Heegaard NHH, Robin Hsiung G-Y, Hyman B, Iqbal K, Lachno DR, Lleó A, Lewczuk P, Molinuevo JL, Parchi P, Regeniter A, Rissman R, et al. CSF biomarker variability in the Alzheimer's Association quality control program. Alzheimers Dement. 2013;9:251–61.
217. Consensus report of the Working Group on: 'Molecular and Biochemical Markers of Alzheimer's Disease'. The Ronald and Nancy Reagan Research Institute of the Alzheimer's Association and the National Institute on Aging Working Group. Neurobiol Aging. 1998;19:109–16.
218. Insel PS, Mattsson N, Donohue MC, Mackin RS, Aisen PS, Jack CR, Shaw LM, Trojanowski JQ, Weiner MW, Alzheimer's Disease Neuroimaging Initiative. The transitional association between β-amyloid pathology and regional brain atrophy. Alzheimers Dement. 2015;11:1171–9.
219. Mattsson N, Brax D, Zetterberg H. To know or not to know: ethical issues related to early diagnosis of Alzheimer's disease. Int J Alzheimers Dis. 2010;2010:841941.

Biomarkers for Alzheimer's Disease and Frontotemporal Lobar Degeneration: Imaging

12

Marco Bozzali and Laura Serra

Abstract

Neuroimaging has become an invaluable tool for the clinical management of patients with cognitive decline and for research purposes. In clinical setting, structural and functional information on the brain tissue damage contributes to define the diagnosis of the major forms of dementia since their early clinical stages. From the research side, quantitative neuroimaging techniques have contributed in clarifying some critical pathophysiological aspects of dementias, playing the unique role of linking together measures of cognitive and behavioural impairment and the presence and distribution of brain tissue abnormalities. Positron emission tomography provides not only information on abnormal brain metabolism, but also on the brain deposition of pathogenic molecules, such as beta-amyloid and tau. On the other hand, quantitative MRI provides information on microstructural brain abnormalities as well as on functional and structural connectivity. In this chapter we review the role of these neuroimaging techniques with a special focus on Alzheimer's disease and frontotemporal dementia.

Keywords

Alzheimer's disease · Frontotemporal dementia · Imaging · Biomarker

Introduction

In recent years, our understanding of neurodegenerative dementias has translated into a change in the clinical approach to patients presenting with impairments in cognition and behaviour. The diagnosis of different forms of neurodegenerative

M. Bozzali (✉) · L. Serra
Neuroimaging Laboratory, Santa Lucia Foundation IRCCS, Rome, Italy
e-mail: m.bozzali@hsantalucia.it

© Springer International Publishing AG 2018
D. Galimberti, E. Scarpini (eds.), *Neurodegenerative Diseases*,
https://doi.org/10.1007/978-3-319-72938-1_12

dementias is currently based not only on their clinical and neuropsychological characterization, but also on the use of biomarkers. Advances in neuroimaging techniques, including magnetic resonance imaging (MRI) and positron emission tomography (PET), have strongly contributed not only in increasing our understanding of clinical and pathophysiological aspects of dementias, but also in improving the diagnostic confidence in clinical settings [1]. MRI, thanks to its ability to image in vivo soft tissues non-invasively and with detailed anatomical resolution, shows high sensitivity in detecting the presence and extension of macroscopic brain abnormalities [2]. In this view, as discussed below, MRI plays the unique role of excluding alternative diagnoses that may mimic a neurodegenerative form of cognitive decline. On the other hand, PET imaging has proven high sensitivity in detecting metabolic abnormalities at a single subject level since early clinical stages of cognitive decline [3]. Additionally, novel tracers, including beta-amyloid and tau protein ligands have become available with the potential of detecting in vivo specific pathological features of brain tissue degeneration [4].

Neurodegenerative dementias, such as Alzheimer's disease (AD) and frontotemporal dementia (FTD), are typically characterized by an insidious onset which is followed by a gradual progression of symptoms. Especially at early clinical stages, the underlying neurodegenerative processes produce selective cognitive dysfunctions that may correspond to the focal distribution of brain damage [1]. As shown in Table 12.1, the combination of biomarker characteristics and neuropsychological

Table 12.1 Neurodegenerative dementia clinical syndromes [1, 3]

Syndrome	CSF characteristics	Key clinical dysfunction	Early cerebral involvement
Amnestic-Alzheimer disease	β-amyloid hyperphosphorylated tau protein	Episodic memory	Medial temporal lobes
Posterior cortical atrophy	β-amyloid hyperphosphorylated tau protein	Visuospatial dysfunctions	Parietal-occipital lobes
Dysexecutive Alzheimer disease	*No pathological correlations available*	Dysexecutive syndrome	Frontal and temporo-parietal lobes
Logopenic PPA	Hyperphosphorylated tau protein	Word retrieval, sentence repetition	Left temporo-parietal lobe
Agrammatic PPA	Hyperphosphorylated tau protein	Agrammatism; Apraxia of Speech	Left posterior frontal lobe and insula
Semantic dementia	Hyperphosphorylated tau protein	Confrontation naming; single word comprehension	Left temporal pole
bv-FTD	Hyperphosphorylated tau protein	Disinhibition; apathy; sleep disorder; perseverative behaviour, dysexecutive syndrome	Frontal, temporal lobes, anterior cingulate, insula

Abbreviations: *PPA* primary progressive aphasia, *bv-FTD* behavioural variant-frontotemporal lobar degeneration. *Modified by McGinnis, 2012* [1]; *Cummings, 2003* [5]

profiles improves the potential of a correct and early diagnosis of neurodegenerative dementias [1, 5]. Additionally, as demonstrated by research evidence mostly based on neuroimaging, cognitive and behavioural disabilities in dementias are not only due to focal brain tissue damage, but also to disconnection mechanisms. In this context, brain connectivity as assessed by functional neuroimaging, has revolutionized our understanding of large-scale neuronal networks and clarified the relationship between their disruption/modifications and the clinical evolution of neurodegenerative diseases [6, 7]. Moreover, the impact of neuropathology may be different across individuals, depending on various genetic and environmental factors. In the so-called sporadic forms of neurodegeneration, the concept of "cognitive reserve" has been put forward to account for inconsistencies between severity of brain tissue damage and symptoms exhibited by patients. Neuroimaging has strongly contributed in supporting the concept of cognitive reserve and in clarifying the potential neurobiological mechanisms by which cognitive reserve mitigates the clinical effect of neurodegeneration.

In this chapter, we will review, for AD and frontotemporal dementia (FTD), the contribution of neuroimaging in supporting a correct clinical diagnosis and the role of advanced neuroimaging techniques in clarifying and monitoring some pathophysiological aspects of disease.

Alzheimer's Disease

Alzheimer's disease (AD) is the most common cause of dementia in elderly populations [8]. Neuropathological studies have identified a sequential accumulation of neurofibrillary tangles and β-amyloid plaques in the brain tissue, as well as the progression of neuronal loss through the cerebral cortex [9]. From a clinical viewpoint, the accumulation of neuropathological abnormalities may precede of many years the clinical onset of AD [10]. In particular, neurofibrillary pathology in the entorhinal cortex, hippocampus and amygdala is considered as the major neurobiological substrate for episodic memory deficits, which are typically observed in AD since early stages (for a review, see [10]). In recent years, the increased knowledge on the neuropathological cascade, occurring in AD brains, and the early cognitive modifications originating from these abnormalities have led to the definition of new diagnostic criteria for preclinical AD [11]. These criteria incorporate several biomarkers, including neuroimaging, to define the presence of AD pathology [11].

Conventional MRI

Conventional MRI has shown the ability to produce brain images with a higher spatial resolution compared to computerized tomography (CT), thus showing much more detailed information about macroscopic brain anatomy. Moreover, MRI is particularly helpful in detecting and excluding other neurological conditions mimicking a neurodegenerative form of cognitive decline, such as brain tumours, normal

Table 12.2 Visual rating of medial temporal lobe atrophy according to MTA scale [10]

Score	Width of choroid fissure	Width of temporal horn	Height of hippocampal formation
0	Normal	Normal	Normal
1	Mild increase	Normal	Normal
2	Moderate increase	Mild increase	Mild decrease
3	Severe increase	Moderate increase	Moderate decrease
4	Severe increase	Severe increase	Severe decrease

pressure hydrocephalus, subdural hematoma and cerebrovascular disease. After exclusion of secondary causes of dementia, conventional MRI may address a correct diagnosis of AD only in a proportion of cases, mainly based on assessment of regional brain atrophy. The simplest approach to determine regional changes of brain volumes is to use rating scales based on visual examination of T1-weighted MR images [12, 13], such as the "medial temporal lobe atrophy" MTA [12] scale. This tool allows a semi-quantitative volumetric assessment of the medial temporal lobe structures (i.e. hippocampus, dentate gyrus, subiculum and parahippocampal gyrus) and enlargements of the temporal horn of the lateral ventricles and choroid fissures (Table 12.2 and Fig. 12.1, panel A). The use of MTA has shown high accuracy in determining the severity of local atrophy in cross-sectional studies that compared AD patients with healthy controls [14]. Conversely, MTA appears to be poorly informative in detecting longitudinal volumetric changes over time [14]. Consistently, a recent study showed a low sensitivity of MTA to detect AD progression in patients with mild cognitive impairment (MCI) [15]. Moreover, new decade-specific MTA cut-off scores for AD have been recently proposed [16]. In these new cut-offs, a MTA score ≥ 1 is sufficient to identify hippocampal atrophy (with 83.3% of sensitivity and 86.4% of specificity) in subjects who are less than 65 years old. A MTA score ≥ 1.5 is necessary to identify clinically relevant atrophy (with 73.7% of sensitivity and 84.6% of specificity) in subjects whose age ranges from 65 to 74 years. A MTA score ≥ 2 is necessary to identify clinically relevant atrophy in subjects over 75 years old (with approximately 75% of sensitivity and 70% of specificity) [16].

There are specific visual rating scales also to quantify the presence and severity of macroscopic white matter (WM) abnormalities. They can be applied to CT images or, with a better definition, to MRI scans (i.e. T2-/proton density [PD]-weighted and/fluid attenuated inversion recovery [FLAIR] images). The age-related white matter changes (ARWMC) [13] and the Fazekas (1987) scales [17], whose application criteria are summarized in Table 12.3 and illustrated in Fig. 12.1 (panel B), allow a simple assessment of macroscopic WM abnormalities. In the diagnostic suspect of AD, taking altogether the information given by MTA and WM lesion assessment, three different patterns may schematically be identified (Fig. 12.1): (1) severe MTA and minimal WM abnormalities; (2) minimal MTA and severe WM abnormalities; (3) moderate MTA and moderate/severe WM abnormalities. In the first two cases, conventional MRI strongly contributes in increasing the diagnostic confidence of degenerative against vascular dementia. In the third case, due to the frequent comorbidity of degenerative and vascular pathology, the contribution of conventional MRI remains limited. Moreover, a recent study showed a strict

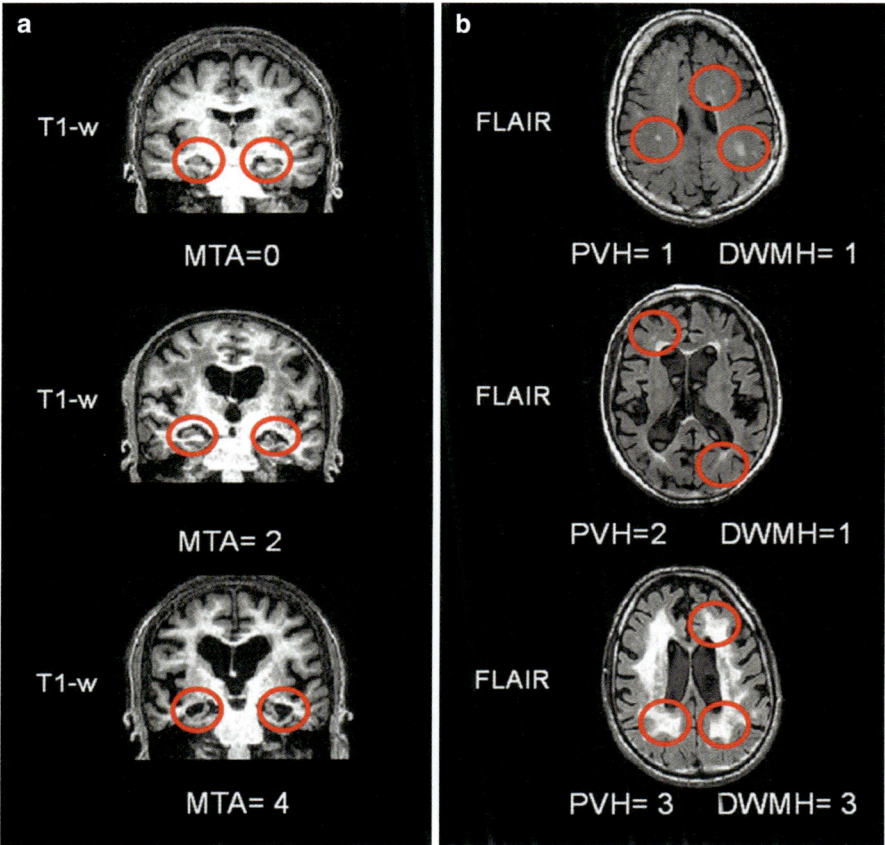

Fig. 12.1 Visual rating scales to assess brain atrophy and white matter hyperintensities. T1-weighted and FLAIR images of patients with AD are shown. Medial temporal lobe atrophy scale (MTA) [10] scores are illustrated in panel **A**, ARWMC [11] and Fazekas scale [15] scores are shown in panel **B**. Red circles highlight the considered medial temporal lobe structures. Abbreviations: *ARWMC* age-related white matter changes, *DWMH* deep white matter hyperintensities, *FLAIR* fluid attenuated inversion recovery images, *MTA* medial temporal lobe atrophy scale, *PVH* periventricular hyperintensities, *T1-w* T1-weighted images

association between brain amyloid deposition and periventricular lesions [18]. Therefore, rating scales can be useful in clinical settings to estimate atrophy and vascular pathology in neurodegenerative disorders.

Advanced MRI Techniques

Brain Volumetrics
Several approaches to quantitative brain volumetrics are currently available, and the simplest methods are those based on manual or semiautomatic delineation of brain

Table 12.3 Visual rating scales to assess white matter hyperintensities

White matter rating scales	Brain area	White matter lesions
ARWMC [11]	Frontal, parieto-occipital, temporal and infratentorial	0 = no lesions 1 = focal lesions 2 = beginning confluence of lesions 3 = diffuse involvement of entire regions
	Basal ganglia	0 = no lesions 1 = one focal lesion 2 = more than one focal lesion 3 = confluent lesions
Fazekas scale [15]	Periventricular (PVH) lesions	0 = absent 1 = caps or pencil-thin lining around ventricles 2 = smooth halo around ventricles 3 = irregular PVH extending into DWM
	Deep (DWM) lesions	0 = absent 1 = discrete diffuse lesions 2 = beginning of confluence of foci 3 = large confluent areas

structures. More recently, the development of more sophisticated registration algorithms has made it possible to bring volumetric images from different subjects into a common space and to identify differences between groups (e.g. patients vs. controls) or correlations with clinical/psychometric measures, on a voxel-by-voxel level basis. The most appropriate MR scans for all types of volumetric assessments are the high-resolution T1-weighted volumes, typically obtained using three-dimensional acquisitions, which provide sufficient anatomical detail, as well as sufficient contrast between grey and WM tissues.

Manual and Semiautomatic Regional Measurements

Given the relevance of MTL atrophy in AD, which corresponds to post-mortem evidence of earlier and predominant neurofibrillary degeneration in this region, first attempts to quantify brain damage (i.e. atrophy) employed manual volumetric assessments of the hippocampus on coronal T1-weigthed images [19]. Figure 12.2 illustrates the steps to obtain manual segmentation of MTL structures.

Comparisons between patients with AD and healthy controls have consistently revealed volumetric reductions of the hippocampus of about 40% [19]. Significant hippocampal reductions have been reported also in patients with MCI [20], thus confirming an involvement of this area as core disease feature since early clinical stages. Interestingly, in clinical follow-up studies, hippocampal volumetrics revealed a more severe atrophy in those MCI patients who converted to AD than in those who remained stable [20]. In terms of potential diagnostic application, hippocampal and entorhinal volumetrics allows a separation of AD and MCI patients from healthy controls with accuracies ranging from 70% (in early MCI) to 100% (in AD patients)

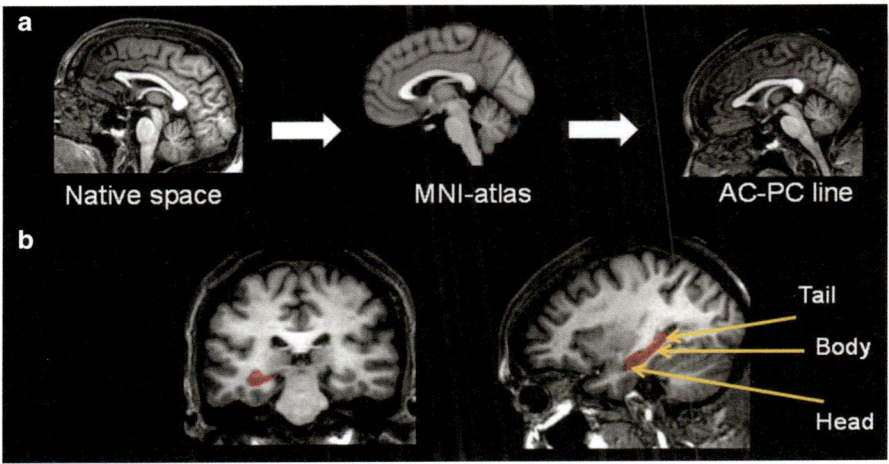

Fig. 12.2 Manual segmentation to obtain hippocampal volumes. Pipeline to rigidly co-register individual T1-weighted images to MNI atlas is shown in panel **A**. Panel **B** shows the anatomical landmarks used to manually segment the hippocampus. Abbreviations: *AC* anterior commissure, *MNI* Montreal Neurological Institute, *PC* posterior commissure

[19]. Additionally, volumetrics of these brain structures has been reported as predictive for a future conversion from MCI to AD with an accuracy of about 80–85% [21]. A recent study showed also a relationship between hippocampal volumes and patients' level of cognitive reserve (CR) [22]. Patients with higher levels of CR accumulated more hippocampal atrophy than those with lower CR to express the same level of cognitive decline. This means that CR helps patients to cope better with the accumulation of brain damage and accounts for variability across individuals between brain damage severity and the level of cognitive impairment.

Nevertheless, manual assessments of MTL volumes are strongly operator dependent, based on different anatomical landmarks across studies, and time consuming. So far, these weaknesses have prevented a wide diffusion of manual assessments in clinical settings, despite the recent on-going efforts of methodological standardization and validation [23, 24].

Automated Methods to Assess Brain Atrophy

For data-driven analyses, voxel-based morphometry (VBM) is one of the most popular techniques to investigate dementias [25]. VBM has proven high reproducibility when using datasets obtained by different MR systems and various optimizations of image processing. This approach is operator independent and does not require any a priori hypothesis on the anatomical localization of the brain tissue loss, as it includes the whole brain (i.e. voxel-wise analysis) [26]. VBM analysis is particularly suitable for grey matter (GM) volume assessments and is based on a series of automatic steps, the main ones including normalization of individual T1-weighted volumes to standard space, brain segmentation and extraction of GM maps, and statistical analyses. Different statistical designs can be employed, which allow to perform between-group comparisons as well as correlations between regional distributions of GM volumes and clinical, neuropsychological and behavioural variables. When applied to AD patients at different clinical stages, VBM has demonstrated a widespread pattern of GM atrophy, including not only the medial temporal lobe structures but also several other areas of the association cortex [27, 28]. Moreover, in AD and amnestic MCI patients, it has been shown a strict association between cognitive profiles and regional patterns of GM atrophy. For instance, hippocampal GM loss has been shown to be associated with patients' episodic memory deficits [27], and posterior cortical atrophy has been found associated with constructional apraxia [28]. Associations between regional GM atrophy and patients' behavioural features have also been demonstrated in AD and MCI, suggesting these symptoms to be part of AD pathophysiology [29]. MCI can also be clinically dominated by neuropsychological deficits other than memory (i.e. non-amnestic MCI). Again, VBM has shown the ability to detect patterns of regional GM loss that fit with the non-amnestic neuropsychological profile, thus allowing a differentiation of MCI patients who are more likely to convert to other forms of dementias [30]. Moreover, VBM has identified different patterns of GM volumes in association with different levels of CR in patients with AD at different clinical stages [31]. Patients with higher CR levels, compared to those with lower CR, showed both decrease and increase of GM volumes in different brain areas [31]. Crucially, when comparing patients with higher against those with lower CR, the former group exhibit more atrophy in areas typically targeted by AD pathology, such as the medial temporal lobes, to express the same level of cognitive decline. Conversely, they are less atrophic in other areas of the association cortex, which might express a CR driven mechanism of compensation.

Diffusion Imaging
Diffusion imaging provides, through the measurement of diffusional motion of water molecules into brain cells, unique information to investigate the WM microarchitecture, connectivity and integrity, documenting the size, shape, orientation and geometry of brain structures [32]. Neurodegenerative processes, such as those occurring in AD, modify tissue integrity, and they can result in an altered diffusion coefficient, which can be measured in vivo by diffusion MRI. The diffusion of water molecules is facilitated along the principal direction of WM fibres, and this allows to reconstruct some WM fibre tracts. The metrics resulting from different steps of diffusion image analysis (e.g. fractional anisotropy, FA; mean diffusivity, MD; radial diffusivity, RD; axial diffusivity AD) can be statistically analysed using both automated voxel-wise methods (e.g. by tract-based spatial statistics—TBSS) [33] or regional approaches (e.g. diffusion tractography reconstruction of WM tracts) [34].

Diffusion imaging has been widely used in studies investigating MCI and AD patients (for a review see [2]). Some of them have reported a widespread alteration of WM tissue integrity in patients with AD at different clinical stages and using both a whole brain analysis [27, 35] or focusing on specific WM tracts [36, 37]. For instance, a study based on diffusion tractography of the cingulum (i.e. the main

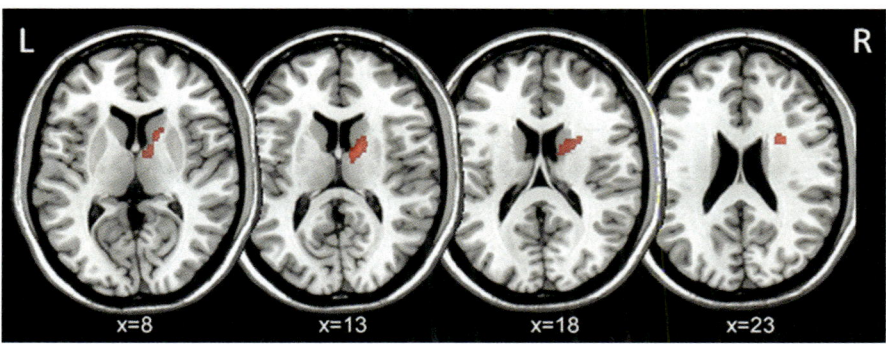

Fig. 12.3 ACM values in patients undergoing AChEIs therapy. Direct associations (red areas) between ACM values and dosage × duration of therapy product in the group of patients under treatment with acetylcholinesterase inhibitors (AChEIs). The area of significant association is located within the anterior limb of the internal capsule. Abbreviations: *ACM* anatomical connectivity mapping, *AChEIs* acetylcholinesterase inhibitor, *L* left, *R* right

pathway of connection between the limbic system and the rest of the brain) shows a progressive disruption of this structure over the transitional stage from MCI to AD [37]. Interestingly, this WM damage accounts, in combination with regional GM loss, for the cognitive features of preclinical and clinical AD stages [37]. Another interesting tract, implicated in AD pathophysiology, is the uncinate fasciculus. It has been shown how damage to this tract accounts for cognitive and behavioural aspects which are typically present at advanced stages of AD [36].

Finally, a novel method of diffusion imaging analysis, called anatomical connectivity mapping (ACM), has been proposed to assess changes in structural brain connectivity across the whole brain [38]. This voxel-wise technique, based on probabilistic tractography, is able to detect in patients with AD, modifications of brain plasticity including those which are likely driven by cholinergic therapy [39] (Fig. 12.3).

Functional MRI

Neuronal activity can be investigated non-invasively, but indirectly, through blood oxygenation level-dependent (BOLD) functional MRI (fMRI). fMRI can be used to assess changes of brain activation in response to patients' performance at cognitive tasks involving specific higher level functions (e.g. memory, visuospatial, executive functions, emotion processing). On the other hand, fMRI can also be used at rest to record coherent fluctuations of brain activity over time, in the so-called resting-state fMRI technique. In this latter case, fMRI provides information on functional brain connectivity within specific networks, some of which have been associated with specific higher level functions. When using fMRI with active tasks, patients' cooperation is essential, and findings obtained in patients with fully developed AD remain for this reason controversial. Investigations based on episodic memory tasks have reported, in AD, reductions of functional activity in the hippocampus and other temporal lobe areas and increased activity in the parietal association cortex [40]. In contrast, other studies have reported a decrease of functional activity (during memory tasks) not only in the temporal lobe but also in parietal and frontal regions [41]. Studies involving patients with MCI have generally reported increased activation in brain areas related to the administered tasks (for a review see [42]). There is some evidence that these increases of functional activity might represent compensatory mechanisms against the incipient occurrence of brain atrophy. In a group of patients with amnestic MCI single domain, it has been shown increased brain activation in a set of tasks exploring memory and visuospatial attention, in the presence of a maintained performance during task execution [43].

As mentioned above, resting-state fMRI does not require any active performance of tasks and allows to record spontaneous brain activity fluctuations when subjects lie in the scanner at rest. Therefore, resting-state fMRI provides information on the integrity of functional brain connectivity [44] and permits to identify different brain networks and to investigate the strength of connectivity within them [45]. Among all brain networks, the default mode network (DMN) has been intensively investigated in patients with dementia. This network includes the posterior cingulated cortex (PCC), the inferior parietal and the medial prefrontal cortex. These regions are believed to be similarly modulated by cognitive tasks [46]. Several studies have been performed on patients at different stages of AD, all documenting an alteration into DMN nodes. A study [6] involving patients with AD, patients with a-MCI and healthy controls, investigated changes in both GM atrophy and functional connectivity into DMN. This study revealed that functional disconnection precedes GM atrophy in the PCC, supporting the hypothesis that GM atrophy in specific regions of AD brains is likely to reflect a long-term effect of brain disconnection and to possibly account for the conversion to AD [6]. In addition, DMN connectivity has been found to be modulated by individual levels of CR [47], thus contributing in clarifying the neurobiological substrate of this compensation mechanisms that helps in delaying the clinical impact of AD pathology.

Recently, a modulation of connectivity due to CR was observed also at larger scale in the brain, based on more sophisticated approaches of image analysis called "brain connectomics". MCI patients with higher CR showed increased functional connectivity in a large network of fronto-parietal nodes (Fig. 12.4A) and decreased connectivity in a network involving fronto-temporo-cerebellar nodes (Fig. 12.4B) [48]. Interestingly, this dichotomy effect was clearly detectable in MCI patients only, suggesting that the CR acts in contrasting AD symptoms in a specific time window of the transitional stage between normal ageing and dementia. This has potential implications for non-pharmacological interventions in AD.

Metabolic Imaging

PET is a sensitive molecular imaging technique for the in vivo quantification of radiotracer concentrations in a picomolar range. PET scanning allows a non-invasive assessment of molecular processes at their sites of action and is in principle capable of detecting disease processes when there is no evidence of structural changes on MRI [49]. ^{18}Fluorodeoxyglucose (^{18}FDG-PET) is a widely available PET tracer that

Fig. 12.4 Functional brain connectivity in patients with high or low cognitive reserve. Networks of higher (panel **A**) and lower (panel **B**) connectivity in patients with a-MCI and high cognitive reserve compared to those with low cognitive reserve. Abbreviations: *a-MCI* amnestic mild cognitive impairment, *R* right

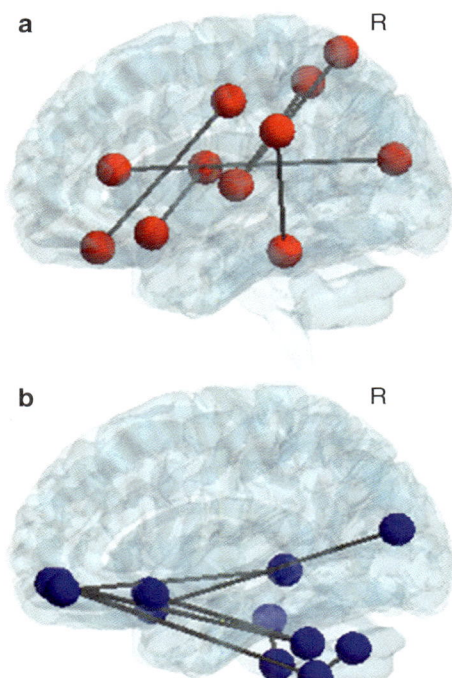

reflects the local glucose metabolism as a proxy index for neuronal activity [50]. Typical ^{18}FDG-PET finding in patients with AD is a pattern of reduced glucose uptake in temporo-parietal association areas, including the precuneus and the posterior cingulate cortex [3, 50, 51]. ^{18}FDG-PET has demonstrated a high specificity in discriminating between patients with AD and healthy subjects (ranging from 70 to 90%) [52] and between patients with AD and those with other forms of degenerative dementia (specificity of 87%) [52]. On the other hand, the ability of ^{18}FDG-PET to identify patients at preclinical AD stages remains a controversial issue [52].

Another useful application in clinical practice is the use of single-photon emission computed tomography (SPECT) after administration of dopamine transporter (DAT) ligands (e.g. [^{123}I]FP-CIT, [^{123}I]β-CIT, [^{99m}Tc]-TRODAT-1), the so-called DAT scan technique. DAT scan allows the detection of striatal dopaminergic dysfunction, which is typically present in patients with Parkinson-related disorders [53] and not in AD.

Available evidence supports the position that an abnormal processing of β-amyloid (Aβ) peptides is the initiating event of AD pathophysiology, which eventually leads to accumulation of Aβ plaques in the brain tissue [54]. This process occurs when individuals are still cognitively intact, many years before the occurrence of clinical manifestations of AD. In this picture, the amyloid PET imaging has been proposed as tool for early detection of AD pathology in vivo, and for the differential diagnosis of dementia

[55]. In AD, PET β-amyloid imaging has shown increased tracer binding in areas known to have high concentrations of amyloid plaques such as medial and orbitofrontal regions, the lateral parietal and temporal cortex, the precuneus and posterior cingulate [55]. Advances in biomarkers for AD pathology have recently led to proposals for more definitive diagnoses in patients with MCI as a prodromal AD stage (International Working Group for New Research Criteria for Diagnosis of AD) [56] or MCI as due to AD (National Institute on Aging and Alzheimer's Association Workgroup) [11]. In the latter case, MCI can be defined as due to AD with "high likelihood" whenever both an amyloid and a neurodegenerative biomarker are positive, with "intermediate likelihood" when one biomarker only is positive and "low likelihood" when both biomarkers are negative for AD pathology. In this perspective, several pharmacological approaches aimed at reducing Aβ levels in the brain tissue are being developed and tested, and many efforts have been focused on generating radiotracers for imaging Aβ in vivo [57]. Currently, the [^{11}C] Pittsburgh compound-B (PIB) is the most popular radiotracer used in AD patients, due to its high affinity and selectivity for fibrillar Aβ in plaques and other Aβ-containing lesions [58]. Most importantly, there are available studies showing that the PIB cortical retention primarily reflects Aβ-related cerebral amyloidosis rather than Lewy bodies or neurofibrillary tangles [59]. This would indicate that PIB can be particularly useful for patients' diagnostic definition since early clinical stages. When considering the prognostic value of PET imaging on the risk of conversion to AD, measures of brain glucose metabolism and amyloid load are both extremely powerful biomarkers [3]. In a longitudinal study, ^{18}FDG-PET positivity performed as the best individual predictor for AD conversion, but the combination of both, ^{18}FDG-PET and ^{11}C-PiB-PET imaging, improved classification accuracy [3].

Finally, although future studied are needed to clarify their specific role, PET imaging shows nowadays the potential of detecting in vivo specific aspects of neurodegeneration, including not only beta-amyloid deposition, but also tau protein accumulation [4].

Frontotemporal Dementias

Frontotemporal dementia (FTD) is the second most common neurodegenerative disease, especially in patients with a pre-senile clinical onset (age < 65 years) [60]. FTD can be defined as a heterogeneous cluster of disorders including two major clinical conditions, one characterized by predominant deficits of language functions (primary progressive aphasia) and one characterized by prominent behavioural symptoms (bv-FTD). Nevertheless, several other cognitive deficits (such as impairment in problem-solving, reasoning, planning, attention and decision-making) can be present in both clinical syndromes, and behavioural disorders can be observed in all clinical variants [61]. In the suspect of FTD, after exclusion of severe macroscopic WM damage, specific patterns of regional brain atrophy may be present and fit with specific clinical/neuropsychological syndromes [60], thus increasing the diagnostic confidence of a correct diagnosis.

Some examples of the neuroradiological aspects observed in FTD variants are illustrated in Fig. 12.5.

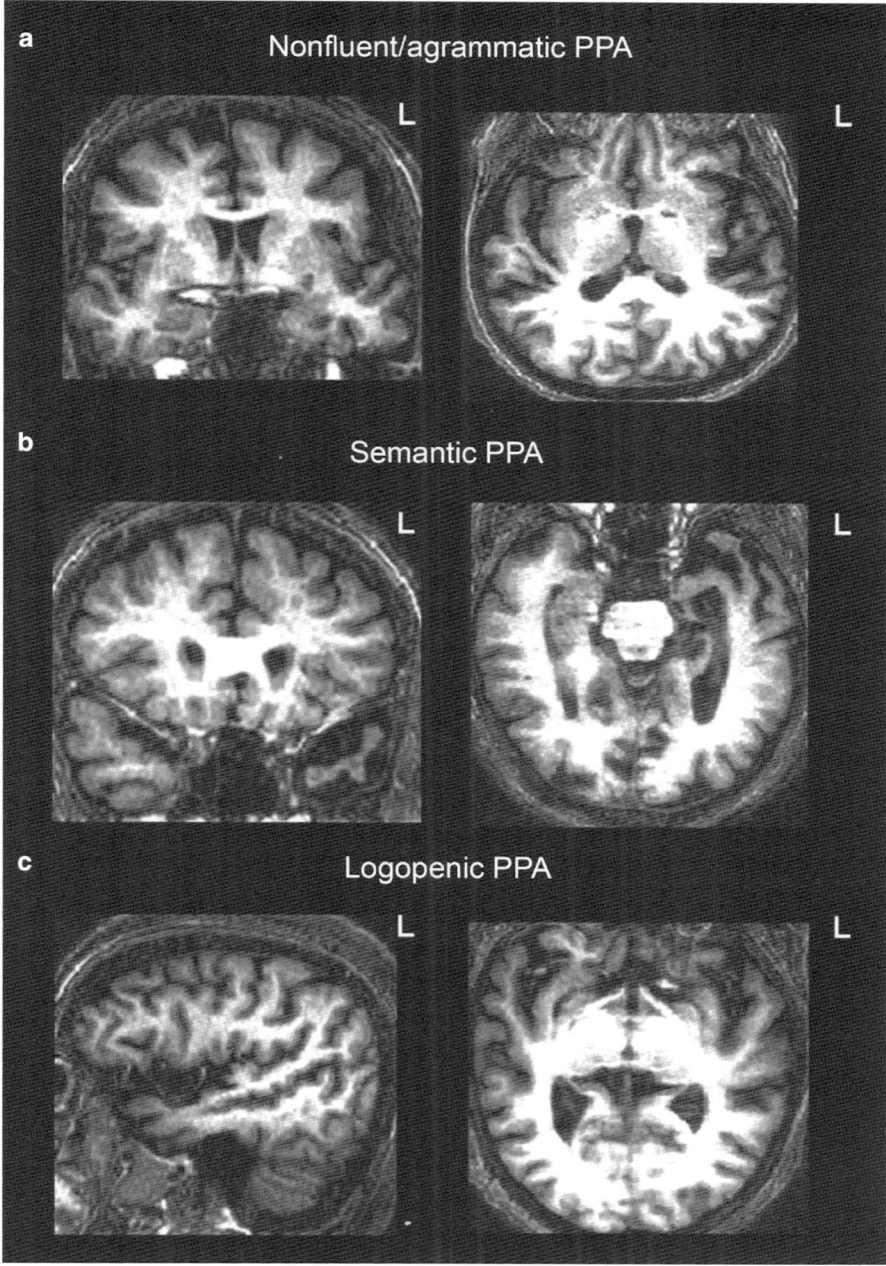

Fig. 12.5 Patterns of brain atrophy in different PPA phenotypes. The typical atrophy of the left insula and perisylvian area in the non-fluent/agrammatic variant of PPA is shown in panel **A**. Panel **B** shows brain atrophy mainly involving the anterior part of the left temporal pole in semantic dementia. Finally, logopenic variant of PPA is characterized by an asymmetric left-side atrophy, involving the perisylvian areas, the posterior part of the superior temporal cortex and the inferior parietal lobes (panel **C**). Abbreviation: *L* left, *PPA* primary progressive aphasia

Histopathological and Genetic Aspects

Over the last few years, many different classifications of FTD variants have been introduced, mainly based on genetic and neuropsychological profiles. Neuroimaging has been used to identify anatomo-functional substrates for these classifications. In all FTD variants, neurodegeneration is mainly due to neuronal loss and gliosis [61, 62], despite a large variety of the underling pathophysiology. Specific protein abnormalities have been identified in a heterogeneous group of diseases, namely, the frontotemporal lobar degeneration (FTLD), which are in most cases associated with a FTD syndrome [62].

The first classification of FTLD has identified two main categories, one associated with the deposition of microtubule-associated protein tau (FTLD-tau) and one associated with deposition of ubiquitin-only immunoreactivity (FTLD-U) (see [63] for a review). In some cases of this latter group of disorders, patients showed also motor neuron disease (FTLD-MND) [63]. More recently, additional sub-classifications have been introduced in the FTLD-U group on the basis of specific molecular features. The presence/absence of transactive response DNA-binding protein of 43 kDa has identified the FTLD-TDP against the aFTLD-U form [64]. In turn, within the aFTLD-U group, another sub-classification has been introduced based on the presence/absence of ubiquitinated protein fused in sarcoma (FUS) [65]. In summary, FTLD currently includes the following forms: FTLD-tau, FTLD-TDP and FTLD-FUS [62].

In most patients with different FTLD forms, several genetic varieties have been identified, the main ones including the following gene mutations: (1) microtubule-associated protein tau (MAPT) [66]; (2) progranulin (GRN) [67]; (3) C9ORF72 [68]. MAPT mutations are commonly observed in the FTLD-tau form [68], while GRN and C9ORF72 mutations are commonly associated to the FTLD-TDP form [68]. Moreover, MAPT mutations have been found in patients with progressive supernuclear palsy, corticobasal syndrome and progressive non-fluent aphasia (PNFA), while GRN mutations [68] have been found associated with semantic dementia (SD), behavioural variant FTD (bv-FTD) and FTLD-MND.

For the purpose of this chapter, which is focused on the neuroimaging, we will limit our description to the clinical classification of FTD as linguistic and behavioural variants.

Clinical Aspects

The Linguistic Variants of FTD

A progressive disorder of language associated with atrophy of the frontal and temporal regions of the left hemisphere was first described in the 1890s by Pick [69]. In the last century, several attempts have been done to further classify the language variant in more specific subtypes. In 1982, Mesulam [70] described a series of cases with "slowly progressive aphasia" and renamed them as primary progressive aphasia (PPA) [71]. In the 1990s, the progressive PNFA and the semantic dementia [72, 73] have been characterized. Each FTD language variety presents with a well-defined pattern of cognitive deficits and brain abnormalities [74]. According to the most recent diagnostic criteria, we can recognize the following PPA subtypes: non-fluent/agrammatic, semantic and logopenic (see Table 12.4).

Table 12.4 Diagnostic features of different variant of PPA [70]

Non-fluent/agrammatic variant of PPA	Semantic variant of PPA	Logopenic variant of PPA
Clinical diagnosis	**Clinical diagnosis**	**Clinical diagnosis**
At least one of the following core features:	Both of the following core features:	Both of the following core features:
– Agrammatism in language production – Effortful, halting speech with inconsistent speech sound errors and distortions (apraxia of speech)	– Impaired confrontation naming – Impaired single-word comprehension	– Impaired single-word retrieval in spontaneous speech and naming – Impaired repetition of sentences and phrases
At least two of three of the following other features:	At least three of the following other features:	At least three of the following other features:
– Impaired comprehension of syntactically complex sentences – Spared single-word comprehension – Spared object knowledge	– Impaired object knowledge (low frequency/low familiar) – Surface dyslexia or dysgraphia – Spared repetition – Spared speech production (grammar and motor speech)	– Speech (phonological) errors in spontaneous speech and naming – Spared single-word comprehension and object knowledge – Spared motor speech – Absence of agrammatism
Imaging-supported diagnosis	**Imaging-supported diagnosis**	**Imaging-supported diagnosis**
Both of the following criteria:	Both of the following criteria:	Both of the following criteria:
– Clinical diagnosis of non-fluent/agrammatic PPA	– Clinical diagnosis of semantic variant of PPA	– Clinical diagnosis logopenic variant of PPA
– Imaging **must** show one or more of the following results: 1. Predominant left posterior fronto-insular atrophy on MRI 2. Predominant left posterior fronto-insular hypoperfusion or metabolism on SPECT or PET	– Imaging **must** show one or more of the following results: 1. Predominant anterior temporal lobe atrophy 2. Predominant anterior temporal hypoperfusion or hypometabolism on SPECT or PET	– Imaging **must** show one or more of the following results: 1. Predominant left posterior perisylvian or parietal atrophy on MRI 2. Predominant left posterior perisylvian or parietal hypoperfusion or hypometabolism on SPECT or PET

(continued)

Table 12.4 (continued)

Non-fluent/agrammatic variant of PPA	Semantic variant of PPA	Logopenic variant of PPA
Non-fluent/agrammatic PPA with definite pathology	**Semantic variant of PPA with definite pathology**	**Logopenic variant of PPA with definite pathology**
Clinical diagnosis and either criterion two or three *must* be present: 1. Clinical diagnosis of non-fluent/agrammatic PPA 2. Histopatologic evidence of specific neurodegenerative pathology 3. Presence of a known pathogenic mutation	Clinical diagnosis and either criterion two or three *must* be present: 1. *Clinical diagnosis of semantic variant of PPA* 2. *Histopatologic evidence of specific neurodegenerative pathology* 3. *Presence of a known pathogenic mutation*	Clinical diagnosis and either criterion two or three *must* be present: 1. *Clinical diagnosis of logopenic variant of PPA* 2. *Histopatologic evidence of specific neurodegenerative pathology* 3. *Presence of a known pathogenic mutation*

Modified by Gorno-Tempini et al., 2011 [74]

The Behavioural Variant of FTD

The behavioural variant of FTD is the most frequent clinical presentation of FTD, whose current diagnostic criteria [75] are summarized in Table 12.5. It is characterized by predominant changes in personality and several behavioural symptoms, including disinhibition, apathy, eating and sleep disorders, lack of empathy and obsessive-compulsive disorders. A progressive deterioration of social behaviour and cognition is also a common feature.

Structural MRI in FTD

The non-fluent/agrammatic PPA is clinically characterized by a prominent impairment in speech production dominated by agrammatism and apraxia [74]. Other typical aphasic deficits include comprehension impairment for sentences, anomia and phonemic errors [74]. Tau but not TDP pathology has been mostly associated to the presence of prominent apraxia of speech [76].

Several studies on the non-fluent/agrammatic variant of PPA have shown a typical atrophy of the left insula and perisylvian area (Fig. 12.5, panel A) [77, 78]. However, an involvement of the left opercular region, Broca's area and motor/premotor cortex has also been reported [77]. Other studies have shown subcortical GM atrophy in the thalamus, the basal ganglia and the amygdala [77]. A recent study based on VBM and diffusion imaging [79] has compared different variants of FTD patients against healthy controls to assess both, GM and WM damage. In patients with the non-fluent/agrammatic variant of PPA, GM and WM loss were found in frontal and temporal language areas, with a selective microscopic damage in the left superior longitudinal/arcuate fasciculus [79]. Patients with the semantic variant of PPA (SD) clinically show fluent aphasia, characterized by anomia, single-word

Table 12.5 Revised criteria for bv-FTD [71]

(I) *Neurodegenerative disease*
The following symptom must be present:
  (A) Progressive deterioration of behaviour and or cognition by observation or history

(II) *Possible bv-FTD*
Three of the following behavioural or cognitive symptoms must be present. The symptoms are persistent or recurrent, rather than sporadic.
  (A) Early behavioural disinhibition (one of the following symptoms must be present):
    1. Socially inappropriate behaviour
    2. Loss of manners or decorum
    3. Impulsive, rash or careless action
  (B) Early apathy or inertia (one of the following symptoms must be present):
    1. Apathy
    2. Inertia
  (C) Early loss or sympathy or empathy (one of the following symptoms must be present):
    1. Diminished response to other people's need and feelings
    2. Diminished social interest, interrelatedness or personal warmth
  (D) Early perseverative, stereotyped or compulsive/ritualistic behaviour (one of the following symptoms must be present):
    1. Simple repetitive movements
    2. Complex, compulsive or ritualistic behaviour
    3. Stereotypy of speech
  (E) Hyperorality and dietary changes (one of the following symptoms must be present):
    1. Altered food preferences
    2. Binge eating, increased consumption of alcohol or cigarettes
    3. Oral exploration on consumption of inedible objects
  (F) Neuropsychological profile: executive/generation deficits with relative sparing of memory and visuospatial functions (all the following symptoms must be present):
    1. Deficits in executive tasks
    2. Relative sparing of episodic memory
    3. Relative sparing of visuospatial skills

(III) *Probable bv-FTD*
All the following symptoms must be present:
  (A) Meets criteria for possible bv-FTD
  (B) Exhibits significant functional decline (by caregiver report or as evidenced by Clinical Dementia Rating Scale or Functional Activities Questionnaire scores
  (C) Imaging results consistent with bv-FTD (one of the following must be present):
    1. Frontal and/or anterior temporal atrophy on MRI or CT
    2. Frontal and/or anterior temporal hypoperfusion or hypometabolism on PET or SPECT

(IV) *Behavioural variant of FTD with definite pathology*
Criterion A and either criterion B or C must be present:
  (A) Meets criteria for possible or probable bv-FTD
  (B) Histopathological evidence of FTD on biopsy or at post-mortem
  (C) Presence of a known pathogenic mutation

(V) *Exclusionary criteria for bv-FTD*
Criteria A and B must be answered negatively for any bv-FTD diagnosis. Criterion C can be positive for possible bv-FTLD but must be negative for probable bv-FTD
  (A) Pattern of deficits is better accounted for by other degenerative nervous system or medical disorders
  (B) Behavioural disturbance is better accounted for by psychiatric diagnosis
  (C) Biomarkers strongly indicative of Alzheimer's disease or other neurodegenerative process

Modified by Rascovsky et al., 2011 [75]

comprehension deficits, impaired object knowledge, surface dyslexia and dysgraphia. In contrast, repetition and speech production are relatively spared (without agrammatism or apraxia of speech [74]). SD is in most cases associated to FTLD-TDP pathology [80], with a pattern of brain atrophy mainly involving the anterior part of the left temporal lobe and, specifically, the temporal pole (Fig. 12.5, panel B) [81]. When considering the WM tissue, microstructural damage is confined to the uncinate fasciculus bilaterally and to the anterior part of the left inferior longitudinal fasciculus [79].

The logopenic variant of PPA is clinically characterized by impairment in the single-word retrieval in spontaneous speech and deficits in sentence repetition [74]. In some cases, phonological errors during spontaneous speech can also be present. Instead, single-word comprehension, object knowledge and motor speech are typically preserved in the absence of agrammatism [74]. This PPA variant is more likely associated to AD pathology [80]. Other studies reported an asymmetric left-side atrophy involving the perisylvian areas, the posterior part of the superior temporal cortex and the inferior parietal lobes (Fig. 12.5, panel C) [74, 76]. Finally, atrophy can also be found in the PCC/precuneus and in the medial temporal lobe [76]. A recent study showed a cortical thinning in the left superior and middle temporal gyrus in patients suffering from the logopenic variant of PPA [82].

Neuroimaging studies on bv-FTD have demonstrated a widespread pattern of brain atrophy including several frontal areas, such as the anterior medial portion of the frontal lobe, the gyrus rectus, the superior frontal gyrus and anterior cingulate [81, 83]. The insula and thalamus can also be affected [83]. Atrophy is typically bilateral with a mild right-side prevalence [83]. The earliest changes occur in the anterior cingulate, orbitofrontal and frontoinsular cortices, and atrophy in these regions may help in differentiating between AD and FTD patients [4].

However, it is possible to distinguish between different subtypes of bv-FTD according to the brain networks which are mainly involved. Whitwell and co-workers (2009) [84] reported four main bv-FTD variants: frontal dominant, temporal dominant, frontotemporal and temporo-fronto-parietal. The only variant showing significant correlations between brain atrophy distribution and its pathological subtype is the temporal-dominant type, which associates with genetic mutation in MAPT [84]. It has also been described a strict association between brain atrophy distribution and tau accumulation in MAPT carriers [85].

It has also been reported that specific brain areas associate with specific clinical symptoms: dorsomedial frontal atrophy is related to apathy and aberrant motor behaviour, while atrophy in the orbitofrontal regions is associated with disinhibition [86]. Moreover, WM atrophy has also been shown, predominantly in the anterior part of the corpus callosum [79].

Functional MRI in FTD

Abnormalities in functional brain connectivity have been reported, within specific networks, in patients with bv-FTD. A peculiar disruption of functional connectivity (as assessed by resting-state fMRI) has been consistently found in the salience network, which includes the frontal and insular cortex and the anterior cingulate [78].

A recent study shows that loss of functional connectivity within the salience network is associated with disease severity [87]. Additionally, alterations in insular or fronto-limbic areas have been found to predict patients' behavioural worsening (i.e. increasing of apathy scores) in both phenotypes, bv-FTD and semantic-PPA [88]. Borroni and co-workers [7] have investigated not only patients with sporadic FTD, but also patients with a genetic granulin variant of FTD along with preclinical mutation carriers. Interestingly, increased functional connectivity was found in the salience network of pre-symptomatic mutation carriers as a compensation mechanism before clinical onset. In contrast, the DMN, which is typically disrupted in AD patients, was found playing an initial compensatory role (increased connectivity) in bv-FTD at early stages.

The DMN has also been investigated in patients with bv-FTD. A recent study demonstrates decreased functional connectivity in the medial prefrontal cortex and lateral temporal lobes into the DMN of both, patients with bv-FTD and, more interestingly, in asymptomatic MAPT carriers when compared to healthy controls [61]. This study highlights an alteration of functional connectivity, which preceded signs of brain atrophy, in asymptomatic subjects destined to develop dementia [61]. Increased functional connectivity has also been reported in several brain networks, including the DMN and the dorsal attention network (DAN) of patients with the behavioural variant of FTD [89], indicating that these changes may be related to patients' apathy symptoms and executive dysfunction. Recently, a study showed divergent patterns of vulnerability in specific functional network components that might account for the clinical heterogeneity observed in bv-FTD [90].

When using a whole brain approach for RS-fMRI data analysis (i.e. graph theoretical analysis) alterations of functional connectivity have been documented in patients with bv-FTD, with some cerebral nodes resulted enhanced (i.e. hub regions in the medial parietal, temporal and occipital cortices) and other nodes resulted decreased (i.e. hub regions in the medial and dorsal frontal cortex, temporal and insular regions, caudate nucleus) [91]. Moreover, functional abnormalities have been significantly found as correlated with executive dysfunctions [91].

Conclusions

Neuroimaging has remarkably changed the clinical approach to neurodegenerative dementias. In clinical settings, MRI, PET and SPECT imaging help addressing the diagnosis of cognitive declines by excluding secondary causes and providing some peculiar features into neurodegenerative disorders. Additionally, quantitative neuroimaging techniques help understanding the pathophysiology of neurodegeneration, linking together quantitative structural and functional information on diseased brains, and clinical, neuropsychological and behavioural characteristics of patients. Overall, combinations of regional brain abnormalities and patterns functional and structural brain disconnection help to characterize the onset and evolution of different neurodegenerative conditions and fit with some major sources on individual variability.

References

1. McGinnis SM. Neuroimaging in neurodegenerative dementias. Semin Neurol. 2012;32(4):347–60.
2. Bozzali M, Serra L, Cercignani M. Quantitative MRI to understand Alzheimer's disease pathophysiology. Curr Opin Neurol. 2016;29(4):437–44.
3. Iaccarino L, Chiotis K, Alongi P, Almkvist O, Wall A, Cerami C, Bettinardi V, Gianolli L, Nordberg A, Perani D. A cross-validation of FDG- and amyloid-PET biomarkers in mild cognitive impairment for the risk prediction to dementia due to Alzheimer's disease in a clinical setting. J Alzheimers Dis. 2017;59(2):603–14.
4. Jack CR Jr, Wiste HJ, Weigand SD, Therneau TM, Knopman DS, Lowe V, Vemuri P, Mielke MM, Roberts RO, Machulda MM, Senjem ML, Gunter JL, Rocca WA, Petersen RC. Age-specific and sex-specific prevalence of cerebral β-amyloidosis, tauopathy, and neurodegeneration in cognitively unimpaired individuals aged 50-95 years: a cross-sectional study. Lancet Neurol. 2017;16(6):435–44.
5. Cummings JL. Toward a molecular neuropsychiatry of neurodegenerative diseases. Ann Neurol. 2003;54(2):147–54.
6. Gili T, Cercignani M, Serra L, Perri R, Giove F, Maraviglia B, Caltagirone C, Bozzali M. Regional brain atrophy and functional disconnection across Alzheimer's disease evolution. J Neurol Neurosurg Psychiatry. 2011;82(1):58–66.
7. Borroni B, Alberici A, Cercignani M, Premi E, Serra L, Cerini C, Cosseddu M, Pettenati C, Turla M, Archetti S, Gasparotti R, Caltagirone C, Padovani A, Bozzali M. Granulin mutation drives brain damage and reorganization from preclinical to symptomatic FTLD. Neurobiol Aging. 2012;33(10):2506–20.
8. Sosa-Ortiz AL, Acosta-Castillo I, Prince MJ. Epidemiology of dementias and Alzheimer's disease. Arch Med Res. 2012;43(8):600–8.
9. Braak H, Braak E. Staging of Alzheimer's disease-related neurofibrillary changes. Neurobiol Aging. 1995;16(3):271–8.
10. Markesbery WR. Neuropathologic alterations in mild cognitive impairment: a review. J Alzheimers Dis. 2010;19(1):221–8.
11. Albert MS, DeKosky ST, Dickson D, Dubois B, Feldman HH, Fox NC, Gamst A, Holtzman DM, Jagust WJ, Petersen RC, Snyder PJ, Carrillo MC, Thies B, Phelps CH. The diagnosis of mild cognitive impairment due to Alzheimer's disease: recommendations from the National Institute on Aging-Alzheimer's Association workgroups on diagnostic guidelines for Alzheimer's disease. Alzheimers Dement. 2011;7(3):270–9.
12. Scheltens P, Leys D, Barkhof F, Huglo D, Weinstein HC, Vermersch P, Kuiper M, Steinling M, Wolters EC, Valk J. Atrophy of medial temporal lobes on MRI in "probable" Alzheimer's disease and normal ageing: diagnostic value and neuropsychological correlates. J Neurol Neurosurg Psychiatry. 1992;55(10):967–72.
13. Wahlund LO, Barkhof F, Fazekas F, Bronge L, Augustin M, Sjögren M, Wallin A, Ader H, Leys D, Pantoni L, Pasquier F, Erkinjuntti T, Scheltens P. European task force on age-related white matter changes. A new rating scale for age-related white matter changes applicable to MRI and CT. Stroke. 2001;32(6):1318–22.
14. Ridha BH, Barnes J, van de Pol LA, Schott JM, Boyes RG, Siddique MM, Rossor MN, Scheltens P, Fox NC. Application of automated medial temporal lobe atrophy scale to Alzheimer disease. Arch Neurol. 2007;64(6):849–54.
15. Persson K, Barca ML, Eldholm RS, Cavallin L, Šaltytė Benth J, Selbæk G, Brækhus A, Saltvedt I, Engedal K. Visual evaluation of medial temporal lobe atrophy as a clinical marker of conversion from mild cognitive impairment to dementia and for predicting progression in patients with mild cognitive impairment and mild Alzheimer's disease. Dement Geriatr Cogn Disord. 2017;44(1–2):12–24.
16. Claus JJ, Staekenborg SS, Holl DC, Roorda JJ, Schuur J, Koster P, Tielkes CE, Scheltens P. Practical use of visual medial temporal lobe atrophy cut-off scores in Alzheimer's disease: validation in a large memory clinic population. Eur Radiol. 2017;27:3147–55.

17. Fazekas F, Chawluk JB, Alvavi A, Hurtig HI, Zimmerman RA. MR signal abnormalities at 1.5T in Alzheimer's disease and normal aging. AJNR. 1987;8:421–6.
18. Marnane M, Al-Jawadi OO, Mortazavi S, Pogorzelec KJ, Wang BW, Feldman HH, Hsiung GY, Alzheimer's Disease Neuroimaging Initiative. Periventricular hyperintensities are associated with elevated cerebral amyloid. Neurology. 2016;86(6):535–43.
19. Seab JP, Jagust WJ, Wong ST, Roos MS, Reed BR, Budinger TF. Quantitative NMR measurements of hippocampal atrophy in Alzheimer's disease. Magn Reson Med. 1988;8(2):200–8.
20. Convit A, De Leon MJ, Tarshish C, De Santi S, Tsui W, Rusinek H, George A. Specific hippocampal volume reductions in individuals at risk for Alzheimer's disease. Neurobiol Aging. 1997;18(2):131–8.
21. Jack CR Jr, Petersen RC, Xu YC, O'Brien PC, Smith GE, Ivnik RJ, Boeve BF, Waring SC, Tangalos EG, Kokmen E. Prediction of AD with MRI-based hippocampal volume in mild cognitive impairment. Neurology. 1999;52(7):1397–403.
22. Mondragón JD, Celada-Borja C, Barinagarrementeria-Aldatz F, Burgos-Jaramillo M, Barragán-Campos HM. Hippocampal volumetry as a biomarker for dementia in people with low education. Dement Geriatr Cogn Dis Extra. 2016;6(3):486–99.
23. Frisoni GB, Jack CR. Harmonization of magnetic resonance-based manual hippocampal segmentation: a mandatory step for wide clinical use. Alzheimers Dement. 2011;7(2):171–4.
24. Boccardi M, Bocchetta M, Ganzola R, Robitaille N, Redolfi A, Duchesne S, Jack CR Jr, Frisoni GB, EADC-ADNI Working Group on The Harmonized Protocol for Manual Hippocampal Segmentation and for the Alzheimer's Disease Neuroimaging Initiative. Operationalizing protocol differences for EADC-ADNI manual hippocampal segmentation. Alzheimers Dement. 2015;11(2):184–94.
25. Ashburner J, Friston KJ. Voxel-based morphometry—the methods. Neuroimage. 2000;11(6):805–21.
26. Bozzali M, Filippi M, Magnani G, Cercignani M, Franceschi M, Schiatti E, Castiglioni S, Mossini R, Falautano M, Scotti G, Comi G, Falini A. The contribution of voxel-based morphometry in staging patients with mild cognitive impairment. Neurology. 2006;67(3):453–60.
27. Serra L, Cercignani M, Lenzi D, Perri R, Fadda L, Caltagirone C, Macaluso E, Bozzali M. Grey and white matter changes at different stages of Alzheimer's disease. J Alzheimers Dis. 2010;19(1):147–59.
28. Serra L, Fadda L, Perri R, Spanò B, Marra C, Castelli D, Torso M, Makovac E, Cercignani M, Caltagirone C, Bozzali M. Constructional apraxia as a distinctive cognitive and structural brain feature of pre-senile Alzheimer's disease. J Alzheimers Dis. 2014;38(2):391–402.
29. Serra L, Perri R, Cercignani M, Spanò B, Fadda L, Marra C, Carlesimo GA, Caltagirone C, Bozzali M. Are the behavioral symptoms of Alzheimer's disease directly associated with neurodegeneration? J Alzheimers Dis. 2010;21(2):627–39.
30. Serra L, Giulietti G, Cercignani M, Spanò B, Torso M, Castelli D, Perri R, Fadda L, Marra C, Caltagirone C, Bozzali M. Mild cognitive impairment: same identity for different entities. J Alzheimers Dis. 2013;33(4):1157–65.
31. Serra L, Cercignani M, Petrosini L, Basile B, Perri R, Fadda L, Spanò B, Marra C, Giubilei F, Carlesimo GA, Caltagirone C, Bozzali M. Neuroanatomical correlates of cognitive reserve in Alzheimer disease. Rejuvenation Res. 2011;14(2):143–51.
32. Basser PJ, Jones DK. Diffusion-tensor MRI: theory, experimental design and data analysis—a technical review. NMR Biomed. 2002;15(7–8):456–67.
33. Smith SM, Jenkinson M, Johansen-Berg H, Rueckert D, Nichols TE, Mackay CE, Watkins KE, Ciccarelli O, Cader MZ, Matthews PM, Behrens TE. Tract-based spatial statistics: voxel-wise analysis of multi-subject diffusion data. Neuroimage. 2006;31(4):1487–505.
34. Jones DK. Studying connections in the living human brain with diffusion MRI. Cortex. 2008;44(8):936–52.
35. Liu Y, Spulber G, Lehtimäki KK, Könönen M, Hallikainen I, Gröhn H, Kivipelto M, Hallikainen M, Vanninen R, Soininen H. Diffusion tensor imaging and tract-based spatial statistics in Alzheimer's disease and mild cognitive impairment. Neurobiol Aging. 2011;32(9):1558–71.

36. Serra L, Cercignani M, Basile B, Spanò B, Perri R, Fadda L, Marra C, Giubilei F, Caltagirone C, Bozzali M. White matter damage along the uncinate fasciculus contributes to cognitive decline in AD and DLB. Curr Alzheimer Res. 2012;9(3):326–33.
37. Bozzali M, Giulietti G, Basile B, Serra L, Spanò B, Perri R, Giubilei F, Marra C, Caltagirone C, Cercignani M. Damage to the cingulum contributes to Alzheimer's disease pathophysiology by deafferentation mechanism. Hum Brain Mapp. 2012;33(6):1295–308.
38. Bozzali M, Parker GJ, Serra L, Embleton K, Gili T, Perri R, Caltagirone C, Cercignani M. Anatomical connectivity mapping: a new tool to assess brain disconnection in Alzheimer's disease. Neuroimage. 2011;54(3):2045–51.
39. Bozzali M, Parker GJ, Spanò B, Serra L, Giulietti G, Perri R, Magnani G, Marra C, G Vita M, Caltagirone C, Cercignani M. Brain tissue modifications induced by cholinergic therapy in Alzheimer's disease. Hum Brain Mapp. 2013;34(12):3158–67.
40. Peters F, Collette F, Degueldre C, Sterpenich V, Majerus S, Salmon E. The neural correlates of verbal short-term memory in Alzheimer's disease: an fMRI study. Brain. 2009;132(7):1833–46.
41. Golby A, Silverberg G, Race E, Gabrieli S, O'Shea J, Knierim K, Stebbins G, Gabrieli J. Memory encoding in Alzheimer's disease: an fMRI study of explicit and implicit memory. Brain. 2005;128(4):773–87.
42. Pihlajamäki M, Jauhiainen AM, Soininen H. Structural and functional MRI in mild cognitive impairment. Curr Alzheimer Res. 2009;6(2):179–85.
43. Lenzi D, Serra L, Perri R, Pantano P, Lenzi GL, Paulesu E, Caltagirone C, Bozzali M, Macaluso E. Single domain amnestic MCI: a multiple cognitive domains fMRI investigation. Neurobiol Aging. 2011;32(9):1542–57.
44. Greicius MD, Krasnow B, Reiss AL, Menon V. Functional connectivity in the resting brain: a network analysis of the default mode hypothesis. Proc Natl Acad Sci U S A. 2003;100(1):253–8.
45. Greicius MD, Supekar K, Menon V, Dougherty RF. Resting-state functional connectivity reflects structural connectivity in the default mode network. Cereb Cortex. 2009;19(1):72–8.
46. Lee ES, Yoo K, Lee YB, Chung J, Lim JE, Yoon B, Jeong Y, Alzheimer's Disease Neuroimaging Initiative. Default mode network functional connectivity in early and late mild cognitive impairment: results from the Alzheimer's disease neuroimaging initiative. Alzheimer Dis Assoc Disord. 2016;30(4):289–96.
47. Bozzali M, Dowling C, Serra L, Spanò B, Torso M, Marra C, Castelli D, Dowell NG, Koch G, Caltagirone C, Cercignani M. The impact of cognitive reserve on brain functional connectivity in Alzheimer's disease. J Alzheimers Dis. 2015;44(1):243–50.
48. Serra L, Mancini M, Cercignani M, Di Domenico C, Spanò B, Giulietti G, Koch G, Marra C, Bozzali M. Network-based substrate of cognitive Reserve in Alzheimer's disease. J Alzheimers Dis. 2017;55(1):421–30.
49. Phelps ME. PET: the merging of biology and imaging into molecular imaging. J Nucl Med. 2000;41(4):661–81.
50. Bohnen NI, Djang DS, Herholz K, Anzai Y, Minoshima S. Effectiveness and safety of 18F-FDG PET in the evaluation of dementia: a review of the recent literature. J Nucl Med. 2012;53(1):59–71.
51. Kato T, Inui Y, Nakamura A, Ito K. Brain fluorodeoxyglucose (FDG) PET in dementia. Ageing Res Rev. 2016;30:73–84.
52. Knopman DS. Diagnostic tests for Alzheimer disease: FDG-PET imaging is a player in search of a role. Neurol Clin Pract. 2012;2(2):151–3.
53. Scherfler C, Schwarz J, Antonini A, Grosset D, Valldeoriola F, Marek K, Oertel W, Tolosa E, Lees AJ, Poewe W. Role of DAT-SPECT in the diagnostic work up of parkinsonism. Mov Disord. 2007;22(9):1229–38.
54. Hardy J, Selkoe DJ. The amyloid hypothesis of Alzheimer's disease: progress and problems on the road to therapeutics. Science. 2002;297(5580):353–6.
55. Rowe CC, Villemagne VL. Amyloid imaging with PET in early Alzheimer disease diagnosis. Med Clin North Am. 2013;97(3):377–98.

56. Visser PJ, Vos S, van Rossum I, Scheltens P. Comparison of International Working Group criteria and National Institute on Aging-Alzheimer's Association criteria for Alzheimer's disease. Alzheimers Dement. 2012;8(6):560–3.
57. Villemagne VL, Rowe CC, Macfarlane S, Novakovic KE, Masters CL. Imaginem oblivionis: the prospects of neuroimaging for early detection of Alzheimer's disease. J Clin Neurosci. 2005;12(3):221–30.
58. Cohen AD, Rabinovici GD, Mathis CA, Jagust WJ, Klunk WE, Ikonomovic MD. Using Pittsburgh Compound B for in vivo PET imaging of fibrillar amyloid-beta. Adv Pharmacol. 2012;64:27–81.
59. Ikonomovic MD, Klunk WE, Abrahamson EE, Mathis CA, Price JC, Tsopelas ND, Lopresti BJ, Ziolko S, Bi W, Paljug WR, Debnath ML, Hope CE, Isanski BA, Hamilton RL, DeKosky ST. Post-mortem correlates of in vivo PiB-PET amyloid imaging in a typical case of Alzheimer's disease. Brain. 2008;131(6):1630–45.
60. Rabinovici GD, Miller BL. Frontotemporal lobar degeneration: epidemiology, pathophysiology, diagnosis and management. CNS Drugs. 2010;24(5):375–98.
61. Whitwell JL, Josephs KA, Avula R, Tosakulwong N, Weigand SD, Senjem ML, Vemuri P, Jones DT, Gunter JL, Baker M, Wszolek ZK, Knopman DS, Rademakers R, Petersen RC, Boeve BF, Jack CR Jr. Altered functional connectivity in asymptomatic MAPT subjects: a comparison to bvFTD. Neurology. 2011;77(9):866–74.
62. Josephs KA, Hodges JR, Snowden JS, Mackenzie IR, Neumann M, Mann DM, Dickson DW. Neuropathological background of phenotypic variability in frontotemporal dementia. Acta Neuropathol. 2011;122(2):137–53.
63. Whitwell JL, Josephs KA. Neuroimaging in frontotemporal lobar degeneration—predicting molecular pathology. Nat Rev Neurol. 2012;8(3):131–42.
64. Neumann M, Sampathu DM, Kwong LK, Truax AC, Micsenyi MC, Chou TT, Bruce J, Schuck T, Grossman M, Clark CM, McCluskey LF, Miller BL, Masliah E, Mackenzie IR, Feldman H, Feiden W, Kretzschmar HA, Trojanowski JQ, Lee VM. Ubiquitinated TDP-43 in frontotemporal lobar degeneration and amyotrophic lateral sclerosis. Science. 2006;314(5796):130–3.
65. Neumann M, Rademakers R, Roeber S, Baker M, Kretzschmar HA, Mackenzie IR. A new subtype of frontotemporal lobar degeneration with FUS pathology. Brain. 2009;132(11):2922–31.
66. Hutton M, Lendon CL, Rizzu P, Baker M, Froelich S, Houlden H, Pickering-Brown S, Chakraverty S, Isaacs A, Grover A, Hackett J, Adamson J, Lincoln S, Dickson D, Davies P, Petersen RC, Stevens M, de Graaff E, Wauters E, van Baren J, Hillebrand M, Joosse M, Kwon JM, Nowotny P, Che LK, Norton J, Morris JC, Reed LA, Trojanowski J, Basun H, Lannfelt L, Neystat M, Fahn S, Dark F, Tannenberg T, Dodd PR, Hayward N, Kwok JB, Schofield PR, Andreadis A, Snowden J, Craufurd D, Neary D, Owen F, Oostra BA, Hardy J, Goate A, van Swieten J, Mann D, Lynch T, Heutink P. Association of missense and 5′-splice-site mutations in tau with the inherited dementia FTDP-17. Nature. 1998;393(6686):702–5.
67. Cruts M, Gijselinck I, van der Zee J, Engelborghs S, Wils H, Pirici D, Rademakers R, Vandenberghe R, Dermaut B, Martin JJ, van Duijn C, Peeters K, Sciot R, Santens P, De Pooter T, Mattheijssens M, Van den Broeck M, Cuijt I, Vennekens K, De Deyn PP, Kumar-Singh S, Van Broeckhoven C. Null mutations in progranulin cause ubiquitin-positive frontotemporal dementia linked to chromosome 17q21. Nature. 2006;442(7105):920–4.
68. Renton AE, Majounie E, Waite A, Simón-Sánchez J, Rollinson S, Gibbs JR, Schymick JC, Laaksovirta H, van Swieten JC, Myllykangas L, Kalimo H, Paetau A, Abramzon Y, Remes AM, Kaganovich A, Scholz SW, Duckworth J, Ding J, Harmer DW, Hernandez DG, Johnson JO, Mok K, Ryten M, Trabzuni D, Guerreiro RJ, Orrell RW, Neal J, Murray A, Pearson J, Jansen IE, Sondervan D, Seelaar H, Blake D, Young K, Halliwell N, Callister JB, Toulson G, Richardson A, Gerhard A, Snowden J, Mann D, Neary D, Nalls MA, Peuralinna T, Jansson L, Isoviita VM, Kaivorinne AL, Hölttä-Vuori M, Ikonen E, Sulkava R, Benatar M, Wuu J, Chiò A, Restagno G, Borghero G, Sabatelli M, ITALSGEN Consortium, Heckerman D, Rogaeva E, Zinman L, Rothstein JD, Sendtner M, Drepper C, Eichler EE, Alkan C, Abdullaev Z, Pack SD, Dutra A, Pak E, Hardy J, Singleton A, Williams NM, Heutink P, Pickering-Brown S, Morris

HR, Tienari PJ, Traynor BJ. A hexanucleotide repeat expansion in C9ORF72 is the cause of chromosome 9p21-linked ALS-FTD. Neuron. 2011;72(2):257–68.
69. Pick A. Ubeer die Beziehungen der senile Hirnatrophie zur aphasie. Prager Medizinische Wochenschr. 1892;17:165–7.
70. Mesulam MM. Slowly progressive aphasia without generalized dementia. Ann Neurol. 1982;11(6):592–8.
71. Mesulam MM, Weintraub S. Spectrum of primary progressive aphasia. Baillieres Clin Neurol. 1992;1(3):583–609.
72. Warrington EK. The selective impairment of semantic memory. Q J Exp Psychol. 1975;27(4):635–57.
73. Snowden JS, Goulding PJ, Neary D. Semantic dementia: a form of circumscribed cerebral atrophy. Behav Neurol. 1989;2:167–82.
74. Gorno-Tempini ML, Hillis AE, Weintraub S, Kertesz A, Mendez M, Cappa SF, Ogar JM, Rohrer JD, Black S, Boeve BF, Manes F, Dronkers NF, Vandenberghe R, Rascovsky K, Patterson K, Miller BL, Knopman DS, Hodges JR, Mesulam MM, Grossman M. Classification of primary progressive aphasia and its variants. Neurology. 2011;76(11):1006–14.
75. Rascovsky K, Hodges JR, Knopman D, Mendez MF, Kramer JH, Neuhaus J, van Swieten JC, Seelaar H, Dopper EG, Onyike CU, Hillis AE, Josephs KA, Boeve BF, Kertesz A, Seeley WW, Rankin KP, Johnson JK, Gorno-Tempini ML, Rosen H, Prioleau-Latham CE, Lee A, Kipps CM, Lillo P, Piguet O, Rohrer JD, Rossor MN, Warren JD, Fox NC, Galasko D, Salmon DP, Black SE, Mesulam M, Weintraub S, Dickerson BC, Diehl-Schmid J, Pasquier F, Deramecourt V, Lebert F, Pijnenburg Y, Chow TW, Manes F, Grafman J, Cappa SF, Freedman M, Grossman M, Miller BL. Sensitivity of revised diagnostic criteria for the behavioural variant of frontotemporal dementia. Brain. 2011;134(9):2456–77.
76. Rohrer JD, Paviour D, Bronstein AM, O'Sullivan SS, Lees A, Warren JD. Progressive supranuclear palsy syndrome presenting as progressive nonfluent aphasia: a neuropsychological and neuroimaging analysis. Mov Disord. 2010;25(2):179–88.
77. Gorno-Tempini ML, Dronkers NF, Rankin KP, Ogar JM, Phengrasamy L, Rosen HJ, Johnson JK, Weiner MW, Miller BL. Cognition and anatomy in three variants of primary progressive aphasia. Ann Neurol. 2004;55(3):335–46.
78. Seeley WW, Crawford RK, Zhou J, Miller BL, Greicius MD. Neurodegenerative diseases target large-scale human brain networks. Neuron. 2009;62(1):42–52.
79. Zhang Y, Tartaglia MC, Schuff N, Chiang GC, Ching C, Rosen HJ, Gorno-Tempini ML, Miller BL, Weiner MW. MRI signatures of brain macrostructural atrophy and microstructural degradation in frontotemporal lobar degeneration subtypes. J Alzheimers Dis. 2013;33(2):431–44.
80. Rohrer JD. Structural brain imaging in frontotemporal dementia. Biochim Biophys Acta. 2012;1822(3):325–32.
81. Hodges JR, Patterson K. Semantic dementia: a unique clinicopathological syndrome. Lancet Neurol. 2007;6(11):1004–14.
82. Leyton CE, Hodges JR, Piguet O, Ballard KJ. Common and divergent neural correlates of anomia in amnestic and logopenic presentations of Alzheimer's disease. Cortex Hodges. 2017;86:45–54.
83. Schroeter ML, Raczka K, Neumann J, von Cramon DY. Neural networks in frontotemporal dementia—a meta-analysis. Neurobiol Aging. 2008;29(3):418–26.
84. Whitwell JL, Przybelski SA, Weigand SD, Ivnik RJ, Vemuri P, Gunter JL, Senjem ML, Shiung MM, Boeve BF, Knopman DS, Parisi JE, Dickson DW, Petersen RC, Jack CR Jr, Josephs KA. Distinct anatomical subtypes of the behavioural variant of frontotemporal dementia: a cluster analysis study. Brain. 2009;132(Pt 11):2932–46.
85. Spina S, Schonhaut DR, Boeve BF, Seeley WW, Ossenkoppele R, O'Neil JP, Lazaris A, Rosen HJ, Boxer AL, Perry DC, Miller BL, Dickson DW, Parisi JE, Jagust WJ, Murray ME, Rabinovici GD. Frontotemporal dementia with the V337M MAPT mutation: tau-PET and pathology correlations. Neurology. 2017;88(8):758–66.
86. Rosen HJ, Allison SC, Schauer GF, Gomo-Tempini ML, Weiner MW, Miller BL. Neuroanatomical correlates of behavioural disorders in dementia. Brain. 2005;128:2612–25.

87. Zhou J, Greicius MD, Gennatas ED, Growdon ME, Jang JY, Rabinovici GD, Kramer JH, Weiner M, Miller BL, Seeley WW. Divergent network connectivity changes in behavioural variant frontotemporal dementia and Alzheimer's disease. Brain. 2010;133(5):1352–67.
88. Day GS, Farb NA, Tang-Wai DF, Masellis M, Black SE, Freedman M, Pollock BG, Chow TW. Salience network resting-state activity: prediction of frontotemporal dementia progression. JAMA Neurol. 2013;70(10):1249–53.
89. Rytty R, Nikkinen J, Paavola L, Abou Elseoud A, Moilanen V, Visuri A, Tervonen O, Renton AE, Traynor BJ, Kiviniemi V, Remes AM. GroupICA dual regression analysis of resting state networks in a behavioral variant of frontotemporal dementia. Front Hum Neurosci. 2013;7:461.
90. Ranasinghe KG, Rankin KP, Pressman PS, Perry DC, Lobach IV, Seeley WW, Coppola G, Karydas AM, Grinberg LT, Shany-Ur T, Lee SE, Rabinovici GD, Rosen HJ, Gorno-Tempini ML, Boxer AL, Miller ZA, Chiong W, DeMay M, Kramer JH, Possin KL, Sturm VE, Bettcher BM, Neylan M, Zackey DD, Nguyen LA, Ketelle R, Block N, Wu TQ, Dallich A, Russek N, Caplan A, Geschwind DH, Vossel KA, Miller BL. Distinct subtypes of behavioral variant frontotemporal dementia based on patterns of network degeneration. JAMA Neurol. 2016;73(9):1078–88.
91. Agosta F, Sala S, Valsasina P, Meani A, Canu E, Magnani G, Cappa SF, Scola E, Quatto P, Horsfield MA, Falini A, Comi G, Filippi M. Brain network connectivity assessed using graph theory in frontotemporal dementia. Neurology. 2013;81(2):134–43.

Genotypic and Phenotypic Heterogeneity in Amyotrophic Lateral Sclerosis

13

Nicola Ticozzi and Vincenzo Silani

Abstract

Genetic risk factors play a major role in the susceptibility to amyotrophic lateral sclerosis (ALS), the most common motor neuron disease of the adult. Although genetic studies have partially elucidated the genetic background of the disease, a large part of ALS heritability is still missing. In this chapter, we discuss the major genes implicated in the pathogenesis of motor neuron diseases; the clinical, pathological, and genetic links between ALS and frontotemporal dementia; and the vast genotypic and phenotypic heterogeneity of ALS. Lastly, we review the most recent strategies for identification of novel genetic risk factors in ALS, detailing their advantages and potential pitfalls.

Keywords

Amyotrophic lateral sclerosis · Frontotemporal dementia · Genotype-phenotype correlation · c9orf72 · Whole genome sequencing

Introduction

Amyotrophic lateral sclerosis (ALS) is a neurodegenerative disorder predominantly affecting the motor system and caused by the progressive loss of motor neurons within the primary motor cortex (upper motor neurons, UMN), the motor nuclei of the brainstem, and the anterior horns of the spinal cord (lower motor neurons, LMN).

N. Ticozzi (✉) • V. Silani
Department of Neurology and Laboratory of Neuroscience, IRCCS Istituto Auxologico Italiano, Milan, Italy

Department of Pathophysiology and Transplantation, 'Dino Ferrari' Center,
University of Milan, Milan, Italy
e-mail: n.ticozzi@auxologico.it

© Springer International Publishing AG 2018
D. Galimberti, E. Scarpini (eds.), *Neurodegenerative Diseases*,
https://doi.org/10.1007/978-3-319-72938-1_13

The incidence of ALS in Europe has been estimated to be ~2 cases per 100,000 population per year, with a prevalence of 5.4 cases per 100,000 population, a lifetime risk of 1:400, and a predominance in men with a 1.5:1 male-female ratio [1, 2]. Age at onset varies greatly, with an average of 60 years (range 55–75), and incidence steadily decreasing in individuals older than 80 years [2].

Classically, ALS manifests insidiously with signs and symptoms reflecting combined UMN (increased deep tendon reflexes, pathological reflexes, spasticity, pyramidal weakness, loss of dexterity) and LMN degeneration (cramps, fasciculations, muscle weakness and atrophy, decreased deep tendon reflexes, and muscle tone) and progresses relentlessly to an invariably fatal outcome, usually due to respiratory failure, with an average survival of 3 years after onset of first symptoms [3]. The clinical heterogeneity is remarkable, evidenced by the large variability in site of onset, mode of presentation, age of onset, and rate of progression. In addition to classic ALS, several clinical patterns can be recognized based on the aforementioned variables, such as bulbar- and respiratory-onset ALS, pyramidal ALS, and flail arm and flail leg syndromes [4, 5]. The correct recognition of these subphenotypes is becoming increasingly important both in the clinical setting for prognostic prediction and patient stratification, as well as in basic research, since they may underlie different etiologies and/or pathomechanisms.

ALS has long been considered a pure motor system disorder, the absence of cognitive impairment, and/or other extramotor signs often being considered a prerequisite for the diagnosis. This has been proven false, as involvement of other neurological functions has been increasingly demonstrated in recent years. In particular, association with subclinical cognitive and/or behavioural dysfunction of frontal type or with full-fledged frontotemporal dementia (FTD) has been recognized as a feature of the disease, to the point that the existence of an ALS-FTD spectrum, with the extremes constituting pure motor and pure cognitive disease, is now broadly accepted [6]. Involvement of other systems, such as extrapyramidal, cerebellar, and autonomic, has also been increasingly recognized in patients with motor neuron diseases, and the notion of ALS as part of a more complex 'multisystem degeneration' is becoming more and more established [7].

Genetics of ALS

Notwithstanding the original description by Charcot of ALS as a non-heritable disease, the subsequent recognition that familial aggregation was present in a portion of ALS cases suggested the importance of genetic factors in the pathogenesis of the disease [8, 9].

Studies of disease concordance rates among monozygotic and dizygotic twins estimate that 53–84% of ALS population risk is genetically determined [10]. Longitudinal analysis of age-adjusted mortality rates in ALS patients provides evidence for the existence of a susceptible population subset [11]. A clear familial aggregation is observed in ~10% of all ALS cases, and the first degree relatives of ALS patients develop disease at ~10 times the rate seen in the general population

[12, 13]. The majority of ALS cases are believed to result from the combined effects of environmental, epigenetic, and other stochastic factors acting upon the background of a genetically predisposed individual [14, 15]. Despite this, much of the current understanding of ALS etiopathogenesis comes from the study of the rare Mendelian subtypes, where disease is primarily, if not exclusively, attributable to single gene defects that segregate with disease in families.

A positive family history for motor neuron diseases has been traditionally used to distinguish ALS into familial (FALS) and sporadic (SALS) forms. This classification, although useful in the clinical setting, is now being challenged as incorrect by growing epidemiological and biological evidence. Several factors, such as small patient family size, advanced age at onset, and reduced penetrance of mutations, may lead to an underestimation of 'true' familial cases [16]. The possibility of ALS phenocopies in the same family is also well documented, resulting into misclassification of sporadic cases as familial [17].

Since the discovery of the first ALS gene in 1993, over 100 disease loci have been proposed to cause the disease or to influence susceptibility and/or clinical phenotypes (http://alsod.iop.kcl.ac.uk/) [18]. Although the pathogenic relevance of several of these genes is well supported (Table 13.1), the contribution of the majority of them to the disease is far from certain. The difficulty in establishing their clinical relevance is primarily caused by the relatively low frequency of the disease, the heterogeneity of causative factors, and the fact that every human genome contains a considerable number of potentially disease-related genetic variants. These issues can also complicate the interpretation of mutations identified at well-established disease genes, where gene size can represent a misleading factor. In fact, the probability of observing entirely incidental variants is not negligible, especially in larger genes [19]. Another issue is that of pleiotropy, which refers to the association of one gene with multiple phenotypes. In the case of ALS, several disease-causing genes have also been associated with FTD, motor neuropathy, spastic paraplegia, progressive bulbar palsy, spinal muscular atrophy, spinocerebellar ataxia, oculomotor apraxia, and schizophrenia [20]. In certain cases even individual mutations can associate with multiple seemingly distinct clinical presentations. A striking example of this phenomenon are mutations within the *VCP* gene, which even within a single family associate with variable combinations of ALS, FTD, Paget's disease of bone, and inclusion body myopathy [21]. Contrary to this, certain ALS mutations associate with very specific clinical profiles [22, 23], reinforcing that much of the clinical heterogeneity seen in the disease may reflect variation in causative as well as modifying factors.

The importance of individual ALS genes varies considerably according to ancestral background. Cumulatively, mutations within four major genes, namely, *SOD1*, *TARDBP*, *FUS*, and *c9orf72*, account for 38–67% of FALS and 5–15% of SALS patients of European ancestry [24–26]. Given the relatively high mutational frequency, a robust genotype-phenotype correlation can be drawn for these genes. The relative prevalence of single gene mutations, however, varies among different populations, with some mutations being extremely common in specific geographic areas and much rarer in others. Conversely, pathogenic mutations in the other ALS-associated genes

Table 13.1 ALS-associated genes

Gene	ALS type	Locus	Encoded protein	Protein function	Mutation	Associated phenotypes
Autosomal dominant						
SOD1	ALS1	21q22.1	Cu/Zn superoxide dismutase	Oxidative stress	MS, ns	ALS
Unknown	ALS3	18q21				ALS
SETX	ALS4	9q34.13	Senataxin	DNA/RNA helicase	MS	ALS; juvenile; AOA-2 (AR)
FUS	ALS6	16p11.2	Fused in sarcoma (FUS)	mRNA processing, transport and metabolism, DNA repair	MS, ns, FS	ALS; het-4
Unknown	ALS7	20p13				ALS
VAPB	ALS8	20q13.33	Vesicle-associated membrane protein-associated protein B	Vesicular trafficking	MS	ALS; SMA, late onset
ANG	ALS9	14q11.1	Angiogenin	Ribonuclease, angiogenesis	MS	ALS; parkinsonism
TARDBP	ALS10	1q36.22	TAR DNA-binding protein (TDP-43)	Transcription regulation, mRNA processing, transport and metabolism	MS	ALS; ALS-FTD; parkinsonism
FIG4	ALS11	6q21	Polyphosphoinositide phosphatase	Endosomal fusion and trafficking	MS	ALS; CMT-4J (AR)
OPTN	ALS12	10p13	Optineurin	Vesicular trafficking, NFkB signalling, autophagy	MS	ALS; open angle glaucoma
VCP	ALS14	9p13.3	Valosin-containing protein	Vesicular trafficking, ubiquitin-dependent protein degradation	MS	ALS; ALS-FTD; IBMPFD-1; CMT-2Y
CHMP2B	ALS17	3p11.2	Chromatin modifying protein 2B	Endosomal trafficking, ubiquitin-dependent protein degradation	MS	ALS; FTD-3
PFN1	ALS18	17p13.2	Profilin 1	Actin polymerization	MS	ALS
ERBB4	ALS19	2q34	V-erb-B2 avian erythroblastic leukaemia viral oncogene homolog 4	Tyrosine kinase, cell migration, and axonal guidance	MS	ALS
hnRNPA1	ALS20	12q13.13	Heterogeneous nuclear ribonucleoprotein A1	mRNA processing, transport, and metabolism	MS	ALS; IBMPFD-3

Gene	ALS type	Locus	Encoded protein	Protein function	Mutation	Associated phenotypes
MATR3	ALS21	5q31.2	Matrin 3	mRNA stabilization	MS	ALS; myopathy
TUBA4A	ALS22	2q35	Tubulin a 4A	Structural component of cytoskeleton	MS, ns	ALS; ALS-FTD
c9orf72	FTDALS1	9p212	Chromosome 9 Open reading frame 72	Endocytosis, endosomal trafficking, autophagy	$(G_4C_2)_n$ repeat	ALS; FTD
CHCHD10	FTDALS2	22q11	Coiled-coil-helix-coiled-coil-helix domain containing 10	Mitochondrial morphology	MS	ALS; FTD; SMAJ
SQSTM1	FTDALS3	5q35	Sequestosome-1 (p62)	NFkB signalling, autophagy	Ns, MS	ALS; FTD; Paget disease of bone
TBK1	FTDALS4	12q14	TANK-binding kinase 1	NFkB signalling	Ns, MS	ALS; FTD
DAO	ALS	12q24	D-amino acid oxidase	Peroxisome function	MS	ALS
BSCL2	ALS	11q13	Seipin	Unknown	MS, ns, FS	SPG17; dHMN-5A; PELD-CGL-2 *(AR)*
CCNF	ALS-FTD	16p13	Cyclin F	Cell proliferation		ALS; FTD
Autosomal recessive						
SOD1	ALS1	21q22.1	Cu/Zn superoxide dismutase	Oxidative stress	p.D90A	ALS
ALS2	ALS2	2q33.2	Alsin	Endosomal trafficking	Ns, FS	ALS; PLS, juvenile; IAHSP
SPG11	ALS5	15q21.1	Spatacsin	Axonal growth and function Intracellular cargo trafficking	Ns, FS	ALS, juvenile; SPG-11, CMT-2X
FUS	ALS6	16p11.2	Fused in sarcoma (FUS)	mRNA processing, transport and metabolism, DNA repair	p.H517Q	ALS
OPTN	ALS12	10p13	Optineurin	Vesicular trafficking, NFkB signalling, autophagy	NS, deletion of exon 15	ALS
SIGMAR1	ALS16	9p13.3	Sigma non-opioid intracellular receptor 1	ER chaperone, proteasome inhibition, mitochondrial retrograde axonal transport	MS, SS	ALS, juvenile; SMA, distal
PARK7	ALS-PD2	1p36.23	Protein deglycase DJ-1	Oxidative stress, ER chaperone	Ns, FS, MS	PD, juvenile ± MND

(continued)

Table 13.1 (continued)

Gene	ALS type	Locus	Encoded protein	Protein function	Mutation	Associated phenotypes
c19orf12	MPAN	9q12	Chromosome 19 Open reading frame 12	Unknown	FS, MS	NBIA-4 ± pyramidal signs – muscle atrophy
VRK1	ALS	14q32	Vaccinia-related kinase 1	Cell proliferation		MND, distal; PCH-1A
PNPLA6	ALS	19p13	Neuropathy target esterase	Neurite outgrowth and elongation	MS, ns	SPG-39 ± muscle atrophy; BNHS; OMCS
X-linked						
UBQLN2	ALS15	Xp11.21	Ubiquilin 2	Ubiquitin-dependent protein degradation	MS	ALS; FTD
Susceptibility genes						
ATXN2	ALS13	12q24.12	Ataxin-2	Ribosomal translation, mitochondrial function, endocytosis, mTOR signalling	$(CAG)_n$ repeat	SCA-2 (*AD*); ALS; parkinsonism;
DCTN1	SALS	2p13	Dynactin 1	Axonal transport	Rare variants	Perry syndrome (*AD*); dHMN-VIIB (*AD*); ALS
NEK1	SALS-FALS	4q33	Never in mitosis gene A-related kinase 1	Cilia formation, DNA-damage response, axon morphology, microtubule stability	Rare variants	SRTD-6 (*AR*); ALS
GLE1	SALS-FALS	9q34.11	RNA export mediator Gle1	mRNA processing, transport and metabolism	Rare variants	LCCS-1-LAAHD (*AR*); ALS
NEFH	SALS	22q12.2	Neurofilament protein, heavy polypeptide	Structural component of cytoskeleton	Rare variants	CMT-2CC (*AD*); ALS
PRPH	SALS	12q13.12	Peripherin	Neurite elongation, axonal outgrowth	Rare variants	ALS
UNC13A	SALS	19p13.11	Unc-13 (*C. elegans*) homolog A	Synaptic vesicle trafficking	Unknown	ALS
SARM1	SALS	17q11.2	Homo Sapiens Sterile alpha and TIR motif containing 1	Activation of axonal degeneration and neuronal death upon stress	Unknown	ALS

13 Genotypic and Phenotypic Heterogeneity in Amyotrophic Lateral Sclerosis

Gene	ALS type	Locus	Encoded protein	Protein function	Mutation	Associated phenotypes
C21orf2	SALS	21q22.3	Chromosome 21 Open reading frame 2	Cilia formation, cell morphology, cytoskeletal maintenance	Rare variants	ALS

MS missense, *NS* nonsense, *SS* splice-site, *FS* frameshift, *ER* endoplasmic reticulum, *mTOR* mechanistic target of rapamycin, *ALS* amyotrophic lateral sclerosis, *FTD* frontotemporal dementia, *AOA* ataxia with ocular apraxia, *HET* hereditary essential tremor, *SMA* spinal muscular atrophy, *CMT* Charcot-Marie-Tooth disease, *IBMPFD* inclusion body myopathy with Paget disease of bone and/or frontotemporal dementia, *SMAJ* SMA-Jokela type, *SPG* spastic paraplegia, *dHMN* distal hereditary motor neuropathy, *PELD* progressive encephalopathy ± lipodystrophy, *CGL* congenital generalized lipodystrophy, *PLS* primary lateral sclerosis, *IAHSP* infantile-onset ascending hereditary spastic paraplegia, *MND* motor neuron disease, *MPAN* mitochondrial membrane protein-associated neurodegeneration, *NBIA* neurodegeneration with brain iron accumulation, *PCH* pontocerebellar hypoplasia, *BNHS* Boucher-Neuhauser syndrome, *OMCS* Oliver-McFarlane syndrome, *SCA* spinocerebellar atrophy, *SRTD* short rib with thoracic dysplasia ± polydactyly, *LCCS* lethal congenital contracture syndrome, *LAAHD* lethal arthrogryposis with anterior horn cell disease are collectively responsible for less than 5% of cases overall. Usually those variants are found in isolated pedigrees, often with atypical ALS phenotypes [27–34], and are private mutations, thus making a clear genotype-phenotype correlation extremely difficult.

Genetic Pleiotropy in ALS

The identification of genetic factors has contributed tremendously to our understanding of ALS biological and clinical complexity.

The first major ALS-associated gene, *SOD1*, has been discovered in 1993 [35]. The gene encodes for a 153 residues long, 32 kDa homodimeric metalloenzyme, which is ubiquitously expressed, highly conserved, and represent ~1% of all cytoplasmic proteins [36]. To date, more than 160 different *SOD1* mutations have been reported, the vast majority of which are missense substitutions distributed throughout the five exons of the gene. Eight frameshift deletions and five insertions, all clustered in exon 4 and 5, and leading to a premature truncation of the protein, have also been described. Collectively, *SOD1* mutations are found in ~20% of all FALS patients and in ~3% of SALS cases [37]. *SOD1* mutations are characterized by a considerable interfamilial and intrafamilial variability of the phenotype with regard to the age, site of onset, and disease duration [38]. A notable exception is represented by A4V, the mutation most frequently observed in ALS1 pedigrees, which is consistently associated with a high penetrance, younger age at onset, prevalence of lower motor neuron signs, and a very rapid disease course, usually less than 12 months [38, 39]. Atypical symptoms, such as external ophthalmoplegia, hyperacusis, and neuralgic pain, have occasionally been reported in A4V long survivors on artificial ventilatory support. Conversely, other mutations, such as G41D, H46R, and G93D, display a very mild phenotype, often with carriers surviving more than 20 years after the onset of the disease [38, 40, 41]. Atypical lower motor neuron phenotypes, such as cramp-fasciculation syndrome, flail limb, and/or pseudopolyneuritic (Patrikios) forms of ALS, are also often observed in patients with less aggressive *SOD1* mutations. The penetrance of mutations is also variable, being almost complete for A4V, and less than 30% at 70 years for I113T [42]. All *SOD1* mutations are inherited as dominant traits, with the exception of the D90A variant, that is observed both in recessive pedigrees in Scandinavia and in dominant pedigrees in the rest of the world [43, 44]. D90A homozygous families consistently display a milder phenotype, characterized by an asymmetrical, slowly progressive, ascending paraparesis with upper motor neuron signs, compared to heterozygous individuals that develop classic ALS.

A major step forward has been the discovery of the protein TDP-43 as the main component of ubiquitinated cytoplasmic inclusions in the nervous tissue of ALS patients, both in sporadic and in familial cases [45]. TDP-43 is a 43 kDa multifunctional DNA-/RNA-binding protein encoded by the *TARDBP* gene, belonging to the heterogeneous ribonucleoprotein (hnRNP) family. After this breakthrough discovery, mutations in the *TARDBP* gene have been found to be a major cause of

Mendelian ALS in several populations of different geographic origin, with a mutational frequency of ~5% in FALS and 1% in SALS [46, 47]. Subsequently, disease causing mutations were identified with a similar frequency also in the *FUS* gene, encoding for another nuclear RNA-binding protein which forms pathological cytoplasmic aggregates when mutated [48, 49]. Genetic and neuropathological findings about TDP-43 and FUS suggest that alterations in RNA homeostasis may represent a key event in ALS pathogenesis.

From a clinical standpoint, *TARDBP*-mutated cases usually display a classic ALS phenotype. Conversely, most patients carrying mutations in the *FUS* gene develop a juvenile ALS characterized by symmetric weakness of the scapular or pelvic girdles and of the proximal muscles of the upper or lower limbs [23]. Also prominent is the involvement of the axial muscles of the neck and trunk [50]. Severe weakness of neck extensor muscles at onset, as frequently seen in *FUS*-mutated patients, is also unusual, being reported in less than 1% of individuals with ALS [51].

The observation of cytoplasmic inclusion immunoreactive for TDP-43 or FUS indicates that mutations decrease their solubility, thus increasing aggregation propensity. Moreover, while in unaffected neurons both proteins localize in the cell nucleus, they are absent from the nuclei of inclusion-bearing cells, suggesting a nucleo-cytoplasmic redistribution. Interestingly, TDP-43-positive inclusions are observed not only in mutation carriers, but also in the vast majority of sporadic ALS and in ~55% of FTD cases, indicating a broad neuropathological overlap between the two diseases. Conversely, FUS aggregates are observed exclusively in ALS patients harbouring mutations and in a minority of non-mutated FTD cases displaying atypical neuropathological phenotypes [52–54].

TDP-43 and FUS share many structural and functional, genetic, and neuropathological similarities. Both proteins have been demonstrated to play a role in several biological processes, including transcriptional regulation, splicing, nucleo-cytoplasmic shuttling, transport, and stabilization of mRNAs [55]. In the central nervous system, both proteins are involved in mRNA transport towards the dendrites and in regulating synaptic plasticity [56–59].

The definitive molecular link between ALS and FTLD-U was however the identification of an expanded $(G_4C_2)_n$ hexanucleotide repeat in the promoter/first intron of the *c9orf72* gene as the major genetic cause of both diseases [60, 61]. In populations of European descent, the mutational frequencies range from 23 to 47% in FALS, 4 to 5% in SALS, 12 to 29% in FTD, and 6 to 86% in ALS-FTD patients [62].

Although the physiological functions of the C9ORF72 protein are currently poorly understood, several possible mechanisms have been proposed to explain the $(G_4C_2)_n$-mediated toxicity. In silico and in vitro models suggest that the protein is involved in regulating endosomal trafficking and that a reduction of mRNA levels due to the presence of the expansion may lead to impaired autophagy and endocytosis (loss-of-function hypothesis) [63, 64]. Increasing evidence, however, indicates that expanded $(G_4C_2)_n$ repeats may cause ALS and FTD through the acquisition of a novel toxic function (gain-of-function hypothesis). Mutant *c9orf72* is bidirectionally transcribed into pre-mRNAs containing sense $(G_4C_2)_n$ and antisense $(C_4G_2)_n$ repeats, with a propensity to form stable RNA G-quadruplex and RNA/DNA hybrid conformations. These abnormal structures, in turn, have a high affinity for RNA-binding proteins which are sequestered into nuclear RNA foci in a fashion similar to myotonic dystrophy [65]. Alternatively, expanded sense and antisense repeats may undergo an abnormal repeat-associated non-ATG translation (RAN translation), thus generating C9RAN proteins composed of repeated GA, GP, GR, PA, and PR dipeptides which in turn form insoluble neuronal inclusions [66, 67].

The motor phenotype of *c9orf72+* patients is often characterized by an earlier age at onset and shorter survival time compared to non-mutated ALS individuals, possibly due to a prominent bulbar involvement in a majority of cases [68, 69]. Patients with concurrent ALS and FTD or with a family history positive for one of both diseases have a significantly higher risk of harbouring expanded $(G_4C_2)_n$ repeats (33–86%), further indicating that the two disorders belong to the same pathogenic continuum [68]. The cognitive deficit of *c9orf72+* patients is usually consistent with a diagnosis of behavioural variant of FTD (bvFTD) and characterized by socially inappropriate, impulsive behaviour and general deterioration in the ability to perform routine daily tasks, or apathy, social isolation, and emotional lability. Patients often display prominent psychiatric features such as visual hallucinations, paranoid behaviour, persecutory delusions, aggressive behaviour, and/or suicidal thoughts [68–71].

Occasionally, mutated patients may display concurrent extrapyramidal and/or cerebellar signs, and in rare instances $(G_4C_2)_n$ repeat expansions have been identified in individuals with corticobasal syndrome, progressive supranuclear palsy, cerebellar ataxia, or olivopontocerebellar degeneration, suggesting that *c9orf72* may contribute to the pathogenesis of a broad spectrum of neurodegenerative diseases beyond ALS and FTD [72–76]. It is worth highlighting that several ALS-associated genes beside *c9orf72* display a high degree of phenotypic heterogeneity, being responsible for several other disorders (Table 13.1). ALS-associated mutations in *ATXN2*, *SPAST*, *FIG4*, *SETX*, *DCTN1*, *VAPB*, *HNRNPA1*, *VCP*, and *OPTN* have been observed in multisystem disorders such as cerebellar ataxias, hereditary spastic paraplegias (HSP), hereditary motor neuropathies (HMN), and IBMPFD [20]. The observation that HSP-associated genes significantly overlap with sets of genes implicated in the pathogenesis of ALS, AD, and PD further suggest the existence of a common genetic background in neurodegeneration [77].

The existence of such etiological relationships has several implications. For genetic research, it suggests that much of the missing heritability in ALS may be explained by variants already associated to related diseases and vice versa. It also supports the potential utility of combining data from individuals with seemingly distinct ailments to gain power in genetic and epidemiological studies. For translational research, the identification of common mechanistic features may provide new targets for drug development and broaden the indications of existing therapies. From a clinical standpoint, it may help interpret the co-occurrence of multiple phenotypes in affected families and the results from next-generation sequencing analysis.

Identification of Novel ALS Genes

Notwithstanding the discoveries of the past 25 years, the genetic determinants of nearly one-third of all FALS cases are still unknown. Mapping the remainder of ALS missing heritability is thus an area of ongoing and active research, and a wide variety of genetic techniques, each one with its own advantages and limitations, has been applied to this task.

Linkage analysis has led to the identification of several ALS genes such as *SOD1*, *ALS2*, *SETX*, *VAPB*, and *UBQLN2* [27, 28, 31, 35, 78, 79] and remains to this day the gold standard for gene hunting. Linkage analysis has proven however to be a difficult task when applied to ALS genetics since it requires large multigenerational pedigrees with multiple affected individuals, which are hard to find in a disease of adult life and with a rapid lethal course.

To overcome these limitations, several other approaches have been used alone or in combination with varying degrees of success. Linkage analysis used in combination with homozygosity mapping led to the identification of mutations in *FUS* and *OPTN* in pedigrees where an autosomal recessive inheritance was suspected [34, 48, 49]. Similarly, the application of next-generation sequencing (NGS) techniques to linkage analysis in small ALS families has been crucial to the discovery of the *VCP* and *PFN1* genes [21, 80].

Until very recently, however, the only method available to identify novel rare causative mutations in large cohorts of unrelated ALS cases has been the 'candidate gene approach', where the gene to be screened was chosen based on some specific characteristic, such as decreased expression in ALS patients or functional relationships to pathways altered in ALS. This approach has intrinsic limitations and pitfalls, especially when applied to the identification of rare variants in complex diseases such ALS. First, it is a hypothesis-driven method, relying on existing knowledge of comparative genomics, physiological properties of candidate protein, and purported pathological pathways. Compared to genome-wide association studies, this a priori bias in selecting a candidate gene may thus reinforce incorrect assumptions about disease pathogenesis. Additionally, any significant association may be actually driven by genetic differences between cases and controls because of population stratification. Lastly, most variants in the genome are actually rare and rather frequent. In fact, over 80% of coding variants have allele frequencies less than 0.5% [81], and many genes demonstrate an increased rate of variation, possibly due to a lack of selective pressure. Based on this knowledge, the possibility of identifying benign case-specific variants is very high, especially in large cohorts. It is thus not surprising that, despite the large number of reports [18], the only ALS gene convincingly identified through the candidate gene approach has been *TARDBP* [46, 47].

The recent development of massive, parallel DNA sequencing technologies has transformed medical genetics, allowing for rapid, high-throughput sequencing of entire exomes (whole exome sequencing, WES) or genomes (whole genome sequencing, WGS) for a modest cost. As such, it is now possible to assess low-frequency and rare variants in an unbiased way and investigate their role in complex diseases. Notwithstanding these technological advancements, however, the detection of rare variants in sequencing-based association studies remains challenging. In particular, the statistical power of classical single-variant association tests is extremely low and would require an unsustainable increase in sample size to be acceptable. To overcome these limitations, several strategies have been developed, alone or in combination [82]. Instead of testing them individually, rare variants can be collapsed in functional units (e.g. exons, domains, genes, etc.) and assigned a different weight based on different characteristics (e.g. number of variants per individual, presence in cases vs. controls, predicted impact on protein function, etc.).

Albeit very recent, the application of these rare variant burden (RVB) analytical strategies to cohorts of unrelated ALS patients has already led to the successful identification of novel genetic factors associated with ALS susceptibility within the *TUBA4A*, *TBK1*, and *NEK1* genes [83–85].

Future Perspectives and Conclusions

Notwithstanding the rapid advancement in the field of ALS genetic, translational efforts to interpret this bulk of information in the clinical setting are lagging behind. For instance, the identification of a novel variant in an ALS-associated gene does not imply that the mutation is pathogenic, especially in the absence of segregation and/or functional data. This is a serious issue in a disease such as ALS, where an abundance of very rare or private mutations has been described. To compound the problem, information about geographic distribution, associated phenotypes, and age-dependent penetrance are available only for a handful of them. Thus, to improve existing genetic counselling protocols, it is not only necessary to identify novel ALS-associated genes, but also to screen large replication cohorts and to systematically collect phenotypic data in order to draw reliable genotype-phenotype correlations.

The growing evidence of a possible clinical and genetic overlap between ALS and other neurodegenerative diseases also warrants further attention, both in the research and the clinical setting. The identification of a common genetic background may help highlight cellular pathways involved in neurodegeneration. At the same time, this knowledge should stimulate clinicians to a deeper investigation of the family history of ALS patients in order to provide better genetic counselling. A relatively uncharted territory, for example, is the degree of genetic overlap between ALS and other motor neuron disorders such as HMNs and HSPs. Although these conditions are usually associated to slow progression and extra-motor signs (e.g. sensory, cognitive, cerebellar and/or extrapyramidal), the phenotypic variability is so wide that it could easily lead to misdiagnoses. In fact, genes such as BSCL2, SPG11, and REEP1 have been associated with ALS phenotypes [30, 86–88], and HSP-associated genes significantly overlap with sets of genes implicated in the pathogenesis of other neurodegenerative diseases, including ALS, further suggesting the existence of a common genetic background in neurodegeneration [77].

Concerning the field of gene hunting in complex diseases, RVB analysis of NGS-generated data has largely supplanted the candidate gene approach, and the development of novel algorithms will likely improve their sensitivity [89]. One key issue will be the choice of WGS over WES. Presently, WES is significantly cheaper than WGS and requires much less storage space and computational power. However, WGS provides a more consistent coverage of the genome than WES, which often fails to capture the entire target region, and can identify rare variants in promoter and enhancer sequences. Notwithstanding the rapidly decreasing price of sequencing, the greatly increased sample sizes required to address power issues, not to mention the hardware and software capabilities needed to store and analyse data, will likely make costs of WGS unbearable for a single research group. As such, international collaborations between researchers will be crucial in promoting successful genetic studies. In the ALS genetic field, Project Mine (http://www.projectmine.com/) is spearheading a massive international effort to sequence the genomes of 15,000 patients with ALS and 7500 controls [90].

In conclusion, genetic studies in ALS have by far given the largest contribution to our understanding of the pathogenesis of the disease, to the recognition of particular subphenotypes within the whole clinical picture of ALS, and currently represent the best chance at identifying novel therapeutic targets.

References

1. Chiò A, Logroscino G, Traynor BJ, et al. Global epidemiology of amyotrophic lateral sclerosis: a systematic review of the published literature. Neuroepidemiology. 2013;41:118–30. https://doi.org/10.1159/000351153.
2. Alonso A, Logroscino G, Jick SS, Hernán MA. Incidence and lifetime risk of motor neuron disease in the United Kingdom: a population-based study. Eur J Neurol. 2009;16:745–51.
3. Logroscino G, Traynor BJ, Hardiman O, et al. Descriptive epidemiology of amyotrophic lateral sclerosis: new evidence and unsolved issues. J Neurol Neurosurg Psychiatry. 2008;79:6–11. https://doi.org/10.1136/jnnp.2006.104828.
4. Statland JM, Barohn RJ, McVey AL, et al. Patterns of weakness, classification of motor neuron disease, and clinical diagnosis of sporadic amyotrophic lateral sclerosis. Neurol Clin. 2015;33:735–48. https://doi.org/10.1016/j.ncl.2015.07.006.
5. Chiò A, Calvo A, Moglia C, et al. Phenotypic heterogeneity of amyotrophic lateral sclerosis: a population based study. J Neurol Neurosurg Psychiatry. 2011;82:740–6. https://doi.org/10.1136/jnnp.2010.235952.
6. Ferrari R, Kapogiannis D, Huey ED, Momeni P. FTD and ALS: a tale of two diseases. Curr Alzheimer Res. 2011;8:273–94.
7. Swinnen B, Robberecht W. The phenotypic variability of amyotrophic lateral sclerosis. Nat Rev Neurol. 2014;10:661–70. https://doi.org/10.1038/nrneurol.2014.184.
8. Kurland LT, Mulder DW. Epidemiologic investigations of amyotrophic lateral sclerosis. 2. Familial aggregations indicative of dominant inheritance. I. Neurology. 1955;5:182–96.
9. Kurland LT, Mulder DW. Epidemiologic investigations of amyotrophic lateral sclerosis. 2. Familial aggregations indicative of dominant inheritance. II. Neurology. 1955;5:249–68.
10. Al-Chalabi A, Fang F, Hanby MF, et al. An estimate of amyotrophic lateral sclerosis heritability using twin data. J Neurol Neurosurg Psychiatry. 2010;81:1324–6. https://doi.org/10.1136/jnnp.2010.207464.

11. Riggs JE. Longitudinal Gompertzian analysis of amyotrophic lateral sclerosis mortality in the U.S., 1977-1986: evidence for an inherently susceptible population subset. Mech Ageing Dev. 1990;55:207–20.
12. Hanby MF, Scott KM, Scotton W, et al. The risk to relatives of patients with sporadic amyotrophic lateral sclerosis. Brain J Neurol. 2011;134:3454–7. https://doi.org/10.1093/brain/awr248.
13. Fang F, Kamel F, Lichtenstein P, et al. Familial aggregation of amyotrophic lateral sclerosis. Ann Neurol. 2009;66:94–9. https://doi.org/10.1002/ana.21580.
14. Andersen PM, Al-Chalabi A. Clinical genetics of amyotrophic lateral sclerosis: what do we really know? Nat Rev Neurol. 2011;7:603–15. https://doi.org/10.1038/nrneurol.2011.150.
15. Renton AE, Chiò A, Traynor BJ. State of play in amyotrophic lateral sclerosis genetics. Nat Neurosci. 2014;17:17–23. https://doi.org/10.1038/nn.3584.
16. Al-Chalabi A, Lewis CM. Modelling the effects of penetrance and family size on rates of sporadic and familial disease. Hum Hered. 2011;71:281–8. https://doi.org/10.1159/000330167.
17. Byrne S, Bede P, Elamin M, et al. Proposed criteria for familial amyotrophic lateral sclerosis. Amyotroph Lateral Scler. 2011;12:157–9. https://doi.org/10.3109/17482968.2010.545420.
18. Abel O, Powell JF, Andersen PM, Al-Chalabi A. ALSoD: a user-friendly online bioinformatics tool for amyotrophic lateral sclerosis genetics. Hum Mutat. 2012;33:1345–51. https://doi.org/10.1002/humu.22157.
19. Kenna KP, McLaughlin RL, Hardiman O, Bradley DG. Using reference databases of genetic variation to evaluate the potential pathogenicity of candidate disease variants. Hum Mutat. 2013;34:836–41. https://doi.org/10.1002/humu.22303.
20. Al-Chalabi A, Jones A, Troakes C, et al. The genetics and neuropathology of amyotrophic lateral sclerosis. Acta Neuropathol. 2012;124:339–52. https://doi.org/10.1007/s00401-012-1022-4.
21. Johnson JO, Mandrioli J, Benatar M, et al. Exome sequencing reveals VCP mutations as a cause of familial ALS. Neuron. 2010;68:857–64. https://doi.org/10.1016/j.neuron.2010.11.036.
22. Conte A, Lattante S, Zollino M, et al. P525L FUS mutation is consistently associated with a severe form of juvenile amyotrophic lateral sclerosis. Neuromuscul Disord. 2012;22:73–5. https://doi.org/10.1016/j.nmd.2011.08.003.
23. Ticozzi N, Silani V, LeClerc AL, et al. Analysis of FUS gene mutation in familial amyotrophic lateral sclerosis within an Italian cohort. Neurology. 2009;73:1180–5. https://doi.org/10.1212/WNL.0b013e3181bbff05.
24. Kenna KP, McLaughlin RL, Byrne S, et al. Delineating the genetic heterogeneity of ALS using targeted high-throughput sequencing. J Med Genet. 2013;50:776–83. https://doi.org/10.1136/jmedgenet-2013-101795.
25. Chiò A, Calvo A, Mazzini L, et al. Extensive genetics of ALS: a population-based study in Italy. Neurology. 2012;79:1983–9. https://doi.org/10.1212/WNL.0b013e3182735d36.
26. Lattante S, Conte A, Zollino M, et al. Contribution of major amyotrophic lateral sclerosis genes to the etiology of sporadic disease. Neurology. 2012;79:66–72. https://doi.org/10.1212/WNL.0b013e31825dceca.
27. Hadano S, Hand CK, Osuga H, et al. A gene encoding a putative GTPase regulator is mutated in familial amyotrophic lateral sclerosis 2. Nat Genet. 2001;29:166–73. https://doi.org/10.1038/ng1001-166.
28. Yang Y, Hentati A, Deng HX, et al. The gene encoding alsin, a protein with three guanine-nucleotide exchange factor domains, is mutated in a form of recessive amyotrophic lateral sclerosis. Nat Genet. 2001;29:160–5. https://doi.org/10.1038/ng1001-160.
29 Yeh T-H, Lai S-C, Weng Y-H, et al. Screening for C9orf72 repeat expansions in parkinsonian syndromes. Neurobiol Aging. 2013;34:1311.e3–4. https://doi.org/10.1016/j.neurobiolaging.2012.09.002.
30. Orlacchio A, Babalini C, Borreca A, et al. SPATACSIN mutations cause autosomal recessive juvenile amyotrophic lateral sclerosis. Brain J Neurol. 2010;133:591–8. https://doi.org/10.1093/brain/awp325.
31. Nishimura AL, Mitne-Neto M, Silva HCA, et al. A mutation in the vesicle-trafficking protein VAPB causes late-onset spinal muscular atrophy and amyotrophic lateral sclerosis. Am J Hum Genet. 2004;75:822–31. https://doi.org/10.1086/425287.

32. Greenway MJ, Andersen PM, Russ C, et al. ANG mutations segregate with familial and "sporadic" amyotrophic lateral sclerosis. Nat Genet. 2006;38:411–3. https://doi.org/10.1038/ng1742.
33. Chow CY, Landers JE, Bergren SK, et al. Deleterious variants of FIG4, a phosphoinositide phosphatase, in patients with ALS. Am J Hum Genet. 2009;84:85–8. https://doi.org/10.1016/j.ajhg.2008.12.010.
34. Maruyama H, Morino H, Ito H, et al. Mutations of optineurin in amyotrophic lateral sclerosis. Nature. 2010;465:223–6. https://doi.org/10.1038/nature08971.
35. Rosen DR, Siddique T, Patterson D, et al. Mutations in Cu/Zn superoxide dismutase gene are associated with familial amyotrophic lateral sclerosis. Nature. 1993;362:59–62. https://doi.org/10.1038/362059a0.
36. McCord JM, Fridovich I. Superoxide dismutase. An enzymic function for erythrocuprein (hemocuprein). J Biol Chem. 1969;244:6049–55.
37. Andersen PM, Sims KB, Xin WW, et al. Sixteen novel mutations in the Cu/Zn superoxide dismutase gene in amyotrophic lateral sclerosis: a decade of discoveries, defects and disputes. Amyotroph Lateral Scler Mot Neuron Disord. 2003;4:62–73.
38. Cudkowicz ME, McKenna-Yasek D, Sapp PE, et al. Epidemiology of mutations in superoxide dismutase in amyotrophic lateral sclerosis. Ann Neurol. 1997;41:210–21. https://doi.org/10.1002/ana.410410212.
39. Juneja T, Pericak-Vance MA, Laing NG, et al. Prognosis in familial amyotrophic lateral sclerosis: progression and survival in patients with glu100gly and ala4val mutations in Cu,Zn superoxide dismutase. Neurology. 1997;48:55–7.
40. Aoki M, Ogasawara M, Matsubara Y, et al. Mild ALS in Japan associated with novel SOD mutation. Nat Genet. 1993;5:323–4. https://doi.org/10.1038/ng1293-323.
41. Luigetti M, Madia F, Conte A, et al. SOD1 G93D mutation presenting as paucisymptomatic amyotrophic lateral sclerosis. Amyotroph Lateral Scler. 2009;10:479–82. https://doi.org/10.3109/17482960802302261.
42. Orrell RW, Habgood JJ, Malaspina A, et al. Clinical characteristics of SOD1 gene mutations in UK families with ALS. J Neurol Sci. 1999;169:56–60.
43. Andersen PM, Forsgren L, Binzer M, et al. Autosomal recessive adult-onset amyotrophic lateral sclerosis associated with homozygosity for Asp90Ala CuZn-superoxide dismutase mutation. A clinical and genealogical study of 36 patients. Brain J Neurol. 1996;119(Pt 4):1153–72.
44. Al-Chalabi A, Andersen PM, Chioza B, et al. Recessive amyotrophic lateral sclerosis families with the D90A SOD1 mutation share a common founder: evidence for a linked protective factor. Hum Mol Genet. 1998;7:2045–50.
45. Neumann M, Sampathu DM, Kwong LK, et al. Ubiquitinated TDP-43 in frontotemporal lobar degeneration and amyotrophic lateral sclerosis. Science. 2006;314:130–3. https://doi.org/10.1126/science.1134108.
46. Sreedharan J, Blair IP, Tripathi VB, et al. TDP-43 mutations in familial and sporadic amyotrophic lateral sclerosis. Science. 2008;319:1668–72. https://doi.org/10.1126/science.1154584.
47. Van Deerlin VM, Leverenz JB, Bekris LM, et al. TARDBP mutations in amyotrophic lateral sclerosis with TDP-43 neuropathology: a genetic and histopathological analysis. Lancet Neurol. 2008;7:409–16. https://doi.org/10.1016/S1474-4422(08)70071-1.
48. Kwiatkowski TJ, Bosco DA, Leclerc AL, et al. Mutations in the FUS/TLS gene on chromosome 16 cause familial amyotrophic lateral sclerosis. Science. 2009;323:1205–8. https://doi.org/10.1126/science.1166066.
49. Vance C, Rogelj B, Hortobágyi T, et al. Mutations in FUS, an RNA processing protein, cause familial amyotrophic lateral sclerosis type 6. Science. 2009;323:1208–11. https://doi.org/10.1126/science.1165942.
50. Ravits J, Paul P, Jorg C. Focality of upper and lower motor neuron degeneration at the clinical onset of ALS. Neurology. 2007;68:1571–5. https://doi.org/10.1212/01.wnl.0000260965.20021.47.
51. Gourie-Devi M, Nalini A, Sandhya S. Early or late appearance of "dropped head syndrome" in amyotrophic lateral sclerosis. J Neurol Neurosurg Psychiatry. 2003;74:683–6.
52. Munoz DG, Neumann M, Kusaka H, et al. FUS pathology in basophilic inclusion body disease. Acta Neuropathol. 2009;118:617–27. https://doi.org/10.1007/s00401-009-0598-9.

53. Neumann M, Rademakers R, Roeber S, et al. A new subtype of frontotemporal lobar degeneration with FUS pathology. Brain J Neurol. 2009;132:2922–31. https://doi.org/10.1093/brain/awp214.
54. Seelaar H, Klijnsma KY, de Koning I, et al. Frequency of ubiquitin and FUS-positive, TDP-43-negative frontotemporal lobar degeneration. J Neurol. 2010;257:747–53. https://doi.org/10.1007/s00415-009-5404-z.
55. Ratti A, Buratti E. Physiological functions and pathobiology of TDP-43 and FUS/TLS proteins. J Neurochem. 2016;138(Suppl 1):95–111. https://doi.org/10.1111/jnc.13625.
56. Fujii R, Okabe S, Urushido T, et al. The RNA binding protein TLS is translocated to dendritic spines by mGluR5 activation and regulates spine morphology. Curr Biol. 2005;15:587–93. https://doi.org/10.1016/j.cub.2005.01.058.
57. Wang I-F, Wu L-S, Chang H-Y, Shen C-KJ. TDP-43, the signature protein of FTLD-U, is a neuronal activity-responsive factor. J Neurochem. 2008;105:797–806. https://doi.org/10.1111/j.1471-4159.2007.05190.x.
58. Liu-Yesucevitz L, Bassell GJ, Gitler AD, et al. Local RNA translation at the synapse and in disease. J Neurosci. 2011;31:16086–93. https://doi.org/10.1523/JNEUROSCI.4105-11.2011.
59. Fallini C, Bassell GJ, Rossoll W. The ALS disease protein TDP-43 is actively transported in motor neuron axons and regulates axon outgrowth. Hum Mol Genet. 2012;21:3703–18. https://doi.org/10.1093/hmg/dds205.
60. DeJesus-Hernandez M, Mackenzie IR, Boeve BF, et al. Expanded GGGGCC hexanucleotide repeat in noncoding region of C9ORF72 causes chromosome 9p-linked FTD and ALS. Neuron. 2011;72:245–56. https://doi.org/10.1016/j.neuron.2011.09.011.
61. Renton AE, Majounie E, Waite A, et al. A hexanucleotide repeat expansion in C9ORF72 is the cause of chromosome 9p21-linked ALS-FTD. Neuron. 2011;72:257–68. https://doi.org/10.1016/j.neuron.2011.09.010.
62. Majounie E, Renton AE, Mok K, et al. Frequency of the C9orf72 hexanucleotide repeat expansion in patients with amyotrophic lateral sclerosis and frontotemporal dementia: a cross-sectional study. Lancet Neurol. 2012;11:323–30. https://doi.org/10.1016/S1474-4422(12)70043-1.
63. Farg MA, Sundaramoorthy V, Sultana JM, et al. C9ORF72, implicated in amyotrophic lateral sclerosis and frontotemporal dementia, regulates endosomal trafficking. Hum Mol Genet. 2014;23:3579–95. https://doi.org/10.1093/hmg/ddu068.
64. Ji YJ, Ugolino J, Brady NR, et al. Systemic deregulation of autophagy upon loss of ALS- and FTD-linked C9orf72. Autophagy. 2017;13:1254–5. https://doi.org/10.1080/15548627.2017.1299312.
65. Haeusler AR, Donnelly CJ, Periz G, et al. C9orf72 nucleotide repeat structures initiate molecular cascades of disease. Nature. 2014;507:195–200. https://doi.org/10.1038/nature13124.
66. Ash PEA, Bieniek KF, Gendron TF, et al. Unconventional translation of C9ORF72 GGGGCC expansion generates insoluble polypeptides specific to c9FTD/ALS. Neuron. 2013;77:639–46. https://doi.org/10.1016/j.neuron.2013.02.004.
67. Mori K, Weng S-M, Arzberger T, et al. The C9orf72 GGGGCC repeat is translated into aggregating dipeptide-repeat proteins in FTLD/ALS. Science. 2013;339:1335–8. https://doi.org/10.1126/science.1232927.
68. Byrne S, Elamin M, Bede P, et al. Cognitive and clinical characteristics of patients with amyotrophic lateral sclerosis carrying a C9orf72 repeat expansion: a population-based cohort study. Lancet Neurol. 2012;11:232–40. https://doi.org/10.1016/S1474-4422(12)70014-5.
69. Gijselinck I, Van Langenhove T, van der Zee J, et al. A C9orf72 promoter repeat expansion in a Flanders-Belgian cohort with disorders of the frontotemporal lobar degeneration-amyotrophic lateral sclerosis spectrum: a gene identification study. Lancet Neurol. 2012;11:54–65. https://doi.org/10.1016/S1474-4422(11)70261-7.
70. Snowden JS, Adams J, Harris J, et al. Distinct clinical and pathological phenotypes in frontotemporal dementia associated with MAPT, PGRN and C9orf72 mutations. Amyotroph Lateral Scler Frontotemporal Degener. 2015;16:497–505. https://doi.org/10.3109/21678421.2015.1074700.
71. Simón-Sánchez J, Dopper EGP, Cohn-Hokke PE, et al. The clinical and pathological phenotype of C9ORF72 hexanucleotide repeat expansions. Brain J Neurol. 2012;135:723–35. https://doi.org/10.1093/brain/awr353.

72. Beck J, Poulter M, Hensman D, et al. Large C9orf72 hexanucleotide repeat expansions are seen in multiple neurodegenerative syndromes and are more frequent than expected in the UK population. Am J Hum Genet. 2013;92:345–53. https://doi.org/10.1016/j.ajhg.2013.01.011.
73. Majounie E, Abramzon Y, Renton AE, et al. Repeat expansion in C9ORF72 in Alzheimer's disease. N Engl J Med. 2012;366:283–4. https://doi.org/10.1056/NEJMc1113592.
74. Lindquist SG, Duno M, Batbayli M, et al. Corticobasal and ataxia syndromes widen the spectrum of C9ORF72 hexanucleotide expansion disease. Clin Genet. 2013;83:279–83. https://doi.org/10.1111/j.1399-0004.2012.01903.x.
75. Lesage S, Le Ber I, Condroyer C, et al. C9orf72 repeat expansions are a rare genetic cause of parkinsonism. Brain J Neurol. 2013;136:385–91. https://doi.org/10.1093/brain/aws357.
76. Hensman Moss DJ, Poulter M, Beck J, et al. C9orf72 expansions are the most common genetic cause of Huntington disease phenocopies. Neurology. 2014;82:292–9. https://doi.org/10.1212/WNL.0000000000000061.
77. Novarino G, Fenstermaker AG, Zaki MS, et al. Exome sequencing links corticospinal motor neuron disease to common neurodegenerative disorders. Science. 2014;343:506–11. https://doi.org/10.1126/science.1247363.
78. Chen Y-Z, Bennett CL, Huynh HM, et al. DNA/RNA helicase gene mutations in a form of juvenile amyotrophic lateral sclerosis (ALS4). Am J Hum Genet. 2004;74:1128–35. https://doi.org/10.1086/421054.
79. Deng H-X, Chen W, Hong S-T, et al. Mutations in UBQLN2 cause dominant X-linked juvenile and adult-onset ALS and ALS/dementia. Nature. 2011;477:211–5. https://doi.org/10.1038/nature10353.
80. C-H W, Fallini C, Ticozzi N, et al. Mutations in the profilin 1 gene cause familial amyotrophic lateral sclerosis. Nature. 2012;488:499–503. https://doi.org/10.1038/nature11280.
81. Tennessen JA, Bigham AW, O'Connor TD, et al. Evolution and functional impact of rare coding variation from deep sequencing of human exomes. Science. 2012;337:64–9. https://doi.org/10.1126/science.1219240.
82. Lee S, Abecasis GR, Boehnke M, Lin X. Rare-variant association analysis: study designs and statistical tests. Am J Hum Genet. 2014;95:5–23. https://doi.org/10.1016/j.ajhg.2014.06.009.
83. Cirulli ET, Lasseigne BN, Petrovski S, et al. Exome sequencing in amyotrophic lateral sclerosis identifies risk genes and pathways. Science. 2015;347:1436–41. https://doi.org/10.1126/science.aaa3650.
84. Smith BN, Ticozzi N, Fallini C, et al. Exome-wide rare variant analysis identifies TUBA4A mutations associated with familial ALS. Neuron. 2014;84:324–31. https://doi.org/10.1016/j.neuron.2014.09.027.
85. Kenna KP, van Doormaal PTC, Dekker AM, et al. NEK1 variants confer susceptibility to amyotrophic lateral sclerosis. Nat Genet. 2016;48:1037–42. https://doi.org/10.1038/ng.3626.
86. James PA, Talbot K. The molecular genetics of non-ALS motor neuron diseases. Biochim Biophys Acta. 2006;1762:986–1000. https://doi.org/10.1016/j.bbadis.2005.04.003.
87. Beetz C, Pieber TR, Hertel N, et al. Exome sequencing identifies a REEP1 mutation involved in distal hereditary motor neuropathy type V. Am J Hum Genet. 2012;91:139–45. https://doi.org/10.1016/j.ajhg.2012.05.007.
88. Züchner S, Wang G, Tran-Viet K-N, et al. Mutations in the novel mitochondrial protein REEP1 cause hereditary spastic paraplegia type 31. Am J Hum Genet. 2006;79:365–9. https://doi.org/10.1086/505361.
89. Auer PL, Lettre G. Rare variant association studies: considerations, challenges and opportunities. Genome Med. 2015;7:16. https://doi.org/10.1186/s13073-015-0138-2.
90. van Rheenen W, Shatunov A, Dekker AM, et al. Genome-wide association analyses identify new risk variants and the genetic architecture of amyotrophic lateral sclerosis. Nat Genet. 2016;48:1043–8. https://doi.org/10.1038/ng.3622.

Lewy Body Dementia

14

L. Bonanni, R. Franciotti, S. Delli Pizzi, A. Thomas, and M. Onofrj

Abstract

Dementia with Lewy bodies (DLB) represents the second most common cause of neurodegenerative dementia after Alzheimer's disease (AD).

The dementia with Lewy bodies (DLB) Consortium has refined its recommendations about the clinical and pathologic diagnosis of DLB, updating the previous report, which has been in widespread use for the last decade. The revised DLB consensus criteria now distinguish clearly between clinical features and diagnostic biomarkers and give guidance about optimal methods to establish and interpret these.

Important new information has been updated about previously reported aspects of DLB, with increased diagnostic weighting given to REM sleep behaviour disorder and ^{123}iodine-metaiodobenzylguanidine (MIBG) myocardial scintigraphy. The diagnostic role of other neuroimaging, electrophysiologic and laboratory investigations is also better specified. Substantial progress has been made since the previous report in the detection and recognition of DLB as a common and important clinical disorder.

Keywords

Dementia with Lewy bodies · Biomarkers · EEG abnormalities · MRI studies · Consensus criteria · Treatment options

L. Bonanni (✉) · R. Franciotti · S. Delli Pizzi · A. Thomas · M. Onofrj
Department of Neuroscience, Imaging and Clinical Sciences,
University G. d'Annunzio of Chieti-Pescara, Pescara, Italy
e-mail: l.bonanni@unich.it

Dementia with Lewy bodies (DLB) represents the second most common cause of neurodegenerative dementia after Alzheimer's disease (AD).

Despite the high prevalence of the disease among neurodegenerative dementia, DLB tends to be underdiagnosed during life and mostly misdiagnosed as AD, due to clinical overlap between the two diseases.

It is important, however, to differentiate between these two forms of dementia since the earliest stages because, compared to patients with AD, those with DLB may be considerably more sensitive to adverse effects of neuroleptics [1] and may exhibit faster disease progression [2] and different response to acetylcholinesterase inhibitors [3]. To reach a satisfactorily accuracy of the diagnosis of DLB, great emphasis has been placed on methods evaluating the uptake of either dopamine transporter (DAT) in basal ganglia [4, 5] or metaiodobenzylguanidine (MIBG) in the myocardium [6]. These methods, respectively, exploring the integrity of the nigrostriatal dopaminergic system and of postganglionic sympathetic cardiac innervation, have been suggested to improve clinical diagnostic accuracy of DLB, but there is a clear need of other biomarkers to assist with accurate identification of this entity.

Cognitively, DLB patients can display marked deficits in executive and visuospatial/visuoperceptual function, as well as marked variations in their level of arousal and attention which are typically known as cognitive fluctuations [7–10]. Clinical features associated with DLB also include spontaneous motor features of parkinsonism [8], but it is the non-motor manifestations including visual hallucinations, autonomic dysfunction, syncope, repeated falls, REM sleep behaviour disorder, delusions and depression, to represent the most disruptive symptoms for patients and their caregivers [8].

There are also a number of treatment challenges. Profound cholinergic deficits occur in DLB, even more than in AD [11]. Restoration of cholinergic function, by the use of cholinesterase inhibitors such as donepezil, rivastigmine and galantamine, may have cognitive and neuropsychiatric benefits including improvements in global cognitive function, attentional function and activities of daily living [12]. However, there are frequently variations in treatment response [13]; thus, responder stratification by use of a cholinergic function biomarker would be beneficial to clinical management. Beyond the cholinesterase inhibitors, the pharmacological repertoire of effective drugs in DLB is small, and agents such as memantine have been tried with mixed success [14–16]. The development of novel therapeutics is required, but this process has been hampered by the lack of DLB biomarkers which are sensitive to treatment response.

The dementia with Lewy bodies (DLB) Consortium has refined its recommendations about the clinical and pathologic diagnosis of DLB, updating the previous report, which has been in widespread use for the last decade. The revised DLB consensus criteria now distinguish clearly between clinical features and diagnostic biomarkers and give guidance about optimal methods to establish and interpret these.

Important new information has been updated about previously reported aspects of DLB, with increased diagnostic weighting given to REM sleep behaviour disorder and ^{123}iodine-metaiodobenzylguanidine (MIBG) myocardial scintigraphy. The diagnostic role of other neuroimaging, electrophysiologic and laboratory investigations is also better specified. Substantial progress has been made since the previous report in the detection and recognition of DLB as a common and important clinical disorder. During that period, DLB has been incorporated as a separate nosological entity into DSM-5, as major neurocognitive disorder with Lewy bodies. 'There remains a pressing need to understand the underlying neurobiology and pathophysiology of DLB, to develop and deliver clinical trials with both symptomatic and disease-modifying agents and to help patients and carers worldwide to inform themselves about the disease, its prognosis, best available treatments, ongoing research and how to get adequate support'.

A collection of studies have been recently performed in order to define possible specific pathophysiological mechanisms underlying the appearance of specific clinical features of DLB, with the double aim to explain clinical presentation and potentially to provide possible diagnostic markers of disease [17].

In this chapter, we will summarize the main results of studies focused on DLB biomarkers, and we will underline the significance of each of these biomarkers in terms of diagnostic accuracy and of pathophysiological mechanisms. Biomarkers will be divided into two sections, as suggested by the last consensus document [17]: indicative and supportive biomarkers. Major emphasis will be given to EEG and MRI studies which are the main fields of contributions by the authors.

Indicative Biomarkers

SPECT-DAT Scan

The functional integrity of dopaminergic nigrostriatal pathway can be studied with single photon emission computed tomography (SPECT) imaging by using ligands of pre-synaptic dopamine transporter (DAT), such as [123I]-N-(3-fluoropropyl)-2-carbomethoxy-3-(4-iodophenyl) nortropane (123I-FP-CIT). A reduction of SPECT ligand binding to DAT correlates with the loss of pre-synaptic dopamine. The rationale supporting the use of 123I-FP-CIT SPECT as a supportive tool in the diagnosis of DLB is represented by the pathological peculiarities of DLB, characterized by abnormal inclusion bodies (Lewy bodies) in limbic, neocortical and brainstem areas with concomitant nigrostriatal degeneration and loss of pre-synaptic dopamine transporters in the striatum [12, 18]. For these reasons, low dopamine transporter uptake in basal ganglia on 123I-FP-CIT SPECT has been listed as an indicative biomarker of DLB in the international consensus criteria for the diagnosis [8, 17] and also included in the guidelines of the National Institute for Health and Clinical Excellence (NICE) and Social Care Institute for Excellence to support the clinical diagnosis of DLB in doubtful cases].

MIBG: Reduced Uptake on Metaiodobenzylguanidine Myocardial Scintigraphy

^{123}Iodine-MIBG myocardial scintigraphy quantifies postganglionic sympathetic cardiac innervation, which is reduced in DLB [19, 20]. Useful sensitivity (69%) and specificity (87%) values for discriminating probable DLB from probable AD rise to

77% and 94% in milder cases (MMSE >21). A caveat to the interpretation of MIBG results should be considered in the light of possible confounding causes, including ischemic heart disease, heart failure, diabetes mellitus, peripheral neuropathies and medications that may cause reduced uptake including labetalol, reserpine, tricyclic antidepressants and sympathomimetics. MIBG imaging was already described in the supportive feature section of the previous version of the Consortium on DLB guidelines [8]. According to this report, a heart/mediastinum ratio (H/M) cut-off point of 1.68 on delayed MIBG images resulted in highly reliable differentiation of DLB from AD with both the sensitivity and the specificity being 100%, regardless of the presence or absence of parkinsonism [21]. Because of its high diagnostic accuracy, MIBG imaging has become one of the essential imaging methods for diagnosing DLB [17].

PSG Confirmation of REM Sleep Without Atonia

PSG demonstration of REM sleep without atonia [22, 23] is a highly specific predictor of Lewy-related pathology. If the PSG shows REM sleep without atonia in a person with dementia and a history of rapid eye movement (REM) sleep behaviour disorder (RBD), there is a 90% likelihood of a synucleinopathy [24] sufficient to justify a probable DLB diagnosis even in the absence of any other core feature or biomarker. RBD is characterized by loss of normal skeletal muscle atonia during REM sleep with prominent motor activity and dreaming [25, 26–29]. The parasomnia occurs more frequently in males and usually begins manifesting after the age of 50 years [27–29]. RBD can occur without any coexisting neurologic disorders or findings (so-called idiopathic RBD) and can be precipitated or aggravated by medications, such as selective serotonin or norepinephrine reuptake inhibitors [30, 31]. RBD can be triggered by structural brain lesions such as brainstem infarcts, tumours, vascular malformations, and demyelinating plaques associated with multiple sclerosis [32, 33]; these findings have provided insights into the location of the networks implicated in human RBD. All structural lesions identified to date have been localized in the dorsal midbrain, pons or medulla. Rare cases of RBD associated to voltage-gated potassium channel complex abnormalities present with abnormalities in the mesial temporal lobe structures and usually not in the brainstem [32]. This finding underline that the precise networks and neurotransmitter systems involved in human RBD remain unclear but most consistently relate to brainstem networks and their efferent or afferent connections.

Supportive Biomarkers

EEG

Resting-state electroencephalographic (rsEEG) rhythms have extensively been used as a possible tool to assess the neurophysiological correlates of dementia [34–36].

For DLB patients, specific and non-specific EEG features were found. However, they can provide an index of the extent to which DLB patients show abnormalities in the structure and function of the brain across the disease progression and therapeutic intervention [37]. RsEEG features which are specific for DLB patients are considered to be 'supportive' in international diagnostic guidelines [8, 17], as the presence of posterior (temporal) transient slow or sharp waves [36, 38]. A characterizing feature of rsEEG in DLB is the fluctuation of global delta and theta power over a few minutes [39–42]. This fluctuation was observed in the vast majority of DLB patients and very few AD patients [42]. The fluctuation of the dominant rsEEG frequency and the power of the alpha and slow frequencies are partially normalized by a short-term administration of acetylcholinesterase inhibitors in DLB patients [43]. A greater degree of parietal delta power band variability has been reported in patients with DLB, compared to AD patients and controls [41]. However, this feature may represent a non-specific EEG feature because increased power in these bands was reported in AD patients in at least four studies [44]. Many EEG studies tried to improve the diagnostic accuracy in differentiating DLB from AD. The use of qualitative EEG analysis as a diagnostic tool to distinguish between DLB and AD remains rare in daily clinical practice because of conflicting studies and the absence of a reliable scoring method [45]. Widespread delta and theta power over the scalp and posterior beta power were found to be higher in DLB than AD patients [41, 46]. Quantitative EEG (QEEG) has demonstrated good discriminative capacity for DLB diagnosis as compared to AD with a predictive value of 100% in cohort studies, even at the stage of MCI [42, 47] and the percentage of 90% in a multicentric cohort study [48]. Specifically, discriminant analysis detected specific cut-offs for every EEG mathematical descriptor; dominant frequency (DF) = 8, dominant frequency variability (DFV) = 2.2 Hz, frequency prevalence (FP) pre-alpha = 33% and FP alpha = 41% for posterior derivations. The occipital low-frequency alpha 2 source activity showed a classification accuracy of 75% in the contrast between the AD and DLB patients [37]. A sensitivity of 79% and a specificity of 76% were obtained by means of grand total EEG (GTE) score applied to distinguish DLB from AD patients [45]. Disrupted alpha band-directed connectivity may underlie the clinical syndrome of DLB and differentiate between DLB and AD. Indeed the common posterior-to-anterior pattern of directed connectivity in controls is disturbed in the alpha band in DLB patients and in the beta band in AD patients [49]. New mathematical approaches on EEG rhythms showed lower connectivity strength in the alpha frequency band in DLB patients compared to both controls and AD. In addition DLB brain network organization was found to be less efficient and contained less hubs [50].

Structural Imaging Studies (Preservation of Medial Temporal GM)

Recent advance in structural MRI allows to perform physical measurements of brain cortical thickness for each individual and to map, within and between groups; the macrostructural changes in grey matter (GM) regions. The measurement of the cortical thickness by methods proposed by Fischl and Dale [51] showed 82% sensitivity and 85% specificity in differentiating AD from DLB [52]. At cortical level, DLB patients show a preservation of medial temporal GM as compared to AD [17, 53, 54] and a thinning in the posterior areas including the precuneus, superior parietal gyrus, cuneus, pericalcarine and lingual gyri [53]. Of note, the posterior atrophy of the cuneus, precuneus and superior parietal cortex has been related to visual deficit and hallucinations in DLB [53, 55]. Moreover, increased rates of cortical thinning in the parietal regions were also correlated with motor deterioration in DLB [56].

Further Evidences Coming from Structural Imaging Studies

At subcortical level, microstructural and macrostructural alterations have been also described in DLB patients. Macrostructural assessment highlighted that the hippocampus, especially in the cornu ammonis and subiculum, is relatively preserved in DLB as compared to AD [56, 57]. GM reduction was also observed in DLB patients in the adjacent extrahippocampal structures including the perirhinal and parahippocampal cortices [57]. Based on studies reporting an involvement of these regions in the visual processing, it has been suggested that an impairment of adjacent extra-hippocampal structures could be contribute to the aetiology of visual hallucinations [58]. Diffusion weighted imaging (DWI) is able to give in vivo microstructural information on the grey and white matters integrity by assessing the Brownian motion of water molecules among neurons. In this context, two parameters are of particular relevance: fractional anisotropy (FA), whose reduction describes axonal degeneration; mean diffusivity (MD), whose increase refers to loss of membrane density and cell loss of both neurons and glia. While the former is a specific index only for axonal integrity, the MD describes both grey and white matters damage [59]. Microstructural damage of GM subcortical nuclei in DLB patients has been observed in the pons, hippocampus and thalamus. In particular, the changes within thalamus were observed for both FA [60] and MD [61, 62]. Additionally, by combining structural MRI and diffusion tensor imaging (DTI) data, the thalamus was further divided into sub-regions according to their structural connectivity to cortex. The assessment of the MD in each thalamic sub-region in DLB has revealed microstructural grey matter preservation of the sub-regions which projects to temporal cortex and [61] increase of MD within the thalamic portions projecting to the prefrontal and parieto-occipital cortices and amygdala [61]. Moreover, DLB patients present reduced structural connectivity within the anterior thalamic radiation, which projects to frontal cortex [62]. These results are in agreement with the role of thalamus in shaping the cortico-cortical control [63] and with emerging hypotheses suggesting that thalamic dysregulation could induce reduced levels of arousal and consciousness state [64]. In this context, we observed reduction of NAA/tCr (marker of axonal density) and increase of tCho/tCr (marker of cholinergic dysfunction) in DLB patients, which correlated with frequency and severity of fluctuating cognition in DLB [62]. These results match with the cholinergic deregulation coexisting with the dopaminergic alteration in DLB patients [65] and with the pharmacological evidence that (1) anticholinergic medication can induce a symptom profile of altered arousal like cognitive fluctuation [66] and (2) administration of cholinesterase inhibitors considerably reduces cognitive fluctuations in DLB [67]. Moreover, it was observed that the microstructural damage of the thalamic portions projecting to cortical posterior regions including parietal and occipital lobes is closely related to the presence and severity of visual hallucinations [61]. These findings are in agreement with the role of the pulvinar in the visual processing [68] and with recent reports from neuropathological studies showing severe neuronal loss in the medial pulvinar in post-mortem brain tissue acquired from patients with DLB [69].

Cingulate Island Sign

The cingulate island sign (CIS), a term referring to sparing of the posterior cingulate relative to the precuneus and cuneus, has been proposed as an FDG-PET imaging feature of DLB [70, 71] due to its good diagnostic power to distinguish DLB patients from AD. The preservation of the CIS is not associated with Aβ load but does predict lower Braak neurofibrillary tangle stage in clinically diagnosed DLB cases [72]. Furthermore, clinical symptoms of DLB (parkinsonism and global cognitive function) were found to be correlated with precuneus plus cuneus hypometabolism but not the CIS [72].

Other Functional Imaging Studies

Functional magnetic resonance imaging (fMRI) was also used to differentiate DLB patients from controls and other dementias. Lower regional homogeneity (ReHo) in sensory-motor cortices and higher ReHo in left middle temporal gyrus was found in DLB when compared with healthy controls [73]. Neuropathological differences between DLB and AD were also found by means of resting state fMRI and measures on network organization. Resting state fMRI evidenced an increased connectivity between the precuneus and regions in the dorsal attention networks and decreased connectivity with prefrontal and visual cortices in DLB compared to the AD group [74]. Greater connectivity between the putamen and frontal, temporal and parietal regions was found in DLB patients compared with AD patients [75]. Right hemisphere functional connectivity was reduced in DLB patients in comparison with control subjects and was correlated with severity of fluctuations [76].

Global network measures showed also significant differences between DLB and AD. DLB group demonstrated a generalized lower synchronization compared with the AD and healthy controls, mainly for edges connecting distant brain regions and higher small worldness [77].

On the other hand, task-based fMRI studies on visuoperceptual impairments in DLB patients reported a relative preservation of function in visual system in DLB [78, 79] and a greater activation in the superior temporal sulcus in DLB compared to AD during the motor part of the tasks [78].

Pharmacological Interventions

Meta-analyses of class I clinical trials of cholinesterase inhibitors (CHEIs), specifically rivastigmine and donepezil, support the use of this class of drugs in DLB for improving cognition, global function and activities of living. There is evidence that even if patients do not improve with CHEIs they are more likely to maintain their cognitive performance stable while taking them [80, 81]. The efficacy of memantine in DLB is less clear, but it is well-tolerated and may have benefits, either as monotherapy or adjunctive to a CHEI [80, 81].

Neuropsychiatric Symptoms

CHEIs may produce substantial reduction in apathy and improve visual hallucinations and delusions in DLB [82]. The use of antipsychotics for the acute management of substantial behavioural disturbance, delusions or visual hallucinations comes with attendant mortality risks in patients with dementia, and particularly in the case of DLB they should be avoided whenever possible, given the increased risk of a serious sensitivity reaction [83]. Low-dose quetiapine may be relatively safer than other antipsychotics and is widely used [84]. There is a positive evidence base for clozapine in PD psychosis, but efficacy and tolerability in DLB have not been established. Newer drugs targeting the serotonergic system, such as pimavanserin [85], may be alternatives, but controlled clinical trial data in DLB are needed. Although depressive symptoms are common in DLB, trial data are insufficient. In alignment with general advice on depression in dementia, selective serotonin reuptake inhibitors, serotonin-norepinephrine reuptake inhibitors and mirtazapine are options in DLB with treatment guided by individual patient tolerability and response.

Motor Symptoms

DLB patients may benefit from levodopa preparations introduced at low doses and increased slowly to the minimum required to minimize motor disability without exacerbating psychiatric symptoms [86, 87].

Akinetic Crisis in DLB

Parkinsonian motor features are listed among the core clinical diagnostic criteria of DLB [8]:

1. Extrapyramidal signs can be severely exacerbated, or appear for the first time, following the administration of typical neuroleptic drugs prescribed to control hallucinatory symptoms and behavioural disturbances.

2. Reactions to neuroleptic agents occur in a high percentage of patients with DLB (30–50%), so that neuroleptic hypersensitivity was already included among the supportive features for the diagnosis of DLB [8]. Responses to neuroleptics vary from mild (worsening of parkinsonian symptoms that resolves on drug withdrawal) to severe reactions, encompassing rigidity, fever, postural hypotension, falls, confusion, collapse and rapid deterioration to death [83].

Aarsland and colleagues [88] quantified the severity of symptoms using a standardized pro forma rating of cognitive symptoms, impairment of consciousness, agitation, worsening of parkinsonism and orthostatic hypotension, and found that 53% of patients with DLB developed severe neuroleptic sensitivity-related symptoms.

The incidence of a potentially lethal condition linked to the use of neuroleptic drugs, defined as neuroleptic malignant syndrome (NMS), which can increase patient mortality threefold, has been recently specifically defined [47]. NMS is a potentially lethal, drug-induced, idiosyncratic condition, first described by Delay et al. in 1960. Five NMS, classically associated with the use of high-potency antipsychotics (AP), such as haloperidol, butyrophenones and phenothiazines, have also been described with newer AP (risperidone, olanzapine), other D2-receptor antagonists (metoclopramide) [89] and following withdrawal of dopaminergic agents [90]. NMS is recognized as part of the so-called akinetic crisis (AC) syndrome, represented by a complication that appears in the course of parkinsonism when infectious diseases, trauma or gastrointestinal tract diseases occur [91]. The symptoms observed in AC overlap those observed in NMS but, unlike the latter, exposure to neuroleptics is not a mandatory causative factor for AC.

AC consists of acute motor symptom worsening characterized by an akinetic state and transient unresponsiveness to current antiparkinsonian treatment and represents an emergency in the management of parkinsonian patients [92].

AC is characterized by a distinctive clinical tetrad of mental status changes, motor abnormalities (bradykinesia and muscle rigidity), autonomic dysfunction (blood pressure instability, diaphoresis and tachycardia) and hyperthermia. Laboratory findings include elevation of serum creatine kinase (CK) and myoglobin [90, 91]. Treatment is mainly supportive and includes withdrawal of the AP or other causative agent and treatment with dopaminergic agonists and dantrolene [90].

Akinetic crisis (AC) is a condition observable in the DLB course, at least as frequently as in that of PD (6.8% vs. 3.9%). It appears to have a more severe outcome, being more frequently fatal in patients with DLB (50% vs. 12.5%), and is independent of clinical variables including severity of either cognitive or motor symptoms and of L-dopa equivalent daily dose. Disease duration before the occurrence of AC has been found to be shorter in patients with DLB than in patients with PD. This is likely due to the fact that DLB clinical expression in the early stage of disease is similar to the phenotype of PD in the advanced stage when patients with PD can present with symptoms requiring neuroleptic treatment.

An alternative hypothesis may be based on the different time of appearance of specific brain area alterations during the course of DLB versus PD. In patients with

DLB, early alterations of insula, which may integrate consciousness and internal homeostasis, has been recently demonstrated [93–95] whereas in patients with PD, the involvement of insula appears late in the course of the disease. Exposure to typical neuroleptics was found to be higher in the DLB group (32.6%) than in the PD group (5.2%), likely due to the more frequent occurrence of severe psychiatric symptoms early in the course of DLB. It is also necessary to point out that the wide exposure to typical neuroleptics, prescribed either by GPs or by primary dementia centres, in a high percentage of patients with DLB, suggests that DLB is an under-recognized clinical entity, often misdiagnosed as Alzheimer's disease. Awareness on the clinical features of DLB needs to be improved, to avoid severe NMS treatment outcomes, including AC [96].

The significant rate of mortality in our cohorts, due to medical complications precipitated by the transient refractoriness to dopaminergic rescue drug administration, suggests that the use of neuroleptics in patients with DLB should be limited to quetiapine or clozapine, which showed an acceptable safety profile [17, 97, 98].

However, the occurrence of AC in those patients with DLB never exposed to neuroleptics suggests that the exposure to neuroleptics and subsequent D2 receptor dysfunction is unlikely to be the only explanation for the development of AC.

Recent reports [99, 100] showed that iodine-123 fluoropropyl-carbomethoxy-3 beta-(4-iodophenyltropane) (FP-CIT) single photon emission CT uptake is extremely reduced in putamen and caudate nuclei in AC and NMS, resulting in the feature of 'burst striatum', with disappearance of the oval-shaped images corresponding to caudate. Interestingly, dopamine transporter (DAT) binding in mitochondrial DNA abnormalities, such as in polymerase (DNA) gamma and catalytic subunit (POLG) mutations, seems to be very low [101].

In a recent cohort study on the occurrence of recurrent and fatal AC in genetic parkinsonism [102], we found that AC was especially common in genetic mutations glucocerebrosidase (GBA), leucine-rich repeat kinase 2 (LRRK2)) involving mitochondrial functions. The hypothesis of genetic mutations associated with mitochondrial dysfunctions could be called into cause in the occurrence of AC in DLB. Further studies are needed, considering wider populations, assessing DLB for genetic mutations and considering the risk of underestimations dependent on reduced penetrance or recessive inheritance.

Further hypotheses may point to a possible link between inflammatory indices and the development of AC, as most patients developed AC after surgery or infections. Acute and chronic systemic inflammation, possibly associated with increases of serum tumour necrosis factor, is associated with a faster progression of cognitive decline in Alzheimer's disease [103]. It could be hypothesized that activation of inflammatory factors could precipitate an unfavourable outcome by inducing AC in patients with DLB/PD.

We hope that by presenting the statistical risk factors for AC–NMS in DLB, the AC issue will be acknowledged and the current practice of prescribing typical neuroleptics in dementia with uncertain phenotype will be discontinued.

References

1. Ballard CG, Holmes C, McKeith IG, O'Brien JT, Ince PG, Perry RH. Clinical symptoms in dementia with Lewy bodies: a prospective clinical and neuropathological comparative study with Alzheimer's disease. Neurology. 1998;50(4):A183.
2. Olichney JM, Galasko D, Salmon DP, Hofstetter CR, Hansen LA, Katzman R, et al. Cognitive decline is faster in Lewy body variant than in Alzheimer's disease. Neurology. 1998;51(2):351–7.
3. Levy R. Alzheimer's disease and Lewy body dementia. Br J Psychiatry. 1994;164(2):268.
4. Walker Z, Costa DC, Walker RW, Shaw K, Gacinovic S, Stevens T, et al. Differentiation of dementia with Lewy bodies from Alzheimer's disease using a dopaminergic presynaptic ligand. J Neurol Neurosurg Psychiatry. 2002;73(2):134–40.
5. O'Brien JT, Colloby S, Fenwick J, Williams ED, Firbank M, Burn D, et al. Dopamine transporter loss visualized with FP-CIT SPECT in the differential diagnosis of dementia with Lewy bodies. Arch Neurol. 2004;61(6):919–25.
6. Yoshita M, Taki J, Yokoyama K, Noguchi-Shinohara M, Matsumoto Y, Nakajima K, et al. Value of 123I-MIBG radioactivity in the differential diagnosis of DLB from AD. Neurology. 2006;66(12):1850–4.
7. Lee DR, Taylor JP, Thomas AJ. Assessment of cognitive fluctuation in dementia: a systematic review of the literature. Int J Geriatr Psychiatry. 2012;27(10):989–98.
8. McKeith IG, Dickson DW, Lowe J, Emre M, O'Brien JT, Feldman H, et al. Diagnosis and management of dementia with Lewy bodies: third report of the DLB Consortium. Neurology. 2005;65(12):1863–72.
9. Mollenhauer B, Forstl H, Deuschl G, Storch A, Oertel W, Trenkwalder C. Lewy body and parkinsonian dementia: common, but often misdiagnosed conditions. Dtsc Arztebl Int. 2010;107(39):684–91.
10. Mosimann UP, Mather G, Wesnes KA, O'Brien JT, Burn DJ, McKeith IG. Visual perception in Parkinson disease dementia and dementia with Lewy bodies. Neurology. 2004;63(11):2091–6.
11. Samuel W, Caligiuri M, Galasko D, Lacro J, Marini M, McClure FS, et al. Better cognitive and psychopathologic response to donepezil in patients prospectively diagnosed as dementia with Lewy bodies: a preliminary study. Int J Geriatr Psychiatry. 2000;15(9):794–802.
12. McKeith I, Mintzer J, Aarsland D, Burn D, Chiu H, Cohen-Mansfield J, et al. Dementia with Lewy bodies. Lancet Neurol. 2004;3(1):19–28.
13. Burn DJ, McKeith IG. Current treatment of dementia with Lewy bodies and dementia associated with Parkinson's disease. Mov Disord. 2003;18(Suppl 6):S72–9.
14. Aarsland D, Ballard C, Walker Z, Bostrom F, Alves G, Kossakowski K, et al. Memantine in patients with Parkinson's disease dementia or dementia with Lewy bodies: a double-blind, placebo-controlled, multicentre trial. Lancet Neurol. 2009;8(7):613–8.
15. Emre M, Tsolaki M, Bonuccelli U, Destee A, Tolosa E, Kutzelnigg A, et al. Memantine for patients with Parkinson's disease dementia or dementia with Lewy bodies: a randomised, double-blind, placebo-controlled trial. Lancet Neurol. 2010;9(10):969–77.
16. Matsunaga S, Kishi T, Iwata N. Memantine for Lewy body disorders: systematic review and meta-analysis. Am J Geriatr Psych. 2015;23(4):373–83.
17. McKeith IG, Boeve BF, Dickson DW, Halliday G, Taylor JP, Weintraub D, et al. Diagnosis and management of dementia with Lewy bodies: fourth consensus report of the DLB Consortium. Neurology. 2017;89(1):88–100.
18. Weisman D, McKeith I. Dementia with Lewy bodies. Semin Neurol. 2007;27(1):42–7.
19. Nakajima K, Okuda K, Yoshimura M, Matsuo S, Wakabayashi H, Imanishi Y, et al. Multicenter cross-calibration of I-123 metaiodobenzylguanidine heart-to-mediastinum ratios to overcome camera-collimator variations. J Nucl Cardiol. 2014;21(5):970–8.

20. Treglia G, Cason E. Diagnostic performance of myocardial innervation imaging using MIBG scintigraphy in differential diagnosis between dementia with Lewy bodies and other dementias: a systematic review and a meta-analysis. J Neuroimaging. 2012;22(2):111–7.
21. Yoshita M, Taki J, Yamada M. A clinical role for [(123)I]MIBG myocardial scintigraphy in the distinction between dementia of the Alzheimer's-type and dementia with Lewy bodies. J Neurol Neurosurg Psychiatry. 2001;71(5):583–8.
22. McCarter SJ, St Louis EK, Duwell EJ, Timm PC, Sandness DJ, Boeve BF, et al. Diagnostic thresholds for quantitative REM sleep phasic burst duration, phasic and tonic muscle activity, and REM atonia index in REM sleep behavior disorder with and without comorbid obstructive sleep apnea. Sleep. 2014;37(10):1649–62.
23. Frauscher B, Iranzo A, Gaig C, Gschliesser V, Guaita M, Raffelseder V, et al. Normative EMG values during REM sleep for the diagnosis of REM sleep behavior disorder. Sleep. 2012;35(6):835–47.
24. Boeve BF, Silber MH, Ferman TJ, Lin SC, Benarroch EE, Schmeichel AM, et al. Clinicopathologic correlations in 172 cases of rapid eye movement sleep behavior disorder with or without a coexisting neurologic disorder. Sleep Med. 2013;14(8):754–62.
25. Ratti PL, Negre-Pages L, Perez-Lloret S, Manni R, Damier P, Tison F, et al. Subjective sleep dysfunction and insomnia symptoms in Parkinson's disease: insights from a cross-sectional evaluation of the French CoPark cohort. Parkinsonism Relat Disord. 2015;21(11):1323–9.
26. Arnulf I. Excessive daytime sleepiness in parkinsonism. Sleep Med Rev. 2005;9(3):185–200.
27. Abbott RD, Ross GW, White LR, Tanner CM, Masaki KH, Nelson JS, et al. Excessive daytime sleepiness and subsequent development of Parkinson disease. Neurology. 2005;65(9):1442–6.
28. Gao J, Huang X, Park Y, Hollenbeck A, Blair A, Schatzkin A, et al. Daytime napping, nighttime sleeping, and Parkinson disease. Am J Epidemiol. 2011;173(9):1032–8.
29. Simuni T, Caspell-Garcia C, Coffey C, Chahine LM, Lasch S, Oertel WH, et al. Correlates of excessive daytime sleepiness in de novo Parkinson's disease: a case control study. Mov Disord. 2015;30(10):1371–81.
30. Breen DP, Williams-Gray CH, Mason SL, Foltynie T, Barker RA. Excessive daytime sleepiness and its risk factors in incident Parkinson's disease. J Neurol Neurosurg Psychiatry. 2013;84(2):233–4.
31. Schrag A, Horsfall L, Walters K, Noyce A, Petersen I. Prediagnostic presentations of Parkinson's disease in primary care: a case-control study. Lancet Neurol. 2015;14(1):57–64.
32. Iranzo A, Fernandez-Arcos A, Tolosa E, Serradell M, Molinuevo JL, Valldeoriola F, et al. Neurodegenerative disorder risk in idiopathic REM sleep behavior disorder: study in 174 patients. PLoS One. 2014;9(2):e89741.
33. Postuma RB, Gagnon JF, Vendette M, Fantini ML, Massicotte-Marquez J, Montplaisir J. Quantifying the risk of neurodegenerative disease in idiopathic REM sleep behavior disorder. Neurology. 2009;72(15):1296–300.
34. Giaquinto S, Nolfe G. The EEG in the normal elderly: a contribution to the interpretation of aging and dementia. Electroencephalogr Clin Neurophysiol. 1986;63(6):540–6.
35. Breslau J, Starr A, Sicotte N, Higa J, Buchsbaum MS. Topographic EEG changes with normal aging and SDAT. Electroencephalogr Clin Neurophysiol. 1989;72(4):281–9.
36. Briel RC, McKeith IG, Barker WA, Hewitt Y, Perry RH, Ince PG, et al. EEG findings in dementia with Lewy bodies and Alzheimer's disease. J Neurol Neurosurg Psychiatry. 1999;66(3):401–3.
37. Babiloni C, Del Percio C, Lizio R, Noce G, Cordone S, Lopez S, et al. Abnormalities of cortical neural synchronization mechanisms in patients with dementia due to Alzheimer's and Lewy body diseases: an EEG study. Neurobiol Aging. 2017;55:143–58.
38. Barber PA, Varma AR, Lloyd JJ, Haworth B, Snowden JS, Neary D. The electroencephalogram in dementia with Lewy bodies. Acta Neurol Scand. 2000;101(1):53–6.
39. Walker MP, Ayre GA, Cummings JL, Wesnes K, McKeith IG, O'Brien JT, et al. Quantifying fluctuation in dementia with Lewy bodies, Alzheimer's disease, and vascular dementia. Neurology. 2000;54(8):1616–25.

40. Walker MP, Ayre GA, Perry EK, Wesnes K, McKeith IG, Tovee M, et al. Quantification and characterization of fluctuating cognition in dementia with Lewy bodies and Alzheimer's disease. Dement Geriatr Cogn Disord. 2000;11(6):327–35.
41. Andersson M, Hansson O, Minthon L, Rosen I, Londos E. Electroencephalogram variability in dementia with Lewy bodies, Alzheimer's disease and controls. Dement Geriatr Cogn Disord. 2008;26(3):284–90.
42. Bonanni L, Thomas A, Tiraboschi P, Perfetti B, Varanese S, Onofrj M. EEG comparisons in early Alzheimer's disease, dementia with Lewy bodies and Parkinson's disease with dementia patients with a 2-year follow-up. Brain. 2008;131(Pt 3):690–705.
43. Onofrj M, Thomas A, Iacono D, Luciano AL, Di Iorio A. The effects of a cholinesterase inhibitor are prominent in patients with fluctuating cognition: a part 3 study of the main mechanism of cholinesterase inhibitors in dementia. Clin Neuropharmacol. 2003;26(5):239–51.
44. Jackson CE, Snyder PJ. Electroencephalography and event-related potentials as biomarkers of mild cognitive impairment and mild Alzheimer's disease. Alzheimers Dement. 2008;4(1 Suppl 1):S137–43.
45. Lee H, Brekelmans GJ, Roks G. The EEG as a diagnostic tool in distinguishing between dementia with Lewy bodies and Alzheimer's disease. Clin Neurophysiol. 2015;126(9):1735–9.
46. Kai T, Asai Y, Sakuma K, Koeda T, Nakashima K. Quantitative electroencephalogram analysis in dementia with Lewy bodies and Alzheimer's disease. J Neurol Sci. 2005;237(1–2):89–95.
47. Bonanni L, Perfetti B, Bifolchetti S, Taylor JP, Franciotti R, Parnetti L, et al. Quantitative electroencephalogram utility in predicting conversion of mild cognitive impairment to dementia with Lewy bodies. Neurobiol Aging. 2015;36(1):434–45.
48. Bonanni L, Franciotti R, Nobili F, Kramberger MG, Taylor JP, Garcia-Ptacek S, et al. EEG markers of dementia with Lewy bodies: a Multicenter Cohort Study. J Alzheimers Dis. 2016;54(4):1649–57.
49. Dauwan M, van Dellen E, van Boxtel L, van Straaten EC, de Waal H, Lemstra AW, et al. EEG-directed connectivity from posterior brain regions is decreased in dementia with Lewy bodies: a comparison with Alzheimer's disease and controls. Neurobiol Aging. 2016;41:122–9.
50. van Dellen E, de Waal H, van der Flier WM, Lemstra AW, Slooter AJ, Smits LL, et al. Loss of EEG network efficiency is related to cognitive impairment in dementia with Lewy bodies. Mov Disord. 2015;30(13):1785–93.
51. Fischl B, Dale AM. Measuring the thickness of the human cerebral cortex from magnetic resonance images. Proc Natl Acad Sci U S A. 2000;97(20):11050–5.
52. Lebedev AV, Westman E, Beyer MK, Kramberger MG, Aguilar C, Pirtosek Z, et al. Multivariate classification of patients with Alzheimer's and dementia with Lewy bodies using high-dimensional cortical thickness measurements: an MRI surface-based morphometric study. J Neurol. 2013;260(4):1104–15.
53. Delli Pizzi S, Franciotti R, Tartaro A, Caulo M, Thomas A, Onofrj M, et al. Structural alteration of the dorsal visual network in DLB patients with visual hallucinations: a cortical thickness MRI study. PLoS One. 2014;9(1):e86624.
54. Colloby SJ, Cromarty RA, Peraza LR, Johnsen K, Johannesson G, Bonanni L, et al. Multimodal EEG-MRI in the differential diagnosis of Alzheimer's disease and dementia with Lewy bodies. J Psychiatr Res. 2016;78:48–55.
55. Blanc F, Colloby SJ, Cretin B, de Sousa PL, Demuynck C, O'Brien JT, et al. Grey matter atrophy in prodromal stage of dementia with Lewy bodies and Alzheimer's disease. Alzheimers Res Ther. 2016;8:31.
56. Mak E, Su L, Williams GB, Watson R, Firbank M, Blamire A, et al. Differential atrophy of hippocampal subfields: a comparative study of dementia with Lewy bodies and Alzheimer disease. Am J Geriatr Psychiatry. 2016;24(2):136–43.
57. Delli Pizzi S, Franciotti R, Bubbico G, Thomas A, Onofrj M, Bonanni L. Atrophy of hippocampal subfields and adjacent extrahippocampal structures in dementia with Lewy bodies and Alzheimer's disease. Neurobiol Aging. 2016;40:103–9.
58. Aminoff EM, Kveraga K, Bar M. The role of the parahippocampal cortex in cognition. Trends Cogn Sci. 2013;17(8):379–90.

59. Canu E, McLaren DG, Fitzgerald ME, Bendlin BB, Zoccatelli G, Alessandrini F, et al. Microstructural diffusion changes are independent of macrostructural volume loss in moderate to severe Alzheimer's disease. J Alzheimers Dis. 2010;19(3):963–76.
60. Mak E, Su L, Williams GB, O'Brien JT. Neuroimaging characteristics of dementia with Lewy bodies. Alzheimers Res Ther. 2014;6(2):18.
61. Delli Pizzi S, Maruotti V, Taylor JP, Franciotti R, Caulo M, Tartaro A, et al. Relevance of subcortical visual pathways disruption to visual symptoms in dementia with Lewy bodies. Cortex. 2014;59:12–21.
62. Delli Pizzi S, Franciotti R, Taylor JP, Thomas A, Tartaro A, Onofrj M, et al. Thalamic involvement in fluctuating cognition in dementia with Lewy bodies: magnetic resonance evidences. Cereb Cortex. 2015;25(10):3682–9.
63. Sherman SM. The thalamus is more than just a relay. Curr Opin Neurobiol. 2007;17(4):417–22.
64. Akeju O, Loggia ML, Catana C, Pavone KJ, Vazquez R, Rhee J, et al. Disruption of thalamic functional connectivity is a neural correlate of dexmedetomidine-induced unconsciousness. elife. 2014;3:e04499.
65. Klein JC, Eggers C, Kalbe E, Weisenbach S, Hohmann C, Vollmar S, et al. Neurotransmitter changes in dementia with Lewy bodies and Parkinson disease dementia in vivo. Neurology. 2010;74(11):885–92.
66. Perry E, Walker M, Grace J, Perry R. Acetylcholine in mind: a neurotransmitter correlate of consciousness? Trends Neurosci. 1999;22(6):273–80.
67. Wesnes KA, McKeith I, Edgar C, Emre M, Lane R. Benefits of rivastigmine on attention in dementia associated with Parkinson disease. Neurology. 2005;65(10):1654–6.
68. Saalmann YB, Pinsk MA, Wang L, Li X, Kastner S. The pulvinar regulates information transmission between cortical areas based on attention demands. Science. 2012;337(6095):753–6.
69. Erskine D, Taylor JP, Firbank MJ, Patterson L, Onofrj M, O'Brien JT, et al. Changes to the lateral geniculate nucleus in Alzheimer's disease but not dementia with Lewy bodies. Neuropathol Appl Neurobiol. 2016;42(4):366–76.
70. Imamura T, Ishii K, Sasaki M, Kitagaki H, Yamaji S, Hirono N, et al. Regional cerebral glucose metabolism in dementia with Lewy bodies and Alzheimer's disease: a comparative study using positron emission tomography. Neurosci Lett. 1997;235(1–2):49–52.
71. Lim SM, Katsifis A, Villemagne VL, Best R, Jones G, Saling M, et al. The 18F-FDG PET cingulate island sign and comparison to 123I-beta-CIT SPECT for diagnosis of dementia with Lewy bodies. J Nucl Med. 2009;50(10):1638–45.
72. Graff-Radford J, Murray ME, Lowe VJ, Boeve BF, Ferman TJ, Przybelski SA, et al. Dementia with Lewy bodies: basis of cingulate island sign. Neurology. 2014;83(9):801–9.
73. Peraza LR, Colloby SJ, Deboys L, O'Brien JT, Kaiser M, Taylor JP. Regional functional synchronizations in dementia with Lewy bodies and Alzheimer's disease. Int Psychogeriatr. 2016;28(7):1143–51.
74. Galvin JE, Price JL, Yan Z, Morris JC, Sheline YI. Resting bold fMRI differentiates dementia with Lewy bodies vs Alzheimer disease. Neurology. 2011;76(21):1797–803.
75. Kenny ER, Blamire AM, Firbank MJ, O'Brien JT. Functional connectivity in cortical regions in dementia with Lewy bodies and Alzheimer's disease. Brain. 2012;135(Pt 2):569–81.
76. Franciotti R, Falasca NW, Bonanni L, Anzellotti F, Maruotti V, Comani S, et al. Default network is not hypoactive in dementia with fluctuating cognition: an Alzheimer disease/dementia with Lewy bodies comparison. Neurobiol Aging. 2013;34(4):1148–58.
77. Peraza LR, Taylor JP, Kaiser M. Divergent brain functional network alterations in dementia with Lewy bodies and Alzheimer's disease. Neurobiol Aging. 2015;36(9):2458–67.
78. Sauer J, ffytche DH, Ballard C, Brown RG, Howard R. Differences between Alzheimer's disease and dementia with Lewy bodies: an fMRI study of task-related brain activity. Brain. 2006;129(Pt 7):1780–8.
79. Taylor JP, Firbank MJ, He J, Barnett N, Pearce S, Livingstone A, et al. Visual cortex in dementia with Lewy bodies: magnetic resonance imaging study. Br J Psychiatry. 2012;200(6):491–8.
80. Wang HF, JT Y, Tang SW, Jiang T, Tan CC, Meng XF, et al. Efficacy and safety of cholinesterase inhibitors and memantine in cognitive impairment in Parkinson's disease, Parkinson's

disease dementia, and dementia with Lewy bodies: systematic review with meta-analysis and trial sequential analysis. J Neurol Neurosurg Psychiatry. 2015;86(2):135–43.
81. Stinton C, McKeith I, Taylor JP, Lafortune L, Mioshi E, Mak E, et al. Pharmacological management of lewy body dementia: a systematic review and meta-analysis. Am J Psychiatry. 2015;172(8):731–42.
82. McKeith I, Del Ser T, Spano P, Emre M, Wesnes K, Anand R, et al. Efficacy of rivastigmine in dementia with Lewy bodies: a randomised, double-blind, placebo-controlled international study. Lancet. 2000;356(9247):2031–6.
83. McKeith I, Fairbairn A, Perry R, Thompson P, Perry E. Neuroleptic sensitivity in patients with senile dementia of Lewy body type. BMJ. 1992;305(6855):673–8.
84. Maust DT, Kim HM, Seyfried LS, Chiang C, Kavanagh J, Schneider LS, Kales HC. Antipsychotics, other psychotropics, and the risk of death in patients with dementia: number needed to harm. JAMA Psychiatry. 2015;72(5):438–45.
85. Cummings J, Isaacson S, Mills R, Williams H, Chi-Burris K, Corbett A, et al. Pimavanserin for patients with Parkinson's disease psychosis: a randomised, placebo-controlled phase 3 trial. Lancet. 2014;383(9916):533–40.
86. Goldman JG, Goetz CG, Brandabur M, Sanfilippo M, Stebbins GT. Effects of dopaminergic medications on psychosis and motor function in dementia with Lewy bodies. Mov Disord. 2008;23(15):2248–50.
87. Molloy S, McKeith IG, O'Brien JT, Burn DJ. The role of levodopa in the management of dementia with Lewy bodies. J Neurol Neurosurg Psychiatry. 2005;76(9):1200–3.
88. Aarsland D, Perry R, Larsen JP, McKeith IG, O'Brien JT, Perry EK, et al. Neuroleptic sensitivity in Parkinson's disease and parkinsonian dementias. J Clin Psychiatry. 2005;66(5):633–7.
89. Trollor JN, Chen X, Sachdev PS. Neuroleptic malignant syndrome associated with atypical antipsychotic drugs. CNS Drugs. 2009;23(6):477–92.
90. Ikebe S, Harada T, Hashimoto T, Kanazawa I, Kuno S, Mizuno Y, et al. Prevention and treatment of malignant syndrome in Parkinson's disease: a consensus statement of the malignant syndrome research group. Parkinsonism Relat Disord. 2003;9(Suppl 1):S47–9.
91. Onofrj M, Thomas A. Acute akinesia in Parkinson disease. Neurology. 2005;64(7):1162–9.
92. Thomas A, Iacono D, Luciano AL, Armellino K, Onofrj M. Acute akinesia or akinetic crisis in Parkinson's disease. Neurol Sci. 2003;24(3):219–20.
93. Blanc F, Colloby SJ, Philippi N, de Petigny X, Jung B, Demuynck C, et al. Cortical thickness in dementia with Lewy bodies and Alzheimer's disease: a comparison of prodromal and dementia stages. PLoS One. 2015;10(6):e0127396.
94. Zhong J, Pan P, Dai Z, Shi H. Voxelwise meta-analysis of gray matter abnormalities in dementia with Lewy bodies. Eur J Radiol. 2014;83(10):1870–4.
95. Koubeissi MZ, Bartolomei F, Beltagy A, Picard F. Electrical stimulation of a small brain area reversibly disrupts consciousness. Epilepsy Behav. 2014;37:32–5.
96. Christopher L, Koshimori Y, Lang AE, Criaud M, Strafella AP. Uncovering the role of the insula in non-motor symptoms of Parkinson's disease. Brain. 2014;137(Pt 8):2143–54.
97. Miyasaki JM, Shannon K, Voon V, Ravina B, Kleiner-Fisman G, Anderson K, et al. Practice parameter: evaluation and treatment of depression, psychosis, and dementia in Parkinson disease (an evidence-based review): report of the Quality Standards Subcommittee of the American Academy of Neurology. Neurology. 2006;66(7):996–1002.
98. Kapur S, Seeman P. Does fast dissociation from the dopamine d(2) receptor explain the action of atypical antipsychotics?: a new hypothesis. Am J Psychiatry. 2001;158(3):360–9.
99. Kaasinen V, Joutsa J, Noponen T, Paivarinta M. Akinetic crisis in Parkinson's disease is associated with a severe loss of striatal dopamine transporter function: a report of two cases. Case Rep Neurol. 2014;6(3):275–80.
100. Martino G, Capasso M, Nasuti M, Bonanni L, Onofrj M, Thomas A. Dopamine transporter single-photon emission computerized tomography supports diagnosis of akinetic crisis of parkinsonism and of neuroleptic malignant syndrome. Medicine (Baltimore). 2015;94(13):e649.

101. Tzoulis C, Tran GT, Schwarzlmuller T, Specht K, Haugarvoll K, Balafkan N, et al. Severe nigrostriatal degeneration without clinical parkinsonism in patients with polymerase gamma mutations. Brain. 2013;136(Pt 8):2393–404.
102. Bonanni L, Onofrj M, Valente EM, Manzoli L, De Angelis MV, Capasso M, et al. Recurrent and fatal akinetic crisis in genetic-mitochondrial parkinsonisms. Eur J Neurol. 2014;21(9):1242–6.
103. Holmes C, Cunningham C, Zotova E, Woolford J, Dean C, Kerr S, et al. Systemic inflammation and disease progression in Alzheimer disease. Neurology. 2009;73(10):768–74.

Rare Dementias

Camilla Ferrari, Benedetta Nacmias, and Sandro Sorbi

Abstract

Dementia is becoming a worldwide phenomenon. Alzheimer's disease represents the first cause of cognitive impairment followed by vascular dementia and fronto-temporal dementia. However, in addition to these well-studied dementia causes there is a wide number of conditions that can cause dementia as infections, toxic-metabolic conditions, inflammatory-autoimmune disorders, or metabolic inborn errors. These uncommon dementia causes, due to the heterogeneous clinical presentation, lack of diagnostic criteria, and rare frequency are often misdiagnosed. Their prevalence has been only partially estimated among young patients (age at onset <65 years), and their management is based only on some expert suggestions. However, a correct diagnosis is of a great importance, since some of them are treatable and reversible dementias.

This chapter presents a comprehensive summary of etiologies, clinical presentation, typical features, diagnostic strategies, and treatments of known uncommon dementias.

C. Ferrari (✉)
IRCCS Don Gnocchi, Florence, Italy
e-mail: camilla.ferrari@unifi.it

B. Nacmias
Department of Neuroscience, Psychology, Drug Research and Child Health (NEUROFARBA), University of Florence, Florence, Italy

S. Sorbi
IRCCS Don Gnocchi, Florence, Italy

Department of Neuroscience, Psychology, Drug Research and Child Health (NEUROFARBA), University of Florence, Florence, Italy

© Springer International Publishing AG 2018
D. Galimberti, E. Scarpini (eds.), *Neurodegenerative Diseases*,
https://doi.org/10.1007/978-3-319-72938-1_15

Keywords

Uncommon dementia · Young-onset dementia · Neurodegenerative disease · Reversible dementia

Introduction

Uncommon dementias indicate a wide heterogeneous group of rare disorders causing cognitive impairment and generally characterized by an early age at onset. Thus, uncommon dementias greatly overlap the concept of young-onset dementia. Conventionally, young-onset dementia includes conditions that afflict patients younger than 65 years of age [1], i.e., early-onset forms of common neurodegenerative dementia as familial Alzheimer's disease cases, dementia associated with other neurological disorders (Huntington's disease, myotonic dystrophies, autosomal dominant cerebellar ataxia, or hereditary spastic paraparesis), or late-onset forms of childhood conditions as mitochondrial disorders, lysosomal storage disorders, and leukodystrophies. Potentially reversible etiologies including inflammatory disorders, infections, toxic/metabolic abnormalities are also part of rare dementia causes. It is to note that information on the frequency of uncommon dementias among elderly are not available, while few epidemiological data on young-onset dementia come from restricted geographical setting. Harvey and colleagues estimated, on a population-based study of two London boroughs, a dementia prevalence of 54 per 100,000 people aged 30–65 years and 98 per 100,000 people aged 45–65 years [2].

Alzheimer's disease represented the most common single diagnosis (34%), and the prevalence of metabolic, infective, and inflammatory/autoimmune diseases was generally estimated in a cumulative percentage of 19 [2]. Among others [3–5], only one study [6], conducted on a population with age ranged between 17 and 45 years, specifically evaluated the prevalence of all uncommon causes of early-onset dementia, producing the following results: neurodegenerative etiology 31.1%, autoimmune and inflammatory 21.3%, metabolic disorders 10.6%, other 7.7%, vascular 6%, and infective 4.7% [6]. The age-stratified analysis showed decreasing frequency of metabolic etiology with aging and an opposite behavior of neurodegenerative etiology.

The diagnosis of uncommon dementias is a challenge, due to the wide number of pathologies with heterogeneous clinical presentation and lack of diagnostic criteria. Suggestions for diagnostic procedure, management, and treatment are generally based on small not controlled study and on expert opinion. To guide clinicians toward differential diagnosis and to avoid misdiagnosis, uncommon dementias have been differently categorized following clinical, pathological, or etiological criteria [1, 6–8]. Rossor and colleagues proposed a flow chart for diagnostic procedure [1]. This chapter presents a comprehensive list of uncommon dementias, grouped in diagnostic categories (Table 15.1), and suggests a two-step diagnostic procedure for the differential diagnosis of rare dementias. A specific protocol for subacute dementia is also provided.

Table 15.1 Diagnostic categories

Neurodegenerative dementias	Lysosomal storage disorders	Infective dementia
Familial Alzheimer's disease Fronto-temporal dementia Lewy body dementia Corticobasal degeneration Progressive supranuclear palsy	Fabry disease Gaucher's disease Niemann-pick type C disease Kuf's disorder (neuronal ceroid lipofuscinosis) Tay-Sachs disease	HIV-related dementia Neurosyphilis Whipple's disease Lyme disease Tuberculosis meningitis
Dementia plus Huntington disease Myotonic dystrophies Autosomal dominant cerebellar ataxia Hereditary spastic paraparesis Fragile X-associated tremor/ataxia syndrome	**Basal ganglia pathologies** Neuroacanthocytoses Neurodegeneration with iron accumulation Fahr's disease Wilson disease **Microgliopathies** Nasu-Hakola disease (polycystic lipomembranous osteodysplasia with sclerosing leukoencephalopathy)	**Inflammatory-autoimmune disorders** Limbic encephalitis Hashimoto encephalopathy Neurosarcoidosis NeuroLES Bechet
Leukoencephalopathies **Leuko (adult-onset)** Adrenoleukodystrophy Krabbe disease Metachromatic leukodystrophy Adult-onset leukoencephalopathy with axonal spheroid and pigmented glia Ovario-leukodystrophy Cerebrotendinous xanthomatosis Pelizaeus-Merzbacher disease Alexander disease Adult polyglucosan body disease Vanishing white matter disease **Familial form of vascular dementia** CADASIL/CARASIL COL4A1 RVCL **Mithocondrial** MELAS MERRF Kearns-Sayre syndrome	**Prion disease** Creutzfeltd-Jakob disease (CJD) sporadic and variant Hereditary prion disorders familial CJD Gestman-Strausler-Schenker disease Familial fatal insomnia	**Toxic metabolic** Alcohol-related dementia B12 deficiency Heavy metal poisoning

CADASIL cerebral autosomal dominant arteriopathy subcortical infarcts and leukoencephalopathy; *CARASIL* cerebral autosomal recessive arteriopathy with subcortical infarcts and leukoencephalopathy; *RVCL* Retinal vasculopathy with cerebral leukodystrophy; *COL4A1* collagen-type(IV)-mutations including pontine autosomal dominant microangiopathy and leukoencephalopathy; *MELAS* mitochondrial encephalopathy, lactic acidosis, and stroke; *MERFF* myoclonic epilepsy and ragged red fibers

Diagnostic Procedure

Uncommon causes of dementia should be suspected in the presence of:

- Young-onset dementia
- Predominance of psychiatric symptoms
- Association with other neurological sign
- Systemic involvement
- Subacute onset and rapid progression
- Positive family history for dementia or other neurological disturbance

All patients should undergo a complete clinical assessment with neurological, neuropsychological, and general examination, basic blood tests, and neuroimaging, preferably magnetic resonance imaging (MRI) [1, 8]. These evaluations represent the first step of the diagnostic procedure. A second diagnostic step, different for each diagnostic category, includes more complex blood or urinary examination, cerebrospinal fluid analysis (CSF), electroencephalography (EEG), electromyography (EMG), fluorodeoxyglucose positron emission tomography (PET-FDG), tissue biopsy, and genetic test.

First diagnostic step: Clinical history includes family history, specific dementia risk as alcohol or heavy metal exposure, description of symptoms by temporal profile of onset, progression and degree of impairment. The objective of the neurological examination is to define the pattern of cognitive and behavioral deficits and to investigate the presence of specific neurological signs (pyramidal, extrapyramidal, cerebellar). General examination is also important in case of systemic illness and can reveal stigmata of some disorders, as Achilles tendon xanthomata in cerebrotendinous xanthomatosis. Basic blood tests should include serum electrolytes, complete blood counts, liver and thyroid function tests, vitamin B12 and folate, cholesterol and triglycerides, glucose, urea and serum creatinine; syphilis and HIV serology in order to detect toxic/metabolic encephalopathy, or infective dementia [1, 7, 8].

Once defined the type of cognitive impairment and the association with specific neurological signs, MRI is of a great utility. Thanks to MRI, it is possible to identify three different types of dementia: dementia due to gray matter degeneration, leukoencephalopathies, and dementia based on basal ganglia disorders.

Gray matter degenerations include two categories: neurodegenerative dementia and dementia plus, while leukoencephalopathies include leukodystrophies, some lysosomal storage disorders, mitochondrial disorders, and familial form of vascular dementia (Fig. 15.1).

Second diagnostic step: Clinical examination and the MRI results lead to the identification of specific subgroups. Thus, different investigations are required as second step in each category (Fig. 15.2).

Subacute Dementia

A specific protocol should be applied in case of dementia with abrupt onset.

15 Rare Dementias

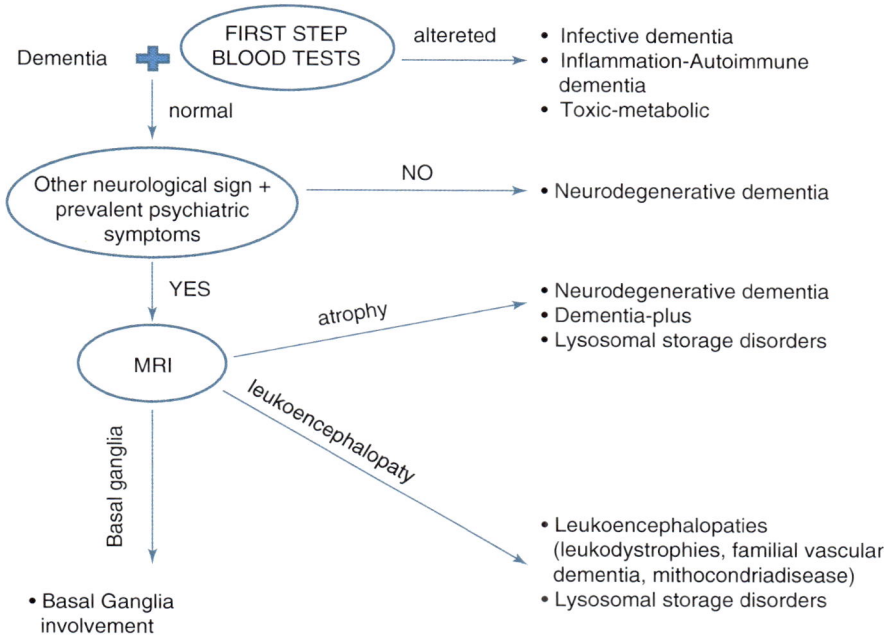

Fig. 15.1 The diagnostic procedure of dementias: first step

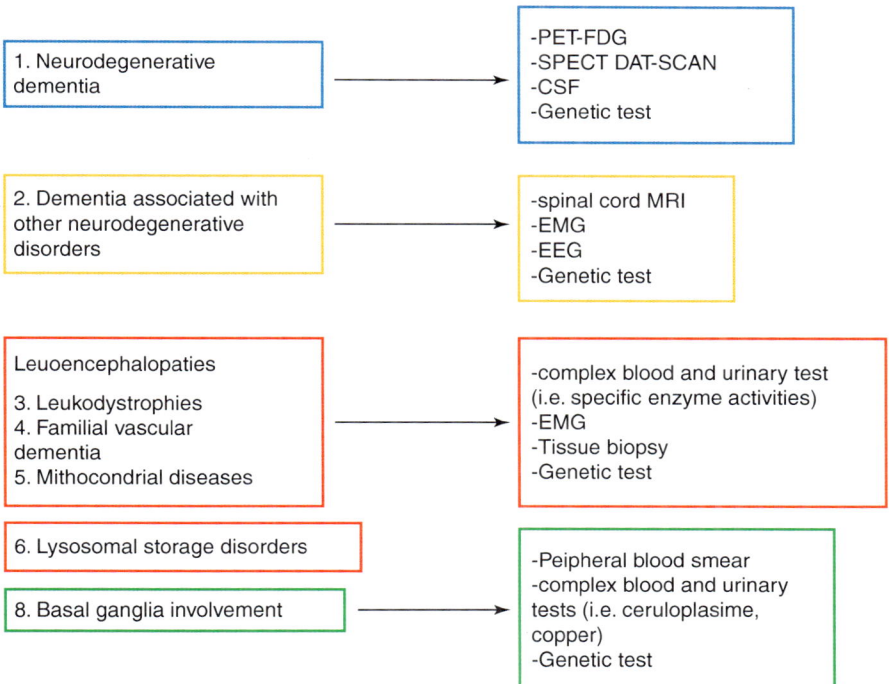

Fig. 15.2 The diagnostic procedure of dementias: second step

Each subject should undergo, in addition to basic blood tests and MRI, an EEG and the CSF examination. Usually, causes of subacute dementia present specific treatment or should be recognized in order to avoid transmission as in case of prion disease.

Table 15.2 summarized clinical features, diagnostic procedure, and treatment of subacute causes of dementias.

Neurodegenerative Dementias

Although rare, represent a group of very well-studied disorders with specific diagnostic criteria (9). Dementia is the first and predominant symptom. For their management, see dedicated chapter of this book [9].

Dementia Plus

Dementia is always associated with other neurological signs, as pyramidal, cerebellar, or muscular signs, and often occurs later in the disease course. Generally, MRI presents cortical atrophy. All the diseases are inherited and genetic test is necessary to confirm the diagnosis.

Huntington Disease [9–11]

Autosomal dominant disease caused by the expansion of CAG trinucleotide repeat sequence on chromosome 4, on a gene encoding for "huntingtin," a protein of unknown function. **Epidemiology:** prevalence in European is 0.5–8/1,000,000. **Clinic:** age at onset 30–50 years, rarely before 20 years (Westphal variant). Symptoms: chorea, psychiatric symptoms, and cognitive decline. Cognitive deficits are mostly in executive function and judgment capacity. Language and semantic memory are generally spared. Disease duration: 15–20 years. **Neuroimaging:** atrophy of caudate and putamen, and frontal lobes. **Diagnosis:** genetic test, CAG expansion (normal <27 repetition, 35–39 incomplete penetrance, >39 pathological).

Therapy. Motor symptoms: tetrabenazine, haloperidol, fluphenazine, risperidone, and olanzapine. **Psychiatric symptoms**: SSRI, clomipramine, atypical antipsychotics. **Cognitive decline:** rivastigmine is possibly the best option; however, data are not sufficient and based on few small open-label trials. Preliminary data on genetic therapy [10] have been reported.

Data from pilot clinical studies on intrastriatal transplantation of striatal neuroblasts from human fetus seems to show a delay in disease progression and transient clinical improvement [12, 13].

Myotonic Dystrophies

CTG trinucleotide repeats expansion disorders transmitted with autosomal dominant inheritance. **Clinic:** five main clinical forms: congenital, childhood-onset, juvenile, adult-onset, and late-onset/asymptomatic [14]. Adult-onset 20–40 years is

Table 15.2 Subacute dementias

Diseases	Clinical features	Blood tests	MRI	EEG	CSF	Therapy
Prion disorders	Dementia Myoclonus Ataxia	Normal	Basal ganglia DWI hyperintensity	Periodic sharp waves	Real-time quaking-induced conversion (PrPSc)	None
Hashimoto's encephalopathy	Dementia Psychosis Myoclonus Epilepsy Tremor Ataxia	Anti-TPO	Aspecific alteration	Slowliness	+protein +OB	Prednisone 50 mg/die or plasmapheresis
Limbic encephalitis	Memory deficits Behavioral changes Temporal epilepsy	< Na+ VES Neoplastic markers Onconeural antibodies Autoantibodies	Bilateral temporal hyperintensity in FLAIR	Temporal epileptic features	Normal or +OB +lymphocytes	Tumor-specific treatment, plasmapheresis
Wernicke's encephalopathy	Ataxia Ophthalmoplegia Disorientation Dementia confabulation	Malnutrition Low vitamin B level Renal, hepatic alteration	T2 hyperintensity thalami, hypothalamus, mammillari corpi, periacqueduttal, BEE damage	Slowliness	Normal	Vitamin replacement Thiamine 500 mg × 3 Magnesium

characterized by weakness, myotonia, cataracts, cardiac conduction defects, insulin resistance, respiratory failure, lower IQ scores, and frontal dysexecutive syndrome. **Neuroimaging:** cortical atrophy, especially in the frontal, temporal, and parietal lobes. **Diagnosis:** EMG testifies dystrophic and myotonic phenomena. Genetic testing [14]. **Therapy:** in 2014, the first phase 1/2a clinical trial on genetic therapy with antisense oligonucleotide [15] has been conducted.

Autosomal Dominant Cerebellar Ataxia

It encompasses a group of neurodegenerative disorders characterized by ataxia and different combination of pyramidal, extrapyramidal signs, and peripheral neuropathy. Dementia is a constant feature of dentatorubral-pallidoluysian atrophy (DRPLA) and SCA17, with behavioral disorders and frontal-type dementia, can also be present in SCA3, SCA8, and SCA13. In other form is more rarely descripted [16].

Hereditary Spastic Paraparesis (SPG)

It is a group of heterogeneous neurodegenerative inherited disorders with the main clinical features of slowly progressive spasticity and weakness of lower limb. Dementia has been reported only in some families with SPG4 form and is specifically associated with the deletion of exon 17 of the spastin gene. The degree of cognitive impairment is not correlated with the severity of spastic symptoms but seems to be related with aging [17].

Fragile X-Associated Tremor/Ataxia Syndrome (FXTAS)

It is a trinucleotide repeat disorder caused by the expansion of CGG on the fragile X mental retardation 1 gene (FMR1) on the X chromosome. Normal alleles present 5–44 repeats, while more than 200 CGG repeats determine the most common inherited form of intellectual disability and autism.

Premutation alleles 55–200 can be associated with FXTAS. Premutation expansions prevalence is 1 per 113–259 female and 1 for 260–800 males [18]. FXTAS affects nearly 40% of permutation males and 8% of permutation female. **Clinic:** age at onset is over 50 years of age. Clinical features are kinetic, intention or postural tremor, cerebellar gait and limb ataxia, parkinsonism and dementia. Patients may have autonomic dysfunction and peripheral neuropathy. Dementia in FXTAS presents memory loss associated with both frontal lobe features (disinhibition, poor executive functioning, perseveration, mood disturbance) and subcortical features (psychomotor slowing, bradyphrenia, attention and concentration difficulties). The onset of cognitive symptoms often follows the onset of movement disorders. **Neuroimaging:** diffuse cerebellar and cerebral atrophy. MRI T2 hyperintensity of middle cerebellar peduncles and subcortical regions. **Diagnosis:** genetic test. It is important to recognize the disorder among the family in order to identify premutant alleles [18]. **Therapy:** treatment of cognitive impairment is based on off-label application of dementia treatments. In early phases of memory impairment, the use of cholinesterase inhibitors can result in short-term improvement. Expert opinion reported some benefit from memantine use [19]. For therapeutic approaches in fragile X syndrome, see review [20].

Leukoencephalopathies

This category includes leukodystrophies and genetic leukoencephalopathies in which the hallmark is the progressive degeneration of CNS white matter. The diagnosis is based on genetic analyses; however, the list of implicated genes already counts more than 60 genes, and isolated cases of novel mutation are frequently described [21, 22]. In addition, despite the large number of genes and cellular pathways, there is an overlapping of clinical and radiological phenotype that made the diagnosis challenging, time consuming, costly, and disappointing. Many efforts have been made to improve diagnostic approach of these disorders [23, 24]; however, knowledge on adult-onset phenotypes is still incomplete. Genetic leukoencephalopathies also include familial form of vascular dementia, mitochondrial diseases, and some lysosomal storage disorders. We will focus on leukodystrophies and genetic leukoencephalopathies, which can cause an adult-onset dementia; nevertheless, the list is continuously increasing.

Leukodystrophies (Adult-Onset)

Represent childhood neurodegenerative disorders, involving myelin development, that can be divided into (1) dysmyelinating (abnormally formed myelin), (2) hypomyelinating (decreased myelin production), and (3) spongiform (cystic degeneration of myelin) [25].

Clinical onset is normally in infancy; however, adult-onset form has been described. Information on epidemiology and clinical data are based on single case report or specific clinical setting. Generally, age at onset is earlier than 45 years of age and is predominantly characterized by psychiatric symptoms. Some disorders present typical body features in the general examination than can help in the diagnosis [24].

Adrenoleukodystrophy

Hereditary X-linked disease caused by mutations of the gene encoding a peroxisomal protein necessary for the metabolization of the very long-chain fatty acids (VLCFA). **Epidemiology:** adrenoleukodystrophy occurs usually in childhood; only 1–3% of cases present an adult-onset. **Clinic:** age at onset of adult form is around 20–30 years of age. Cognitive disorders are characterized by psychotic symptoms, character changes, hyperorality, tendency to wander, stereotypical vocal expression, and by subcortical signs as bradyphrenia and concentration deficits. Impaired vision and hearing are characteristic. Paraparesis, sphincteric disturbance, and sexual dysfunction can be present [26]. **Neuroimaging:** T2 white matter hyperintensity especially in parieto-occipital regions. **Diagnosis:** high plasma level of VLCFA (C24-C30 chain length). Genetic test. **Therapy:** dietary treatment with the use of glycerol trioleate and glycerol trierucate that lower VCLFA in culture. Unfortunately, the use of these oils does not improve clinical course for symptomatic patients, however in asymptomatic cases, genetically detected, determines a less severe illness course [26]. Genetic and hematopoietic strategies are under evaluation [27].

Krabbe Disease (Globoid Cell Leukodystrophy)

Autosomal recessive lysosomal storage disease caused by a deficiency of the lysosomal enzyme galactocerebrosidase (GALC). The GALC deficiency impaired the degradation of galactosylceramide, a major myelin lipid, whose excess elicits formation of multinucleated macrophages, the globoid cells [28]. However, recently it has been demonstrated that GALC is also responsible for the degradation of galactosylsphingosine (psychosine) that is a highly cytotoxic glycolipid that preferentially kills oligodendrocytes in the central nervous system and Schwann cells in the peripheral nervous system [29]. **Epidemiology:** adult-onset is rare. A recent review of published cases reported 28 adult-onset cases [28]. **Clinic:** age at onset of adult form is between 25 and 72 years. Symptoms progression is slow and disease duration can be more than 10 years. Patients present pyramidal signs in 96% of cases, dysarthria (31%), cerebellar ataxia (27%), deep sensory signs (23%), tongue atrophy (15%), and optic neuropathy (12%). Cognitive decline is described in 12% of cases [28]. **Neuroimaging:** MRI T2 hyperintensity of optic radiations, posterior part of corpus callosum, and of cortico-spinal tracts.

Diagnosis: deficient galactocerebrosidase activity in leucocytes or fibroblasts. Psychosine concentration on dried blood spot [29]. **Therapy:** hematopoietic stem cell transplantation [28].

Metachromatic Leukodystrophy (MLD) [30]

Autosomal recessive lysosomal sphingolipid storage disorder. It is caused by a deficiency of the enzyme arylsulfatase A resulting in the accumulation of non-degradated sulfatide in oligodendrocytes, Schwann cells, and some neurons.

Sulfatide accumulation is the trigger to demyelination. **Epidemiology:** incidence in Europe is 1 per 100,000 live births. Even if more than 100 mutations have been described, among Caucasian only three are frequent (splice donor site mutation of the exon 2/intron 2; missense mutations causing Pro-426Leu substitution, missense mutation causing Ile-179Ser substitution). **Clinic:** there is genotype-phenotype correlation with disease severity based on the amount of residual enzyme activity. The adult form, with an onset beyond the age of 16 years, corresponds to 18–20% of MLD cases. The adult form shows two possible distinct phenotypes: one with a predominant cerebello-pyramidal presentation and the other with predominantly psychiatric features. The psychiatric presentation is often associated with a specific mutation in Caucasian (Ile170S). Neurological signs appear later with seizures, chorea or dystonia. Cognitive impairment is characterized by attentional disturbance, reduced speed of processing, and executive functions impairment. In some cases, patients present fronto-temporal-like dementia symptoms.

Neuroimaging: symmetric confluent T2 MRI high signal in the periventricular regions and corpus callosum. Within the abnormal white matter, low-density tigroid stripes are present. The tigroid stripes are typical of MLD but not specific. **Diagnosis:** arylsulfatase A enzyme activity reduced in blood leukocytes; increased sulfatide excretion in 24-h urine sample. **Therapy:** no available treatment. Hematopoietic stem cell transplantation is beneficial only in late juvenile or adult patients in the early stages of the disease. Recently genetic strategies have been developed [31].

Adult-Onset Leukoencephalopathy with Axonal Spheroids and Pigmented Glia (ALSP) [32, 33]

This term includes two previous distinct autosomal dominant disorders: hereditary diffuse leukoencephalopathy with axonal spheroids (HDLS) and pigmentary orthochromatic leukodystrophy (POLD), both caused by a colony stimulating factor 1 receptor (CSF1R) mutation first described in 2012 [32]. **Epidemiology:** first description was among a Swedish family in 1989, and now 122 cases from 90 families were identified worldwide [33]. **Clinic:** the mean age of onset is 43 years (range 18–78 years); core symptoms are behavioral changes, depression, frontal-type dementia, motor impairment as parkinsonism, spastic paresis, or ataxia. **Neuroimaging:** dilation of the lateral ventricles, white matter lesions predominantly frontal and slightly asymmetric, cortical atrophy, thinning and abnormal signaling of the corpus callosum.

Diagnosis: genetic testing. **Therapy**: in the literature is described only one case of stem cell transplantation demonstrating a halt of disease progression; the patient was treated based on wrong diagnosis [34].

Ovario-Leukodystrophies

Biallelic mutations in the mitochondrial alanyl-tRNA synthetase 2 gene (AARS2), previously associated with infancy cardiomiopathy, were first described in 2014 associated with leukodystrophy and ovarian failure in female [35]. Ten cases in the literature. Clinic: symptoms are a variable combination of dementia, upper motor neuron signs, ataxia, and ovarian failure in females. MRI revealed slightly asymmetric abnormal T2 hyperintense signal in the frontoparietal and periventricular white matter, with white matter rarefaction, involvement of the corpus callosum and pyramidal tracts and punctate areas of restricted diffusion. Clinical and radiological presentation overlaps often with ALSP [36].

Cerebrotendinous Xanthomatosis [25, 37]

Autosomal recessive disorder due to mutations on the gene for the mitochondrial enzyme sterol 27-hydroxylase responsible for the production of bile acids. The deficiency of that mitochondrial enzyme determines increased plasma level of cholestanol and deposition in different body's tissue, as Achilles tendon, nervous system, and lungs. **Epidemiology:** few cases described in the literature, probably underdiagnosed [38]. **Clinic:** peculiar triad of the disorder is tendon xanthomata (especially of Achilles tendons), juvenile ocular cataracts, and nervous system dysfunction. Nervous system symptoms consist in behavioral problems, dementia, psychiatric disorders, pyramidal weakness, cerebellar ataxia, and seizures. **Neuroimaging:** white matter hyperintensity above and especially below the tentorium and in some cases focal lesions. Diffuse brain and spine atrophy. **Diagnosis:** increased plasma level of cholestanol associated with low or normal level of cholesterol. **Therapy:** since the disease results from a defect of bile acid synthesis, treatment consists in assumption of chenodeoxycholic acid. Chenodeoxycholic acid can reverse encephalopathy in early stages. Long-term treatment with chenodeoxycholic acid (750 mg/day) suppresses abnormal bile acid synthesis [37].

Pelizaeus-Merzbacher Disease

X-linked hypomyelinating leukoencephalopathy due to mutations in the proteolipid gene (PLP). The PLP gene encodes for two proteolipid proteins, PLP that is the prominent protein of the CNS myelin and DM 20 that is involved in olygodendrocyte differentiation. **Epidemiology:** only rare single cases are reported. **Clinical:** classic presentation is before 5 years of age and consists in nystagmus, stridor, hypotonia, spasticity, ataxia, and choreoathetosis. Rare adult-onset cases present spastic paraparesis and sometimes tremor ataxia and dementia. **Neuroimaging:** central white matter is reduced in volume and present diffuse hyperintensity (cerebral hemispheres, cerebellum, and brainstem) with thin corpus callosum. Some preserved myelin islands are present living a "tigroid" appearance. **Diagnosis:** genetic test. **Therapy:** no treatment. Interestingly PLP gene mutations are also associated with spastic paraplegia type 2 (SPG2) [25].

Alexander Disease

Autosomal dominant disorder caused by mutations on glial fibrillary acidic protein gene (GFAP). GFAP gene mutations cause an overexpression of abnormal protein.

Epidemiology: few cases reported in the literature [39]. **Clinic:** age at onset is between 13 and 62 years. Disease duration can be few years or decades. Symptoms are dysarthria, dysphonia, dysphagia (bulbar and pseudobulbar signs), pyramidal signs, ataxia, and palatal myoclonus. Cognitive decline occurs late in the disease course. **Neuroimaging:** paucity of myelin especially in the frontal lobe, cystic degeneration, and cavitation of white matter are frequently present. Basal ganglia and thalami are also affected as well as medulla oblongata and cervical spinal cord.

Diagnosis: genetic test. Peculiar histological finding: rosenthal fibers that are eosinophilic inclusion localized in astrocytes cytoplasm. **Therapy:** no treatment.

Adult Polyglucosan Body Disease [40]

Autosomal recessive polyglucosane storage disorder caused by mutations of glycogen branching enzyme gene (GBE1). This disorder is often observed among Ashkenazi Jewish families. **Clinic:** onset is in fifth or sixth decades of life with myelopathy signs, peripheral axonal sensory-motor neuropathy, and neurogenic bladder. Weakness and sensory loss typically start in lower limbs. Around 2/3 of patients have cognitive impairment at onset, with cortical and subcortical deficits. **Neuroimaging:** periventricular, subcortical, and deep white matter changes that extend to cervical-medullary junction. Brain and spinal cord atrophy. **Diagnosis:** decreased GBE1 activity on skin fibroblasts or muscle. Intra-axonal polyglucosan bodies on sural nerve biopsy. Genetic test [40]. **Therapy:** no treatment.

Vanishing White Matter Disease [41]

Autosomal recessive disorder caused by mutations on one of the five genes encoding subunits of eukaryotic translation initiation factor 2B (EIF2B). Mutations in the EIF2B gene disrupted the normal stress-elicited compensatory mechanisms (synthesis of new protein and signals promoting either cellular survival or apoptosis). **Epidemiology:** case reports [41]. **Clinic:** based on a recent review (24 cases), the mean age at onset results 30 years (range 122–62). Characteristic of the disease are episodes of acute deterioration with hypotonia, irritability, vomiting, seizures, unconsciousness after minor head trauma, febrile infections, sun exposure, or fear. Extracranial involvement often includes ovarian dysgenesis. Cognitive decline is described in 62% of cases. **Neuroimaging:** MRI shows diffuse abnormalities in white matter that can have signal intensity near that of CSF, diffuse disappearance of cerebral white matter. Relative sparing of temporal lobes. **Diagnosis:** genetic test. **Therapy:** no treatment.

Familial Form of Vascular Dementia

There is a group of monogenic causes of strokes and a major cause of vascular dementia characterized by lacunar infarcts, white matter hyperintensities on T2/FLAIR MRI, asymptomatic cerebral microbleeds on gradient echo MRI, and symptomatic subcortical hemorrhage [42].

CADASIL (cerebral autosomal dominant arteriopathy with subcortical infarcts and leukoencephalopathy): it is an autosomal dominant disorder due to mutations in the NOTCH3 gene. Epidemiology: estimated population prevalence in the UK of about 2 per 100,000 [43]. Clinic: migraine, lacunar strokes, and cognitive impairment. Migraine, usually with aura, is the first symptom in 60–75%, with onset usually in the 20s or early 30s. Strokes are lacunar with a mean age of onset of 46 ± 9.7 years. Psychiatric disturbances particularly depression and apathy but also anxiety are common, and onset of depression may precede any other symptoms. Cognitive impairment, with early involvement of selective executive dysfunction and impaired processing speed. Diagnosis: genetic test. Therapy: no effect of donepezil on cognitive decline [44].

CARASIL (cerebral autosomal recessive arteriopathy with subcortical infarcts and leukoencephalopathy): [42] autosomal recessive disease due to mutation of HTRA1 gene. Clinic: recurrent subcortical ischemic strokes, cognitive impairment, motor impairment, and early-onset diffuse alopecia. Diagnosis: genetic testing.

Retinal vasculopathy with cerebral leukodystrophy (RVCL): TREX1 gene mutation with autosomal dominant inherited. Clinic is characterized by cognitive impairment, visual loss (capillary obliteration, avascular areas in retina, microaneurysms, telangiectatic capillaries), migraines, and subcortical ischemic or hemorrhagic strokes [42].

For the other rare monogenic vascular disorder, see review [42, 44].

Mitochondrial Diseases

It is a group of progressive neurodegenerative disorders associated with polygenetic, maternally inherited, mitochondrial DNA mutation. Prevalence studies report that mitochondrial disease affects 9 in 100,000 adults aged less than 65 years. The clinical presentation of mitochondrial disease is varied and can occur almost in any stage of life. Dementia is generally of a subcortical type, and neuroimaging shows characteristic involvement of white matter, especially in MELAS [45]. See a comprehensive review [46].

Lysosomal Storage Disorders

It is a group of metabolic inborn errors with usually clinical onset in infancy and childhood. Psychiatric disorders are predominant as well as clinical signs of diffuse nervous system involvement (pyramidal, extrapyramidal, cerebellar). Rare late-onset cases can present cognitive decline and less devastating neurological deficits. Some disorders are characteristically associated with leukoencephalopathy while other with gray matter alterations. Diagnosis is based on the demonstration of decreased activity of specific metabolic enzyme. Some of these disorders can be treated with enzyme replacement therapy (ERT) [8] Table 15.3.

Basal Ganglia Pathologies

This category includes disorders with specific involvement of the basal ganglia, degeneration, or material accumulation, which accounts for typical neurological features: chorea, dystonia or parkinsonism, psychiatric symptoms (obsessive-compulsive disorder, depression, schizophrenia-like psychosis), dementia (Table 15.4).

Neuroacanthocytosis (NA) [56]
NA can be divided in four groups: (1) classic NA involving basal ganglia degeneration (2). Neurodegeneration with only occasionally acanthocytosis (about 10%of cases). (3) Paroxysmal dyskinetic disorders. (4). Low lipoprotein blood level associated with ataxia, sensory neuropathy, and no movement disorder. Group 1 and 2 present dementia.

Neurodegeneration with Brain Iron Accumulation (NBIA) [57]
It comprises a group of inherited entities, with a cumulative prevalence of less than 1/1,000,000. So far ten causative genes have been described: PANK2, PLA2G6, C19orf12, COASY, FA2H, ATP13A2, WDR45, FTL, CP, DCAF17. However numerous patients remain without genetic diagnosis. Pantothenate kinase-associated neurodegeneration (PKAN) represents the 35–50% of NBIA cases; PLA2G6-associated neurodegeneration (PLAN) is the second most common, around 20% of cases.

Fahr Disease [58, 59]
A physiological calcification in brain can be seen in up to 20% of CT scan and can also be secondary to parathyroid disorders, phacomatosis, infections, inflammation, or hemorrhage. Idiopathic brain calcification presents so far four causative genes: SCL20A2, PDGFB, PDGFRB, and XPR1. SCL20A2 is the most common and account for about 60% of cases.

Table 15.3 Lysosomal storage disorders

Type of inheritance	Epidemiology	Clinical feature	MRI characteristics	Diagnosis	Enzyme replace therapy
Fabry's disease [47]					
X-linked	1 in 40,000/1 in 117,000 male live births. Heterozygous females can present some symptoms	Small fiber neuropathy, juvenile posterior stroke, depression,cognitive impairment (deficits in executive functioning, information processing speed and attention) Systemic involvement: Renal and cardiac disorders, angiokeratoma, hypohidrosis, corneal and lenticular opacities	Periventricular and deep white matter hyperintensity, and lacunars strokes	Alfa galactosidase A enzyme decreased activity	Yes. Efficacy [48].
Gaucher's disease [49]					
AR heterozygous subjects show idiopathic Parkinson disease (it is the single largest risk factor for PD) [50].	General incidence (type1–3) is approximately 1/40,000 to 1/60,000 births, rising to 1/800 in Ashkenazi Jews	Adult-onset (17–55 years) type 3: Akinetic-rigid syndrome poorly responsive to dopa therapy, supranuclear palsy, seizures, cerebellar ataxia. Cognitive and psychotic disturbances Type 2 juvenile presentation: Predominantly bulbar signs Type 1: No neurological signs hepatosplenomegaly, anemia, thrombocytopenia	Involvement of basal ganglia	Glucocerebrosidase enzyme, decreased activity	YES but no efficacy on neurological signs, does not cross blood-brain barrier

(continued)

Table 15.3 (continued)

Type of inheritance	Epidemiology	Clinical feature	MRI characteristics	Diagnosis	Enzyme replace therapy
Niemann-Pick type C disease [51]					
AR, NPC1, NPC2 genes mutation. Intracellular transport of endocytosed cholesterol	Reports from European countries describe an incidence of 1/120,000 live births.	Cerebellar ataxia, dysarthria, dysphagia, supranuclear gaze palsy, seizure, dystonia and progressive dementia. Cognitive impairment has been described as initial manifestation in 25–40% of adult-onset cases, 20–30 years of age. Often psychiatric symptoms, as psychosis, are present alone. Splenomegaly is present in more than half adult patients	Severe atrophy of white matter tracts, huge neuronal loss in corpus callosum. Atrophy of cerebellum (loss of Purkinje neurons), striatum, thalamus, and hippocampus	Accumulation of unesterified cholesterol in perinuclear vesicles (lysosomes) of skin fibroblasts. Genetic test	No ERT. Miglustat, inhibitor of glucosylceramide synthase
Kuf's disorder (neuronal ceroid lipofuscinosis) [52, 53]					
AR or AD	General incidence 1/100,000 live births	9 clinical forms Adult-onset: 12–60 years, at least 2 of dementia(fronto-temporal type), ataxia, pyramidal or extrapyramidal signs, seizure or myoclonus	Cerebral (hippocampus) and cerebellar atrophy, callosal thinning, altered signal in basal ganglia	Peripheral lymphocytes or skin biopsy One or more -granular osmiophilic deposits [GRODs], -fingerprint profiles, -curvilinear profiles, or rectilinear complexes in more than one cell type 14 genes In adult form:CLN6, CTSF, DNAJC5, GRN, and PPT1 [52]	Carbamazepine, phenytoin and lamotrigine may increase seizure activity and myoclonus and result in clinical deterioration. Actually 8 clinical trials. Anti-inflammatory, gene therapy, stem cells, ERT [53]

Type of inheritance	Epidemiology	Clinical feature	MRI characteristics	Diagnosis	Enzyme replace therapy
Tay-Sachs disease [54, 55]					
AR HEXA gene	Common in Ashkenazi Jews until carrier screening was introduced in the 1970s and disease incidence was reduced by up to 90%	Spinocerebellar ataxia, psychosis, progressive muscular atrophy, cognitive impairment. In infancy presents hypotonia and early vegetative state. Cherry-red macular spots	Cerebellar atrophy	Hexosaminidase A enzyme activity, decreased or absent	NO

Table 15.4 Basal ganglia pathology

Neuroacanthocytosis (NA) [56] (basal ganglia degeneration)
Chorea-acanthocytosis
AR, vacuolar protein sorting 13 homolog A (VPS13A) gene
30–50 years, limb and orobuccal chorea, psychiatric symptoms, dementia
Caudate atrophy, increased T2 signal in basal ganglia
Blood acanthocytosis, reduced chorein on red blood cells
McLeod syndrome
X-linked, XK gene
25–60 years, 80% schizophrenia-like psychosis, dementia elevated CK, myopathy, 60% cardiomyopathy
Striatal atrophy
Blood acanthocytosis, elevated CK
Neurodegeneration with brain iron accumulation (NBIA) [57] (iron accumulation)
Pantothenate kinase-associated neurodegeneration (PKAN) or Hallevorden Spatz syndrome
AR, pantothenate kinase 2 (PANK2) gene
20–30 years, duration up to 15 years.
Cognitive decline 100%, psychiatric symptoms >50%. Dystonia: Eyes, mouth, and neck. MRI: eye of the tiger sign," (bilateral areas of hyperintensity iron deposition within a region of hypointensity in the globus pallidus on T2)
PLA2G6
AR, PLA2G6
Early adulthood form: subacute dystonia-parkinson, pyramidal signs, cognitive and psychiatric features
Iron in basal ganglia and substantia nigra; cerebellar atrophy
Kufor Rakeb disease
AR, ATP13A2
Young-onset pallido-pyramidal syndrome with oculogyric dystonic spasms, supranuclear gaze palsy, hypometric saccades, facial-faucial-finger mini-myoclonus, autonomic dysfunction and psychiatric features (visual hallucinations) oculogyric dystonic spasms, facial-faucial-finger mini-myoclonus, supranucelar gaze palsy, autonomic dysfunction, psychiatric and cognitive features
Neuropherritinopathy
AD, FTL (ferritin light chain) gene
40 years, chorea, oro-facial action-specific dystonia, parkinsonism, cognitive deficits, behavioral abnormalities
Low ferritin concentration in males
Iron in caudate, globus pallidus, putamen, substantia nigra
Followed by cystic degeneration in putamen
Aceruloplasminemia
AR, CP (ceruloplasmine) gene
Retinal degeneration, diabetes mellitus, neurological symptoms, 40–50 years ataxia, involuntary movement, parkinsonism, cognitive decline
Undetectable serum ceruloplasmina, elevated ferritin
Decreased iron and microcytic anemia
Iron in liver and brain (caudate, putamen, pallidum, thalamus)

Table 15.4 (continued)

Fahr disease [58, 59] **(calcium accumulation)**
AD, SLC20A2 gene (sodium dependent phosphate transporter)
20–50 years, extrapyramidal symptoms, cerebellar dysfunction
Dementia neuropsychiatric symptoms.
Bilateral calcification of basal ganglia, thalamus, dentate nucleus
Secondary causes of altered calcium metabolism should be investigated (i.e., hypo or hyper-parathyrodism)
Wilson disease [60, 61] **(copper accumulation)**
AR, ATP7B gene (copper-transporting ATPase)
Late-onset presentation 20–30 years (European mutation H1069Q)
Neurological, hepatic, osteomuscular (copper accumulation)
Involuntary movements, dystonia, psychiatric symptoms
Antisocial behaviors, cognitive decline
Elevation of liver enzyme, joint pain and swelling
Kaiser-flash rings cornea
Low serum ceruloplasmin, increased 24-h urinary copper
Excretion
Therapy: Penicillamine (decoppering treatment).
Initial dose 125–250 mg per day, gradually increasing up to 1–3 g per day. Maintenance phase 250–750 mg per day

Wilson Disease [60–62]

It is a rare inherited disorder of copper metabolism that primarily cause hepatic, neurologic, and ophthalmic manifestation, however also includes musculoskeletal manifestations (synovitis, early osteoarthritis, osteoporosis, rickets, spontaneous fracture). Therapy: clinically improvement is seen after few months of decoppering therapy (penicillamine) even in patients with severe neurological disability. Decoppering ensures that presymptomatic individuals remain symptom free [62].

Others: Microgliopathies

Nasu-Hakola disease (NHD) or polycystic lipomembranous osteodysplasia with sclerosing leukoencephalopathy and the TREM2 spectrum [63, 64]. NHD is an autosomal recessive inherited disorder characterized by progressive dementia and repeated fractures during adolescence. It has been supposed that microglial dysfunction and activation is the primary cause of the disorders, which has been defined "microgliopathy." NHD phenotype is associated with mutation in TREM2 gene (triggering receptor expressed on myeloid cells 2), TYROBP (TYRO protein tyrosine kinase binding protein) or DAP12 (DNAX-activating protein 12), that are implicated in microglia activation for phagocytosis of apoptotic neuronal debris and inflammatory resolution.

The number of described mutation on TREM2 is increasing; the last was discovered few months ago [63]. It is interestingly to note that loss of function of TREM2 was initially associated with NHD phenotype, but later on other mutations were associated with different clinical phenotype: late-onset Alzheimer disease, behavioral variant of fronto-temporal dementia, semantic variant of primary progressive aphasia. For a comprehensive review see [63].

Clinic of NHD phenotype [64]. NHD was first reported at the same time in Finland and Japan in the 1970s. So far, there have been described approximately 200 cases worldwide. Disease onset ranges from 10 to 46 years. Average disease duration 16 years. Disease starts with foot and knee pains and repeated pathological fractures. Dementia is characterized by personality changes, memory disorder, apraxia, agnosia, acalculia, and disorientation. Some patients can have urinary incontinence, seizures, and pyramidal signs. Bone X-rays show cystic lesion in epiphyses of long bones. **Neuroimaging:** can be variable, characterized by diffuse atrophy of gray and white matter, and sometimes basal ganglia calcification can be associated.

Prion Disease

It is a group of disease characterized by a spongiform degeneration of the whole brain due to the deposition of misfolded prion protein, normal component of neurons cells [65]. The most common form of prion disorder is Creutzfeldt-Jakob disease that is sporadic and occurs with a frequency of 1 per million inhabitants. It is a devastating subacute dementia with ataxia and myoclonus.

Infective Dementia [8, 66–70]

Cognitive decline is associated with other systemic symptoms: mood disorder, frequent infectious, systemic illness in HIV [66]; meningitis and tabes dorsalis in neurosyphilis [67]; lymphocytic meningitis, arthralgia, peripheral or facial neuropathies in Lyme disease [68, 69]; arthralgia, gastrointestinal symptoms, ataxia in Whipple's disease [70].

Inflammatory-Autoimmune Disorders

Limbic Encephalitis [1, 9, 71, 72]
See Table 15.2.

Hashimoto Encephalopathy
See Table 15.2 [9, 73].

Toxic Metabolic [1, 7, 8, 74–76]

Alcohol-related dementia represents one of the most frequent causes of dementia in young population; Rossor reported a prevalence of 10% [1]. Dementia is associated with cerebellar signs. In case of thiamine deficiency dementia could be subacute with confusion and ophthalmoplegia (Wernicke–Korsakoff encephalopathy) (Table 15.2) [1, 7, 8, 74].

Conclusions

Uncommon causes of dementia comprise a wide number of very rare and often misdiagnosed disorders, including late-onset forms of childhood metabolic inborn errors, inflammatory disorders, infectious disease, and toxic/metabolic abnormalities.

Clinical data on the most of them are based only on single case report, and often the diagnosis is challenging due to the clinical heterogeneity among and within disorders.

Thus, a complete list of uncommon dementia is not possible.

The creation of diagnostic categories, even if arbitrary, can help clinicians toward differential diagnosis, and may reduce diagnostic errors, that is of great importance since disease modified therapies are available in some cases.

A creation of regional or national registry may be useful to real estimate the prevalence of uncommon dementias and to improve our clinical knowledge.

References

1. Rossor MN, Fox NC, Mummery CJ, et al. The diagnosis of young-onset dementia. Lancet Neurol. 2010;9:793–806.
2. Harvey RJ, Skelton-Robinson M, Rossor MN, et al. The prevalence and causes of dementia in people under the age of 65 years. J Neurol Neurosurg Psychiatry. 2003;74:1206–9.
3. Fujihara S, Brucki SM, Rocha MS, et al. Prevalence of presenile dementia in a tertiary outpatient clinic. Arq Neuropsiquiatr. 2004;62(3):592–5.
4. Panegyres PK, French K. Course and causes of suspected dementia in young adults: a longitudinal study. Am J Alzheimers Dis Other Demen. 2007;22(1):48–56.
5. Shinagawa S, Ikeda M, Toyota Y, et al. Frequency and clinical characteristics of early-onset dementia in consecutive patients in a memory clinic. Dement Geriatr Cogn Disord. 2007;24(1):42–7.
6. Kelley BJ, Boeve BF, Josephs KA. Young-onset dementia: demographic and etiologic characteristics of 235 patients. Arch Neurol. 2008;65(11):1502–8.
7. Sampson EL, Warren JD, Rossor MN. Youn onset dementia. Postgrad Med J. 2004;80:125–39.
8. Kuruppu DK, Matthews BR. Young-Onset Dementia. Semin Neurol. 2013;33:365–85.
9. Roos RA. Huntington's disease: a clinical review. Orphanet J Rare Dis. 2010;5:40–8.
10. Wyant KJ, Ridder AJ, Dayalu P. Huntington's disease—update on treatments. Curr Neurol Neurosci Rep. 2017;17:33.
11. Sorbi S, Hort J, Erkinjuntti T, et al. EFNS-ENS guidelines on the diagnosis and management of disorders associated with dementia. Eur J Neurol. 2012;19:1159–79.
12. Gallina P, Paganini M, Lombardini L, et al. Progress in restorative neurosurgery: human fetal striatal transplantation in Huntington's disease. Review. J Neurosurg Sci. 2011;55(4):371–81.
13. Paganini M, Biggeri A, Romoli AM et al. Fetal striatal grafting slows motor and cognitive decline of Huntington's disease. J Neurol Neurosurg Psychiatry. 2014.
14. De Antonio M, Dogan C, Hamroun D, et al. Unravelling the myotonic dystrophy type 1 clinical spectrum: a systematic registry-based study with implications for disease classification. Rev Neurol. 2016;172:572–80.

15. Gourdon G, Meola G. Myotonic dystrophies: state of the art of new therapeutic developments for the CNS. Front Cell Neurosci. 2017;11:101. https://doi.org/10.3389/fncel.2017.00101.
16. Rossia M, Perez-Lloretb S, Doldand L, et al. Autosomal dominant cerebellar ataxias: a systematic review of clinical features. Eur J Neurol. 2014;21:607–15.
17. Murphy S, Gorman G, Beetz C, et al. Dementia in SPG4 hereditary spastic paraplegia: clinical, genetic, and neuropathologic evidence. Neurology. 2009;73(5):378–84.
18. Bourgeois J, Coffey S, Rivera SM, et al. Fragile X premutation disorders – expanding the psychiatric perspective. J Clin Psychiatry. 2009;70(6):852–62.
19. Hagerman RJ, Hall DA, Coffey S, et al. Treatment of fragile X-associated tremor ataxia syndrome (FXTAS) and related neurological problems. Clin Interv Aging. 2008;3(2):251–62.
20. Mila M, Alvarez-Mora MI, Madrigal I, Rodriguez-Revenga L. Fragile X syndrome: an overview and update of the FMR1 gene. Clin Genet 2017 . doi: https://doi.org/10.1111/cge.13075.
21. Parikh S, Bernard G, Leventer RJ, et al. A clinical approach to the diagnosis of patients with leukodystrophies and genetic leukoencephalopathies. Mol Genet Metab. 2015;114:501–15.
22. Lynch DS, Zhang WJ, Lakshmanan R, et al. Analysis of mutations in AARS2 in a series of CSF1R-negative patients with adult-onset leukoencephalopathy with axonal spheroids and pigmented glia. JAMA Neurol. 2016;73(12):1433–9.
23. Vanderver A. Genetic leukoencephalopathies in adults. Continuum (Minneap Minn). 2016;22(3):916–42.
24. Nannucci S, Donnini I, Pantoni L. Inherited leukoencephalopathies with clinical onset in middle and old age. J Neurol Sci. 2014;347(1–2):1–13.
25. Köhler W. Leukodystrophies with late disease onset: an update. Curr Opin Neurol. 2010;23(3):234–41.
26. Luda E, Barisone MG. Adult-onset adrenoleukodystrophy: a clinical and neuropsychological study. Neurol Sci. 2001;22:21–5.
27. Turk BR, Moser AB, Fatemi A. Therapeutic strategies in adrenoleukodystrophy. Wien Med Wochenschr. 2017;167:219–22.
28. Debs R, Froissart R, Aubourg P, et al. Krabbe disease in adults: phenotypic and genotypic update from a series of 11 cases and a review. J Inherit Metab Dis. 2013;36:859–68.
29. Lim SM, Choi B-O, Oh S-i, et al. Patient fibroblasts-derived induced neurons demonstrate autonomous neuronal defects in adult-onset Krabbe disease. Oncotarget. 2016;7(46):74496–509.
30. Gieselmann V, Krägeloh-Mann I. Metachromatic leukodystrophy – an update. Neuropediatrics. 2010;41:1–6.
31. Penati R, Fumagalli F, Calbi V, et al. Gene therapy for lysosomal storage disorders: recent advances for metachromatic leukodystrophy and mucopolysaccharidosis I. J Inherit Metab Dis. https://doi.org/10.1007/s10545-017-0052-4.
32. Rademakers R, Baker M, Nicholson AM, et al. Mutations in the colony stimulating factor 1 receptor (CSF1R) gene cause hereditary diffuse leukoencephalopathy with spheroids. Nat Genet. 2012;44:200–5.
33. Konnoa K, Yoshidac T, Mizuno T, et al. Clinical and genetic characterization of adult-onset leukoencephalopathy with axonal spheroids and pigmented glia associated with CSF1R mutation. Eur J Neurol. 2017;24:37–45.
34. Eichler FS, Li J, Guo Y, et al. CSF1R mosaicism in a family with hereditary diffuse leukoencephalopathy with spheroids. Brain. 2016;139:1666–72.
35. Dallabona C, Diodato D, Kevelam SH, Haack TB, Wong LJ, Salomons GS, et al. Novel (ovario) leukodystrophy related to AARS2 mutations. Neurology. 2014;82:2063–71.
36. Lakshmanan R, Adams ME, Lynch DS, et al. Redefining the phenotype of ALSP and AARS2 mutation-related leukodystrophy. Neurol Genet. 2017;3(2):e135.
37. Fraidakis MJ. Psychiatric manifestations in cerebrotendinous xanthomatosis. Transl Psychiatry. 2013:1–11.
38. Appadurai V, DeBarber A, Chiang P-W, et al. Apparent underdiagnosis of Cerebrotendinous Xanthomatosis revealed by analysis of ~60,000 human exomes. Mol Genet Metab. 2015;116(4):298–304.

39. Pareyson D, Fancellu R, Mariotti C, et al. Adult-onset Alexander disease: a series of eleven unrelated cases with review of the literature. Brain. 2008;131:2321–31.
40. Mochel F, Schiffmann R, Steenweg ME, et al. Adult polyglucosan body disease: natural history and key magnetic resonance imaging findings. Ann Neurol. 2012;72(3):433–41.
41. Carra-Dalliere C, Horzinski L, Ayrignac X, et al. Natural history of adult-onset eIF2B-related disorders: a multicentric survey of 24 cases. Rev Neurol (Paris). 2011;167(11):802–11.
42. Tan RY, Markus HS. Monogenic causes of stroke: now and the future. J Neurol. 2015;262:2601–16.
43. Narayan SK, Gorman G, Kalaria RN, et al. The minimum prevalence of CADASIL in northeast England. Neurology. 2012;78(13):1025–7.
44. Søndergaarda CB, Nielsenb JE, Hansena CK, Christensena H. Hereditary cerebral small vessel disease and stroke. Clin Neurol Neurosurg. 2017;155:45–57.
45. Kaufman KR, Zuber N, Rueda-Lara MA, et al. MELAS with recurrent complex partial seizures, non convulsive status epilepticus, psychosis, and behavioral disturbances: case analysis with literature review. Epilepsy Behav. 2010;18(4):494–7.
46. McFarland R, Taylor RT, Turnbull DM. A neurological perspective on mitochondrial disease. Lancet Neurol. 2010;9:829–40.
47. Sigmundsdottir L, Tchan MC, Knopman AA, et al. Cognitive and psychological functioning in Fabry disease. Arch Clin Neuropsychol. 2014;29(7):642–50.
48. El Dib R, Gomaa H, Ortiz A, et al. Enzyme replacement therapy for AndersonFabry disease: a complementary overview of a Cochrane publication through a linear regression and a pooled analysis of proportions from cohort studies. PLoS One. 2017;12(3):e0173358. https://doi.org/10.1371/journal.pone.0173358.
49. Stirnemann J, Belmatoug N, Camou F, et al. A review of Gaucher disease pathophysiology, clinical presentation and treatments. Int J Mol Sci. 2017;18(2):pii: E441. https://doi.org/10.3390/ijms18020441.
50. O'Regan G, deSouza RM, Balestrino R, Schapira AH. Glucocerebrosidase mutations in Parkinson disease. J Parkinsons Dis. 2017. https://doi.org/10.3233/JPD-171092.
51. Vanier MT. Niemann-pick disease type C. Orphanet J Rare Dis. 2010;5:16.
52. Berkovic SF, Staropoli JF, Carpenter S, et al. Diagnosis and misdiagnosis of adult neuronal ceroid lipofuscinosis (Kufs disease). Neurology. 2016;87(6):579–84.
53. Geraets RD, Koh SY, Hastings ML, et al. Moving towards effective therapeutic strategies for neuronal Ceroid Lipofuscinosis. Orphanet J Rare Dis. 2016;11:40.
54. Neudorfer O, Pastores GM, Zeng BJ, et al. Late-onset Tay-Sachs disease: phenotypic characterization and genotypic correlations in 21 affected patients. Genet Med. 2005;7(2):119–23.
55. Patterson MC. Gangliosidoses. Handb Clin Neurol. 2013;113:1707–8. https://doi.org/10.1016/B978-0-444-59565-2.00039-3.
56. Walterfanga M, Evansa A, Chee Leong Looid J, et al. The neuropsychiatry of neuroacanthocytosis syndrome. Neurosci Biobehav Rev. 2011;35:1275–83.
57. Tello C, Darling A, Lupo V, et al. On the complexity of clinical and molecular bases of neurodegeneration with brain iron accumulation. Clin Genet. 2017. https://doi.org/10.1111/cge.13057.
58. Saleem S, Aslam HM, Anwar M. Fahr's syndrome: literature review of current evidence. Orphanet J Rare Dis. 2013;8:156.
59. Batla A, Tai XY, Schottlaender L, et al. Deconstructing Fahr's disease/syndrome of brain calcification in the era of new genes. Parkinsonism Relat Disord. 2017;37:1–10.
60. Taly AB, Meenakshi-Sundaram S, Sinha S, Swamy HS, Arunodaya GR. Wilson disease: description of 282 patients evaluated over 3 decades. Medicine (Baltimore). 2007;86(2):112–21.
61. Bhatnagar N, Lingaiah P, Lodhi JS, Karkhur Y. Pathological fracture of femoral neck leading to a diagnosis of Wilson's disease: a case report and review of literature. J Bone Metab. 2017 May;24(2):135–9. https://doi.org/10.11005/jbm.2017.24.2.135.
62. Aggarwal A, Bhatt M. Recovery from severe neurological Wilson's disease with copper chelation. Int Rev Neurobiol. 2013;110:313–48.

63. Dardiotis E, Siokas V, Pantazi E, et al. A novel mutation in TREM2 gene causing Nasu-Hakola disease and review of the literature. Neurobiol Aging. 2017;53:194.e13–22.
64. Kaneko M, Sano K, Nakayama J, Amano N. Nasu-Hakola disease: the first case reported by Nasu and review: the 50th anniversary of Japanese society of neuropathology. Neuropathology. 2010;30(5):463–70.
65. Kang HE, Mo Y, Abd Rahim R, Lee HM, Ryou C. Prion diagnosis: application of real-time quaking-induced conversion. Biomed Res Int 2017;2017:5413936.
66. Valcour V, Paul R, Chiao S, Wendelken LA, Miller B. Screening for cognitive impairment in human immunodeficiency virus. Clin Infect Dis. 2011;53(8):836–42.
67. Read PJ, Donovan B. Clinical aspects of adult syphilis. Intern Med J. 2012;42(6):614–20.
68. Halperin JJ. Nervous system lyme disease: diagnosis and treatment. Curr Treat Options Neurol. 2013;15(4):454–64.
69. Halperin JJ. Lyme disease: a multisystem infection that affects the nervous system. Continuum (Minneap Minn). 2012;18(6 Infectious Disease):1338–50.
70. Puéchal X. Whipple's disease. Ann Rheum Dis. 2013;72(6):797–803.
71. Graus F, Saiz A. Limbic encephalitis: an expanding concept. Neurology. 2008;70:500–1.
72. Serratrice G, Pellissier JF, Serratrice J, De Paula A. Limbic encephalitis – evolving concepts. Bull Acad Natl Med. 2008;192(8):1531–41.
73. Mocellin R, Walterfang M, Velakoulis D. Hashimoto's encephalopathy: epidemiology, pathogenesis and management. CNS Drugs. 2007;21(10):799–811.
74. Victor M. Alcoholic dementia. Can J Neurol Sci. 1994;21:88–99.
75. Chalouhi C, Faesch S, Anthoine-Milhomme MC, et al. Neurological consequences of vitamin B12 deficiency and its treatment. Pediatr Emerg Care. 2008;24(8):538–41.
76. Ibrahim D, Froberg B, Wolf A, Rusyniak DE. Heavy metal poisoning: clinical presentations and pathophysiology. Clin Lab Med. 2006;26(1):67–97.

Neurodevelopmental and Neurodegenerative Alterations in the Pathophysiology of Schizophrenia: Focus on Neuro-Immuno-Inflammation

Bernardo Dell'Osso, M. Carlotta Palazzo, and A. Carlo Altamura

Abstract

Schizophrenia is a severely impairing psychiatric disorder with a precocious onset, accounting for a conspicuous burden of disability worldwide. With respect to the etiology of schizophrenia, as for other major psychoses, the gene–environment interaction seems to be the most accredited model. In particular, alterations in the immune system have been repeatedly reported, involving both the unspecific and specific pathways of the immune system and suggesting that inflammatory/autoimmune processes might play an important role in the development of the disorder. Relating to this hypothesis, an imbalance in the inflammatory cytokines has been associated with schizophrenia and, more broadly, alterations in the inflammatory and immune systems seem to be already present in the early stages of the disorder. Such phenomenon could be responsible of specific neurodevelopmental abnormalities, which identify the roots of the disorder during brain development, with consequences that do not become clinically evident until adolescence or early adulthood. On the other hand, longitudinal cohort studies on schizophrenic patients demonstrated a progressive loss of grey matter, more evident in the frontal and temporal lobes of the brain. These two perspectives, the neurodevelopmental and neurodegenerative one, are thought to coexist in the complex and still unravelled etiology of schizophrenia, with studies supporting both of them. This chapter aims at providing the state of the art in the field.

B. Dell'Osso (✉) • M. Carlotta Palazzo • A. Carlo Altamura
Department of Psychiatry, University of Milan, Fondazione IRCCS Ca' Granda, Ospedale Maggiore Policlinico, Milan, Italy
e-mail: bernardo.dellosso@policlinico.mi.it

© Springer International Publishing AG 2018
D. Galimberti, E. Scarpini (eds.), *Neurodegenerative Diseases*,
https://doi.org/10.1007/978-3-319-72938-1_16

Keywords

Schizophrenia · Neurodegeneration · Neurodevelopment · Immunity · Inflammation

Background

Schizophrenia is a major psychosis accounting for conspicuous burden of disability worldwide, due to its early onset, chronic and severe evolution. Kraepelin and Bleuler already recognized that a significant part of schizophrenic subjects had previously shown behavioural abnormalities over childhood [1]. However, schizophrenia rarely develops in preadolescents [2]: its prevalence increases from age 14 with a peak in the late teens/beginning of 20s [3–5]. An equal gender ratio has been described for schizophrenic patients with onset during adolescence, while paediatric onset is more common in male patients [5, 6].

Genetic studies conducted during the 1990s reported differences in neurological development in high-risk children [7–9]. Indeed, neurodevelopmental abnormalities, occurring throughout childhood, have been reported in up to 50% of high-risk children, born from schizophrenic mothers [9], comprising hypoactivity, hypotonia, soft neurological signs—poor motor coordination, in particular—and deficits in attention and information processing in late childhood.

Taken as a whole, converging evidence supports the hypothesis that at least part of the genetic vulnerability to schizophrenia involves abnormal neurodevelopment [1].

From a clinical perspective, late childhood and adolescence onset is prognostic of a worse functioning in real-life setting: these patients are usually more severe from a clinical point of view, treatment resistant (when treated pharmacologically) and more keen to develop side effects. A major concern for these patients is the presence of a relevant cognitive impairment that makes some of the available psychotherapeutic and rehabilitative approaches poorly effective. Family support and easy access to treatment are crucial but not always possible. Furthermore, the actual duration of untreated psychosis in these patients is likely to be longer.

It has been pointed out that many environmental risk factors seem to operate before, around or immediately after birth, including pregnancy and birth complications, perinatal and early-childhood brain damages, altered foetal development, season of birth and drug abuse, including cannabinoid intake [1]. Therefore, up to one-third of the variance in liability to schizophrenia can be attributed to non-genetic factors.

Despite consistent evidence supporting the presence of neurodevelopmental alterations in schizophrenia, many authors have put emphasis on the neurodegenerative processes occurring over the course of the illness [10]. Currently, however, the traditional neurodegenerative hypothesis has been largely questioned and, at least to some extent, revisited [11].

As a matter of fact, the debate, as to whether there is an abnormal developmental or degenerative process in the natural history of schizophrenia, likely stems from a spurious dichotomy and depends on the stage at which its observation begins. However, the effect of long-term, albeit necessary, pharmacological treatment in relation to brain neurotrophic factor is a major concern nowadays, and it represents an important bias in evaluating the pathophysiological underpinnings of the disorder. Regardless of its neurodevelopmental versus neurodegenerative nature, an imbalance in inflammatory markers has been intensively studied in the past 20 years, and it is in fact largely documented [11–13]. Evidence of immune activation was obtained from the detection of abnormal levels of pro-inflammatory cytokines and their receptors in the peripheral blood and cerebrospinal fluid of schizophrenic patients [14].

Cytokines are involved in normal central nervous system (CNS) development and consequently result involved in the pathogenesis of many neuropsychiatric disorders, acting directly on neural cells or modulating neurotransmitter and peptidergic pathways [14]. In such perspective, neurobiological hypotheses linking the neurodevelopmental alterations occurring in schizophrenia with the inflammatory processes, largely documented over the course of the illness, have been put forward [15].

Neurodevelopmental Hypothesis of Schizophrenia

Several lines of evidence strongly indicate that Schizophrenia may be a neurodevelopmental disorder [15]. The "neurodevelopmental model" of schizophrenia postulates that the disorder represents the result of an aberrant neurodevelopmental process starting much earlier than the onset of clinical symptoms, caused by a combination of genetic and environmental factors [16, 17], producing a functional impairment in the long-term course of the disease [18, 19] as well as a cerebral damage (still not specific). Subtle changes in the cognitive performances are in fact detectable in many psychiatric conditions, including major psychoses, since their early stage. For instance, first-episode psychosis (FEP) patients show a cognitive impairment across several domains, supporting the neurodevelopmental hypothesis [20], including theories on how early stress with or without genetic vulnerability may moderate cognitive function in psychosis.

In particular, several investigators believe that the damage occurs during brain development, over the intrauterine period and the first few years after birth [21]. Main neurodevelopmental abnormalities in schizophrenia consist of changes in the expression of proteins involved in early migration of neurons and glial cells, their proliferation, axonal outgrowth, synaptogenesis, connectivity and apoptosis [22].

The "neurodevelopmental model" seems to be based on reports of an excess of adverse events occurring during the pre- and perinatal periods, which would lead to the presence of cognitive and behavioural signs, starting in adolescence and childhood, and becoming clinically evident in early adulthood. The lack of clear neurodegenerative patterns (as mentioned imaging findings are more accurate nowadays but not specific in many patients affected by schizophrenia), however, limits the support to this theory [23]. Nonetheless, multiple markers of congenital anomalies, indicative of neurodevelopmental insults, have been indicated as supportive for the neurodevelopmental model of schizophrenia [24, 25], including agenesis of corpus callosum, stenosis of Sylvian aqueduct, cerebral hematomas and cavum septum pellucidum. The presence of low-set ears, epicanthal eye folds, wide spaces between the first and second toes and abnormal dermatoglyphics are, in turn, suggestive of both first and second trimester abnormalities [22] but are present in a minor percentage of patients. Multiple records, moreover, indicate the presence of premorbid neurological soft signs in children, who had subsequently developed schizophrenia [7, 26]. Additionally, children at high risk for schizophrenia were found to show a broad range of abnormalities, the most prominent of which seemed to occur in attention, motor function, coordination, sensory integration, mood and social behaviours [27]. Indeed, such abnormalities may have predictive value in determining which children will later keep on showing overt signs of either schizophrenia spectrum disorders or schizophrenia itself [7], but this hypothesis still lacks appropriate studies to be supported.

With respect to the role of genetic factors, this has been investigated by several studies, with polygenic model acting additively or multiplicatively. Linkage and association studies [28, 29] have shown 12 chromosomal regions, containing 2181 known genes [230] and 9 specific genes, involved in the possible etiology of the disorder [29]. On the other hand, environmental factors, including pre- and perinatal complications, as well as maternal infections occurring during pregnancy, were found to play an important role in the pathogenesis of schizophrenia. In particular, a meta-analysis of population-based data found significant estimates for three main categories of pre- and perinatal complications: (1) complications of pregnancy (e.g. bleeding, pre-eclampsia, diabetes), (2) abnormal foetal growth and development (e.g. low birth weight, congenital malformations, small head circumference) and (3) complications of delivery (e.g. asphyxia, uterine atony, emergency caesarean section) [30].

Obstetric complications are supposed to increase the risk of developing schizophrenia in two main ways: acting alone and/or interacting with genetic risk factors [30, 31]. In fact, it has been suggested that specific susceptibility genes for schizophrenia may be regulated by hypoxia/ischaemia [32] occurring during birth.

Other environmental factors, potentially causing abnormal neurodevelopment, include possible infective processes occurring during pregnancy. Maternal infections can, for instance, increase the risk for the offspring to develop schizophrenia during adulthood [33, 34].

The available body of research in the field suggests that pre−/perinatal infections (including viruses as influenza, measles, polio, herpes simplex type 2 or bacteria like diphtheria and pneumonia) and other environmental insults, that adversely affect infant brain development, may increase the likelihood to develop schizophrenia in later life, particularly in genetically susceptible individuals [35–40]. Association studies regarding the influenza A virus showed that the maximum risk for the embryonic brain is represented by the exposure to the infective agent during the fourth and seventh month of gestation [41]. Subsequent studies have shown that rubella may increase the risk for the development of schizophrenia in the progeny of exposed mothers by 10- to 20-fold [42, 43]. Finally, prenatal exposure to influenza in the first trimester increased the risk of developing schizophrenia by sevenfold, and infection in early to mid-gestation increased the risk by threefold. Also the presence of maternal antibodies against *Toxoplasma gondii* can lead to a 2.5-fold increased risk.

Alterations of Inflammatory Pathways in Schizophrenia

The dysregulation of the inflammatory response system represents a major piece of evidence in the pathophysiology of schizophrenia, along with genetic and environmental factors, ultimately affecting the neurodevelopmental process [44, 45]. Neuroimmunology is a recent yet rapidly growing field of research investigating the interface between immunology and development of chronic mental illness, including areas such as stress, neuroplasticity, genetics and cytokines [46]. The latter ones, in particular, play a pivotal role in infectious and inflammatory processes and mediate the crosstalk between the brain and the immune system. Therefore, cytokines are supposed to be the main actors of the immune and inflammatory abnormalities, documented in schizophrenia [47].

Because cytokines are large hydrophilic polypeptides, their ability to cross the brain–blood barrier is reduced, at least under physiologic conditions. The presence of abnormal circulating levels of pro-inflammatory cytokines, therefore able to trespass the hematoencephalic barrier, and their receptors is well established in schizophrenic patients [48–50] and their first-degree relatives [48, 51], thus confirming the presence of immune abnormalities that could develop from a pre-inflammatory state in the CNS, which has been always considered as an immune-sequestered district [52, 53].

In the last two decades, different hypotheses in relation to the cytokine-mediated development of schizophrenia have been proposed.

As a matter of fact, cytokines play an important role during neurodevelopment and in CNS functions at all stages, starting with the induction of neuroepithelium [55]. Subsequently, cytokines monitor the renewal of neuroepithelial cells, which act as precursors for all neurons, microglia and adult progenitors, as well as framework for radially migrating neurons [56]. Such processes are orchestrated by cytokines and related responses of their target cells [57]. As a general rule, there is an overproduction of neurons and glia and cytokines are pivotal to either promote survival of cells or to induce apoptosis of cells with impaired connections [58]. Therefore, even minimal variations on cytokine levels could result in subsequent functional impairment [59].

An increase of cytokines, following maternal infection, may alter the immune status of the brain, causing abnormal cells development with subsequent brain damage [60]. It is clear that maternal immune activation (MIA) induces cytokines increase in the placenta (IL-1beta, IL-6, TNF-alpha) and amniotic fluid (IL-6, TNF-alpha) [61]. The action of cytokines on the placenta might alter the transfer of cells, nutrient, oxygen, growth factors and maternal antibodies, each of which with potential crucial effect on foetal development [61].

Besides affecting neurodevelopment, some cytokines (i.e. IL-2 and IL-6) appear to have a role in the progression of schizophrenic illness. For instance, IL-2 stimulates the proliferation of T lymphocytes and its inhibition contributes to humoral immunity enhancement [62]. Kim and colleagues found lower IL-2 serum levels in schizophrenics with long duration of illness [63]. Such findings suggest that IL-2 may be a key modulator of dopaminergic metabolism and psychotic symptoms in schizophrenia [64].

Another possible contribution to the progression of the illness might be due to a hyper-activation of humoral immunity, which stimulates the tryptophan 2,3-dioxygenase enzyme, with an increased transformation of the amino acid tryptophan in kynurenic acid, that acts as a N-methyl-D-aspartate (NMDA) receptor antagonist [64].

Among cytokines, IL-6 potentiates B lymphocyte proliferation, and it seems to play a key role in the immunological abnormalities observed in schizophrenic patients [53]. It is also interesting to note that several studies showed that a long duration of illness in schizophrenia is associated with higher serum levels of IL-6 [53]. Moreover, elevated IL-6 serum concentrations have been proposed as key factors, responsible for cerebral atrophy observed in schizophrenic patients with long duration of illness [65, 66].

Neurodegeneration in Schizophrenia

Neuroanatomical abnormalities are common in schizophrenic patients yet unspecific and largely thought to originate from a neurodevelopmental defect [21], therefore possibly representing a structural substrate for the disorder. However, there is growing evidence that the magnitude and pattern of such abnormalities could progress over time [67], involving a proper neurodegenerative process, biased, as stated above, by many factors, including ageing, treatment, treatment discontinuation and co-occurring neurological conditions including vascular disorders. The combination of neurodevelopmental and neurodegenerative processes in the pathogenesis of schizophrenia is not surprising but likely plausible [10]. Tissue losses in the brain can involve different areas: for example, decreases in the volume of the temporal lobe [68], in the hippocampal volume [69] and in the volume of parahippocampal gyrus [70] were reported. Similarly, several studies have shown reductions in the grey matter of volume of cortical structures in schizophrenic patients.

The molecular basis of grey matter volume losses in schizophrenic subjects is still poorly understood, even though such anomalies seem to be more likely connected to the loss of organization of neuronal processes more than to the actual loss of cell bodies. In fact, post-mortem studies in schizophrenic brains showed abnormal neuronal organization within corticolimbic structures [71, 72]. For instance, a MR imaging study reported that schizophrenic patients showed vertical sulcal patterns more frequently than healthy controls [73], while other studies also demonstrated distortions of normal patterns of cortical asymmetries in schizophrenia and hippocampal volume reductions only on the left side [74, 75]. Even though some studies reported the progression of neuroanatomical abnormalities in schizophrenic patients, the point of whether such alterations are static or dynamic is still open to argument. Indeed, post-mortem studies also reported larger abnormalities in the left temporal lobes of patients with schizophrenia, i.e. temporal horn enlargement [76] and neuronal heterotopia. These results, however, remain obscure and need further explanation [72].

Some studies report that ventricular enlargement and grey matter volume losses are progressive over periods of 1–5 years in schizophrenic subjects [77, 78], while other studies describe that such structural measures are highly stable over time [79, 80]. Some investigators reported the presence of cortical thickness reductions in schizophrenic patients; in particular, the absence of widespread cortical thinning before disease onset implies that the cortical thinning is unlikely to simply reflect genetic liability to schizophrenia but is predominantly driven by disease-associated factors [81].

On the other hand, recent studies of individuals with "prodromal" schizophrenia showed that relatively rapid changes in neuroanatomical structure early in the course of illness can be found [67]. Several different mechanisms of neuronal injury are now under investigation in relation to the pathogenesis of schizophrenia. Some investigators suggested that a developmental deficit of NMDA receptor-bearing GABAergic interneurons would place an individual at increased risk for excitotoxic neuronal injury later in life [82]. Excitotoxicity (i.e. neurodegeneration via the overactivity of excitatory neurotransmission) represents an interesting mechanism to explain neuronal injury in schizophrenia, because it could be initiated and maintained through the action of neurotransmitter systems, such as the monoamines, that have long been implicated in schizophrenia [83]. Another intriguing theory to explain neuronal injury in schizophrenia is the dysregulation of apoptosis [84], a process normally associated with the elimination of redundant neurons during development [85] and evaluated in the past decades as a mechanism deeply involved in neoplastic processes. Also, glucocorticoid hormones [86], triggered by environmental stressors, including those (e.g. famine) associated with an increased risk for schizophrenia [87], have been implicated as factors contributing to neurodegenerative impairment.

Conclusions

This chapter sought to summarize the most intriguing models and evidence linking abnormalities in the neurodevelopment with altered immune/inflammatory mechanisms in schizophrenic patients. However, such perspective does not exclude the possibility to consider also the presence of progressive neurodegeneration as a prominent biological feature of the disorder. In fact, it seems likely that what we currently diagnose as a unitary disorder includes, actually, highly heterogeneous entities, in terms of pathophysiology [88]. These would include forms predominantly characterized by neurodevelopmental alterations (e.g. inflammatory features), as well as others with minor or absent neurodevelopmental aspects, but marked and progressive neurodegeneration, starting from the early adolescence, as main biological feature. Therefore, the attempt to solve the question whether schizophrenia is or is not a neurodevelopmental disorder or a progressive neurodegenerative seems to be outdated and needs to overcome by recent biological acquisitions [1]. Differences in the genetic background could, moreover, give account of these two different timing and patterns of illness evolution and presentation.

Conflict of Interest Authors declare they have no conflict of interest with the content of the present article.

References

1. Jones PB, Buckley PF. Schizophrenia. Amsterdam: Elsevier; 2006.
2. Gillberg C, Steffenburg S. Outcome and prognostic factors in infantile autism and similar conditions: a population-based study of 46 cases followed through puberty. J Autism Dev Disord. 1987;17:273–87.
3. Thomsen PH. Schizophrenia with childhood and adolescent onset: a nationwide register-based study. Acta Psychiatr Scand. 1996;94:187–93.
4. Amminger GP, Harris MG, Conus P, et al. Treated incidence of first episode psychosis in the catchment area of EPPIC between 1997 and 2000. Acta Psychiatr Scand. 2006;114:337–45.
5. British Psychological Society. Psychosis and schizophrenia in children and young people: recognition and management. Leicester: British Psychological Society; 2013.
6. Hollis C. Adult outcomes of child and adolescent onset schizophrenia: diagnostic stability and predictive validity. Am J Psychiatry. 2000;157:1652–9.
7. Fish B, Marcus J, Hans S, Auerbach JG, Perdue S. Infants at risk for schizophrenia: sequelae of a genetic neurointegrative defect. Arch Gen Psychiatry. 1992;49:221–35.
8. Altamura AC, Pozzoli S, Fiorentini A, Dell'Osso B. Neurodevelopment and inflammatory patterns in schizophrenia in relation to pathophysiology. Prog Neuro-Psychopharmacol Biol Psychiatry. 2013;42:63–70.
9. Walker E, Lewine RJ. Prediction of adult-onset schizophrenia from childhood home movies of the patients. Am J Psychiatry. 1990;147(8):1052–6.
10. Csernansky JG. Neurodegeneration in schizophrenia: evidence from in vivo neuroimaging studies. Sci World J. 2007;7:135–43.
11. Rund BR. Is schizophrenia a neurodegenerative disorder? Nord J Psychiatry. 2009;63(3):196–201.
12. Boin F, Zanardini R, Pioli R, Altamura CA, Maes M, Gennarelli M. Association between − G308A tumor necrosis factor alpha gene polymorphism and schizophrenia. Mol Psychiatry. 2001;6(1):79–82.
13. Lin A, Kenis G, Bignotti S, Tura GJ, De Jong R, Bosmans E, et al. The inflammatory response system in treatment-resistant schizophrenia: increased serum interleukin-6. Schizophr Res. 1998;32(1):9–15.
14. Maes M, Bocchio Chiavetto L, Bignotti S, Battisa Tura GJ, et al. Increased serum interleukin-8 and interleukin-10 in schizophrenic patients resistant to treatment with neuroleptics and the stimulatory effects of clozapine on serum leukemia inhibitory factor receptor. Schizophr Res. 2002;54(3):281–91.
15. Gourion D, Gourevitch R, Leprovost JB, Olié H Iôo JP, Krebs MO. Neurodevelopmental hypothesis in schizophrenia. L'Encéphale. 2004;30:109–18.
16. Rapoport JL, Addington AM, Frangou S, Psych MR. The neurodevelopmental model of schizophrenia: update 2005. Mol Psychiatry. 2005;10:434–49.

17. Singh SM, McDonald P, Murphy B, O'Reilly R. Incidental neurodevelopmental episodes in the etiology of schizophrenia: an expanded model involving epigenetics and development. Clin Genet. 2004;65:435–40.
18. Buckley P. The clinical stigmata of aberrant neurodevelopment in schizophrenia. J Nerv Ment Dis. 1998;186(2):79–86.
19. Keshavan MS, Murray RM. Neurodevelopment and adult psychopathology. Cambridge: Cambridge University Press; 1997.
20. Aas M, Dazzan P, Mondelli V, Melle I, Murray RM, Pariante CM. A systematic review of cognitive function in first-episode psychosis, including a discussion on childhood trauma, stress, and inflammation. Front Psych. 2014;4:182.
21. Weinberger DR. Implications of normal brain development for the pathogenesis of schizophrenia. Arch Gen Psychiatry. 1987;44:660–9.
22. Fatemi SH, Folsom TD. The neurodevelopmental hypothesis of schizophrenia, revisited. Schizophr Bull. 2009;35(3):528–48.
23. Lewis DA, Levitt P. Schizophrenia as a disorder of neurodevelopment. Annu Rev Neurosci. 2002;25:409–32.
24. Lloyd T, Dazzan P, Dean K, Park SB, Fearon P, Doody GA, et al. Minor physical anomalies in patients with first-episode psychosis: their frequency and diagnostic specificity. Psychol Med. 2008;38:71–7.
25. Meltzer HY, Fatemi SH. Schizophrenia and other psychotic disorders. In: Ebert MH, Loosen PT, Nurcombe B, editors. Current diagnosis and treatment in psychiatry. Norwalk, CT: Appleton and Lange; 2000. p. 260–77.
26. Barkus E, Stirling J, Hopkins R, Lewis S. The presence of neurological soft signs along the psychosis proneness continuum. Schizophr Bull. 2006;32:573–7.
27. Niemi LT, Suvisaari JM, Tuulio-Henriksson A, Lonnqvist JK. Childhood developmental abnormalities in schizophrenia: evidence from high-risk studies. Schizophr Res. 2003;60:239–58.
28. Lewis CM, Levinson DF, Wise LH, DeLisi LE, Straub RE, Hovatta I, et al. Genome scan meta-analysis of schizophrenia and bipolar disorder, part II: schizophrenia. Am J Hum Genet. 2003;73:34–48.
29. Sullivan PF, Eaves LJ, Kendler KS, Neale MC. Genetic case–control association studies in neuropsychiatry. Arch Gen Psychiatry. 2001;58:1015–24.
30. Cannon TD, van Erp TG, Rosso IM, Huttunen M, Lönnqvist J, Pirkola T, et al. Fetal hypoxia and structural brain abnormalities in schizophrenic patients, their siblings, and controls. Arch Gen Psychiatry. 2002;59:35–41.
31. Boog G. Obstetrical complications and subsequent schizophrenia in adolescent and young adult offsprings: is there a relationship? Eur J Obstet Gynecol Reprod Biol. 2004;114:130–6.
32. Schmidt-Kastner R, van Os J, Steinbusch H WM, Schmitz C. Gene regulation by hypoxia and the neurodevelopmental origin of schizophrenia. Schizophr Res. 2006;84:253–71.
33. Karlsson H, Bachmann S, Schroder J, McArthur J, Torrey EF, Yolken RH. Retroviral RNA identified in the cerebrospinal fluids and brains of individuals with schizophrenia. Proc Natl Acad Sci U S A. 2001;98:4634–9.
34. Lewis DA. Retroviruses and the pathogenesis of schizophrenia. Proc Natl Acad Sci U S A. 2001;94:4293–4.
35. Jones P, Cannon M. The new epidemiology of schizophrenia. Psychiatr Clin North Am. 1998;21(1):1–25.
36. Mednick SA, Machon RA, Huttunen MO, Bonett D. Adult schizophrenia following prenatal exposure to an influenza epidemic. Arch Gen Psychiatry. 1988;45:189–92.
37. Torrey EF, Rawlings R, Waldman IN. Schizophrenic births and viral diseases in two states. Schizophr Res. 1988;1:73–7.
38. Suvisaari J, Haukka J, Tanskanen A, Hovi T, Lönnqvist J. Association between prenatal exposure to poliovirus infection and adult schizophrenia. Am J Psychiatry. 1999;156:1100–2.
39. Buka SL, Tsuang MT, Torrey EF, Klebanoff MA, Bernstein D, Yolken RH. Maternal infections and subsequent psychosis among offspring. Arch Gen Psychiatry. 2001;58:1032–7.

40. Watson CG, Kucala T, Tilleskjor C, Jacobs L. Schizophrenic birth seasonality in relation to the incidence of infectious diseases and temperature extremes. Arch Gen Psychiatry. 1984;41:85–90.
41. Brown AS, Begg MD, Gravenstein S, Schaefer CA, Wyatt RJ, Bresnahan M, et al. Serologic evidence of prenatal influenza in the etiology of schizophrenia. Arch Gen Psychiatry. 2004;61:774–80.
42. Brown AS, Schaefer CA, Wyatt RJ, Goetz R, Begg MD, Gorman JM, Susser ES. Maternal exposure to respiratory infections and adult schizophrenia spectrum disorders: a prospective birth cohort study. Schizophr Bull. 2000;26:287–95.
43. Brown AS. Prenatal infection as a risk factor for schizophrenia. Schizophr Bull. 2006; 32:200–2.
44. Altamura AC, Boin F, Maes M. HPA axis and cytokines dysregulation in schizophrenia: potential implications for the antipsychotics treatment. Eur Neuropsychopharmacol. 1999;10:1–4.
45. Jablensky A. Epidemiology of schizophrenia: the global burden of disease and disability. Eur Arch Psychiatry Clin Neurosci. 2000;250(6):274–85.
46. Mundo E, Altamura AC, Vismara S, Zanardini R, Bignotti S, Randazzo R, et al. MCP-1 gene (SCYA2) and schizophrenia: a case–control association study. Am J Med Genet B Neuropsychiatr Genet. 2005;132B:1–4.
47. Altamura AC, Pozzoli S, Fiorentini A, Dell'Osso B. Neurodevelopment and inflammatory patterns in schizophrenia in relation to pathophysiology. Prog Neuro-Psychopharmacol Biol Psychiatry. 2013;5(42):63–70.
48. Garver DL, Tamas RL, Holcomb JA. Elevated interleukin-6 in the cerebrospinal fluid of a previously delineated schizophrenia subtype. Neuropsychopharmacology. 2003;28:1515–20.
49. Miller BJ, Buckley P, Seabolt W, Mellor A, Kirkpatrick B. Meta-analysis of cytokine alterations in schizophrenia: clinical status and antipsychotic effects. Biol Psychiatry. 2011;70:663–71.
50. Potvin S, Stip E, Sepehry AA, Gendron A, Bah R, Kouassi E. Inflammatory cytokines alterations in schizophrenia: a systematic quantitative review. Biol Psychiatry. 2008;63:801–8.
51. Nunes SO, Matsuo T, Kaminami MS, Watanabe MA, Reiche EM, Itano EN. An autoimmune or an inflammatory process in patients with schizophrenia, schizoaffective disorder, and their biological relatives. Schizophr Res. 2006;84:180–2.
52. Ganguli R, Yang Z, Shurin G, Chengappa KN, Brar JS, Gubbi AV, Rabin BS. Serum interleukin-6 concentration in schizophrenia: elevation associated with duration of illness. Psychiatry Res. 1994;51:1–10.
53. Naudin J, Mege JL, Azorin JM, Dassa D. Elevated circulating levels of IL-6 in schizophrenia: an overview. Eur Arch Psychiatry Clin Neurosci. 1996;20(3):269–73.
54. Müller N, Riedel M, Scheppach C, Brandstätter B, Sokullu S, Krampe K, et al. Beneficial antipsychotic effects of celecoxib add-on therapy compared to risperidone alone in schizophrenia. Am J Psychiatry. 2002;159:1029–34.
55. Gulden J, Reiter JF. Neur-ons and neur-offs: regulators of neural induction in vertebrate embryos and embryonic cells. Hum Mol Genet. 2008;17:R60–6.
56. Pinto L, Gotz M. Radial glial cell heterogeneity—the source of diverse progeny in the CNS. Prog Neurobiol. 2007;7:797–805.
57. Lee RH, Mills EA, Schwartz N, Bell MR, Deeg KE, Ruthazer ES, et al. Neurodevelopmental effects of chronic exposure to elevated levels of pro-inflammatory cytokines in a developing visual system. Neural Dev. 2010;5:2.
58. Deverman BE, Patterson P. Cytokines and CNS development. Neuron. 2009;64:61–77.
59. Chklovskii DB. Exact solution for the optimal neuronal layout problem. Neural Comput. 2004;16(10):2067–78.
60. Braun AS, Derkits EJ. Prenatal infection and schizophrenia: a review of epidemiologic and translational studies. Am J Psychiatry. 2010;167:261–80.
61. Patterson PH. Immune involvement in schizophrenia and autism. Etiology, pathology and animal models. Behav Brain Res. 2009;204:313–21.
62. Bresee C, Rapaport MH. Persistently increased serum soluble interleukin-2 receptors in continuously ill patients with schizophrenia. Int J Neuropsychopharmacol. 2009;12(6):861–5.

63. Kim YK, Kim L, Lee MS. Relationships between interleukins, neurotransmitters and psychopathology in drug-free male schizophrenics. Schizophr Res. 2000;44:165–75.
64. Muller N, Schwarz M. Schizophrenia as an inflammation-mediated dysbalance of glutamatergic neurotransmission. Neurotox Res. 2006;10(2):131–48.
65. Akiyama K. Serum levels of soluble IL-2 receptor alpha, IL-6 and IL-1 receptor antagonist in schizophrenia before and during neuroleptic administration. Schizophr Res. 1999; 37:97–106.
66. Waddington JL. Neurodynamics of abnormalities in cerebral metabolism and structure in schizophrenia. Schizophr Bull. 1993;19:55–69.
67. Keshavan MS, Berger G, Zipursky RB, Wood SJ, Pantelis C. Neurobiology of early psychosis. Br J Psychiatry Suppl. 2005;48:s8–18.
68. Suddath RL, Casanova MF, Goldberg TE, et al. Temporal lobe pathology in schizophrenia: a quantitative MRI study. Am J Psychiatry. 1989;146:464–72.
69. Breier A, Buchanan RW, Elkashef A, Munson RC, Kirkpatrick B, Gellad F. Brain morphology and schizophrenia. A magnetic resonance imaging study of limbic, prefrontal cortex, and caudate structures. Arch Gen Psychiatry. 1992;49:921–6.
70. McCarley RW, Shenton ME, O'Donnell BF, Faux SF, Kikinis R, Nestor PG, Jolesz FA. Auditory P300 abnormalities and left posterior superior temporal gyrus volume reduction in schizophrenia. Arch Gen Psychiatry. 1993;50:190–7.
71. Falkai P, Bogerts B. Cell loss in the hippocampus of schizophrenics. Eur Arch Psychiatry Neurol Sci. 1986;236:154–61.
72. Jakob H, Beckmann H. Prenatal developmental disturbances in the limbic allocortex in schizophrenics. J Neural Transm. 1986;65:303–26.
73. Kikinis R, Shenton ME, Gerig G, Hokama H, Haimson J, O'Donnell BF, et al. Temporal lobe sulco-gyral pattern anomalies in schizophrenia: an in vivo MR three-dimensional surface rendering study. Neurosci Lett. 1994;182:7–12.
74. Luchins DJ, Weinberger DR, Wyatt RJ. Schizophrenia and cerebral asymmetry detected by computed tomography. Am J Psychiatry. 1982;139:753–7.
75. Johnstone EC, Crow TJ, Frith CD, Husband J, Kreel L. Cerebral ventricular size and cognitive impairment in chronic schizophrenia. Lancet. 1976;2(7992):924–6.
76. Heckers S, Heinsen H, Heinsen Y, Beckmann H. Morphometry of the parahippocampal gyrus in schizophrenics and controls. Some anatomical considerations. J Neural Transm. 1990;80:151–5.
77. Davis KL, Buchsbaum MS, Shihabuddin L, Spiegel-Cohen J, Metzger M, Frecska E, et al. Ventricular enlargement in poor-outcome schizophrenia. Biol Psychiatry. 1998;43:783–93.
78. Nair TR, Christensen JD, Kingsbury SJ, Kumar NG, Terry WM, Garver DL. Progression of cerebroventricular enlargement and the subtyping of schizophrenia. Psychiatry Res. 1997;74(3):141–50.
79. Illowsky BP, Juliano DM, Bigelow LB, Weinberger DR. Stability of CT scan findings in schizophrenia: results of an 8 year follow-up study. J Neurol Neurosurg Psychiatry. 1998;51:209–13.
80. Nasrallah HA, Olson SC, McCalley-Whitters M, Chapman S, Jacoby CG. Cerebral ventricular enlargement in schizophrenia: a preliminary follow-up study. Arch Gen Psychiatry. 1986;43:157–9.
81. Sprooten E, Papmeyer M, Smyth AM, Vincenz D, Honold S, Conlon GA, et al. Cortical thickness in first-episode schizophrenia patients and individuals at high familial risk: a cross-sectional comparison. Schizophr Res. 2013;151(1–3):259–264. pii: S0920-9964(13)00523-9 [Epub ahead of print]. https://doi.org/10.1016/j.schres.2013.09.024.
82. Olney JW, Newcomer JW, Farber NB. NMDA receptor hypofunction model of schizophrenia. J Psychiatr Res. 1999;33:523–33.
83. Farber NB, Hanslick J, Kirby C, McWilliams L, Olney JW. Serotonergic agents that activate 5HT2A receptors prevent NMDA antagonist neurotoxicity. Neuropsychopharmacology. 1998;18:57–62.
84. Glantz LA, Gilmore JH, Lieberman JA, Jarskog LF. Apoptotic mechanisms and the synaptic pathology of schizophrenia. Schizophr Res. 2006;81:47–63.

85. De Zio D, Giunta L, Corvaro M, Ferraro E, Cecconi F. Expanding roles of programmed cell death in mammalian neurodevelopment. Sem cell developmental. Biol. 2005;16:281–94.
86. Sapolsky RM. Glucocorticoids and hippocampal atrophy in neuropsychiatric disorders. Arch Gen Psychiatry. 2000;57:925–35.
87. Susser E, Hoek HW, Brown A. Neurodevelopmental disorders after prenatal famine: the story of the Dutch famine study. Am J Epidemiol. 1998;147:213–6.
88. Altamura AC. A multidimensional (pharmacokinetic and clinical–biological) approach to neuroleptic response in schizophrenia. With particular reference to drug resistance. Schizophr Res. 1993;8(3):187–98.

Parkinson's Disease: Contemporary Concepts and Clinical Management

17

Vanessa Carvalho, Carlota Vicente Cunha, and João Massano

Abstract

Parkinson's disease (PD) is a common neurodegenerative disorder causing a remarkable burden at the individual, family, social, and economic levels. Several risk and protective factors have been recently identified, providing potential for the research and implementation of preventive strategies. Although most cases remain sporadic, various monogenic forms of PD have been described, including autosomal dominant (e.g., LRRK2, SNCA, VPS35, EIF4G1, CHCHD2), autosomal recessive (e.g., parkin, PINK1, DJ-1, DNAJC6), and X-linked (e.g., RAB39B). The pathophysiology of PD is still intriguing, with several recent concepts and theories, including evidence that disease pathology might spread along the various neural systems and regions as a prion protein. Thorough scientific knowledge and clinical experience are required to establish the diagnosis correctly, and novel criteria have been freshly proposed to aid clinicians in this task. This process implies also effectively distinguishing PD from less common parkinsonian disorders including Kufor-Rakeb syndrome, Perry syndrome, a few degenerative ataxias and spastic paraplegias, and several forms of neurodegeneration with brain iron accumulation (NBIA), among others. Treating PD is a challenging enterprise, as the various options should be considered, and often

---

V. Carvalho
Department of Neurology, Hospital Pedro Hispano, Matosinhos Local Health Unit, Matosinhos, Portugal

C.V. Cunha
Department of Neurology, Hospital de Santo António/Porto Hospital Center, Porto, Portugal

J. Massano (✉)
Department of Neurology, Centro Hospitalar São João, Porto, Portugal

Department of Clinical Neurosciences and Mental Health, Faculty of Medicine University of Porto, Porto, Portugal
e-mail: jmassano@med.up.pt

© Springer International Publishing AG 2018
D. Galimberti, E. Scarpini (eds.), *Neurodegenerative Diseases*,
https://doi.org/10.1007/978-3-319-72938-1_17

rerouted, taking into account disease stage, motor and non-motor symptoms, and non-PD concomitant patient features. Although general guidelines and strategies are available, it is essential to tailor therapy to each patient, so that quality of life is maximized for many years, while minimizing risks and adverse effects. In carefully selected patients, deep brain stimulation, subcutaneous apomorphine, and levodopa/carbidopa intestinal gel should be considered whenever optimized noninvasive strategies are insufficient to guarantee these goals.

Keywords

Parkinson's disease · Epidemiology · Genetics · Diagnosis · LRRK2 · Parkin · Levodopa · Dopamine receptor agonists · Deep brain stimulation · Treatment

Introduction

Two hundred years have gone by since the publication of *An Essay on the Shaking Palsy* by James Parkinson [1]. Yet, Parkinson's disease (PD) continues to fascinate clinicians and researchers alike, while troubling a growing number of patients and their families. Currently, there are no efficacious preventive or disease-modifying strategies, and much less a cure for PD, which is frustrating for all stakeholders in the field, but also inspiring for those pursuing innovative protective or therapeutic interventions. Many important developments emerged in recent years with regard to the knowledge on the different aspects of PD, including epidemiology, genetics, pathophysiology, diagnosis, and treatment. These will be outlined in this chapter in order to provide the readership with updated data and pragmatic advice on PD and its optimal management.

Epidemiology

Incidence and Prevalence

Parkinson's disease is the second most common neurodegenerative disorder, after Alzheimer's disease. Its overall estimated prevalence is 0.3%, and the number of individuals afflicted by this disorder is expected to double by 2030, imposing an increasing social and economic burden on societies as populations grow older [2]. The median age of onset is between 60 and 65 years of age and both incidence and prevalence of PD rise with increasing age, peaking between 70 and 79 years old in most studies [3, 4]. PD is rare before the fifth decade, with an overall prevalence of 1 per 100,000 in individuals with age between 40 and 49 years, increasing up to around 1900 per 100,000 above the age of 80 [5]. Age-adjusted prevalence seems to be lower in Africa, when compared to Europe or North America. Incidence in Asia is apparently similar to Europe and America, although data on race or ethnicity is scarce and inconsistent [4, 6].

In most studies the incidence of PD varies from 17 to 19 per 100,000 individuals per year, and there is some evidence that it has been increasing in the past few decades, particularly in men aged 70 years or older [3]. This trend can, however, be partially explained by an increased awareness of signs and symptoms of parkinsonism. Most studies show that both incidence and prevalence are 1.5–2.0 times higher in men than women. Furthermore, onset occurs on average 2 years later in women, who more often present a tremor dominant motor phenotype of PD, which has been associated with slower progression of disability [7].

However, it should be taken into account that both prevalence and incidence are highly variable among epidemiologic surveys. These differences are largely explained by diverse survey designs, methodology, and various diagnostic criteria, but can also reflect dissimilar susceptibilities between different populations [4, 8–11].

Risk and Protective Factors

PD is likely a multifactorial disorder with a strong environmental component. In fact, several studies have reported associations of several factors and the risk of PD (Table 17.1). Most data come from case-control studies, in which exposure is measured using questionnaires, an approach prone to recall and selection bias, and only a few exposures have shown consistent correlation in large population-based studies [4–12].

Smoking is one of the most studied variables, lying among those with consistent data. There is a robust inverse association between smoking and the risk of PD, not explained by common biases such as selection or confounding exposures [13–15]. Past smokers and current smokers have a lower risk of PD when compared to never-smokers, and disease becomes apparent at an older age [13–17]. There is also a dose-dependent lower risk of PD linked to more years of tobacco use, higher

Table 17.1 Exposures associated with increased or decreased risk of PD

Risk factors	Protective factors
Age	Tobacco smoking
Male gender	Coffee consumption
Exposure to pesticides	Elevated blood urate levels
Farmers	NSAIDs
Rural living	Exercise
Well water drinking	Mediterranean diet
Traumatic brain injury	
Hepatitis C infection	
Dairy intake	
Personal or family history of melanoma	

cigarette burden, and fewer years since quitting [16, 17]. However, whether this association is causal remains controversial, and some hypothesize that this is a mere epiphenomenon, given that PD patients tend to have a non-novelty-seeking personality and seem to be less inclined to initiate or continue to smoke, drink alcohol or coffee [18]. For instance, PD patients are less likely to have ever smoked, and those who develop a smoking habit quit at an earlier age than controls, which might reflect a less rewarding experience from smoking, and might indicate ease to quit as a prodromal event to PD [19]. Nevertheless, the correlation of smoking and PD remains statistically significant after adjusting for novelty-seeking assessment scales [17], and this theory would not explain the risk difference between ever-smokers and never-smokers (as smoking typically starts before the fourth decade of life, usually long before prodromal PD), the inverse association between passive and parental smoking, and the time trends variation in the incidence of PD [20–22].

Coffee consumption also has a strong inverse correlation with the risk of PD, both in case-control and prospective studies, not attenuated when adjusted for smoking. This association seems to be stronger in men than women. Caffeine is thought to be the responsible component, since other non-coffee sources of caffeine, but not decaffeinated coffee, also correlate inversely with PD risk [12, 23–26].

Data are still inconsistent regarding alcohol. Several case-control studies report either a moderate inverse association or no change in risk, and this relation is attenuated when adjusted to the effect of smoking and coffee intake [27–29]. Conceivably the effects of alcohol could be mediated by serum uric acid levels [27, 28], as there is an inverse association between blood urate levels and the risk of PD [30]. Population-based studies reported a 30% lower risk of PD in patients with gout in both sexes [31]. Disease progression also seems to correlate negatively with urate levels [32] although this association is less clear in some groups such as women, smokers, and individuals younger than 60 years. The antioxidant properties of urates are thought to mediate the beneficial effects [33, 34].

In PD neuronal degeneration is frequently accompanied by microglial activation. Hence, it seems possible that anti-inflammatory drugs, like NSAIDs, could prevent or at least delay disease onset. In 2003, a prospective study with two large cohorts reported a lower risk of PD in regular users of non-aspirin NSAIDs but not in those using aspirin, with a greater reduction of risk associated with regular use and long-term use, consistent with a dose–response relationship [35]. However, subsequent research failed to find this association in NSAIDs other than ibuprofen, with 30% lower risk of PD in users when compared to nonusers [36, 37].

Diets with high intake of vegetables, whole grains, poultry, and fish, such as the Mediterranean diet, are associated with a lower risk of PD [38, 39]. This could be explained by the high concentrations of complex phenols and vitamins, which can serve as antioxidants, and the lower consumption of compounds associated with higher PD risk, such as animal fat or dairy. Similarly, high intakes of flavonoids, such as anthocyanin-rich foods [40], and frequent moderate to vigorous physical exercise in mid or later life are also associated with lower PD risk [41, 42].

Interest on environmental exposure and PD risk heightened in the 1980s, with the identification of MPTP, a neurotoxin causing acute dopaminergic neuron loss and parkinsonism. It has been hypothesized that environmental exposures could facilitate nigrostriatal degeneration, either alone or in combination with genetic predispositions, and could cause PD [12]. For example, most studies show a positive association between pesticide exposure and PD [43, 44]. Farming, living in rural areas, or drinking well water also correlates positively with increased PD risk [43, 45, 46]. Welding is associated with parkinsonism, due to the exposure to manganese (manganism), but it is not clear whether it can specifically increase the risk for PD [47].

Several long-term population-based studies have shown an association between the intake of dairy products and the future risk of PD [25, 48]. This could be explained by the bioconcentration in milk of certain organic compounds such as organochlorine pesticides or by the lower levels of urates in high consumers of milk [49].

The possible role of traumatic brain injury (TBI) gained attention after the world-renowned boxer and activist Muhammad Ali (aka Cassius Marcellus Clay Jr.) was diagnosed with PD, back in the 1980s. It is hypothesized that TBI can cause damage to the blood–brain barrier, disrupt mitochondrial function, and increase α-synuclein accumulation in the brain [4]. A recent meta-analysis of 22 studies (19 case-control studies, 2 nested case-control studies, and 1 cohort study) reported a pooled OR of 1.57 for the association of PD and head trauma [50]. This risk seems to be particularly high within the first year following TBI, with declining risk in subsequent years [51, 52]. These results are explained by many authors through reverse causation (i.e., patients with subclinical PD are more likely to fall). However, a recent study reported 44% increased risk of PD diagnosis in individuals with TBI when compared with individuals with non-TBI trauma (i.e., fractures) [53].

There is also higher occurrence of PD in patients with melanoma and vice versa [54, 55]. One explanatory theory is that melanoma and PD share environmental or genetic risk factors or pathogenic pathways. For instance, individuals with red hair color or homozygous for the melanocortin-1-receptor Arg51Cys variant (i.e., melanoma risk factors), or family history of melanoma, display greater PD risk [56–58].

Lastly, recent studies [59–61] have shown a relationship between hepatitis C virus (HCV) infection and PD. One of them suggests that there is also an increased risk of PD following hepatitis B virus infection [60], without association between PD and autoimmune hepatitis, chronic hepatitis, or HIV, which favors a specific aspect of the viral infection, rather than a general inflammatory process or the use of antivirals. Whether such association reflects shared disease mechanism, shared genetic or environmental susceptibility, sequelae of the hepatitis or a consequence of treatment remains to be determined.

Etiology, Genetics, and Pathophysiology

Genetics

Causative Genes
Around 10% of PD cases have an identifiable genetic cause. The identification of mutations in the *SNCA* gene as a cause of autosomal-dominant parkinsonism [62]

marked the beginning of a fruitful era in PD genetics, bringing new insights into pathophysiology. Several PD-related gene loci have been identified to date, with at least eight causative genes of dopa-responsive parkinsonism described so far (Table 17.2) [62–80]. Other Mendelian disorders display parkinsonism as a predominant characteristic, bearing also atypical features for PD, such as Perry syndrome (parkinsonism, severe depression, weight loss, central hypoventilation) or Kufor-Rakeb syndrome (juvenile-onset parkinsonism, dementia, pyramidal signs, supranuclear gaze palsy), and should be considered, when clinically appropriate, in the differential diagnosis of early-onset PD or when there is a family history of parkinsonism (Table 17.3) [80–87].

Table 17.2 Monogenic forms of Parkinson's disease

Disorder, inheritance pattern (former designation)	Gene, locus	Clinical features	Neuropathology	Additional comments
PARK-parkin, AR (PARK2)	*Parkin*, 6q26	Early-onset PD, with slow motor progression. Early motor complications from levodopa treatment. Mild dysautonomia; frequently lower limb dystonia and hyperreflexia	Absence of LB. Loss of pigmented neurons in the SNpc and in the *locus coeruleus*	Up to 50% of AR JP cases and 15% of sporadic PD Second most common cause of monogenic parkinsonism Heterozygous status might be a risk factor for PD
PARK-PINK1, AR (PARK6)	*PINK1*, 1p36.12	Early-onset PD with slow motor progression. Patients can also present with atypical features such as prominent dystonia and cognitive and psychiatric disturbances, or early gait impairment without cognitive disturbance	Loss of neurons in the SNpc and LB pathology in selected nuclei of the brainstem, SNpc, and nucleus basalis of Meynert. Absence of involvement of the LC	Second most common cause of AR JP 2–4% of the sporadic early-onset PD cases Heterozygous status might be a risk factor for PD
PARK-DJ-1, AR (PARK7)	*DJ1*, 1p36.23	Early-onset PD with slow progression. Lower limb dystonia, dementia, bulbar, and motor neuron signs have been reported	Neuronal loss in SNpc and LB pathology	Rare, ≤1% of AR JP cases

Table 17.2 (continued)

Disorder, inheritance pattern (former designation)	Gene, locus	Clinical features	Neuropathology	Additional comments
PARK-DNAJC6, AR (PARK19)	*DNAJC6*, 1p31.3	Early-onset PD, with slow progression and good response to levodopa. Patients can also present with atypical parkinsonism in childhood, with rapid progression and little or no response to levodopa, accompanied by pyramidal signs, dystonia, seizures, and cognitive impairment	Unknown	–
PARK-SNCA, AD (PARK1/PARK4)	*SNCA*, 4q22.1	Early-onset PD, with fast progression. Behavioral and cognitive impairment, dementia is a common feature	Widespread LB pathology in the brainstem and cerebral cortex	First described gene for monogenic PD
PARK-LRRK2, AD (PARK8)	*LRRK2*, 12q12	Generally classical PD phenotype. Patients more often present with tremor and more frequently suffer from dystonia when compared to sporadic PD. Abduction-adduction tremor of the lower limbs is a typical feature	Highly variable. Most have LB pathology, with neuronal loss in the SNpc	Most common cause of genetic PD: 2% of sporadic and 5% of familial PD in Northern Europe and North America. Highly prevalent in some populations (10% of Portuguese, 20% of Ashkenazi Jewish ancestry, and 40% of North African Berber Arab patients with PD)
PARK-VPS35, AD (PARK17)	*VPS35*, 16q11.2	Tremor-dominant classical PD, with good response to levodopa	Unknown	–
PARK-EIF4G1, AD (PARK18),	*EIF4G*, 3q27.1	Classical PD, with preserved cognition	LB pathology	–

(continued)

Table 17.2 (continued)

Disorder, inheritance pattern (former designation)	Gene, locus	Clinical features	Neuropathology	Additional comments
PARK-CHCHD2, AD	*CHCHD2*, 7p11.2	Classical PD	Unknown	–
PARK-RAB39B, X-linked	*RAB39B*, Xq28	Early-onset PD, with good response to levodopa. Cognitive impairment in childhood and early-onset PD has been reported	LB pathology. Loss of pigmented neurons in SNpc and LC. Abundant cortical LB, tau NFT, and axonal spheroids in the basal ganglia	–

AD autosomal dominant, *AR* autosomal recessive, *AR JP* autosomal recessive juvenile parkinsonism, *LB* Lewy bodies, *LC* locus coeruleus, *NFT* neurofibrillary tangles, *PD* Parkinson's disease, *SNpc* substantia nigra pars compacta (based on information from [62–80])

Table 17.3 Non-PD monogenic disorders that can present with prominent parkinsonism and clinically resemble PD in some cases

Disorder (alternative or former designations)	Gene, *locus*	Inheritance pattern
DYT/PARK-ATP13A2 (Kufor-Rakeb syndrome, PARK9)	*ATP13A2*, 1p36.13	AR
PARK-FBXO7 (PARK15)	*FBXO7*, 22q12.3	AR
NBIA/DYT-PANK2 (PKAN)	*PANK2*, 20p13	AR
NBIA/DYT/PARK-PLA2G6 (PLAN, PARK14)	*PLA2G6*, 22q13.1	AR
Gaucher disease	*GBA*, 1q22	AR
HSP-KIAA1840 (spastic paraplegia type 11)	*SPG11*, 15q21.1	AR
HSP-ZFYVE26 (spastic paraplegia type 15)	*ZFYVE26*, 14q24.1	AR
HSP/NBIA-FA2H (FAHN, SPG35)	*FA2H*, 16q23.1	AR
Spastic paraplegia type 38	*AP5Z1*, 7p22.1	AR
Polymerase gamma (POLG)	*POLG1*, 15q26.1	AD or AR
SCA-ATXN2 (spinocerebellar ataxia type 2)	*ATXN2*, 12q24.12	AD
SCA-ATXN3 (spinocerebellar ataxia type 3, Machado-Joseph disease)	*ATXN3*, 14q32.12	AD
Progranulin-associated frontotemporal dementia	*GRN*, 17q21.32	AD
MAPT-associated frontotemporal dementia	*MAPT*, 17q21.31	AD
C9orf72-associated frontotemporal dementia	*C9orf72*, 9p21.2	AD
Perry syndrome	*Dynactin 1 (DCTN1)*, 2p13.1	AD
DYT/PARK-ATP1A3 (rapid-onset dystonia–parkinsonism)	*ATP1A3*, 19q13.2	AD
Lubag disease (X-linked dystonia–parkinsonism)	*TAF1*, Xq13.1	X-linked
Fragile X tremor/ataxia syndrome or FXTAS	*FMR1*, Xq27.3	X-linked

AD autosomal dominant, *AR* autosomal recessive, *PKAN* pantothenate kinase-associated neurodegeneration, *PLAN* PLA2G6-associated neurodegeneration, *FAHN* fatty acid hydrolase-associated neurodegeneration (based on information from [80–87])

Susceptibility Genes

Research efforts have identified genetic variants that alter the risk of PD, rather than causing it. For instance, heterozygous mutations in the *GBA* gene (encoding lysosomal enzyme glucocerebrosidase that causes Gaucher disease when homozygous mutations are present) are a well-established genetic risk factor for PD, with carriers having a fivefold greater risk [78, 80, 88].

While the first studies consisted of small case-control studies using candidate genes, more recently genome-wide association studies (GWAS) have provided a comprehensive study of the genome, identifying loci that contain genetic variants that confer risk for a certain disease. This analysis is based in the common disease/common variant theory, postulating that, for common diseases, the risk is likely to be conferred by a constellation of common variants that individually increase risk by a relatively small amount. So, the risk could be derived from the joint, the risk of the identified loci in combination, tracing a risk profile for PD. They have provided potential associations between common genes, single-variant polymorphisms (SNPs) and PD, and both PARK-designated and non-PARK-designated genes (including *BST1, CCDCC2/HIP1R, DGKQ/GAK, GBA, LRRK2, MAPT, MCCC1/LAMP3, SNCA, STK39,* and *SYT11/ RAB25SNCA, UCHL1, LRRK2, PARK 16, GAK, MAPT, GBA, NAT2, INOS2A, GAK, HLA-DRA,* and *APOE*) [89–92].

Neuropathological Correlates

PD, like most neurodegenerative disorders, is caused by pathological accumulation of abnormal proteins within vulnerable neuronal populations, leading to cell death. Neuropathologically PD is characterized by the presence of round eosinophilic inclusions in the neuronal perikarya called Lewy bodies (LB), with similar inclusions within the cell processes referred to as Lewy neurites (LN). LB and LN are mainly composed of alpha-synuclein, but also contain neurofilaments and ubiquitin [93]. Although LB deposition is the neuropathological hallmark of PD, it occurs also in a number of other diseases such as Dementia with Lewy bodies (DLB), multiple system atrophy (MSA), and PLAN-associated neurodegeneration with brain iron accumulation (NBIA type 2) together referred to as synucleinopathies [93].

In PD (both sporadic and most inherited types, Table 17.2), there is neuronal loss within the *substantia nigra pars compacta* (SNpc), which projects to the putamen (the dopaminergic nigrostriatal pathway), and of the noradrenergic neurons in *locus coeruleus*. Interestingly, in the SNpc, this neuronal loss is not random but rather occurs in a region-specific manner, with remarkable loss of the ventrolateral tier, while cell loss from the medial tier does not differ significantly from normal aging. Parkinsonian features emerge when there is moderate to severe loss of these neurons [93]. Nonetheless, LB pathology is not limited to these two nuclei. In fact, it is found in several vulnerable central nervous system (CNS) regions and also in the peripheral autonomic nervous system, including the enteric plexus, paravertebral autonomic ganglia, and sympathetic nerve fibers in the adrenal gland and heart [93, 94].

Prion Propagation of PD Pathology

In 2003, Braak and co-workers have proposed a neuropathological staging system for LB pathology in PD (Table 17.4) [95]. According to these authors, there is caudo-rostral progression of pathology, which is first seen in the dorsal motor nucleus of the vagus nerve and olfactory bulb, with subsequent loss of pigmented neurons in the SNpc, which relates to the first recognizable motor symptoms of PD noticeable only at pathological stage IV. Expanding on this concept, in 2008, Halliday and co-workers have identified three main pathological patterns: that seen in younger patients, who follow the slow progression proposed by the Braak staging; another one typical of patients with older-onset disease (PD with dementia, PDD), who have shorter survival and higher cortical LB loads (both limbic and neocortical) earlier in their disease course, eventually coexisting with Alzheimer's pathology (AP); and those presenting with typical dementia with Lewy bodies (DLB), who have considerable diffuse neocortical pathology at the onset, often coexisting with AP [96, 97].

This theory of disease propagation gained further support when, in 2008, autopsies of PD patients who had received implanted embryonic tissue in the 1980s to 1990s revealed LB pathology, not only in the patients' brains but also in the grafted neurons [98, 99]. These grafted neurons seemed too young to have developed alpha-synuclein aggregates through independent cell-autonomous processes, and several cellular mechanisms were proposed to explain these finding, including neuroinflammation, oxidative stress, or lack of neurotrophic support. However, the theory that has gained greater support hypothesizes that these aggregates result from protein transfer from the host brain cells to graft neurons, behaving like a prion protein [100, 101]. In animal models, injections of alpha-synuclein into animals result in neurons with intracellular inclusions at the injection sites, from where they can spread to distant locations [102–104]. Once inside a new neuron, this exogenous alpha-synuclein can oligomerize with the endogenous protein and seed the formation of aggregates [100–102].

Table 17.4 Neuropathological staging of PD proposed by Braak et al. (see [95])

Parkinson's disease Braak staging	Anatomical site of pathological findings
Stage 1 and 2	Brainstem (medulla oblongata and pontine tegmentum) – Dorsal motor nucleus of the vagus nerve – Intermediate reticular zone – *Locus coeruleus*
Stage 3 and 4	Midbrain, basal prosencephalon, and mesocortex – *Substantia nigra pars compacta* – Transentorhinal cortex – Hippocampal CA2 sector
Stage 5 and 6	Neocortex – Prefrontal cortex – Temporal cortex – Insular cortex – Anterior cingulate cortex

The Gut as Starting Point

As mentioned later in the text, constipation and olfactory impairment are two of the most common early non-motor manifestations of PD, which is also consistent with the Braak staging (see previous section) [105]. Also, LB pathology has been detected in the gastrointestinal tract up to 20 years prior to the clinical diagnosis of PD [106]. Bridging clinical and experimental observations with pathological findings, Braak and co-workers postulated the dual-hit hypothesis, which states that a neurotropic pathogen/toxin could enter the brain via two routes: (1) nasal, with anterograde progression into the temporal lobe and (2) enteric, with retrograde progression to the dorsal motor nucleus of the vagus nerve (DMNV), from where they could reach the medulla and spread [107]. Evidence from animal models also supports this theory, as chronic intragastric administration of low doses of rotenone in wild-type mice triggered neuropathological changes typical of PD, with spatiotemporal progression. In another experiment, mice undergoing hemivagotomy and then treated with rotenone showed significantly lower accumulation of alpha-synuclein in the DMNV deprived of nervous connection with the digestive tract due to hemivagotomy, as compared to the side remaining anatomically intact [105]. Adding to this body of evidence, human research findings have been reported, with lower PD risk associated with truncal vagotomy, but not superselective vagotomy [108].

This hypothesis could also explain some of the environmental risks. One of the proposed explanations for this effect is the change in the composition of the gut microbiome, by modulation of intestinal inflammation, leading to less alpha-synuclein aggregation, and lower risk of PD. Consistent with this theory, data regarding intestinal microbiota in PD patients suggest that there are changes in this population and even correlations with the motor phenotype. Nevertheless, reverse causality could also explain these findings, and whether these are cause or consequence remains to be determined [105]. Of note, both the Braak staging system and the dual-hit hypothesis are controversial, and there is published divergent literature [101].

Clinical Features, Diagnosis, and Disease Course

Parkinsonism refers to the clinical constellation of bradykinesia, rigidity, resting tremor, and postural and gait impairment [109]. Defining PD as the cause of this syndrome requires careful history taking and examination, followed by a few paraclinical tests, to exclude other causes of parkinsonism. There is currently no single biomarker that can accurately define or refute the presence of PD [110].

Motor Symptoms

Establishing the presence of *bradykinesia* is the first step in diagnosing PD. Movements become slower and of smaller amplitude with repeated tasks (decremental slowness). Early in disease course, bradykinesia can manifest as decreased facial expression (hypomimia) and soft speech (hypophonia) [111].

Rigidity is the resistance of muscles to passive movement around a joint. Unlike spasticity or paratonia, it is not altered by the amplitude or velocity of the maneuver but can be enhanced by asking the patient to perform voluntary movements of the contralateral limb (Froment's sign). Patients can report rigidity as decreased range of movement and shoulder pain (often misdiagnosed as orthopedic or rheumatologic).

Rest tremor of the arms and legs usually starts unilaterally and distally, with a typical frequency of 4–6 Hz. When affecting the thumb and forefinger, it can acquire the classical "pill rolling" characteristics. It then progresses more proximally and to the contralateral side and frequently affects the jaw and tongue. Head tremor in PD is very rare and should prompt careful reconsideration of the diagnosis. Asymmetric postural tremor occurs at times; it is usually faster (6–7 Hz) than rest tremor and occurs immediately on stretching out the arms, but it is not useful to support the diagnosis of PD. If the posture is maintained, re-emergent tremor can occur with the same frequency of rest tremor [109].

Other motor features of PD, which may not develop until well into the disease course, include postural instability, gait initiation difficulties, freezing, progressively flexed posture, and dysphagia.

Non-motor Symptoms

Although the diagnosis of PD currently relies on the typical motor features, various non-motor symptoms (NMS) emerge in PD and become increasingly prevalent over the course of the disease [112]. The onset of some of them can precede the classic clinical motor picture by years or even decades. They are a major determinant of quality of life, progression of overall disability, and of nursing home placement. NMS include disorders of sleep–wake cycle regulation, cognitive impairment, disorders of mood and affect, autonomic dysfunction, and sensory symptoms [110]. The neuroanatomical basis of NMS remains largely undefined. Current dopaminergic treatment usually causes little benefit on those features, because most of them are related to non-dopaminergic changes.

Sensory Features

Hyposmia or anosmia is present in 90% of patients with PD, usually bilaterally and may precede the onset of motor features [112]. Alterations in olfaction in PD are most likely due to changes in central olfactory processing in the olfactory bulb and amygdala, related to substance P and acetylcholine deficiency. This theory is supported by evidence of normal biopsy samples of the olfactory epithelium [113], the presence of Lewy bodies and Lewy neurites in the olfactory bulb and cortex, other brain regions related to olfaction [114], and atrophy of the olfactory bulb in MRI studies of PD patients [115]. The failure of olfactory deficits to respond to dopaminergic medications also supports the lack of involvement of dopaminergic systems.

Visual disturbances in PD manifest with impaired acuity and color vision, diplopia, and positive phenomena (e.g., illusions, hallucinations). Their incidence increases with disease progression [112] and does not improve with dopaminergic therapy (indeed visual hallucinations may worsen, particularly with dopamine receptor agonist treatment). Impaired acuity and color vision may be related to changes in dopaminergic transmission in the retina. Indeed, retinal thinning is found in both patients with idiopathic PD and carriers of mutations in *GBA* with and without parkinsonism [116]. The old concept that visual hallucinations in PD came only as a consequence of dopaminergic therapy is incorrect, as such symptoms are present in untreated patients and may occur as early as the prodromal phase. Visual hallucinations in PD have been linked with perceptual, executive, and sleep dysfunction, are a predictor of cognitive decline in later disease, and probably reflect the distribution of Lewy body pathology in the occipital cortex [117].

Changes in sensory function and onset of pain are a common and frequently underreported feature of PD, affecting up to 80% of patients [118]. Several methods have been proposed to classify the complex pain syndrome of PD. A recent clinical tool is the King's Parkinson's disease pain scale, which classifies pain in PD in musculoskeletal, fluctuation related, central, nocturnal, orofacial, and peripheral pain [119]. The loss of dopaminergic input to the basal ganglia alters sensory perception, and in fact pain fluctuates with the motor function, often worsening during the off state. However, there are both dopaminergic and non-dopaminergic pain pathways, as well as neuropathic and nociceptive pain. The exact contribution of each system to pain in PD is unclear, as multiple neurotransmitter pathways are involved, making specific pain-relieving treatments challenging.

Autonomic Dysfunction

Autonomic dysfunction in PD encompasses bladder, bowel, and sexual dysfunction, as well as cardiovascular complications such as postural hypotension. Urinary dysfunction in PD is associated with detrusor muscle hyperreflexia due to a centrally mediated mechanism related to the loss of the inhibitory role of the basal ganglia. No specific alterations have been found to occur in the bladder of patients with PD [120]. Urinary symptoms include nocturia, increased frequency, and urgency of micturition. Dopaminergic therapy improves bladder symptoms in a minority of PD patients, but no specifically aimed intervention towards this comorbidity has been proven effective to this day [112]. Dysfunction occurs along the entire length of the gastrointestinal tract in PD, translated clinically in excessive salivation, dysphagia, impaired gastric emptying, constipation, and impaired defecation. The deposition of Lewy bodies at almost every level of the gastrointestinal tract in PD [121] and early pathological involvement of the dorsal motor nucleus of the vagus nerve (important in autonomic control of the bowel) may explain why gastrointestinal disturbances such as constipation occur earlier than alterations in motor function. The usually pronounced time lapse between these symptoms and the diagnosis of PD has raised the possibility that constipation is a risk factor for PD as well as a prodromal marker [122]. Apomorphine and levodopa infusions improve gastrointestinal motility. The gastrointestinal effects of apomorphine are not prevented by the peripheral dopamine antagonist domperidone, suggesting that they have a central origin. As for the majority of non-motor symptoms, other neurotransmitters may also be involved—there is preliminary evidence of a serotoninergic component to impaired motility [123].

Cardiovascular autonomic dysfunction is common in PD, causing both orthostatic hypotension and labile hypertension. Orthostatic hypotension is defined as a decrease in systolic blood pressure of 20 mmHg or a decrease in diastolic blood pressure of 10 mmHg within 3 min of standing when compared with blood pressure from the sitting or supine position. It results from an inadequate sympathetic response to postural changes in blood pressure and occurs in PD with a reported frequency of 30–58% [124]. Being one of the earliest premotor symptoms, it has been related to an increased risk of degenerative synucleinopathies within 10 years of diagnosis [125]. Labile hypertension includes supine hypertension (associated with target organ damage, as well as an increased risk of cardiovascular events in PD patients) and postprandial hypotension. Changes in heart rate have also been documented in untreated PD patients [126]. Cardiovascular function seems to be related to motor fluctuations. During off periods, patients have a higher resting heart rate and both a greater orthostatic fall of blood pressure and supine hypertension.

Sleep Disorders

Most patients with PD suffer from disturbances in sleep and wakefulness. These can be divided into two categories: daytime somnolence and sleep attacks, and nocturnal sleep disturbances. The latter encompasses not only rapid eye movement behavior disorder (RBD) but also insomnia (disease-related or drug-induced), periodic limb movements in sleep, and restless legs syndrome. The prevalence of sleep-related disturbances increases with disease duration.

RBD refers to the enactment of dreams during REM sleep, enabled by the loss of the normal REM sleep atonia. History taking is usually enough to presume RBD; however, its definite diagnosis requires polysomnography to document the absence of normal REM sleep atonia. The median estimate time between RBD symptom onset and the development of degenerative disease is 13 years.

Arousal and wakefulness are maintained by a complex neuronal network that connects several brain structures. Dopaminergic neurons in the ventral tegmental area and *substantia nigra* (SN) receive inputs from hypothalamic projections and form loops that ascend through the thalamus and cortex and descend through the pedunculopontine nucleus and the reticular formation. The early involvement of the brainstem in PD is the reason why sleep disturbances frequently precede motor symptoms. The multiple neurotransmitters involved in these pathways (dopamine, 5HT, and noradrenaline) may explain why some sleep disturbances respond better to dopaminergic treatment than others. For instance, restless leg syndrome and periodic limb movements seem to improve, while excessive daytime sleepiness and sleep attacks may be worsened by dopaminergic drugs. Insomnia is especially difficult to manage in PD patients, as it may be exacerbated by the return of motor symptoms during the night, or conversely by drug-related features such as nightmares, hallucinations, dyskinesia, and/or dystonia [125].

Neuropsychiatric Disorders

Neuropsychiatric features can occur in PD from the prodromal premotor phase to the late stages of the disease.

Anxiety is felt by up to 60% of PD patients and is more frequent in young women, people with young-onset PD, and in later stages of the disease. Generalized anxiety, panic attacks, and social phobias usually (but not always) occur in association with depression. Interestingly, anxiety fluctuates with the motor status, increasing during off periods.

Clinical significant depression affects 35% of patients [127], but milder symptoms are more frequent, in particular anhedonia and apathy. The early onset of neuropsychiatric features is thought to reflect its relation to pathology outside the nigrostriatal pathway. Recent evidence suggests that noradrenergic function is particularly compromised in PD-associated depression in comparison with endogenous depression [122]. The biochemical and pathological basis of depression and anxiety in PD is complex, as both can be related to the underlying neurodegenerative process or come as a pathological reaction to the perceived disability. Improvement of mood disorders with dopaminergic treatment and deep brain stimulation (DBS) could either point to a dopaminergic component of such symptoms or be secondary to improved motor function.

Apathy and fatigue occur in over half of individuals with PD and are increasingly recognized as independent non-motor features of the disease. Fatigue is not only a consequence of motor dysfunction but is rather related to the disease itself, as it occurs in patients with good motor function [112].

Psychotic symptoms in PD include illusions, hallucinations, and delusions that form a continuum progressing over the course of the disease [128]. In early stages of PD, milder phenomena occur, usually with preserved insight. Examples include visual illusions (misperception of actual stimuli, e.g., a lamp may look like a person), presence hallucinations (perceptual experience of someone else being in the room, without visual content), visual hallucinations "de passage" (unspecific shadows or ill-formed human shaped images quickly showing up and fading in the periphery of the visual field), and even fully formed colorful and detailed visual hallucinations, typically people, animals, and objects. Such episodes are usually short (seconds to minutes) and may occur several times a day, particularly when the patient is alone or in dim light and quiet environment. In clinical practice, it is important to question the patient openly about these symptoms, as they may not be spontaneously voiced, out of fear or shame. In later PD stages, insight dwindles and delusions (false beliefs) and hallucinations in other modalities, such as auditory phenomena, may occur. Psychotic symptoms predict worsening cognitive function [121, 129].

Medication onset is a modifier rather than a necessary feature for the occurrence of psychotic symptoms. Current evidence indicates that the evolution of psychotic symptoms is related to the progression of brain LB pathology [95].

Cognitive Impairment

Cognitive decline and dementia are a major cause of disability in PD patients. Prevalence of dementia ranges from 15–20% after 5 years to 46% at 10 years, with individuals with a predominantly bradykinetic-rigid phenotype bearing a greater risk. However, some level of cognitive dysfunction is present in a much larger percentage of patients. Early PD patients often present with executive function impairment, related to frontostriatal pathology, which may be dopamine responsive. The onset of cortical posterior cognitive deficits such as visual spatial construction, language, and memory heralds the progression of mild cognitive impairment (MCI) to dementia. This "dual syndrome hypothesis" of cognition in PD is supported by clinical evidence that MCI can either progress to dementia, or remain stable, or even revert in some patients [125]. The pathophysiology of cognitive impairment in PD is complex. Cortical LB pathology seems to play a major role, but amyloid plaque pathology may also contribute.

Clinical Diagnosis and the New Diagnostic Criteria

The accurate clinical diagnosis of PD can be challenging. Early diagnostic errors can rate as high as 24% in specialized centers. Common misclassifications in clinicopathological series are atypical parkinsonian syndromes, in particular multiple system atrophy and progressive supranuclear palsy, whereas clinically based studies find essential tremor, drug-induced parkinsonism, and vascular parkinsonism as the main diagnostic caveats [93]. Diagnostic criteria designed in 1988 by the UK Brain Bank to assign diagnosis in a pathological series became commonly used in clinical practice. Even though there is a high positive predictive value at the time of death, sensitivity and reliability of early diagnosis were only slightly above 80% at first visit, since many exclusionary features take time to emerge [130, 131].

A recent change of diagnostic criteria has been proposed by the International Parkinson and Movement Disorder Society [122], designed to render the diagnostic process as close to evaluation by an expert as possible and to enhance diagnostic confidence in early disease stages. The proposed diagnostic flow is outlined as follows: motor parkinsonism is the core feature of the disease, defined as bradykinesia plus rest tremor or rigidity. After documentation of parkinsonism, determination of PD as the cause relies on three categories of diagnostic features: absolute exclusion criteria (which rule out PD), red flags (which must be counterbalanced by additional supportive criteria to allow diagnosis of PD), and supportive criteria (positive features that increase confidence of the PD diagnosis). Two levels of certainty are delineated: clinically established PD (maximizing specificity at the expense of reduced sensitivity) and probable PD (which balances sensitivity and specificity).

When thinking about the pathological basis of PD, it seems logical that a negative presynaptic dopamine transporter scan (i.e., normal functioning dopaminergic system on SPECT) should be considered an absolute exclusion criterion. The focus on early diagnosis and the advances on diagnostic testing translate in the inclusion of positive meta-iodobenzylguanidine cardiac scintigraphy (MIGB) and olfactory loss as supportive criteria.

The expanding knowledge on the vast non-motor profile of PD has led to a significant change in diagnostic criteria. Dementia is regarded as a non-motor feature of PD, frequently present early in the disease. It has been argued that dismissing the "one year rule" to differentiate PD from DLB could be premature, as it remains useful in clinical practice and research studies [132].

Parkinson's Disease Timeline

An International Parkinson Disease and Movement Disorder Society (MDS) task force recently suggested new terminology for the various stages of PD. In a system similar to other models, three stages were identified: preclinical, prodromal, and clinical. Early and mid-stage clinical PD encompasses the premotor, motor, and motor complications which culminate in the late stage of disease [122].

Preclinical
Preclinical disease refers to a state in which neurodegeneration has started, but no symptoms or signs are evident. Its diagnosis remains dependent upon reliable biomarkers that precede clinical signs, which have not currently been identified [133].

Prodromal
Overt clinical PD is preceded by a prodromal phase of years or even decades, characterized by specific non-motor symptoms and subtle motor dysfunction. Prodromal disease can be understood by interpreting the neuropathological staging of PD while taking into account the high threshold of dopaminergic function loss before motor symptoms appear. Efforts to develop neuroprotective therapies are focusing on the early stages of disease, which offer the best opportunity to intervene [125].

MDS has also published research diagnostic criteria to estimate the probability that an individual has prodromal PD [133]. Prodromal and risk markers are combined into an evidence-based and adaptable Bayesian model whose predictive validity and, therefore, usefulness for selecting populations for "disease prevention" trials are still under research [109]. The strongest marker of prodromal PD is rapid eye movement (REM) sleep behavior disorder. Other markers supported by strong evidence include subtle motor dysfunction, olfactory loss, autonomic dysfunction, and affective disorders. Diagnostic testing is cautiously included in this model. Hyperechogenicity of SN, PD-related pattern on SPECT/PET, dopaminergic PET/SPECT abnormalities, hippocampal hyperperfusion, electrocardiogram beat-to-beat variability, and alpha-synuclein gastrointestinal biopsy still yield a low predictive value [133].

Clinical
Traditionally, progression of PD is regarded as an increase in severity of motor symptoms with the emergence of levodopa-induced motor complications and mounting disability. This motor progression is nonlinear, with a more rapid decline

in motor function in earlier stages compared with later stages. Most patients with PD who receive dopaminergic therapy go on to develop motor complications, whose occurrence is related to disease duration and to the duration and cumulative dose of dopaminergic drugs. An estimated 40–50% of patients will develop motor complications after 4–6 years of treatment [109]. On top of the disability caused by motor symptoms and motor complications, there is also the burden of NMS (Fig. 17.1).

Wearing-off is usually the first motor complication surfacing in the course of disease and refers to the re-emergence of dopamine deficiency-related symptoms. In these periods, patients present with stiffness, slowness, or tremor. Greater symptom severity occurs during "off periods," while "on periods" occur when effective dopamine replacement restores motor function.

Dyskinesias are involuntary choreiform or dystonic movements related to the variable levels of dopamine resulting in abnormal patterns of basal ganglia activity. Peak-dose dyskinesias occur when levodopa levels are at their highest. On the other end of the spectrum are off period dyskinesias, typically painful abnormal posturing (dystonia). Biphasic dyskinesias occur when the levels of levodopa are rising, then reduce or disappear when a certain threshold is reached, and finally return when levels fall again; these are particularly difficult to manage. There are other motor complications and fluctuations such as *delayed on* (longtime interval until a levodopa dose kicks in), *no-on* (failure of a single dose to bring any benefits), and *sudden off* states.

The term "late stage PD" concerns the clinical phenotype in which disability is mostly associated with non-motor symptoms such as dementia, psychosis, and

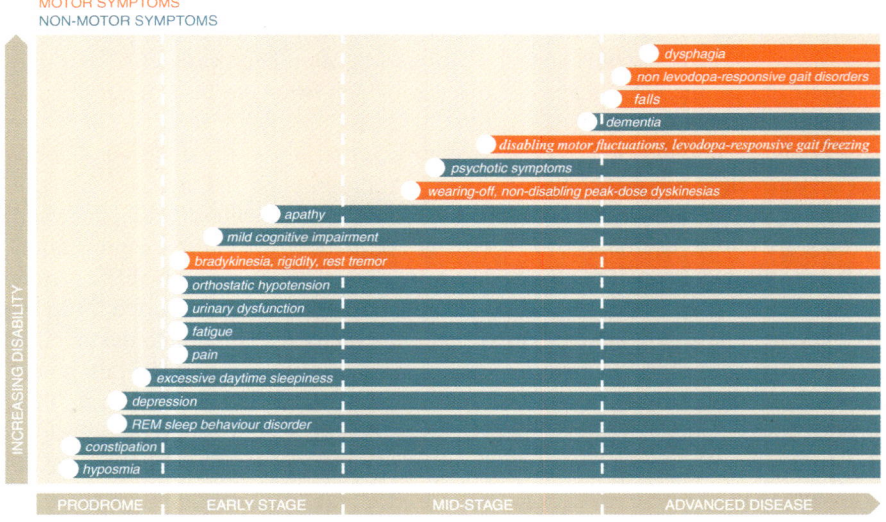

Fig. 17.1 Typical Parkinson's disease clinical progression, throughout the several stages (prodrome, early stage, mid-stage, and advanced PD). Please note the various motor and non-motor symptoms, as well as their approximate timing of appearance, taking into account that variations among individual patients exist. Every symptom contributes cumulatively to increasing disability and loss of quality of life

motor symptoms that are unresponsive to levodopa, such as postural instability, severe gait changes, and falls. These symptoms are the strongest independent predictors of institutionalization and death, and effective treatments are still lacking [134]. Late stages of PD are of increasing clinical relevance owing to improved treatment and survival. Unlike other disease stages, the duration of late stage PD is relatively homogeneous [135].

Parkinson's Disease Therapy

The effective treatment of PD warrants both solid scientific knowledge and clinical experience. Therapeutic strategies should be defined according to disease stage, the type and severity of motor and non-motor symptoms, extent of functional impairment determined by each disease manifestation, and PD-unrelated circumstances such as comorbidities, concomitant medications, and the patient's social and economic setting. The treatment of PD has been thoroughly reviewed in several evidence-based publications [136–139]. In this section we outline the principles of PD therapy from a practical standpoint.

Treatment of Motor Symptoms

Parkinsonian features often emerge subtly and can go unnoticed or remain misinterpreted for several years until expert advice is finally sought. Disability may be inexistent or mild in earlier disease stages, and there is no mandatory reason to start therapy at this point, as no disease-modifying therapy is currently available [136, 137, 140]. In practice, therapy is actively offered to patients when symptoms cause discomfort, disability, or interfere with the person's lifestyle.

Early Disease Stages

After deciding to start therapy, an intricate choice process is carried out, as there are different treatment options. These are explained to the patient, including the advantages and pitfalls of each alternative. A number of patients with mild symptoms fare quite well for some time on monotherapy with a monoamine oxidase B (MAO-B) inhibitor. MAO-B inhibitors are taken once (e.g., rasagiline 1 mg tablets, selegiline 1.25 mg orally disintegrating tablets) or twice (e.g., selegiline 5 mg tablets) a day. They are usually well tolerated, although caution should be taken to avoid concomitant administration of drugs that increase the risk of serotonergic syndrome, such as antidepressants. MAO-B inhibitors provide significant, although modest, symptomatic benefit, and patients will eventually need add-on or alternative drugs at some point—dopamine receptor agonists and/or levodopa. MAO-B inhibitors are associated with a lower risk of motor complications as compared to levodopa, but the benefits concerning PD symptoms and quality of life are also lower.

Dopamine receptor agonists (DRAs) can be of the ergoline (e.g., bromocriptine, cabergoline, dihydroergocryptine, pergolide) or non-ergoline (e.g., piribedil, pramipexole, ropinirole, rotigotine) types. Ergoline DRAs have been associated with fibrotic reactions affecting the heart, lungs, and retroperitoneal space, thus warranting special vigilance, and some of them have been withdrawn in many countries. Once daily controlled release formulations of pramipexole, ropinirole, and rotigotine are available. Rotigotine is the only drug available as transdermal patches. Common side effects of DRAs include nausea, vomiting, hypotension, lower limb edema, and triggering or worsening of hallucinations or delusions. Excessive daytime sleepiness and sudden sleep "attacks" may cause important problems. Impulse control disorders (e.g., pathological gambling, compulsive shopping, hypersexuality, binge eating) may be seen in more than 10% of patients, particularly in younger men, or those with specific previous personality traits (e.g., novelty seekers) or behaviors (e.g., tobacco, drug, or alcohol dependence). Patients must be warned about these possible events before choosing to have DRAs and should be carefully monitored after these have been started. DRAs provide a significant symptomatic relief and offer the possibility of a once-daily regimen, which is an advantage over levodopa, especially in less compliant patients or those with an active life. On the other hand, DRAs have been associated with a lower risk of motor complications in comparison to levodopa. Nonetheless it is predictable that, at some point, many patients on DRAs will be switched to levodopa, either partially or completely, given the advantages of levodopa. Although time to onset of motor complications from the point at which therapy is started may be longer if DRAs are used, patients treated with levodopa display a better symptomatic relief; [141] thus, delaying levodopa may not be the wisest strategy for every patient. DRAs are usually used in patients younger than 70 years old, due to the lower risk of serious cognitive and behavioral adverse effects in this age range. Nonetheless, many older patients tolerate DRAs well and benefit from their use.

Immediate-release levodopa is the single most effective drug in the treatment of PD, namely, concerning the improvement of motor symptoms and quality of life. It is administered with a dopa decarboxylase inhibitor (e.g., carbidopa, benserazide) in order to avoid peripheral dopaminergic action and adverse effects, while increasing levodopa bioavailability beyond the blood–brain barrier. A practical golden rule with DRAs and levodopa is to "start low and go slow," in order to lessen adverse effects while trying to find the lowest effective dose, which is important due to the fact that the cumulative dopaminergic dose correlates with the risk of motor complications [142]. It is important to find an acceptable balance between clinical benefits and risks, in order to allow patients to have an independent life without troublesome parkinsonian symptoms, while keeping them at lower risk of important motor complications and non-motor adverse effects as much as possible. Levodopa is usually the top choice in older patients or those who have predictably lower survival either due to PD or comorbidities, given the lower risk of intolerable adverse effects, while maximizing clinical benefit. It is usually started at 50 mg one to three times daily, and a slow dose titration is performed, with the usual target therapeutic dose ranging from 150 to 400 mg per day in early disease stages. Prolonged-release levodopa should not be used during daytime due to its erratic drug delivery. Combining levodopa and/or DRAs and/or MAO-B inhibitors is common practice. Levodopa should be administered without protein-rich food, as absorption will likely be impaired and serum levels will become unpredictable. In practice, we advise patients to have levodopa at least 20–30 min before any meal, and intervals between administrations should be more or less the same (e.g., every 3–5 h, depending on the number of daily doses).

Typically, a "honeymoon" period will last for a few years (usually 2–5) after levodopa and/or DRAs have been started, with patients experiencing significant improvement of symptoms while living a fully independent life, until motor complications finally emerge.

Anticholinergics (benztropine, trihexyphenidyl) should not be used routinely due to the risk of important adverse effects associated with a marginal ability to improve PD symptoms, with the exception of rest tremor. These drugs should be reserved for younger patients without cognitive impairment and only if there is upsetting rest tremor not improved by other therapeutic approaches.

Motor Complications

After a few years of dopaminergic therapy, particularly after levodopa has been introduced, motor complications emerge in a considerable proportion of patients, albeit of variable severity. Wearing-off and peak-dose dyskinesias are usually seen first. For thorough evidence-based reviews on the treatment of motor complications in PD, please refer to the guidelines issued by the European Federation of Neurological Societies/Movement Disorders Society European Section [136] and the Movement Disorder Society review on this subject [137]. Table 17.5 offers a few practical strategies on how to manage motor complications in PD.

Expert optimized medical therapy is often not enough to provide the desired quality of life, either due to motor complications or drug-induced intolerable adverse effects. In this case deep brain stimulation (DBS) might be considered in selected patients whose symptoms respond to levodopa (assessed by formal levodopa challenge). An age limit of 70 years old was established in many centers, but this is usually flexible and discussed on a case-by-case basis within the multidisciplinary team. In order to be offered the procedure, patients should not be demented, must not have a severe or uncontrolled psychiatric disorder, and cannot have structural brain lesions that prevent correct electrode placement or contraindications for intracranial surgery. Several randomized controlled trials have shown the clinical benefits of DBS in PD, even in patients showing early motor complications [144–146].

Further options for managing patients with severe motor complications not amenable by optimized medical treatment are subcutaneous apomorphine and the intestinal infusion of levodopa/carbidopa gel. There is evidence of efficacy in PD for any of these options [147, 148].

Treatment of Non-motor Symptoms

In general there is a relative lack of evidence to guide the treatment of non-motor symptoms (NMS) in PD. While there are some controlled data regarding a few of them [138, 139], many strategies are chosen on a purely empirical basis.

Table 17.5 Practical drug-based strategies to manage motor complications in Parkinson's disease

Motor complication	Drug-based management strategies
Wearing-off	• Add DRA (if not part of current therapy) • Increase levodopa or DRA dose • Increase number of levodopa intakes while decreasing time intervals between each one • Add MAO-B inhibitor (e.g., rasagiline, selegiline) • Add COMT inhibitor (e.g., entacapone, opicapone, tolcapone) • Add safinamide • DBS of the STN or GPi
Peak-dose dyskinesias	• Add amantadine • Decrease amount of daily levodopa • Stop MAO-B inhibitor or COMT inhibitor if part of current therapy • DBS of the STN or GPi
Nocturnal and morning off	• Add prolonged-release levodopa at nighttime • Add MAO-B inhibitor • Add prolonged-release DRA at nighttime or 24-h rotigotine transdermal patch (or increase dose if already part of regimen) • Liquid levodopa in the morning • Subcutaneous apomorphine (morning dose, nighttime infusion) • Intestinal levodopa/carbidopa gel infusion • DBS of the STN or GPi
Severe fluctuations, including biphasic dyskinesias	• Consider all the strategies mentioned above
Gait freezing	• "Off" state freezing: same strategies as for wearing-off • "On" state freezing: unresponsive to dopaminergic therapies • Cueing (visual, auditory)

COMT catechol-*O*-methyltransferase, *DRA* dopamine receptor agonist, *DBS* deep brain stimulation, *GPi* globus pallidus internus, *MAO-B* monoamine oxidase B, *STN* subthalamic nucleus (based on data from [136, 137, 143])

In practice, psychotic symptoms can be treated with quetiapine, given that this drug has a low potential to exacerbate parkinsonism, although there is no real evidence of efficacy in PD. Typical (e.g., haloperidol, chlorpromazine) and many second-generation (e.g., risperidone, olanzapine) antipsychotics should be avoided, due to the high risk of worsening parkinsonian symptoms. Clozapine is efficacious but carries the need for frequently monitoring the blood cell count due to the risk of agranulocytosis.

Dementia may be improved with cholinesterase inhibitors, which may also exert beneficial effects on hallucinations (e.g., donepezil, rivastigmine, galantamine), or memantine.

Depression and anxiety are usually treated with serotonin selective reuptake inhibitors (SSRIs, e.g., sertraline, escitalopram), or serotonin and noradrenaline reuptake inhibitors (SNRIs, e.g., venlafaxine). If needed to reduce acute significant anxiety, benzodiazepines should be used exceptionally and only during short periods of time (less than 4 weeks).

REM sleep behavior disorder usually improves under clonazepam taken at nighttime. Melatonin is an alternative. Periodic limb movements in sleep also improve under clonazepam.

Constipation is treated with laxatives, particularly macrogol, and patients should be advised to increase water and vegetable consumption, along with higher levels of physical activity. Droxidopa and midodrine may be considered for troublesome orthostatic hypotension. Phosphodiesterase 5 (PDE5) inhibitors such as sildenafil and tadalafil may improve erectile dysfunction. Urinary dysfunction is a particularly complex issue, warranting specialized urological assessment and management.

One should always keep in mind that NMS may worsen under drugs used to treat PD. Whenever possible it is recommended that these should be stopped (or their dose decreased, as appropriate), in the following circumstances:

- Dementia, delirium, hallucinations, delusions: amantadine, MAO-B inhibitors, COMT inhibitors, DRAs, anticholinergics, benzodiazepines, tricyclic antidepressants, oxybutynin
- Orthostatic hypotension: levodopa, DRAs
- Constipation, urinary retention: amantadine, anticholinergics, tricyclic antidepressants
- Erectile dysfunction, anorgasmia: SSRIs, NSRIs

Other Therapeutic Interventions

Non-pharmacological and non-surgical therapies can be useful and should be offered to PD patients whenever appropriate and available, such as physiotherapy, physical exercise, speech and language therapy, and occupational therapy [137, 149, 150]. The detailed discussion of these interventions falls beyond the scope of this text.

In conclusion, there is an array of effective symptomatic therapies for PD, particularly regarding motor symptoms. Clinical expertise and thorough knowledge on available options are essential to achieve optimized results. Future research will focus on developing effective disease-modifying therapies, a goal for which precision medicine might be fundamental [81, 140, 151].

Acknowledgements The authors thank Rosa Bandeirinha for drawing Fig. 17.1.

References

1. Parkinson J. An essay on the shaking palsy. London: Sherwood, Neely, and Jones; 1817.
2. Dorsey E, Constantinescu R, Thompson J, Biglan K, Holloway R, Kieburtz K. Projected number of people with Parkinson disease in the most populous nations, 2005 through 2030. Neurology. 2007;68:384–6.
3. Twelves D, Perkins KSM, Uk M, Counsell C. Systematic review of incidence studies of Parkinson's disease. Mov Disord. 2003;18(1):19–31.
4. Ascherio A, Schwarzschild MA. The epidemiology of Parkinson's disease: risk factors and prevention. Lancet Neurol. 2016;15(12):1257–72.

5. Pringsheim T, Jette N, Frolkis A, Steeves TDL. The prevalence of Parkinson's disease: a systematic review and meta-analysis. Mov Disord. 2014;29(13):1583–90.
6. Okubadejo NU, Bower JH, Rocca WA, Maraganore DM. Parkinson's disease in Africa: a systematic review of epidemiologic and genetic studies. Mov Disord. 2006;21(12):2150–6.
7. Haaxma CA, Bloem BR, Borm GF, Oyen WJG, Leenders KL, Eshuis S, et al. Gender differences in Parkinson's disease. J Neurol Neurosurg Psychiatry. 2007;78(8):819–24.
8. Pupillo E, Cricelli C, Mazzoleni F, Cricelli I, Pasqua A, Pecchioli S, et al. Epidemiology of Parkinson's disease: a population-based study in primary care in Italy. Neuroepidemiology. 2016;47(1):38–45.
9. Blin P, Dureau-Pournin C, Foubert-Samier A, Grolleau A, Corbillon E, Jové J, et al. Parkinson's disease incidence and prevalence assessment in France using the national healthcare insurance database. Eur J Neurol. 2015;22(3):464–71.
10. Caslake R, Taylor K, Scott N, Gordon J, Harris C, Wilde K, et al. Age-, gender-, and socioeconomic status-specific incidence of Parkinson's disease and parkinsonism in North East Scotland: the PINE study. Parkinsonism Relat Disord. 2013;19(5):515–21.
11. Von Campenhausen S, Bornschein B, Wick R, Bötzel K, Sampaio C, Poewe W, et al. Prevalence and incidence of Parkinson's disease in Europe. Eur Neuropsychopharmacol. 2005;15(4):473–90.
12. Kieburtz K, Wunderle KB. Parkinson's disease: evidence for environmental risk factors. Mov Disord. 2013;28(1):8–13.
13. Hernán MA, Takkouche B, Caamaño-Isorna F, Gestal-Otero JJ. A meta-analysis of coffee drinking, cigarette smoking, and the risk of Parkinson's disease. Ann Neurol. 2002;52(3):276–84.
14. Ritz B. Pooled analysis of tobacco use and risk of Parkinson disease. Arch Neurol. 2007;64(7):990.
15. Calle EE, Thun MJ, Ascherio A, Thacker EL, Reilly EJO, Weisskopf MG, et al. Temporal relationship between cigarette smoking and risk of Parkinson disease. Neurology. 2007;68:764–9.
16. Chen H, Huang X, Guo X, Mailman RB, Park Y, Kamel F, et al. Smoking duration, intensity, and risk of Parkinson disease. Neurology. 2010;74(11):878–84.
17. Evans AH, Lawrence AD, Potts J, MacGregor L, Katzenschlager R, Shaw K, et al. Relationship between impulsive sensation seeking traits, smoking, alcohol and caffeine intake, and Parkinson's disease. J Neurol Neurosurg Psychiatry. 2006;77(3):317–21.
18. Menza M. The personality associated with Parkinson's disease. Curr Psychiatry Rep. 2000;2:421–6.
19. Ritz B, Lee P, Lassen CF, Lee P, Lassen CF. Parkinson disease and smoking revisited ease of quitting is an early sign of the disease. Neurology. 2014;83:1396–402.
20. Savica R, Grossardt BR, Bower JH, Ahlskog JE, Rocca WA. Time trends in the incidence of Parkinson disease. JAMA Neurol. 2016;73(8):981–9.
21. O'Reilly J, Chen H, Gardener H, Gao X, Schwarzschild MA, Ascherio A. Smoking and Parkinson's disease: using parental smoking as a proxy to explore causality. Am J Epidemiol. 2009;169(6):678–82.
22. Mellick GD, Gartner CE, Silburn PA, Battistutta D. Passive smoking and Parkinson disease. Neurology. 2006;67:179–81.
23. Palacios N, Gao X, Mccullough ML, Schwarzschild MA. Caffeine and risk of Parkinson's disease in a large cohort of men and women. Mov Disord. 2012;27(10):1276–82.
24. Ascherio A, Chen H, Schwarzschild MA, Zhang SM. Caffeine, postmenopausal estrogen, and risk of Parkinson's disease. Neurology. 2003;60:790–5.
25. Ross GW, Abbott RD, Petrovitch H, Morens DM, Curb JD, Popper JS. Association of coffee and caffeine intake with the risk of Parkinson disease. JAMA. 2000;283(20):2674–9.
26. Costa J, Lunet N, Santos C, Santos J, Vaz-Carneiro A. Caffeine exposure and the risk of Parkinson's disease: a systematic review and meta-analysis of observational studies. J Alzheimers Dis. 2010;20(20):221–S238.

27. Palacios N, Gao X, Reilly EO, Schwarzschild M, Mccullough ML, Mayo T, et al. Alcohol and risk of Parkinson's disease in a large, prospective cohort of men and women. Mov Disord. 2012;27(8):980–7.
28. Zhang D, Jiang H. Alcohol intake and risk of Parkinson's disease: a meta-analysis of observational studies. Mov Disord. 2014;0(0):1–4.
29. Fukushima W, Miyake Y, Tanaka K, Sasaki S, Kiyohara C, Tsuboi Y, et al. Alcohol drinking and risk of Parkinson's disease: a case-control study in Japan. BMC Neurol. 2010;10(1):111.
30. Weisskopf MG, Reilly EO, Chen H, Schwarzschild MA, Ascherio A. Plasma urate and risk of Parkinson's disease. Am J Epidemiol. 2007;166(5):561–7.
31. Vera MDE, Rahman MM, Rankin J, Kopec J, Gao X, Choi H. Gout and the risk of Parkinson's disease: a cohort study. Arthritis Rheum. 2008;59(11):1549–54.
32. Schwarzschild MA, Schwid SR, Marek K, Watts A, Lang AE, Oakes D, et al. Serum urate as a predictor of clinical and radiographic progression in Parkinson disease. Arch Neurol. 2010;65(6):716–23.
33. Alonso A, Rodríguez LAG, Logroscino G, Hernán MA. Gout and risk of Parkinson disease: a prospective study. Neurology. 2010;69:1696–700.
34. Andreadou E, Nikolaou C, Gournaras F, Rentzos M, Boufidou F, Tsoutsou A, et al. Serum uric acid levels in patients with Parkinson's disease: their relationship to treatment and disease duration. Clin Neurol Neurosurg. 2009;111:724–8.
35. Gagne JJ, Power MC. Anti-inflammatory drugs and risk of Parkinson disease. Neurology. 2010;74:995–2001.
36. Chen H, Jacobs E. Nonsteroidal antiinflammatory drug use and the risk for Parkinson's disease. Ann Neurol. 2005;58:963–7.
37. Gao X, Chen H, Schwarzschild MA. Use of ibuprofen and risk of Parkinson disease. Neurology. 2011;76:863–9.
38. Alcalay RN, Gu Y, Mejia-santana H, Cote L, Marder KS, Scarmeas N. The association between Mediterranean diet adherence and Parkinson's disease participants and methods. Mov Disord. 2012;27(6):771–4.
39. Gao X, Chen H, Fung TT, Logroscino G, Schwarzschild MA, FB H, et al. Prospective study of dietary pattern and risk of Parkinson disease 1 – 3. Am J Clin Nutr. 2007;86:1486–94.
40. Gao X, Cassidy A, Schwarzschild MA. Habitual intake of dietary flavonoids and risk of Parkinson disease. Neurology. 2013;78:1138–45.
41. Yang F, Lagerros YT, Bellocco R, Adami H, Fang F, Pedersen NL, et al. Physical activity and risk of Parkinson's disease in the Swedish National March Cohort. Brain. 2015;138:269–75.
42. Xu Q, Park Y, Blair A. Physical activities and future risk of Parkinson disease. Neurology. 2010;75:341–8.
43. Lai BCL, Marion SA, Teschke K, Tsui JKC. Occupational and environmental risk factors for Parkinson's disease. Parkinsonism Relat Disord. 2002;8:297–309.
44. Tanner CM, Ross GW, Jewell SA, Hauser RA, Jankovic J, Factor SA, et al. Occupation and risk of Parkinsonism. Arch Neurol. 2015;66(9):1106–13.
45. Petrovitch H, Ross GW, Abbott RD, Sanderson WT. Plantation work and risk of Parkinson disease in a population-based longitudinal study. Arch Neurol. 2002;59:1787–92.
46. Firestone JA, Smith-weller T, Franklin G, Swanson P, Longstreth WT, Checkoway H. Pesticides and risk of Parkinson disease. Arch Neurol. 2015;62:91–5.
47. Racette BA, Nielsen SS, Sheppard L, Seixas N, Warden MN. Dose-dependent progression of parkinsonism in manganese-exposed welders. Neuroepidemiology. 2016;88:344–51.
48. Abbott RD, Ross GW, White LR, Sanderson WT, Burchfiel CM, Sharp DS, et al. Environmental, life-style, and physical precursors of clinical Parkinson's disease: recent findings from the Honolulu-Asia aging study. J Neurol. 2003;250(3):30–9.
49. Abbott RD, Ross GW, Petrovitch H, Masaki KH, Launer LJ, Nelson JS, et al. Midlife milk consumption and substantia nigra neuron density at death. Neurology. 2015;86:512–9.
50. Jafari S, Etminan M, Aminzadeh F, Samii A. Head injury and risk of Parkinson disease: a systematic review and meta-analysis. Mov Disord. 2013;28(9):1222–9.

51. Rugbjerg K, Ritz B, Korbo L, Martinussen N. Risk of Parkinson's disease after hospital contact for head injury: population based case control study. BMJ. 2008;337(a2494):1–10.
52. Fang F, Chen H, Ascertainment C. Head injury and Parkinson's disease: a population-based study. Mov Disord. 2012;27(13):1632–5.
53. Gardner RC, Burke JF, Nettiksimmons J, Goldman S, Tanner CM. Traumatic brain injury in later life increases risk for Parkinson's disease. Ann Neurol. 2015;77(6):987–95.
54. Liu R, Lu Y. Meta-analysis of the relationship between Parkinson disease and melanoma. Neurology. 2011;76:2002–9.
55. Olsen JH, Friis S, Frederiksen K. Malignant melanoma and other types of cancer preceding Parkinson disease. Epidemiology. 2006;17(5):582–7.
56. Gao X, Simon KC, Han J, Schwarzschild MA, Ascherio A. Genetic determinants of hair color and Parkinson's disease risk. Ann Neurol. 2009;65(1):76–82.
57. Kareus SA, Figueroa KP, Cannon-Albright LA, Pulst SM. Shared predispositions of parkinsonism and cancer. Arch Neurol. 2012;69(12):1572–7.
58. Gao X, Simon KC, Han J, Schwarzschild MA, Ascherio A. Family history of melanoma and Parkinson disease risk. Neurology. 2009;73:1286–91.
59. WY W, Kang K, Chen SL, Chiu SY, Yen AM, Fann JC, et al. Hepatitis C virus infection: a risk factor for Parkinson's disease. J Viral Hepat. 2015;22:784–91.
60. Pakpoor J, Noyce A, Selkihova M, Lees A. Viral hepatitis and Parkinson disease. A national record-linkage study. Neurology. 2017;88:1–5.
61. Tsai H, Liou H, Muo C, Lee C, Yen R, Kao C. Hepatitis C virus infection as a risk factor for Parkinson disease. Neurology. 2015;86:1–7.
62. Polymeropoulos MH, Lavedan C, Leroy E, Ide SE, Dehejia A, et al. Mutation in the α-Synuclein gene identified in families with Parkinson's disease. Science. 1997;276:2045–7; (80) 2012
63. Coelln R, von Dawson VL, Dawson TM. Parkin-associated Parkinson's disease. Cell Tissue Res. 2004;27:175–84.
64. Lesage S, Brice A, Curie-paris PM, Umr S. Parkinson's disease: from monogenic forms to genetic susceptibility factors. Hum Mol Genet. 2009;18(1):48–59.
65. Taipa R, Melo-pires M, Magalha M, Alonso I. DJ-1 linked parkinsonism (PARK7) is associated with Lewy body pathology. Brain. 2016;139(6):1680–7.
66. Christoph B, Lücking CB, Dürr A, Bonifati V, Vaughan J, De Michele G, Gasser T, Harhangi BS, Meco G, Denèfle P, Wood NW, Agid Y, Brice A, French Parkinson's Disease Genetics Study Group, European Consortium on Genetic Susceptibility in Parkinson's Disease. Association between early-onset Parkinson's disease and mutations in the parkin gene. N Engl J Med. 2000;342(21):1560–7.
67. Kumazawa R, Tomiyama H, Li Y, Imamichi Y, Funayama M, Yoshino H, et al. Mutation analysis of the PINK1 gene in 391 patients with Parkinson disease. Arch Neurol. 2008;65(6):802–8.
68. Olgiati S, Quadri M, Fang M, Rood JPMA, Saute JA, Chien HF, et al. DNAJC6 mutations associated with early-onset Parkinson's disease. Ann Neurol. 2016;79:244–56.
69. Marras C, Lang A, van de Warrenburg BP, Sue CM. Nomenclature of genetic movement disorders: recommendations of the International Parkinson and Movement Disorder Society Task Force. Mov Disord. 2016;31(4):436–57.
70. Wider C, Skipper L, Solida A, Brown L, Farrer M, Dickson D. Autosomal dominant dopa-responsive parkinsonism in a multigenerational Swiss family. Parkinsonism Relat Disord. 2008;14:465–70.
71. Lesage S, Bras J, Cormier- F, Condroyer C, Nicolas A, Darwent L, et al. Loss-of-function mutations in RAB39B are associated with typical early-onset Parkinson disease. Neurol Genet. 2015;1:1–3.
72. Samaranch L, Lorenzo-Betancor O, Arbelo JM, Ferrer I, Lorenzo E, Irigoyen J, et al. PINK1-linked parkinsonism is associated with Lewy body pathology. Brain. 2010;133:1128–43.
73. Köroglu Ç, Baysal L, Cetinkaya M, Karasoy H, Tolun A. DNAJC6 is responsible for juvenile parkinsonism with phenotypic variability. Parkinsonism Relat Disord. 2013;19:320–4.

74. Edvardson S, Cinnamon Y, Ta-shma A, Shaag A, Yim Y, Zenvirt S, et al. A deleterious mutation in DNAJC6 encoding the Auxilin, is associated with juvenile parkinsonism. PLoS One. 2012;7(5):4–8.
75. Vilariño-Guell C, Wider C, Ross OA, Dachsel JC, Kachergus JM, Lincoln SJ, et al. VPS35 mutations in Parkinson disease. Am J Hum Genet. 2011;89:162–7.
76. Funayama M, Ohe K, Amo T, Furuya N, Yamaguchi J, Saiki S, et al. Articles CHCHD 2 mutations in autosomal dominant late-onset Parkinson's disease: a genome-wide linkage and sequencing study. Lancet Glob Health. 2015;4422(14):1–9.
77. Wilson GR, Sim JCH, Mclean C, Giannandrea M, Galea CA, Riseley JR, et al. Mutations in RAB39B cause X-linked intellectual disability and early-onset Parkinson disease with a -Synuclein pathology. Am J Hum Genet. 2014;95(6):729–35.
78. Houlden H, Singleton AB. The genetics and neuropathology of Parkinson's disease. Acta Neuropathol. 2013;124(3):325–38.
79. Chartier-Harlin M-C, Dachsel JC, Vilariño-Guell C, Lincoln SJ, Lepretre F, Hulihan MM, et al. Translation initiator EIF4G1 mutations in familial Parkinson disease. Am J Hum Genet. 2011;89:398–406.
80. Ferreira M, Massano J. An updated review of Parkinson's disease genetics and clinicopathological correlations. Acta Neurol Scand. 2017;135(3):273–84.
81. von Coelln R, Shulman LM. Clinical subtypes and genetic heterogeneity of lumping and splitting in Parkinson disease. Curr Opin Neurol. 2016;29:727–34.
82. Bonifati V. Autosomal recessive parkinsonism. Parkinsonism Relat Disord. 2012;18:4–6.
83. Puschmann A. Parkinsonism and related disorders monogenic Parkinson's disease and parkinsonism: clinical phenotypes and frequencies of known mutations. Parkinsonism Relat Disord. 2013;19(4):407–15.
84. Hirst J, Madeo M, Edgar JR, Yarrow A, Deconinck T, Baets J, et al. Complicated spastic paraplegia in patients with AP5Z1 mutations (SPG48). Neurol Genet. 2016;2(5):e98.
85. Baizabal-carvallo JF, Jankovic J. Parkinsonism, movement disorders and genetics in frontotemporal dementia José. Nat Rev Neurol. 2016;12(3):175–85.
86. Mallaret M, Lagha-Boukbiza O, Biskup S, Jacques I, Gabrielle N, Anheim M, et al. SPG15: a cause of juvenile atypical levodopa responsive parkinsonism. J Neurol. 2014;261:435–7.
87. Stephanie T. Hirschbichler. Classic PD-like rest tremor associated with the tau p.R406W mutation. Parkinsonism Relat Disord. 2015;21:1002–4.
88. Sidransky E, Nalls MA, Aasly JO, Aharon-Peretz J, Annesi G, Barbosa ER, et al. Multicenter analysis of glucocerebrosidase mutations in Parkinson's disease. N Engl J Med. 2009;361:1651–61.
89. Singleton AB, Farrer MJ, Bonifati V. The genetics of Parkinson's disease: progress and therapeutic implications monogenic loci. Mov Disord. 2013;28(1):14–23.
90. Klein C, Westenberger A. Genetics of Parkinson's disease. Cold Spring Harb Perspect Med. 2012;2(a008888):1–15.
91. Delamarre A, Meissner WG. Epidemiology, environmental risk factors and genetics of Parkinson's disease. Presse Med. 2017;46(2 Pt 1):175–81.
92. International Parkinson Disease Genomics Consortium. Imputation of sequence variants for identification of genetic risks for Parkinson's disease: a meta-analysis of genome-wide association studies. Lancet. 2013;377(9766):641–9.
93. Dickson DW, Braak H, Duda JE, Duyckaerts C, Gasser T, Halliday GM, et al. Neuropathological assessment of Parkinson's disease: refining the diagnostic criteria. Lancet Neurol. 2009;8(12):1150–7.
94. Braak H, de Vos RA, Bohl J, Del Tredic K. Gastric alpha-synuclein immunoreactive inclusions in Meissner's and Auerbach's plexuses in cases staged for Parkinson's disease-related brain pathology. Neurosci Lett. 2006;396:67–72.
95. Braak H, Del Tredic K, Rüb U, de Vos RA, Jansen Steur EN, Braak E. Staging of brain pathology related to sporadic Parkinson' s disease. Neurobiol Aging. 2003;24:197–211.
96. Halliday G, Hely M, Reid W. The progression of pathology in longitudinally followed patients with Parkinson's disease. Acta Neuropathol. 2008;115:409–15.

97. Halliday GM, Mccann H. The progression of pathology in Parkinson s disease. Ann N Y Acad Sci. 2010;1184:188–95.
98. Li J, Englund E, Holton JL, Soulet D, Hagell P, Lees AJ, et al. Lewy bodies in grafted neurons in subjects with Parkinson' s disease suggest host-to-graft disease propagation. Nat Med. 2008;14(5):501–3.
99. Kordower JH, Chu Y, Hauser RA, Freeman TB, Olanow CW. Lewy body – like pathology in long-term embryonic nigral transplants in Parkinson' s disease. Nat Med. 2008;14(5):504–6.
100. Dunning CJR, Reyes JF, Steiner JA, Brundin P. Progress in neurobiology can Parkinson's disease pathology be propagated from one neuron to another? Prog Neurobiol. 2012;97(2):205–19.
101. Brundin P, Ma J, Kordower JH. How strong is the evidence that Parkinson's disease is a prion disorder? Curr Opin Neurol. 2016;29:459–66.
102. Goedert M, Masuda-suzukake M, Falcon B. Like prions: the propagation of aggregated tau and a-synuclein in neurodegeneration. Brain. 2016;140(2):266–78.
103. Masuda-suzukake M, Nonaka T, Hosokawa M, Oikawa T, Arai T, Akiyama H, et al. Prion-like spreading of pathological a-synuclein in brain. Brain. 2013;136(4):1128–38.
104. Desplats P, Lee H, Bae E, Patrick C, Rockenstein E, Crews L, et al. Inclusion formation and neuronal cell death through neuron-to-neuron transmission of alpha-synuclein. Proc Natl Acad Sci U S A. 2009;106(32):13010–5.
105. Klingelhoefer L, Reichmann H. Pathogenesis of Parkinson disease — the gut – brain axis and environmental factors. Nat Rev Neurol. 2014;11:1–12.
106. Stokholm MG, Danielsen EH, Hamilton-dutoit SJ. Pathological a-Synuclein in gastrointestinal tissues from prodromal Parkinson disease patients. Ann Neurol. 2016;79:940–9.
107. Hawkes CH, Del Tredici K, Braak H. Review: Parkinson's disease: a dual-hit hypothesis. Neuropathol Appl Neurobiol. 2007;33:599–614.
108. Svensson E, Thomsen RW, Djurhuus JC, Pedersen L, Borghammer P, Sørensen HT. Vagotomy and subsequent risk of Parkinson's disease. Ann Neurol. 2015;78:522–9.
109. Bhatia KP, Massano J. Clinical approach to Parkinson's disease. Cold Spring Harb Perspect Med. 2012;2(6):1–16.
110. Poewe W, Seppi K, Tanner CM, Halliday GM, Brundin P, Volkmann J, et al. Parkinson disease. Nat Rev Dis Primers. 2017;3:17013.
111. Hess CW, Okun MS. Diagnosing Parkinson disease. Neurol Contin. 2016;22(4):1047–63.
112. Schapira AHV, Chaudhuri KR, Jenner P. Non-motor features of Parkinson disease. Nat Rev Neurosci. 2017;18(7):435–50. https://doi.org/10.1038/nrn.2017.62
113. Witt M, Bormann K, Gudziol V, Pehlke K, Barth K, Reichmann H, et al. Biopsies of olfactory epithelium in patients with Parkinson's disease. Mov Disord. 2009;24(6):906–14.
114. Harding AJ, Broe GA, Halliday GM. Visual hallucinations in Lewy body disease relate to Lewy bodies in the temporal lobe. Brain. 2002;125:391–403.
115. Wang J, You H, Liu J. Association of olfactory bulb volume and olfactory sulcus depth with olfactory function in patients with Parkinson disease. Am J Neuroradiol. 2011;32:677–81.
116. Tsironi EE, Dastiridou A, Katsanos A, Dardiotis E, Veliki S, Patramani G, et al. Perimetric and retinal nerve fiber layer findings in patients with Parkinson's disease. BMC Ophthalmol. 2012;12:54.
117. Archibald NK, Hutton SB, Clarke MP, Mosimann UP, Burn DJ. Visual exploration in Parkinson's disease and Parkinson's disease dementia. Brain. 2013;136:739–50.
118. Wasner G, Deuschl G. Pains in Parkinson disease — many syndromes. Nat Rev Neurol. 2012;8(5):284–94.
119. Chaudhuri KR, Rizos A, Trenkwalder C, Rascol O, Pal S. King's Parkinson's disease pain scale, the first scale for pain in PD: an international validation. Mov Disord. 2015;30(12):1623–31.
120. Mcdonald C, Winge K, Burn DJ. Lower urinary tract symptoms in Parkinson's disease: prevalence, aetiology and management. Parkinsonism Relat Disord. 2016;35:8–16.
121. de Riva P, Smith K, Xie SX, Weintraub D. Course of psychiatric symptoms and global cognition in early Parkinson disease. Neurology. 2014;83:1096–103.

122. Postuma RB, Berg D. The new diagnostic criteria for Parkinson's disease. In: Parkinson's disease, vol. 132. 1st ed: Elsevier; 2017. p. 55–78.
123. Sakakibara R, Uchiyama T, Yamanishi T, Shirai K, Hattori T. Review article. Bladder and bowel dysfunction in Parkinson's disease. J Neural Transm. 2008;115:443–60.
124. Goldstein DS. Orthostatic hypotension as an early finding in Parkinson's disease. Clin Auton Res. 2006;16:46–54.
125. Postuma RB, Berg D. Advances in markers of prodromal Parkinson disease. Nat Rev Neurol. 2016;12(11):622–34.
126. Pilleri M, Levedianos G, Weis L, Gasparoli E, Facchini S, Biundo R, et al. Parkinsonism and related disorders heart rate circadian profile in the differential diagnosis between Parkinson disease and multiple system atrophy. Parkinsonism Relat Disord. 2013;2:217–21.
127. Reijnders JS, Ehrt U, Weber WE, Aarsland D, Leentjens AF. A systematic review of prevalence studies of depression in Parkinson's disease. Mov Disord. 2008;23(2):183–9.
128. Dominic H, Creese B, Politis M, Chaudhuri KR, Weintraub D, Ballard C, et al. The psychosis spectrum in Parkinson disease. Nat Rev Neurol. 2017;13(2):81–95.
129. Anang JBM, Bertrand J, Romenets SR, Latreille V, Panisset M, Montplaisir J, et al. Predictors of dementia in Parkinson disease. A prospective cohort study. Neurology. 2014;83:1253–60.
130. Hughes AJ, Daniel SE, Kilford L, Lees AJ. Accuracy of clinical diagnosis of idiopathic Parkinson's disease: a clinico-pathological study of 100 cases. J Neurol Neurosurg Psychiatry. 1992;55:181–4.
131. Rizzo G, Copetti M, Arcuti S, Martino D. Accuracy of clinical diagnosis of Parkinson disease. A systematic review and meta-analysis. Neurology. 2016;87(2):237–8.
132. Boeve BF, Dickson DW, Duda JE, Ferman TJ, Galasko DR, Galvin JE, et al. Arguing against the proposed definition changes of PD. Mov Disord. 2016;31(11):1619–22.
133. Berg D, Postuma RB, Adler CH, et al. MDS research criteria for prodromal Parkinson's disease. Mov Disord. 2015;30(12):1600–9.
134. Miyasaki JM. Treatment of advanced Parkinson disease and related disorders. Continuum (Minneap Minn). 2016;22(4 Movement Disorders):1104–16.
135. Coelho M, Ferreira JJ. Late-stage Parkinson disease. Nat Rev Neurol. 2012;8(8):435–42.
136. Ferreira JJ, Katzenschlager R, Bloem BR, et al. Summary of the recommendations of the EFNS/MDS-ES review on therapeutic management of Parkinson's disease. Eur J Neurol. 2013;20(1):5–15.
137. Fox SH, Katzenschlager R, Lim SY, et al. The Movement Disorder Society evidence-based medicine review update: treatments for the motor symptoms of Parkinson's disease. Mov Disord. 2011;26(Suppl 3):S2–41.
138. Seppi K, Weintraub D, Coelho M, et al. The Movement Disorder Society evidence-based medicine review update: treatments for the non-motor symptoms of Parkinson's disease. Mov Disord. 2011;26(Suppl 3):S42–80.
139. Zesiewicz TA, Sullivan KL, Arnulf I, et al. Practice parameter: treatment of nonmotor symptoms of Parkinson disease: report of the Quality Standards Subcommittee of the American Academy of Neurology. Neurology. 2010;74(11):924–31.
140. Espay AJ, Brundin P, Lang AE. Precision medicine for disease modification in Parkinson disease. Nat Rev Neurol. 2017;13(2):119–26.
141. Espay AJ, Lang AE. Common myths in the use of levodopa in Parkinson disease: when clinical trials misinform clinical practice. JAMA Neurol. 2017;74(6):633–4.
142. Scott NW, Macleod AD, Counsell CE. Motor complications in an incident Parkinson's disease cohort. Eur J Neurol. 2016;23(2):304–12.
143. Rascol O, Perez-Lloret S, Ferreira JJ. New treatments for levodopa-induced motor complications. Mov Disord. 2015;30(11):1451–60.
144. Perestelo-Pérez L, Rivero-Santana A, Pérez-Ramos J, et al. Deep brain stimulation in Parkinson's disease: meta-analysis of randomized controlled trials. J Neurol. 2014;261:2051–60.
145. Schuepbach WMM, Rau J, Knudsen K, et al. Neurostimulation for Parkinson's disease with early motor complications. N Engl J Med. 2013;368(7):610–22.

146. Mansouri A, Taslimi S, Badhiwala JH. et al, Deep brain stimulation for Parkinson's disease: meta-analysis of results of randomized trials at varying lengths of follow-up. J Neurosurg. 2017. https://doi.org/10.3171/2016.11.JNS16715.
147. Volkmann J, Albanese A, Antonini A, et al. Selecting deep brain stimulation or infusion therapies in advanced Parkinson's disease: an evidence-based review. J Neurol. 2013;260(11):2701–14.
148. Clarke CE, Worth P, Grosset D, Stewart D. Systematic review of apomorphine infusion, levodopa infusion and deep brain stimulation in advanced Parkinson's disease. Parkinsonism Relat Disord. 2009;15(10):728–41.
149. Abbruzzese G, Marchese R, Avanzino L, Pelosin E. Rehabilitation for Parkinson's disease: current outlook and future challenges. Parkinsonism Relat Disord. 2016;22(Suppl 1):S60–4.
150. Grazina R, Massano J. Physical exercise and Parkinson's disease: influence on symptoms, disease course and prevention. Rev Neurosci. 2013;24(2):139–52.
151. Monteiro A, Massano J. Parkinson's disease cluster: the wind of change. Int J Clin Neurosci Mental Health. 2014;1:7. https://doi.org/10.21035/ijcnmh.2014.1.7.

Neurodegeneration and Multiple Sclerosis

Axel Petzold

Abstract
Neurodegeneration causes inexorable loss of neurons and function in both diseases and aging. Neurodegeneration damage produces a range of progressive disabilities from cognitive decline, behavioral and mood disorders to problems with movement, co-ordination, and sensory dysfunction. Neurodegeneration is a major and growing public health issue which in its broadest sense embraces classical neurodegenerative disorders such as Alzheimer's disease and Parkinson's disease, as well as multiple sclerosis (MS), diabetes, and acute brain injury among many other conditions. This chapter discusses the clinical and pathophysiological features of neurodegeneration in MS.

Keywords
Demyelinating disease · Multiple sclerosis · Neurodegeneration · Trans-synaptic axonal degeneration · Protein biomarker · Cerebrospinal fluid · Retina · Optical coherence tomography

---

A. Petzold
Moorfields Eye Hospital, London, UK

The Neuroimmunology and CSF Laboratory, London, UK

MS Centre and Dutch Expertise Centre for Neuro-ophthalmology, VUmc, Amsterdam, The Netherlands
e-mail: a.petzold@ucl.ac.uk

Introduction

Neurodegeneration causes inexorable loss of neurons and function in both diseases and aging [1]. Neurodegeneration damage produces a range of progressive disabilities from cognitive decline, behavioral and mood disorders to problems with movement, co-ordination, and sensory dysfunction. Neurodegeneration is a major and growing public health issue which in its broadest sense embraces classical neurodegenerative disorders such as Alzheimer's disease and Parkinson's disease, as well as multiple sclerosis (MS), diabetes, and acute brain injury among many other conditions. This chapter discusses the clinical and pathophysiological features of neurodegeneration in MS.

The historical context will be discussed first, because our understanding of MS pathology has been much influenced by demyelination and a concept of dissemination in time and space [2, 3]. Next, the classical pathological features of neurodegeneration in MS are reviewed in more detail [4]. Axonal loss will be placed centrally because of the important link to irreversible loss of function [1, 4, 5]. The resulting disability has a major impact on an individual patient's life [5]. Here limitations will be reviewed of those clinical and paraclinical assessments which were predominantly focused on demyelination and/or evidence for dissemination in time and space [2, 6]. It is against this backdrop that biomarkers for neurodegeneration will be presented [7]. The chapter closes with an outlook on how this knowledge may be applied to future treatment trials targeted at halting neurodegeneration in MS [5].

Historical Context

Most of the credited clinico-pathological descriptions of MS date back to the mid-nineteenth century. The classical pathological features embrace inflammation, demyelination, and gliosis [1, 4, 8].

Jean Marin Charcot, who pioneered the pathophysiological explanation of the symptoms observed in patients distinguished three steps in the pathology of MS, which he called *la sclérose en plaques disseminée, la sclerose generalisée et la sclerose multiloculaire*. First, astrocytic and microglial activation: "la multiplication des noyaux et l'hypertroplasie concomitante des fibres réticulées de la névroglie sont le fait initial." Second, neuro-axonal degeneration: "l'atrophie dégénerative des éléments nerveux est secondaire." The interested reader is referred to a wonderful historical account on axonal pathology for more details [9]. And third, astrogliosis: "la névroglie fait place au tissu fibrillaire." Ultimately, it was demyelination ("dépouillés de leur myéline" [10]) which became the key pathological feature of the disease, here depicted in a frequently cited sketch (Fig. 18.1).

The cause for these features has remained enigmatic ever since James Dawson's dichotomization into "inflammatory" and "developmental" concepts [11].

While pathologically succinct, the difficulty for the treating physician remains to recognize and communicate a diagnosis of MS to the patient. Historically, MS was

18 Neurodegeneration and Multiple Sclerosis

Fig. 18.1 The figure shows the original sketch of an MS lesion from the landmark paper of Charcot [10]. The image depicts a fresh MS plaque colored with carmine. Charcot's text implies the presence of axonal pathology based on morphological observations of diameter and continuity. His interpretation is careful as he does not exclude possible preparation-related artifacts. The original text reads as "Elle représente une préparation frâche, provenant du centre d'une plaque scléreuse, colorié par le carmin et traité e par delacération. Au centre, vaisseau capillaire portant plusieurs noyaux. A droite et à gauche, cylindres d'axe, les uns volumineux, les autres d'un très–petit diamètre, tous dé pouillés de leur myéline. Le vaisseau capillaire et les cylindres d'axe étaient fortement colorés par le carmin. Les cylindres d'axe ont des bords parfaitement lisses, ne presentant aucune ramification. Dans l'intervalle des cylindres d'axe, membranes fibrilles de formation récente, à peu près parallèles les unes aux autres dans la partie droite de la préparation, formant à gauche et au centre, une sorte de réseau résultant, soit de l'enchevêment, soit de l'anastomose des fibrilles. Celles–ci se distinguent des cylindres d'axe, 1 par leur diamètre qui est beaucoup moindre; 2 par les ramifications qu'elles offrent dans leur trajet; 3 parce qu'elles ne se colorent pas par le carmin. — C á et là , noyaux disséminés. Quelques–uns paraissent en connexion avec les fibrilles conjonctives; d'autres ayant pris une forme irre gulière, due à l'action de la solution ammoniacale du carmin." [10]

recognized in the pre-antibiotic area where inflammatory diseases such as syphilis presented major public health issues. Separating one from the other was not always straightforward. Not surprisingly, given the multitude of symptoms and signs mimicking other diseases, MS was also considered a chameleon. In absence of a diagnostic test, the clinical judgement cannot be substituted for. This notion is reflected in a series of diagnostic criteria, all more or less stating that the patient's symptoms and signs ought to be compatible with the characteristics of MS [12–14]. The careful and systematic, evidence-based approach on which these criteria rest distilled a conceptual framework which may be phrased as "dissemination in time and space" [2].

Dissemination in time (DIT) and dissemination in space (DIS) are well suited to describe the occurrence of radiologically recognizable MS lesions in the brain and spinal cord [2].

It was precisely the absence of clear evidence for these characteristic features which made it so challenging to develop diagnostic criteria for primary progressive multiple sclerosis (PPMS) [15]. Later, Thompson and colleagues phrased this as "Neither set of criteria is appropriate to PPMS, since the basic requirement of two discrete episodes of neurological dysfunction cannot by definition be fulfilled." [16]. The clinical cornerstone of what emerged in International Panel diagnostic criteria was the documented clinical progression for more than 1 year [13].

Paradoxically, the first in vivo observation of axonal loss in MS was difficult to publish at all, according to anecdotal reports from the authors. Hoyt and colleagues had observed retinal nerve-fiber bundle defects in the eyes of patients with MS [17]. Much more frequently cited is the follow-up paper on this observation by Frisen et al. stating the presence of "insidious atrophy" of retinal nerve fibers in the eyes of patients with multiple sclerosis [18]. The second case reported by Frisen and Hoyt was a 15-year-old student athlete with a clinical diagnosis of "multifocal demyelinating disease," but without any history of optic neuritis. One may speculate that one argument for rejection at the time might have been that multiple sclerosis was a demyelinating disease, and the question was raised: why should there be at all atrophy of the non-myelinated axons in the eye of a patient who did not even suffer from optic neuritis?

Axonal loss was only some 24 years later firmly put on the MS research agenda by the American cell biologist Bruce Trapp and the Norwegian pathologist Lars Bo [19]. The conceptional change this influential pathological study had will be discussed in the next section.

Pathological Features

Axonal Loss in Multiple Sclerosis

In order to put the observation by Trapp et al. into context, one needs to recall that axonal pathology may not be the most striking feature in the MS brain but certainly is the one with the highest impact for the patient [19–23]. Historically, axonal loss in MS has been associated with the "burnt-out" phase of the disease [24, 25]. Only with the wide availability of immunohistological techniques it was possible to demonstrate axonal pathology in *active* MS lesions [26]. There was extensive staining for amyloid precursor protein (APP), and the APP-positive structures resembled transected axons. It was however, the three-dimensional reconstruction of these axonal ovoids, using confocal microscopy, which conclusively demonstrated axonal transections within acute MS lesions [19]. Interestingly, an accumulation of neurofilament protein was observed in the so-called end-bulbs. In vivo imaging of the development of axonal degeneration is available for experimental models [27–29].

In other words, the important new insight from this work was that a high number of transected axons were already present in acute lesions [19, 26] and in patients with a short clinical course [19]. This data changed the earlier perception of axonal loss in MS [30, 31].

The data from Trapp et al. is consistent with the concept that an important trigger for axonal loss are MS lesions [3]. But because disability continued to progress even after successful suppression of the inflammatory part of the disease, other aspects of axonal pathology were discussed [32]. Axons might be driven into a fatal energy deficit [4, 33, 34]. There is good evidence that mitochondrial pathology and sodium channel redistribution contribute to an "ATP penalty" [35–40]. Axonal transport might be impaired [41–44]. Next, there might be loss of trophic support or increase of inhibitory substances such as Nogo [45]. A barrier may result from astrogliosis. A low-grade inflammatory process might persist [46]. There is the problem of failure to remyelinate. There may be acceleration of physiological processes of aging-related neurodegeneration. Endogenous capacities of repair might have their limits [46]. In sum, those factors causing axonal degeneration might eventually outnumber those which were protective [47].

It is worthwhile to remember some limitations, axonal injury remains a dynamic process and quantification of axonal loss in histological material might be complicated by tissue edema, the presence of inflammatory cells, and the problem of establishing a relationship with the number of healthy axons. There is a crucial dependence on well-preserved tissue with limited capacities of the existing brain banks. Most postmortem studies were biased to tissue from patients with long-standing disease duration, and there is a lack of representative tissue from the clinically and therapeutically relevant early disease phase. Some early tissue might be available through biopsy, but again questions might be asked how representative such tissue really is if taken because the presentation was very atypical. Finally, there are shortcomings to the analytical methods, dyes, and antibodies used.

Concepts of Axonal Degeneration

Like axonal injury, axonal degeneration is also a dynamic process. Most recent insights come from experimental studies in mice on fluorescently labeled axons [27, 48]. It may be opportune to go back in time and revisit the first systematic description of axonal injury by Waller which gave rise to the eponym "Wallerian degeneration" [49].

In brief, Wallerian degeneration is a complex process which describes the degeneration of the *distal* axonal stump after axonal transection from the neuron. Wallerian degeneration begins with the enzymatic proteolysis of the axonal cytoskeleton [50]. Additionally, Wallerian degeneration affects also the sheathing glial cells, causes alterations in the adjacent blood-tissue barriers, and stimulates cells of macrophage lineage. From a mechanistic point of view, Wallerian degeneration is of anterograde direction.

Wallerian degeneration has to be distinguished from *dying-back* neuropathy, defined as the slow *proximal* spread of nerve fiber breakdown and ultimate apoptosis of the neuron [51]. The term *dying back* was introduced to describe the spatio-temporal pattern of central and peripheral nerve fiber pathology in degenerative diseases. Contemporary understanding is that axonal degeneration is defined by direction into anterograde and retrograde.

An important, mechanistic question to be asked is how the process of neurodegeneration can spread from a sick to a healthy neuron/axon? One attractive concept is *trans-synaptic* axonal degeneration [52, 53]. These authors used a noninvasive, utrarapid imaging technique, readily tolerated by patients, retinal optical coherence tomography (OCT) [54]. The study design was elegant and simple by focusing on neurodegeneration in the visual pathways. Following a stroke in the posterior visual pathways, dying-back neuropathy spread (trans-synaptic) from the second-order neuron located in the lateral geniculate nucleus (LGN) to the axons (retinal nerve fiber layer, RNFL) of the first-order neuron (retinal ganglion cell, RGC) [52, 53]. These studies have advanced the understanding of acquired axonal degeneration [55].

In addition to retrograde trans-synaptic axonal degeneration, there is evidence for anterograde trans-synaptic axonal degeneration from a postmortem study of the visual system of patients with multiple sclerosis [56].

Taken together, these data suggests a concept of *bidirectional (trans-synaptic) axonal degeneration* [57] (Fig. 18.2).

The attraction of this unified concept of bidirectional (trans-synaptic) axonal degeneration is that not only it is convenient to explaining how neurodegeneration spreads in MS, but more importantly it may contribute to opening a therapeutic window for future neuroprotective strategies in MS. The aim here will be to prevent the trans-synaptic part of the degenerative process and thereby at least limit the impairment for the patient.

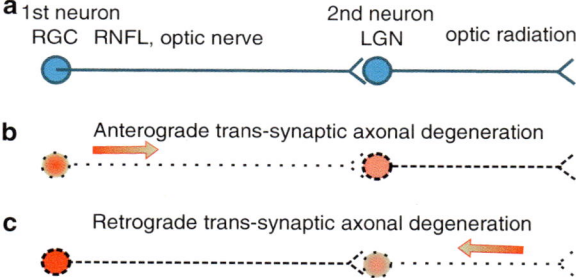

Fig. 18.2 A simplified and uniform mechanistic concept of axonal degeneration. (**a**) The normal situation is here shown for the visual system. The first-order neuron is represented by the retinal ganglion cell (RGC). The first axon is represented by the retinal nerve fiber layer (RNFL) which is named optic nerve after the axons passed through the lamina cribrosa. Here an axon is shown to synapse in the lateral geniculate nucleus (LGN) with the second-order neuron. Next, the second neuron sends its axon through the optic radiations to the occipital cortex. (**b**) Anterograde axonal degeneration starts at the RGC/RNFL/optic nerve (e.g., with optic neuritis). Once anterograde axonal degeneration reaches the LGN, it continues as trans-synaptic anterograde axonal degeneration. (**c**) Retrograde axonal degeneration starts with axonal transections in the optic radiations (e.g., with eloquently placed white matter lesions). Once retrograde axonal degeneration reaches the LGN, the process continues as trans-synaptic retrograde axonal degeneration. Ultimately this leads to loss of retinal nerve fibers and apoptosis of the RGC. Longitudinally, the trans-synaptic part of this concept of bidirectional axonal degeneration will always have to occur with a time lag. Understanding this time lag may potentially open a new therapeutic window for future neuroprotective strategies in MS

The Patient

The use and definition of terms to describe a patient's impairment, disability, and handicap in this section were based on the recommendations of the system adopted by the World Health Organization (WHO).

Impairment describes the "loss or abnormality…of structure of function." *Disability* describes "a restriction or lack…of ability to perform an activity in the manner of within the range considered normal for a human being." *Handicap* describes "the disadvantage for an individual…that prevents or limits the performance of a role that is normal…for that individual." To be more specific, handicap represents the effects of impairments or disabilities in a wide social context and may be substantially influenced by the cultural background.

By definition (DIS and DIT [2]), a patient will suffer from MS-related symptoms causing potentially reversible impairment in different parts of his/her body. From a patient's perception, gait and vision are the two most valuable functions [58]. Both gait and vision topped a list of 13 bodily functions during the early (<5 years) and late (>15 years) disease course. Importantly, early in the disease where patients were still ambulatory, gait was rated more valuable compared to visual function, but there was a crossover with long-disease duration. With the ever-increasing use of visual communication channels (e.g., smart phones, tablets, social media), it can be anticipated that from a patients point of view, the value and dependence on the visual system will continue to increase in the near future. This may be particularly true for those handicapped patients who crucially depend on the visual system for social interaction. Not surprisingly all of above is related to a patient's quality of life [59].

Two questions are frequently asked by patients: "Will this happen again?" (relapse) and "Will I end up in a wheelchair?" (neurodegeneration). The first one may, with caution, be answered based on the momentary clinical and radiological disease activity. Addressing the second question is more challenging because of a relative lack of longitudinal data from well-validated outcome measures for neurodegeneration.

Clinical and Paraclinical Assessments

> "There are few neurological diseases in which the diagnosis depends so much upon the skill of the examiner in knowing what questions to ask and how to interpret the replies." [60]

Clinical Scales

Impairment or loss of function is quantified by clinical scales. The paradox between clinical examination and each clinical scale is that normal functioning is tested, but loss of function is quantified. Because of the potential of CNS regeneration and plasticity, the clinical appearance of disability is a dynamic process. This forms the basis on which MS patients had been classified [61]. A more recent approach separated an "active" from a "non-active" subtype based on clinical and MRI data [62].

A range of validated clinical scales is now in use. For MS the most widely applied scale is the extended disability status scale (EDSS) for multiple sclerosis developed by Kurtzke in 1983 [63]. The EDSS combines a disability status scale [64] with functional systems [65]. For a comprehensive up-to-date review of outcome measures in MS, the reader is referred to van Munster and Uitdehaag [66].

Psychometry is tested by the Paced Auditory Serial Addition Test (PASAT) [67]. The National Adult Reading Test (NART) is used to give an estimate of the premorbid IQ [68]. Current intellectual function is assessed by the Advanced Progressive Matrices, Set 1 (Ravens). Memory is assessed by recognition of words and faces [69]. The paired-associate learning test estimates learning abilities. Attention is readily quantified by the speed of letter counting [70]. Tests of executive function include the Wisconsin Card Sorting Test (Nelson) and the Cambridge Neuropsychological Test Automated Battery (CANTAB) [68, 71]. Fatigue is commonly estimated by Krupp's fatigue rating scale [72]. Anxiety and depression have been measured using the National Hospital Anxiety and Depression Scale (HAD) measuring quality of life and measures for outcome of neuro-rehabilitation [73].

The timed walk test (TWT), 9-hole Peg test (9HPT) and Paced Auditory Serial Addition Test (PASAT) have been combined mathematically to give the multiple sclerosis functional composite (MSFC) [66, 74]. The MSFC has the potential to provide a more reliable measure of changes of function in MS than the EDSS, which is nonlinear and biased toward locomotion [75]. In addition, the MSFC may be perceived as a "melting pot" which permits to embrace other relevant clinical measures within a statistically valid concept. One potential extension of the MSFC may be low-contrast letter acuity [76]. One advantage of such multidimensional measures relates to the potential to cover both disease activity and progression in MS [66].

A cross-sectional measure of disease severity in individual patients is provided by the global Multiple Sclerosis Severity Score (MSSS) [77]. The global MSSS is taken from a statistically constructed "look-up table." This table provides normally distributed disease severity scores for patients with an EDSS between 0 and 9.5 and a disease duration between 1 and 30 years.

Newer developments include patient-reported outcome measures (POM) [66]. A well-established example for a POM is the MSIS-29 [78].

The advantages of clinical scales (and questionnaires) are that they may provide a more holistic view of an individual patient's disability compared to paraclinical tests. But there are also limitations to be considered:

1. Psycho-physiological testing heavily depends on the patient's co-operation and motivation.
2. Biased to data from the system tested. This has been a frequently discussed limitation of the EDSS which is biased to the pyramidal system.
3. Learning effects. This is particularly challenging for testing cognition longitudinally.

4. Challenges of validation across cultural and language-barriers. This may impact on the use as an outcome measure in multicenter studies.
5. Multiple biological causes for poor performance. In MS this includes:
   (a) Conduction block
   (b) Demyelination
   (c) Axonal loss

Paraclinical Tests

"The technological advances that have contributed to a better understanding of the pathophysiology and pathogenesis of MS have resulted in a disturbing increase in the number of false diagnoses of MS based exclusively on the results of test procedures." [60]

Paraclinical tests are a double-edged sword, but do have their merits in experienced hands if used as an extension of the clinical reasoning. The four most frequently used paraclinical tests over the past 50 years comprise in alphabetical order: cerebrospinal fluid (CSF), computed tomography (CT), MRI, and visual evoked potentials (VEP), acknowledging that MRI has become the sole paraclinical test of the 2010 revision of the McDonald criteria for RRMS [13]. A historical head-to-head comparison based on the earlier Poser criteria is presented in Table 18.1.

Of note, none of these studies investigated the relevance of any of these tests for axonal loss, which as pointed out earlier was not the main focus of MS research at the time.

While sensitive for diagnostic purposes, the limitations of MRI to *predict* development disability were elegantly summarized by Kappos and colleagues in a thoroughly conducted meta-analysis: "Neither the initial scan nor monthly scans over six months were predictive of change in the EDSS in the subsequent 12 months or 24 months. The mean of gadolinium-enhancing-lesion counts in the first six monthly scans was weakly predictive of EDSS change after 1 year (odds ratio = 1.34, $p = 0.082$) and 2 years (odds ratio = 1.65, $p = 0.049$)" [82].

Table 18.1 Paraclinical tests used in MS

Reference	Test	Sensitivity (%)	Conclusion
Polman et al. [79]	CSF	72.2	Diagnostic classification
	CT2	17.0	Differential diagnosis
	VEP	62.0	Diagnostic classification
Beer et al. [80]	CSF	77	Best reclassification specificity
	MRI	84	Highly sensitive, demonstrates DIS
	VEP	37	Useful if MRI and CSF are not diagnostic
Filippini et al. [81]	CSF		
	MRI	70	Most sensitive test
	VEP		

For each test, the diagnostic sensitivity of the respective study is presented alongside the author's main conclusions

This meta-analysis demonstrates the difficulties in predicting accumulation of irreversible disability, which is related to neurodegeneration, based on a paraclinical test focused on inflammatory disease activity. In contrast, MRI data on CNS atrophy are much better correlated to sustained disability [83, 84]. There is data on perfusion, functional MRI, high-field MRI, new sequences specifically addressing iron storage, double inversion recovery (DIR), and MR spectroscopy (MRS). For in-depth review of these and other MRI techniques, the reader is referred to recent reviews on the issue [85–88].

Likewise, for the CSF there is conflicting evidence on the relationship of CSF oligoclonal bands (OCBs) and disability [89]. There are some reports suggesting that the absence of OCBs in the CSF of patient with MS may be a good prognostic sign [90–95]. Others did not find any prognostic value of either presence or absence of CSF OCBs [96–98].

There may also be leverage using VEPs (and other evoked potentials) as a paraclinical test for neurodegeneration in MS [6].

It may be suggested to separate those paraclinical tests which permit detection of axonal loss (and neurodegeneration) in the acute phase from those which are superior for documenting axon loss after some time has elapsed. Tentatively, retinal OCT was added to this list as an emerging paraclinical test for retinal layer atrophy:

1. Early phase of ensuing axonal injury and loss:
   (a) Biomarkers for acute axonal damage [99–101]
   (b) Imaging markers for neuronal dysfunction and apoptosis [102–104]
2. Late phase of axonal loss having resulted in manifest atrophy:
   (a) MRI atrophy markers [105, 106]
   (b) OCT [107, 108]
   (c) VEP and motor evoked potentials (MEP) [6, 109]

Acute Neurodegeneration in MS: Body Fluid Biomarkers

In MS, disintegration of the axonal membrane causes release of biomarkers from injured axons and neurons in the surrounding extracellular fluid (ECF) [110]. These biomarkers diffuse from the brain ECF into the CSF and blood. Sampling from each of these body fluid compartments is possible with related advantages and disadvantages.

A review of the biomarker literature in MS shows that most early studies were cross-sectional and frequently of limited sample size [100, 111–115]. This radically changed in the past 2 years. Pioneering studies relied on in-house developed immunoassays for the quantification of biomarkers. With availability of commercial tests for quantification of key biomarkers such as the neurofilament proteins from the blood the literature on the subject has increased exponentially [99, 116, 117].

18 Neurodegeneration and Multiple Sclerosis

Table 18.2 Blood biomarkers in MS and their cellular sources

Blood biomarker	Neuron and axon	Astrocyte	Microglia	Oligodendrocyte	Other cells
14-3-3γ	+	+	+	+	+
Amyloid β42	+				
Apo-E	+	+	+		
FABPs	+	+	+	+	+
FFA	+	+	+	+	+
Ferritin			+		+
GAP-43	+				
Gelsolin	+				+
GFAP		+			
HNE	+	+	+	+	+
NSE	+				+
Neurofilaments	+				
S100B		+		+	+
Tau	+	+	+	+	+
UCHL-1	+				

Because of the essentially correlative nature of clinical biomarker investigations, only a snapshot in time is provided by cross-sectional studies. Not surprisingly, some studies find a clinical relevant correlation for a particular biomarker, while others do not. Some of these issues can be addressed by a meta-analysis. It will however be much more important to obtain high-quality long-term data. Therefore, Table 18.2 summarizes blood biomarkers categorized to their cell-type specificity. For an extended biomarker table and in-depth review on CSF biomarkers for neurodegeneration, see [110, 114, 118].

The measurement of cell-type-specific biomarkers indirectly permits to estimate the degree of damage to the respective cellular source. For example, an increase of blood neurofilament (Nf) levels gives indirect evidence for neuro-axonal damage. Neurofilaments have consistently found to be of prognostic value in MS [99, 119–130].

Importantly, there has been convincing analytical and experimental work to substantiate the hypothesis that Nf levels are related to neurodegeneration [116, 130–139]. Tests are now commercially available with the most sensitive technology being Simoa [140].

Newly Validated Atrophy-Related Imaging Biomarkers for Neurodegeneration: Optical Coherence Tomography

An emerging imaging technology for neurodegeneration in MS is retinal optical coherence tomography (OCT) [108]. The results of the early time-domain OCT meta-analysis have now been repeated for spectral-domain OCT. The results of the two meta-analyses were almost identical underlining the robustness of the method.

While it is well known that optic neuritis causes loss of the retinal nerve fiber layer [18], it only recently emerged that such atrophy can also be present in eyes not affected by optic neuritis [108, 141–153]. Because retinal nerve fiber layer (RNFL) thickness also correlated with clinical scales and MRI measures, there is a need to test the reliability and validity of OCT in a multicenter setting.

Outlook

Taken together, neurodegeneration is an important feature of MS pathology because it is responsible for irreversible disability in patients. The dynamic nature of neurodegeneration poses challenges to the techniques used for monitoring. Some methods have their strengths in the acute phase; others only become reliable once neurodegeneration becomes manifest as atrophy. A holistic model combining the respective strength and weaknesses is presented in Fig. 18.3.

This may be an opportune moment to end this chapter with an open question building on an analogy. In diabetes mellitus, patients measure several times per day their blood glucose levels to optimize individual treatment. Additional paraclinical tests are used to closely monitor related organ damage with the aim to further guide patient management. How can we combine our respective expertise and methods to achieve a similar feat in MS?

Fig. 18.3 A holistic model combining the strength of biomarkers suited for diagnosis (whole brain and spinal cord MRI) of the acute phase of neurodegeneration (e.g., body fluid neurofilament levels) with those more reliable during the later phase of neurodegeneration-related atrophy measures (retinal OCT). A fundamental problem of imaging techniques is that any inflammation-related edema in the acute phase will mask neurodegeneration-related atrophy. Likewise, body fluid biomarkers such as neurofilaments will predominantly be released from disintegrating axons/neurons during the acute phase and only to a smaller degree during the "burnt-out phase." A logical combination of these two distinct methodological approaches would be to have them integrated in longitudinal studies on neurodegeneration in MS

18 Neurodegeneration and Multiple Sclerosis

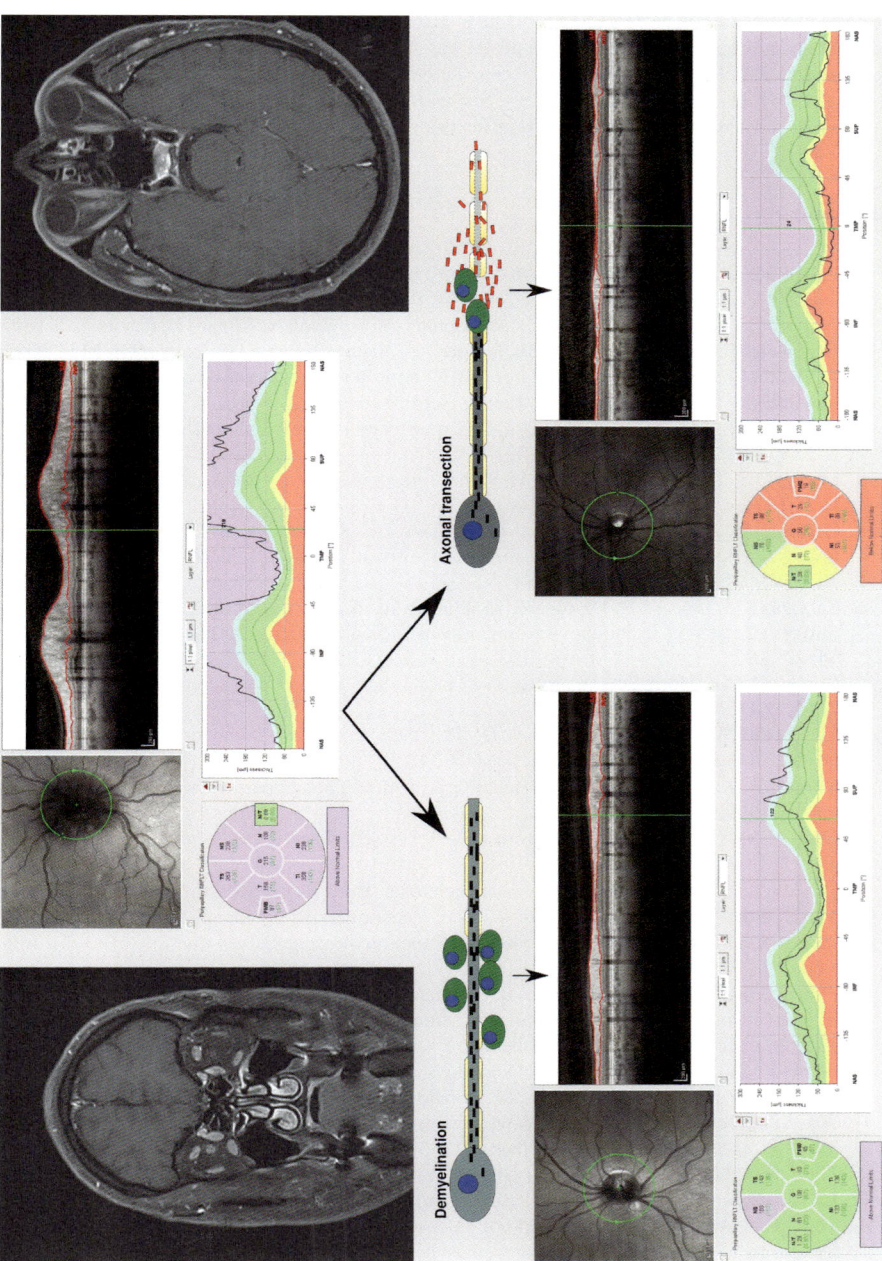

Acknowledgements The MS Center VUmc is partially funded by a program grant of the Dutch MS Research Foundation. The research of AP was supported by the National Institute for Health Research (NIHR) Biomedical Research Centre based at Moorfields Eye Hospital NHS Foundation Trust and UCL Institute of Ophthalmology. The views expressed are those of the author(s) and not necessarily those of the NHS, the NIHR, or the Department of Health.

References

1. Salapa HE, Lee S, Shin Y, Levin MC. Contribution of the degeneration of the neuro-axonal unit to the pathogenesis of multiple sclerosis. Brain Sci. 2017;7(6). ISSN: 2076-3425
2. Filippi M, Rocca MA, Ciccarelli O, De Stefano N, Evangelou N, Kappos L, Rovira A, Sastre-Garriga J, Tintorè M, Frederiksen JL, Gasperini C, Palace J, Reich DS, Banwell B, Montalban X, Barkhof F, MAGNIMS Study Group. MRI criteria for the diagnosis of multiple sclerosis: MAGNIMS consensus guidelines. Lancet Neurol. 2016;15:292–303.
3. Lassmann H, Bradl M. Multiple sclerosis: experimental models and reality. Acta Neuropathol. 2017;133(2):223–44. ISSN: 1432-0533
4. Mahad DH, Trapp BD, Lassmann H. Pathological mechanisms in progressive multiple sclerosis. Lancet Neurol. 2015;14:183–93.
5. Ontaneda D, Thompson AJ, Fox RJ, Cohen JA. Progressive multiple sclerosis: prospects for disease therapy, repair, and restoration of function. Lancet (London, England). 2017;389(10076):1357–66. ISSN: 1474-547X
6. Lascano AM, Lalive PH, Hardmeier M, Fuhr P, Seeck M. Clinical evoked potentials in neurology: a review of techniques and indications. J Neurol Neurosurg Psychiatry. 2017.; ISSN: 1468-330X
7. Comabella M, Montalban X. Body fluid biomarkers in multiple sclerosis. Lancet Neurol. 2014;13:113–26.
8. McDonald WI, Miller DH, Barnes D. The pathological evolution of multiple sclerosis. Neuropathol Appl Neurobiol. 1992;18:319–34.
9. Kornek B, Lassmann H. Axonal pathology in multiple sclerosis. A historical note. Brain Pathol. 1999;9:651–6.
10. Charcot M. Histologie de la sclérose en plaques (II). Gazette des hopitaux, vol 14. 1868; p. 557–8.
11. Dawson J. The histology of disseminated sclerosis. Trans Royal Soc Edin. 1916;50:517–740.
12. McDonald WI, Compston A, Edan G, et al. Recommended diagnostic criteria for multiple sclerosis: guidelines from the International Panel on the diagnosis of multiple sclerosis. Ann Neurol. 2001;50:121–7.
13. Polman CH, Reingold SC, Banwell B, Clanet M, Cohen JA, Filippi M, Fujihara K, Havrdova E, Hutchinson M, Kappos L, Lublin FD, Montalban X, OĆonnor P, Sandberg-Wollheim M, Thompson AJ, Waubant E, Weinshenker B, Wolinsky JS. Diagnostic criteria for multiple sclerosis: 2010 revisions to the McDonald criteria. Ann Neurol. 2011;69:292–302.
14. Poser CM, Paty DW, Scheinberg L, McDonald WI, Davis FA, Ebers GC, Johnson KP, Sibley WA, Silberberg DH, Tourtellotte WW. New diagnostic criteria for multiple sclerosis: guidelines for research protocols. Ann Neurol. 1983;13:227–31.
15. Thompson AJ, Polman CH, Miller DH, McDonald WI, Brochet B, Montalban X FM, De Sa J. Primary progressive multiple sclerosis. Brain. 1997;120:1085–96.
16. Thompson AJ, Montalban X, Barkhof F, Brochet B, Filippi M, Miller DH, Polman CH, Stevenson VL, McDonald WI. Diagnostic criteria for primary progressive multiple sclerosis: a position paper. Ann Neurol. 2000;47:831–5.
17. Hoyt WF, Schlicke B, Eckelhoff RJ. Fundoscopic appearance of a nerve-fibre-bundle defect. Br J Ophthalmol. 1972;56:577–83.
18. Frisen L, Hoyt WF. Insidious atrophy of retinal nerve fibers in multiple sclerosis. Funduscopic identification in patients with and without visual complaints. Arch Ophthalmol. 1974;92:91–7.

19. Trapp BD, Peterson JP, et al. Axonal transection in the lesions of multiple sclerosis. N Engl J Med. 1998;338:278–85.
20. McDonald WI. Relapse, remission, and progression in multiple sclerosis. N Engl J Med. 2000;343:1486–7.
21. Trapp BD, Ransohoff RM, Fisher E, Rudick RA. Neurodegeneration in multiple sclerosis, relationship to neurological disability. Neuroscientist. 1999;5:48–57.
22. Waxman SG. Demyelinating diseases — new pathological insights, new therapeutic targets. N Engl J Med. 1998;338:323–5.
23. Wujek JR, Bjartmar C, Richer E, et al. Axon loss in the spinal cord determines permanent neurological disability in an animal model of multiple sclerosis. J Neuropathol Exp Neurol. 2002;61:23–32.
24. Greenfield JG, King LS. Observations on the histopathology of the cerebral lesions in desseminated sclerosis. Brain. 1936;59:445–58.
25. Putnam TJ. Studies in multiple sclerosis VII similarities between some forms of "encephalomyelitis" and multiple sclerosis. Arch Neurol Psychiatr. 1935:1289–308.
26. Ferguson B, Matyszak MK, Esiri MM, Perry VH. Axonal damage in acute multiple sclerosis lesions. Brain. 1997;120:393–9.
27. Kerschensteiner M, Schwab ME, Lichtman JW, Misgeld T. In vivo imaging of axonal degeneration and regeneration in the injured spinal cord. Nat Med. 2005;11:572–7.
28. Marinkovic P, Reuter MS, Brill MS, Godinho L, Kerschensteiner M, Misgeld T. Axonal transport deficits and degeneration can evolve independently in mouse models of amyotrophic lateral sclerosis. Proc Natl Acad Sci U S A. 2012;109:4296–301.
29. Misgeld T, Kerschensteiner M. In vivo imaging of the diseased nervous system. Nat Rev Neurosci. 2006;7:449–63.
30. Charcot M. Histologie de la sclérose en plaques (I). Gazette des hopitaux, vol 14. 1868; p. 554–5.
31. Rindfleisch E. Histologisches Detail zur grauen Degeneration von Gehirn und RÃ1 4ckenmark. Arch Pathol Anat Physiol Klin Med (Virchow). 1863;26:474–83.
32. Stadelmann C. Multiple sclerosis as a neurodegenerative disease: pathology, mechanisms and therapeutic implications. Curr Opin Neurol. 2011;24:224–9.
33. Su KG, Banker G, Bourdette D, Forte M. Axonal degeneration in multiple sclerosis: the mitochondrial hypothesis. Curr Neurol Neurosci Rep. 2009;9:411–7.
34. Trapp BD, Stys PK. Virtual hypoxia and chronic necrosis of demyelinated axons in multiple sclerosis. Lancet Neurol. 2009;8:280–91.
35. Black JA, Newcombe J, Trapp BD, Waxman SG. Sodium channel expression within chronic multiple sclerosis plaques. J Neuropathol Exp Neurol. 2007;66:828–37.
36. Cambron M, DHaeseleer M, Laureys G, Clinckers R, Debruyne J, De Keyser J. White-matter astrocytes, axonal energy metabolism, and axonal degeneration in multiple sclerosis. J Cereb Blood Flow Metab. 2012;32:413–24.
37. Herrero-Herranz E, Pardo LA, Gold R, Linker RA. Pattern of axonal injury in murine myelin oligodendrocyte glycoprotein induced experimental autoimmune encephalomyelitis: implications for multiple sclerosis. Neurobiol Dis. 2008;30:162–73.
38. Lazzarino G, Amorini AM, Eikelenboom MJ, Killestein J, Belli A, Di Pietro V, Tavazzi B, Barkhof F, Polman CH, Uitdehaag BMJ, Petzold A. Cerebrospinal fluid ATP metabolites in multiple sclerosis. Mult Scler. 2010;16:549–54.
39. Mahad DJ, Ziabreva I, Campbell G, Lax N, White K, Hanson PS, Lassmann H, Turnbull DM. Mitochondrial changes within axons in multiple sclerosis. Brain. 2009;132:1161–74.
40. Witte ME, Lars BÃ, Rodenburg RJ, Belien JA, Musters R, Hazes T, Wintjes LT, Smeitink JA, Geurts JJG, De Vries HE, van der Valk P, van Horssen J. Enhanced number and activity of mitochondria in multiple sclerosis lesions. J Pathol. 2009;219:193–204.
41. van den Berg R, Hoogenraad CC, Hintzen RQ. Axonal transport deficits in multiple sclerosis: spiraling into the abyss. Acta Neuropathol. 2017;134(1):1–14. ISSN: 1432-0533

42. Kreutzer M, Seehusen F, Kreutzer R, Pringproa K, Kummerfeld M, Claus P, Deschl U, Kalkul A, Beineke A, Baumgärtner W, Ulrich R. Axonopathy is associated with complex axonal transport defects in a model of multiple sclerosis. Brain Pathol. 2012;22(4):454–71.
43. Petzold A, Gveric D, Groves M, Schmierer K, Grant D, Chapman M, Keir G, Cuzner L, Thompson EJ. Phosphorylation and compactness of neurofilaments in multiple sclerosis: indicators of axonal pathology. Exp Neurol. 2008;213:326–35.
44. Schirmer L, Merkler D, König FB, Brück W, Stadelmann C. Neuroaxonal regeneration is more pronounced in early multiple sclerosis than in traumatic brain injury lesions. Brain Pathol. 2013;23:2–12.
45. Ineichen BV, Kapitza S, Bleul C, Good N, Plattner PS, Seyedsadr MS, Kaiser J, Schneider MP, Zörner B, Martin R, Linnebank M, Schwab ME. Nogo-a antibodies enhance axonal repair and remyelination in neuro-inflammatory and demyelinating pathology. Acta Neuropathol. 2017;134(3):423–40. ISSN: 1432-0533
46. Hemmer B, Kerschensteiner M, Korn T. Role of the innate and adaptive immune responses in the course of multiple sclerosis. Lancet Neurol. 2015;14:406–19.
47. Simons M, Misgeld T, Kerschensteiner M. A unified cell biological perspective on axon-myelin injury. J Cell Biol. 2014;206:335–45.
48. Romanelli E, Sorbara CD, Ivana NÄ, Dagkalis A, Misgeld T, Kerschensteiner M. Cellular, subcellular and functional in vivo labeling of the spinal cord using vital dyes. Nat Protoc. 2013;8:481–90.
49. Waller A. Experiments on the section of glossopharyngeal and hypoglossal nerves of the frog and observations of the alternatives produced thereby in the structure of their primitive fibres. Philos Trans R Soc Lond A. 1850;140:423–9.
50. George R, Griffin JW. Delayed macrophage responses and myelin clearance during Wallerian degeneration in the central nervous system: the dorsal radiculotomy model. Exp Neurol. 1994;129:225–36.
51. Spencer PS, Schaumburg HH. Ultrastructural studies of the dying-back process IV differential vulnerability of PNS and CNS fibers in experimental central-peripheral distal axonopathies. J Neuropathol Exp Neurol. 1977;36:300–20.
52. Jindahra P, Petrie A, Plant GT. Retrograde trans-synaptic retinal ganglion cell loss identified by optical coherence tomography. Brain. 2009;132:628–34.
53. Jindahra P, Petrie A, Plant GT. The time course of retrograde trans-synaptic degeneration following occipital lobe damage in humans. Brain. 2012;135:534–41.
54. Jindahra P, Hedges TR, Mendoza-Santiesteban CE, Plant GT. Optical coherence tomography of the retina: applications in neurology. Curr Opin Neurol. 2010;23:16–23.
55. Dinkin M. Trans-synaptic retrograde degeneration in the human visual system: slow, silent, and real. Curr Neurol Neurosci Rep. 2017;17(2):16. ISSN: 1534-6293
56. Evangelou N. Size-selective neuronal changes in the anterior optic pathways suggest a differential susceptibility to injury in multiple sclerosis. Brain. 2001;124:1813–20.
57. Balk LJ, Steenwijk MD, Tewarie P, Daams M, Killestein J, Wattjes MP, Vrenken H, Barkhof F, Polman CH, Uitdehaag BMJ, Petzold A. Bidirectional trans-synaptic axonal degeneration in the visual pathway in multiple sclerosis. J Neurol Neurosurg Psychiatry. 2015;86:419–24.
58. Heesen C, Böhm J, Reich C, Kasper J, Goebel M, Gold SM. Patient perception of bodily functions in multiple sclerosis: gait and visual function are the most valuable. Mult Scler. 2008;14:988–91.
59. Lisanne J. Balk, Danko Coric, Jenny A. Nij Bijvank, Joep Killestein, Bernard Mj Uitdehaag, and Axel Petzold. Retinal atrophy in relation to visual functioning and vision-related quality of life in patients with multiple sclerosis. Mult Scler (Houndmills, Basingstoke, England). 2017:1352458517708463. ISSN: 1477–0970.
60. Poser CM. The unfortunate triumph of mechanodiagnosis in multiple sclerosis: a clinicians lament. Clin Neurol Neurosurg. 1992;94(Suppl):S139–42.
61. Lublin FD, Reingold SC. Defining the clinical course of multiple sclerosis: results of an international survey National Multiple Sclerosis Society (USA) Advisory Committee on Clinical Trials of New Agents in Multiple Sclerosis. Neurology. 1996;46:907–11.

62. Lublin FD, Reingold SC, Cohen JA, Cutter GR, Sørensen PS, Thompson AJ, Wolinsky JS, Balcer LJ, Banwell B, Frederik Barkhof BB Jr, Calabresi PA, Clanet M, Comi G, Fox RJ, Freedman MS, Goodman AD, Inglese M, Kappos L, Kieseier BC, Lincoln JA, Lubetzki C, Miller AE, Montalban X, OĆonnor PW, Petkau J, Pozzilli C, Rudick RA, Sormani MP, Stüve O, Waubant E, Polman CH. Defining the clinical course of multiple sclerosis: the 2013 revisions. Neurology. 2014;83:278–86.
63. Kurtzke JF. Rating neurological impairment in multiple sclerosis: an expanded disability status scale (EDSS). Neurology. 1983;33:1444–52.
64. Kurtzke JF. A new scale for evaluating disability in multiple sclerosis. Neurology. 1955;5:580–3.
65. Kurtzke JF. Natural history and clinical outcome measures for multiple sclerosis studies. Why at the present time does EDSS scale remain a preferred outcome measure to evaluate disease evolution? Neurol Sci. 2000;21:339–41.
66. van Munster CEP, Uitdehaag BMJ. Outcome measures in clinical trials for multiple sclerosis. CNS Drugs. 2017;31(3):217–36. [Epub ahead of print]. ISSN: 1179-1934
67. Gronwall DM. Paced auditory serial-addition task: a measure of recovery from concussion. Percept Mot Skills. 1977;44:367–73.
68. Nelson HE, editor. National adult reading test: manual. Windsor: NFER-Nelson; 1982.
69. Warrington EK, editor. Recognition memory tests. NFER Nelson: Windsor; 1984.
70. Willison JR, Thomas DJ, du Boulay GH, et al. Effect of high haematocrit on alertness. Lancet. 1980;19:846–8.
71. Sahakian BJ, Owen MA. Computerized assessment in neuropsychiatry using CANTAB: discussion paper. J R Soc Med. 1992;85:399–402.
72. Krupp LB, LaRocca NG, Muir-Nash J, Steinberg AD. The fatigue severity scale application to patients with multiple sclerosis and systemic lupus erythematosus. Arch Neurol. 1989;46:1121–03.
73. Thompson AJ. Multiple sclerosis: rehabilitation measures. Semin Neurol. 1998;18:397–403.
74. Cutter GR, Baier ML, Rudick RA, Cookfair DL, Fischer JS, Petkau J, Syndulko K, Weinshenker BG, Antel JP, Confavreux C, Ellison GW, Lublin F, Miller AE, Rao SM, Reingold S, Thompson A, Willoughby E. Development of a multiple sclerosis functional composite as a clinical trial outcome measure. Brain. 1999;122:871–82.
75. Barkhof F. The clinico–radiological paradox in multiple sclerosis. Curr Opin Neurol. 2002;15:239–45.
76. Balcer LJ, Baier ML, Cohen JA, Kooijmans MF, Sandrock AW, Nano-Schiavi ML, Pfohl DC, Mills M, Bowen J, Ford C, Heidenreich FR, Jacobs DA, Markowitz CE, Stuart WH, Ying G-S, Galetta SL, Maguire MG, Cutter GR. Contrast letter acuity as a visual component for the multiple sclerosis functional composite. Neurology. 2003;61:1367–73.
77. Roxburgh RHSR, Seaman SR, Masterman T, Hensiek AE, Sawcer SJ, Vukusic S, et al. Multiple sclerosis severity score. Using disability and disease duration to rate disease severity. Neurology. 2005;64:1144–51.
78. Hawton A, Green C, Telford C, Zajicek J, Wright D. Using the multiple sclerosis impact scale to estimate health state utility values: mapping from the MSIS-29, version 2, to the EQ-5D and the SF-6D. Value Health. 2012;15:1084–91.
79. Polman CH, Koetsier JC, Wolters EC. Multiple sclerosis: incorporation of results of laboratory techniques in the diagnosis. Clin Neurol Neurosurg. 1985;87:187–92.
80. Beer S, Rösler KM, Hess CW. Diagnostic value of paraclinical tests in multiple sclerosis: relative sensitivities and specificities for reclassification according to the poser committee criteria. J Neurol Neurosurg Psychiatry. 1995;59:152–9.
81. Filippini G, Comi GC, Cosi V, Bevilacqua L, Ferrarini M, Martinelli V, Bergamaschi R, Filippi M, Citterio A, DÍncerti L. Sensitivities and predictive values of paraclinical tests for diagnosing multiple sclerosis. J Neurol. 1994;241:132–7.
82. Kappos L, Moeri D, et al. Predictive value of gadolinium–enhanced magnetic resonance imaging for relapse rate and changes in disability or impairment in multiple sclerosis: a meta–analysis. Lancet. 1999;353:964–9.

83. Bonati U, Fisniku LK, Altmann DR, Yiannakas MC, Furby J, Thompson AJ, Miller DH, Chard DT. Cervical cord and brain grey matter atrophy independently associate with long-term MS disability. J Neurol Neurosurg Psychiatry. 2011;82:471–2.
84. Fisniku LK, Chard DT, Jackson JS, Anderson VM, Altmann DR, Miszkiel KA, Thompson AJ, Miller DH. Gray matter atrophy is related to long-term disability in multiple sclerosis. Ann Neurol. 2008;64:247–54.
85. Gass A, Rocca MA, Agosta F, Ciccarelli O, Chard D, Valsasina P, Brooks JCW, Bischof A, Eisele P, Kappos L, Barkhof F, Filippi M, et al. MRI monitoring of pathological changes in the spinal cord in patients with multiple sclerosis. Lancet Neurol. 2015;14:443–54.
86. Kaunzner UW, Gauthier SA. MRI in the assessment and monitoring of multiple sclerosis: an update on best practice. Ther Adv Neurol Dis. 2017;10(6):247–61. ISSN: 1756-2856
87. Rovira À, Wattjes MP, Tintoré M, Tur C, Yousry TA, Sormani MP, De Stefano N, Filippi M, Auger C, Rocca MA, Barkhof F, Fazekas F, Kappos L, Polman C, Miller D, Montalban X, MAGNIMS study group. Evidence-based guidelines: MAGNIMS consensus guidelines on the use of MRI in multiple sclerosis-clinical implementation in the diagnostic process. Nat Rev Neurol. 2015;11:471–82.
88. Wattjes MP, Barkhof F. High field MRI in the diagnosis of multiple sclerosis: high field-high yield? Neuroradiology. 2009;51:279–92.
89. Petzold A. Intrathecal oligoclonal IgG synthesis in multiple sclerosis. J Neuroimmunol. 2013;262:1–10.
90. Farina G, Magliozzi R, Pitteri M, Reynolds R, Rossi S, Gajofatto A, Benedetti MD, Facchiano F, Monaco S, Calabrese M. Increased cortical lesion load and intrathecal inflammation is associated with oligoclonal bands in multiple sclerosis patients: a combined CSF and MRI study. J Neuroinflammation. 2017;14(1):40. ISSN: 1742-2094
91. Joseph FG, Hirst CL, Pickersgill TP, Ben-Shlomo Y, Robertson NP, Scolding NJ. CSF oligoclonal band status informs prognosis in multiple sclerosis: a case control study of 100 patients. J Neurol Neurosurg Psychiatry. 2009;80:292–6.
92. Kuhle J, Disanto G, Dobson R, Adiutori R, Bianchi L, Topping J, Bestwick JP, Meier U-C, Marta M, Dalla Costa G, Runia T, Evdoshenko E, Lazareva N, Thouvenot E, Iaffaldano P, Direnzo V, Khademi M, Piehl F, Comabella M, Sombekke M, Killestein J, Hegen H, Rauch S, DÁlfonso S, Alvarez-Cermeño JC, Kleinová P, Horáková D, Roesler R, Lauda F, Llufriu S, Avsar T, Uygunoglu U, Altintas A, Saip S, Menge T, Rajda C, Bergamaschi R, Moll N, Khalil M, Marignier R, Dujmovic I, Larsson H, Malmestrom C, Scarpini E, Fenoglio C, Wergeland S, Laroni A, Annibali V, Romano S, Martínez AD, Carra A, Salvetti M, Uccelli A, Torkildsen Ø, Myhr KM, Galimberti D, Rejdak K, Lycke J, Frederiksen JL, Drulovic J, Confavreux C, Brassat D, Enzinger C, Fuchs S, Bosca I, Pelletier J, Picard C, Colombo E, Franciotta D, Derfuss T, Lindberg R, Yaldizli Ö, Vécsei L, Kieseier BC, Hartung HP, Villoslada P, Siva A, Saiz A, Tumani H, Havrdová E, Villar LM, Leone M, Barizzone N, Deisenhammer F, Teunissen C, Montalban X, Tintoré M, Olsson T, Trojano M, Lehmann S, Castelnovo G, Lapin S, Hintzen R, Kappos L, Furlan R, Martinelli V, Comi G, Ramagopalan SV, Giovannoni G. Conversion from clinically isolated syndrome to multiple sclerosis: a large multicentre study. Mult Scler. 2015;21:1013–24.
93. Lechner-Scott J, Spencer B, de Malmanche T, Attia J, Fitzgerald M, Trojano M, Grandaison F, Gomez JAC, Izquierdo G, Duquette P, Girard M, Grammond P, Oreja-Guevara C, Hupperts R, Bergamaschi R, Boz C, Giuliani G, van Pesch V, Iuliano G, Fiol M, Cristiano E, Verheul F, Laura Saladino M, Slee M, Barnett M, Deri N, Fletcher S, Vella N, Shaw C, Herbert J, Moore F, Petkovska-Boskova T, Jokubatis V, Butzkueven H. The frequency of CSF oligoclonal banding in multiple sclerosis increases with latitude. Mult Scler. 2011;18(7):974–82.
94. Moulin D, Paty DW, Ebers GC. The predictive value of cerebrospinal fluid electrophoresis in possible multiple sclerosis. Brain. 1983;106(Pt 4):809–16.
95. Zeman AZ, Kidd D, McLean BN, Kelly MA, Francis DA, Miller DH, Kendall BE, Rudge P, Thompson EJ, McDonald WI. A study of oligoclonal band negative multiple sclerosis. J Neurol Neurosurg Psychiatry. 1996;60:27–30.

96. Imrell K, Landtblom A-M, Hillert J, Masterman T. Multiple sclerosis with and without CSF bands: clinically indistinguishable but immunogenetically distinct. Neurology. 2006;67:1062–4.
97. Koch M, Heersema D, Mostert J, Teelken A, De Keyser J. Cerebrospinal fluid oligoclonal bands and progression of disability in multiple sclerosis. Eur J Neurol. 2007;14:797–800.
98. Lourenco P, Shirani A, Saeedi J, Oger J, Schreiber WE, Tremlett H. Oligoclonal bands and cerebrospinal fluid markers in multiple sclerosis: associations with disease course and progression. Mult Scler (Houndmills, Basingstoke, England). 2013;19(5):577–84. ISSN: 1477-0970
99. Disanto G, Barro C, Benkert P, Naegelin Y, Schädelin S, Giardiello A, Zecca C, Blennow K, Zetterberg H, Leppert D, Kappos L, Gobbi C, Kuhle J, Swiss MS Cohort Study (SMSC) Group. Serum neurofilament light: a biomarker of neuronal damage in multiple sclerosis. Ann Neurol. 2017;81(6):857–70. ISSN: 1531-8249
100. Petzold A. Neurofilament phosphoforms: surrogate markers for axonal injury, degeneration and loss. J Neurol Sci. 2005;233:183–98.
101. Stangel M, Fredrikson S, Meinl E, Petzold A, Stüve O, Tumani H. The utility of cerebrospinal fluid analysis in patients with multiple sclerosis. Nat Rev Neurol. 2013;9:267–76.
102. Cordeiro MF, Normando EM, Jorge Cardoso M, Miodragovic S, Jeylani S, Davis BM, Guo L, Ourselin S, A'Hern R, Bloom PA. Real-time imaging of single neuronal cell apoptosis in patients with glaucoma. Brain. 2017;140:1757–67.
103. Narayana PA, Doyle TJ, Lai D, Wolinsky JS. Serial proton magnetic resonance spectroscopic imaging, contrast-enhanced magnetic resonance imaging, and quantitative lesion volumetry in multiple sclerosis. Ann Neurol. 1998;43:56–71.
104. Wattjes MP, Harzheim M, Lutterbey GG, Bogdanow M, Schmidt S, Schild HH, TrÃber F. Prognostic value of high-field proton magnetic resonance spectroscopy in patients presenting with clinically isolated syndromes suggestive of multiple sclerosis. Neuroradiology. 2008;50:123–9.
105. Leocani L, Rocca MA, Comi G. MRI and neurophysiological measures to predict course, disability and treatment response in multiple sclerosis. Curr Opin Neurol. 2016;29:243–53.
106. Wattjes MP, Steenwijk MD, Stangel M. MRI in the diagnosis and monitoring of multiple sclerosis: an update. Clin Neuroradiol. 2015;25(Suppl 2):157–65.
107. Britze J, Pihl-Jensen G, Frederiksen JL. Retinal ganglion cell analysis in multiple sclerosis and optic neuritis: a systematic review and meta-analysis. J Neurol. 2017;264(9):1837–53. [epub ahead of print]. ISSN: 1432-1459
108. Petzold A, de Boer JF, Schippling S, Vermersch P, Kardon R, Green A, Calabresi PA, Polman C. Optical coherence tomography in multiple sclerosis: a systematic review and meta-analysis. Lancet Neurol. 2010;9:921–32.
109. Klistorner A, Garrick R, Barnett MH, Graham SL, Arvind H, Sriram P, Yiannikas C. Axonal loss in non-optic neuritis eyes of patients with multiple sclerosis linked to delayed visual evoked potential. Neurology. 2013;80:242–5.
110. Petzold A. Biomarkers of disease progression. In: Wilkins A, editor. Progressive multiple sclerosis. Berlin: Springer; 2013. p. 115–46.
111. Awad A, Hemmer B, Hartung H-P, Kieseier B, Bennett JL, Stuve O. Analyses of cerebrospinal fluid in the diagnosis and monitoring of multiple sclerosis. J Neuroimmunol. 2010;219:1–7.
112. Berger T, Reindl M. Multiple sclerosis: disease biomarkers as indicated by pathophysiology. J Neurol Sci. 2007;259:21–6.
113. Bielekova B, Martin R. Development of biomarkers in multiple sclerosis. Brain. 2004;127:1463–78.
114. Dujmovic I. Cerebrospinal fluid and blood biomarkers of neuroaxonal damage in multiple sclerosis. Mult Scler Int. 2011:767–83.
115. Kuhle J, Petzold A. What makes a prognostic biomarker in CNS diseases: strategies for targeted biomarker discovery? Part 2: chronic progressive and relapsing diseases. Expert Opin Med Diagn. 2011;5:393–410.

116. Bacioglu M, Maia LF, Preische O, Schelle J, Apel A, Kaeser SA, Schweighauser M, Eninger T, Lambert M, Pilotto A, Shimshek DR, Neumann U, Kahle PJ, Staufenbiel M, Neumann M, Maetzler W, Kuhle J, Jucker M. Neurofilament light chain in blood and CSF as marker of disease progression in mouse models and in neurodegenerative diseases. Neuron. 2016;91(2):494–6. ISSN: 1097-4199
117. Gaiottino J, Norgren N, Dobson R, Topping J, Nissim A, Malaspina A, Bestwick JP, Monsch AU, Regeniter A, Lindberg RL, Kappos L, Leppert D, Petzold A, Giovannoni G, Kuhle J. Increased neurofilament light chain blood levels in neurodegenerative neurological diseases. PLoS One. 2013;8:e75091.
118. Olsson B, Lautner R, Andreasson U, Öhrfelt A, Portelius E, Bjerke M, Hölttä M, Rosén C, Olsson C, Strobel G, et al. CSF and blood biomarkers for the diagnosis of Alzheimer's disease: a systematic review and meta-analysis. Lancet Neurol. 2016;15:673–84.
119. Arrambide G, Espejo C, Eixarch H, Villar LM, Alvarez-Cermeño JC, Picón C, Kuhle J, Disanto G, Kappos L, Sastre-Garriga J, Pareto D, Simon E, Comabella M, Río J, Nos C, Tur C, Castilló J, Vidal-Jordana A, Galán I, Arévalo MJ, Auger C, Rovira A, Montalban X, Tintore M. Neurofilament light chain level is a weak risk factor for the development of MS. Neurology. 2016;87:1076–84. ISSN: 1526-632X
120. Brettschneider J, Petzold A, Junker A, Tumani H. Axonal damage markers in the cerebrospinal fluid of patients with clinically isolated syndrome improve predicting conversion to definite multiple sclerosis. Mult Scler. 2006;12:143–8.
121. Gunnarsson M, Malmeström C, Axelsson M, Sundström P, Dahle C, Vrethem M, Olsson T, Piehl F, Norgren N, Rosengren L, Svenningsson A, Lycke J. Axonal damage in relapsing multiple sclerosis is markedly reduced by natalizumab. Ann Neurol. 2011;69:83–9.
122. Kuhle J, Leppert D, Petzold A, Regeniter A, Schindler C, Mehling M, Anthony DC, Kappos L, Lindberg RLP. Neurofilament heavy chain in CSF correlates with relapses and disability in multiple sclerosis. Neurology. 2011;76:1206–13.
123. Kuhle J, Barro C, Disanto G, Mathias A, Soneson C, Bonnier G, Yaldizli Ö, Regeniter A, Derfuss T, Canales M, Schluep M, Du Pasquier R, Krueger G, Granziera C. Serum neurofilament light chain in early relapsing remitting MS is increased and correlates with CSF levels and with MRI measures of disease severity. Mult Scler (Houndmills, Basingstoke, England). 2016;22(12):1550–9. ISSN: 1477-0970
124. Kuhle J, Nourbakhsh B, Grant D, Morant S, Barro C, Yaldizli Ö, Pelletier D, Giovannoni G, Waubant E, Gnanapavan S. Serum neurofilament is associated with progression of brain atrophy and disability in early MS. Neurology. 2017;88(9):826–31. ISSN: 1526-632X
125. Lycke JN, Karlsson JE, Andersen O, Rosengren LE. Neurofilament protein in cerebrospinal fluid: a potential marker of activity in multiple sclerosis. J Neurol Neurosurg Psychiatry. 1998;64:402–4.
126. Lycke J, Andersen O, Rosengren L. Neurofilament in cerebrospinal fluid: a potential marker of activity in multiple sclerosis. Eur J Neurol. 1996;3:100.
127. Malmeström C, Haghighi S, Rosengren L, Andersen O, Lycke J. Neurofilament light protein and glial fibrillary acidic protein as biological markers in MS. Neurology. 2003;61:1720–5.
128. Petzold A, Rejdak K, Plant GT. Axonal degeneration and inflammation in acute optic neuritis. J Neurol Neurosurg Psychiatry. 2004;75:1178–80.
129. Petzold A, Mondria T, Kuhle J, Rocca MA, Cornelissen J, Te Boekhorst P, Lowenberg B, Giovannoni G, Filippi M, Kappos L, Hintzen R. Evidence for acute neurotoxicity after chemotherapy. Ann Neurol. 2010;68:806–15.
130. Teunissen CE, Iacobaeus E, Khademi M, Brundin L, Norgren N, Koel-Simmelink MJA, Schepens M, Bouwman F, Twaalfhoven HAM, Blom HJ, Jakobs C, Dijkstra CD. Combination of CSF N-acetylaspartate and neurofilaments in multiple sclerosis. Neurology. 2009;72:1322–9.
131. Abdo WF, van de Warrenburg BP, Munneke M, van Geel WJ, Bloem BR, Kremer HP, Verbeek MM. CSF analysis differentiates multiple-system atrophy from idiopathic late-onset cerebellar ataxia. Neurology. 2006;67:474–9.

132. Hares K, Kemp K, Gray E, Scolding N, Wilkins A. Neurofilament dot blot assays: novel means of assessing axon viability in culture. J Neurosci Methods. 2011;198:195–203.
133. Jackson SJ, Baker D, Louise Cuzner M, Diemel LT. Cannabinoid-mediated neuroprotection following interferon-gamma treatment in a three-dimensional mouse brain aggregate cell culture. Eur J Neurosci. 2004;20:2267–75.
134. Jackson SJ, Diemel LT, Pryce G, Baker D. Cannabinoids and neuroprotection in CNS inflammatory disease. J Neurol Sci. 2005;233:21–5.
135. Kuhle J, Regeniter A, Leppert D, Mehling M, Kappos L, Lindberg RLP, Petzold A. A highly sensitive electrochemiluminescence immunoassay for the neurofilament heavy chain protein. J Neuro-Oncol. 2010;220:114–9.
136. Lu C-H, Kalmar B, Malaspina A, Greensmith L, Petzold A. A method to solubilise protein aggregates for immunoassay quantification which overcomes the neurofilament "hook" effect. J Neurosci Methods. 2011;195:143–50.
137. Petzold A, Baker D, Pryce G, Keir G, Thompson EJ, Giovannoni G. Quantification of neurodegeneration by measurement of brain-specific proteins. J Neuroimmunol. 2003;138:45–8.
138. Pryce G, Ahmed Z, Hankey DJ, et al. Cannabinoids inhibit neurodegeneration in models of multiple sclerosis. Brain. 2003;127:2191–202.
139. Shaw G, Yang C, Ellis R, Anderson K, et al. Hyperphosphorylated neurofilament NF-H is a serum biomarker for axonal injury. Biochem Biophys Res Commun. 2005;336:1268–77.
140. Kuhle J, Barro C, Andreasson U, Derfuss T, Lindberg R, Sandelius Å, Liman V, Norgren N, Blennow K, Zetterberg H. Comparison of three analytical platforms for quantification of the neurofilament light chain in blood samples: ELISA, electrochemiluminescence immunoassay and Simoa. Clin Chem Lab Med. 2016;54(10):1655–61. ISSN: 1437-4331
141. Albrecht P, Fröhlich R, Hartung H-P, Kieseier BC, Methner A. Optical coherence tomography measures axonal loss in multiple sclerosis independently of optic neuritis. J Neurol. 2007;254:1595–6.
142. Bock M, Brandt AU, Dörr J, Kraft H, Weinges-Evers N, Gaede G, Pfueller CF, Herges K, Radbruch H, Ohlraun S, Bellmann-Strobl J, Kuchenbecker J, Zipp F, Paul F. Patterns of retinal nerve fiber layer loss in multiple sclerosis patients with or without optic neuritis and glaucoma patients. Clin Neurol Neurosurg. 2010;112:647–52.
143. Fisher JB, Jacobs DA, Markowitz CE, Galetta SL, Volpe NJ, Ligia Nano-Schiavi M, Baier ML, Frohman EM, Winslow H, Frohman TC, Calabresi PA, Maguire MG, Cutter GR, Balcer LJ. Relation of visual function to retinal nerve fiber layer thickness in multiple sclerosis. Ophthalmology. 2006;113:324–32.
144. Frohman EM, Dwyer MG, Frohman T, Cox JL, Salter A, Greenberg BM, Hussein S, Conger A, Calabresi P, Balcer LJ, Zivadinov R. Relationship of optic nerve and brain conventional and non-conventional MRI measures and retinal nerve fiber layer thickness, as assessed by OCT and GDx: a pilot study. J Neurol Sci. 2009;282:96–105.
145. Gundogan FC, Demirkaya S, Sobaci G. Is optical coherence tomography really a new biomarker candidate in multiple sclerosis?–a structural and functional evaluation. Invest Ophthalmol Vis Sci. 2007;48:5773–81.
146. Jeanjean L, Castelnovo G, Carlander B, Villain M, Mura F, Dupeyron G, Labauge P. Retinal atrophy using optical coherence tomography (OCT) in 15 patients with multiple sclerosis and comparison with healthy subjects. Rev Neurol (Paris). 2008;164:927–34.
147. Klistorner A, Arvind H, Nguyen T, Garrick R, Paine M, Graham S, ODay J, Grigg J, Billson F, Yiannikas C. Axonal loss and myelin in early ON loss in postacute optic neuritis. Ann Neurol. 2008;64:325–31.
148. Pulicken M, Gordon-Lipkin E, Balcer LJ, Frohman E, Cutter G, Calabresi PA. Optical coherence tomography and disease subtype in multiple sclerosis. Neurology. 2007;69:2085–92.
149. Quelly A, Cheng H, Laron M, Schiffman JS, Tang RA. Comparison of optical coherence tomography and scanning laser polarimetry measurements in patients with multiple sclerosis. Optom Vis Sci. 2010;87:576.

150. Ratchford JN, Quigg ME, Conger A, Frohman T, Frohman E, Balcer LJ, Calabresi PA, Kerr DA. Optical coherence tomography helps differentiate neuromyelitis optica and MS optic neuropathies. Neurology. 2009;73:302–8.
151. Siger M, Dziegielewski K, Jasek L, Bieniek M, Nicpan A, Nawrocki J, Selmaj K. Optical coherence tomography in multiple sclerosis: thickness of the retinal nerve fiber layer as a potential measure of axonal loss and brain atrophy. J Neurol. 2008;255:1555–60.
152. Anand Trip S, Schlottmann PG, Jones SJ, Altmann DR, Garway-Heath DF, Thompson AJ, Plant GT, Miller DH. Retinal nerve fiber layer axonal loss and visual dysfunction in optic neuritis. Ann Neurol. 2005;58:383–91.
153. Zaveri MS, Conger A, Salter A, Frohman TC, Galetta SL, Markowitz CE, Jacobs DA, Cutter GR, Ying G-S, Maguire MG, Calabresi PA, Balcer LJ, Frohman EM. Retinal imaging by laser polarimetry and optical coherence tomography evidence of axonal degeneration in multiple sclerosis. Arch Neurol. 2008;65:924–8.

Index

A
Adrenoleukodystrophy, 321
Adult polyglucosan body disease, 324
Adult-onset leukoencephalopathy with axonal spheroids and pigmented glia (ALSP), 323
Aβ generating enzymes, 227
Akinetic crisis (AC) syndrome, 305
Alcohol consumption, 103
Alcohol intake, 81
Alcohol-related dementia, 332
Alexander disease, 324
ε4 alleles, 53
ALS, *see* Amyotrophic lateral sclerosis (ALS)
Alzheimer Disease Genetics Consortium (ADGC), 60
Alzheimer's disease (AD), 5, 6, 22–25, 55–62, 95–106, 257–260, 262
    advanced MRI techniques
        brain volumetrics, 257–258
        diffusion imaging, 260–261
        functional MRI, 261–262
        manual and semiautomatic regional measurements, 258–259
        VBM, 259–260
    APOE gene, 52
    atypical form, 25–27
    atypical variants of, 233
    candidate gene association studies, 54
    clinical presentation, 22
    clinical stage, 232
    conventional MRI, 255
    cost of care for, 30
    DIAN cohort, 41
    definition, 148
    diagnostic criteria
        DSM-V, 25
        IWG-2, 24
        NIA-AA, 24
        NINCDS-ADRDA, 22–23
    drug development, 42
    dynamic biomarker model, 230
    early genome-wide linkage analyses, 54
    early stages, 21
    extreme phenotype study, 43
    familial early-onset, 30
    fluid biomarkers, 223–225
    forms, 52
    genes, 31
    genome-wide association studies
        early small-scale, 56
        gene–gene interactions, 62
        large meta-analyses, 57–59
        large small-scale, 56–61
        limitation, 55
        non-European populations, 61–62
        SNP genotype, 55
    genomic studies, 31
    metabolic imaging, 262–264
    modifiable risk factors
        depression, 103
        diet, 102–103
        lifestyle, 100–101
        OSA syndrome, 104–106
        traumatic brain injuries, 104
        vascular and metabolic risk factors, 96–100
    mutation analyses, 40–41
    neuropathological studies, 255
    next-generation sequencing technology, 42
    non-modifiable risk factors
        age, 95–96
        familial aggregation, 96
        gender, 96
    post-GWAS era, 64–66
    preclinical, 231
    searched for rare variant associations, 63–64
    sequencing-based study, 63, 64
Amyloid angiopathy, 34

β-amyloid (Aβ) peptides, 263
Amyloid plaques (Aβ(beta)), 33
Amyloid-β precursor protein (APP) models, 189–190
Amyloid-β-related biomarkers
  Aβ oligomers, 227
  blood-based measures, 227–228
  generating enzymes, 227
  peptides, 226–227
  sAPP peptides, 227
Amyotrophic lateral sclerosis (ALS), 9, 10, 124
  classic, 280
  genetics, 280–286
  identification, 289–290
  incidence, 280
  pleiotropy, 286–288
Anosmia, 360
Anticholinergics, 369
Anxiety, 363, 370
Aβ oligomers, 227
Apathy, 363
Aβ peptides, 226–227
Apolipoprotein E (*APOE*), 52, 97
  alleles, 53
  discovery, 52
APP
  AD mutation, 34
  Ala713Val variant, 35
  amyloid plaques, 33
  biology, 33
  CNVs, 36
  codon Glu693, 34
  composed, 33
  mutations, 35
Autosomal dominant cerebellar ataxia, 320
Autosomal dominant disorder, 324
Autosomal recessive disorder, 323, 324
Autosomal recessive lysosomal sphingolipid storage disorder, 322
Autosomal recessive lysosomal storage disease, 322
Autosomal recessive polyglucosane storage disorder, 324
Autosomal-dominant AD, 232
Axonal injury, 382

B
Behavioural-variant frontotemporal dementia (bvFTD), 114–115
Benson's syndrome, 25
Blood pressure, 98–99
Bradykinesia, 359

Brain volumetrics, 257
BTNL2, 176
bvFTD, *see* Behavioural-variant frontotemporal dementia (bvFTD)

C
CADASIL (cerebral autosomal dominant arteriopathy with subcortical infarcts and leukoencephalopathy), 325
CARASIL (cerebral autosomal recessive arteriopathy with subcortical infarcts and leukoencephalopathy), 325
Cardiovascular autonomic dysfunction, 362
Cardiovascular Risk Factors, Aging, and Dementia (CAIDE) risk score, 104
CEP131, 172, 176
Cerebral microbleeds, 84
Cerebrotendinous xanthomatosis, 323
Cerebrovascular disease, 78, 94
Charged multivesicular body protein 2B (CMBP2B) models, 204–205
CHCHD10, 135–136
CHMP2B, 133, 134
Cholinesterase inhibitors (CHEIs), 304
Cingulate island sign (CIS), 303
Coffee consumption, 352
Cognition
  small vessel disease
    cerebral microbleeds, 84
    clinical expression, 82
    different expressions, 83
    lacunes, 83
    white matter changes, 82
  vascular risk factors, 78
    alcohol intake, 81
    diabetes, 79
    hypertension, 80–81
    midlife, 78
    smoking, 81
    stroke, 79
Cognitive impairment, 30
Cognitive reserve, 255
Cohorts for Heart and Aging Research in Genomic Epidemiology (CHARGE), 56
Combinatorial AD models, 190–192
Complex disorders, *see* Genetics
Constipation, 371
Copy-neutral variations, 2
Copy number variations (CNVs), 36
C17orf89, 176
C9orf72, 131–133, 150, 168, 169, 206, 207, 281, 287, 288

CpG islands, 10
CRISPR/Cas9, 189
CTSC, 170, 174
Cytokines, 339

D

Dementia, 30, 95–106
  cause of, 94
  modifiable risk factors
    depression, 103
    diet, 102
    lifestyle, 100–101
    OSA syndrome, 105–106
    traumatic brain injuries, 104
    vascular and metabolic risk factors, 96–100
  non-modifiable risk factors
    age, 95
    familial aggregation, 96
    gender, 96
  protective factors, 95
  risk factors, 95
  risk scores, 104–105
  in stroke survivors, 79
  types, 148
Dementia with Lewy bodies (DLB)
  akinetic crisis, 304–306
  causes, 298
  clinical features, 298
  cognitive, 298
  consortium, 298
  indicative biomarkers, 299–300
  motor symptoms, 304
  neuropsychiatric symptoms, 304
  pharmacological interventions, 304
  supportive biomarkers
    cingulate island sign, 303
    EEG, 300
    functional imaging study, 303
    structural imaging studies, 301, 302
Depression, 103, 370
Diabetes, 79
Diabetes mellitus (DM), 97–98
Diagnostic and statistical manual of mental disorders (DSM), 148
Diet, 352
  alcohol consumption, 103
  antioxidant vitamins, 102
  dietary patterns, 102
  folate, 103
  vitamin B6, B12, 103
DLB, *see* Dementia with Lewy bodies (DLB)
DNA methylation, 10–11
  classical method, 14

Dominant disease, 318
Dominantly Inherited Alzheimer Network (DIAN), 41
Dopamine receptor agonists (DRAs), 367
Dutch Prevention of Dementia by Intensive Vascular Care (PreDIVA) study, 105
Dyskinesias, 366

E

Education, 101
EEG, 300
ENTHD2, 172, 176
Enzyme-linked immunosorbent assay (ELISA), 236
Epigenetics, 10
  chromatin remodeling, 13–14
  DNA methylation, 10–11
  ncRNA, 11–13
  technologies used in studies, 14
European Alzheimer's Disease Initiative (EADI), 56

F

Fahr disease, 326
Fatigue, 363
Finnish Geriatric Intervention Study to Prevent Cognitive Impairment and Disability (FINGER), 105
Fluid biomarkers, 226–237
  AD, 223–225
    atypical variants, 233
    autosomal-dominant, 232
    clinical stage, 232
    dynamic biomarker model, 230
    preclinical, 231
  amyloid-β-related
    Aβ peptides, 226–227
    blood-based measures, 227–228
    generating enzymes, 227
    oligomers, 227
    sAPP peptides, 227
  clinical trials, 233
    enrich study populations, 234
    surrogate, 235
    toxicity and side effects, 234
    treatment effects, 234
  CSF, 222, 223
  FTD, 225, 233
  genetics, 235
  imaging, 235
  inflammatory activity, 230
  microglial and astrocytic activation, 230

Fluid biomarkers (*cont.*)
  NFL, 229
  pathological process, 225
  progranulin, 230
  research, 236
    ELISA, 236
    measurements, 237
  Tau-related, 228
    blood-based measurements, 229
    isoforms, 228
    phospho-tau, 228
    synaptic loss, 229
    total, 228
  TDP-43, 229
  ubiquitin, 230
^{18}Fluorodeoxyglucose (^{18}FDG-PET), 262
Fluorodeoxyglucose positron emission tomography (FDG-PET) imaging, 117
Fragile X-associated tremor/ataxia syndrome (FXTAS), 320
Frontal variant of AD (fvAD), 26
Frontotemporal dementia (FTD), 8, 116, 124, 162–172, 192, 266, 268–271, 280
  biomarkers, 233
  bvFTD, 114–115
  categorize, 149
  characteristics, 149
  clinical aspects
    behavioural variants, 268
    functional MRI, 270–271
    linguistic variants, 266
    structural MRI, 268–270
  clinical features, 114
  clinical variants, 225
  cognitive deficits, 264
  comprehensive assessment, 153
  definition, 113
  description, 114
  diagnostic criteria, 116–117
  differential diagnosis, 119
  fluid biomarkers, 225
  genetics, 149–153
  GWAS, 162
    international clinical, 169–171
    international FTLD-TDP, 162–169
    Italian clinical, 171–172
  histopathological and genetic aspects, 266
  imaging, 117–118
  PPA, 115, 116
  prevalence, 114
  workout, 118–119
Frontotemporal lobar degeneration (FTLD), 8, 9, 126–138, 150, 192
  clinical syndrome, 124–125

CMBP2B models, 204
C9orf72 models, 206
definition, 124
FUS models, 205
genetics, 126
  major causal genes, 126–133
  modifiers, 138
  rare causal genes, 133–137
neuropathology, 125
PGRN models, 200–202
Tau models, 192–196
TDP-43 models, 196–200
VCP models, 202
FTD, *see* Frontotemporal dementia (FTD)
Functional magnetic resonance imaging (fMRI), 303
Fused in sarcoma (*FUS*), 137
Fused in sarcoma (FUS) models, 205–206

G
Genetic(s), 126, 353
  biomarkers, 235
  candidate genes study, 3
  familial cases, 153, 154
  GWAS, 3
  linkage analysis study, 3
  NGS technology study, 4
  of FTLD (*see* Frontotemporal lobar degeneration (FTLD))
  PD, causative genes, 353
  sporadic cases, 154
  variations, 2
Genetic and Environmental Risk in Alzheimer's Disease (GERAD), 56
Genetic test, 119
Genetic variability
  basics, 155–157
  study, 157–158
Genome, 188
Genome-wide association study (GWAS), 3, 32, 162, 163, 169, 171, 172
  design, 154, 158
  early small-scale, 56
  FTD (*see* Frontotemporal dementia (FTD))
  international clinical
    follow-up study, 171
    lessons learn, 169
  international FTLD-TDP
    follow-up study, 163
    lessons learn, 162
  interpretation, 161–162
  Italian clinical
    follow-up study, 172
    lessons learn, 171

large small-scale, 56–61
non-European populations, 61–62
practice, 159–160
Genotype-phenotype correlation, 281
GRN, 128–130
GWAS, *see* Genome-wide association studies (GWAS)

H
Hereditary spastic paraparesis, 320
Hereditary X-linked disease, 321
Homo sapiens, 155
Hypercholesterolaemia, 96–97
Hypertension, 80–81
Hyposmia, 360

I
IBMPFD, 202, 203
Inducible mouse models, 188
International clinical FTD
GWAS, 169–171, 174
International FTLD-TDP GWAS, 162–169
International Genomics of Alzheimer's Project (IGAP) consortium, 60
International Working Group (IWG-2) Criteria, 24
Italian clinical FTD GWAS, 171, 172, 176

K
Krabbe disease, 322

L
Lacunes, 83
Late-onset AD (LOAD), 52
Leukodystrophy, 321
adult polyglucosan body disease, 324
Alexander disease, 324
ALSP, 323
cerebrotendinous xanthomatosis, 323
Krabbe disease, 322
metachromatic leukodystrophy, 322
Ovario-Leukodystrophy, 323
Pelizaeus-Merzbacher disease, 324
vanishing white matter disease, 324
Leukoencephalopathy, 321
Levodopa, 368
Lewy bodies (LB), 357
Logopenic aphasia, 27
Logopenic progressive aphasia (LPA), 149
Lysosomal storage disorders, 326

M
MAPT, 62, 126–128
Mass spectrometry-based methods, 237
Mendelian FTD, 150
Mentally stimulating activity, 101
Metabolic syndrome, 100
Metachromatic leukodystrophy (MLD), 322
Metaiodobenzylguanidine (MIBG), 299–300
Microglial-mediated innate immunity, 65
Microgliopathy, 331
Mitochondrial diseases, 325
Monoamine oxidase B (MAO-B) inhibitor, 367
Motor neuron diseases, 281
Mouse models
Alzheimer's disease, 189
APP models, 189–190
combinatorial models, 190–192
conditional, 188
FTLD, 192
C9orf72 models, 206–207
CMBP2B models, 204
FUS models, 205–206
PGRN models, 200–202
Tau models, 192–196
TDP-43 models, 196–200
VCP models, 202–204
sophisticated, 188
transgenesis techniques, 188
Movement Disorder Society (MDS), 365
Multidomain Alzheimer Preventive Trial (MAPT), 105
Multiple sclerosis (MS), 385, 387, 388
clinical and paraclinical assessments, 385
acute neurodegeneration, 388–389
clinical scale, 385–387
paraclinical tests, 387–388
disability, 385
historical context, 380–382
impairment, 385
pathological features, 382–384
Mutation Database, 34
Myopathy, 134
Myotonic dystrophies, 318

N
Nasu-Hakola disease (NHD), 331
National Institute of Neurological and Communicative Disorders and Stroke-Alzheimer's Disease and Related Disorders Association (NINCDS-ADRDA), 22–23
National Institute on Aging (NIA) and Alzheimer's Association (AA) criteria, 24

Neuroacanthocytosis (NA), 326
Neurocognitive disorder (NCD), 25, 148
Neurodegeneration, 380, 389
Neurodegeneration with brain iron
   accumulation (NBIA), 326
Neurodegenerative dementia, 318
Neurofilament light (NFL), 229
Neuroleptic malignant syndrome (NMS), 305
Neurological disorders, 2
Next-generation sequencing (NGS), 4, 42
nfPPA, *see* Non-fluent variant primary
   progressive aphasia (nfPPA)
Noncoding RNA (ncRNA), 11–13
Non-fluent variant primary progressive aphasia
   (nfPPA), 116
Null hypothesis, 161

O
Obesity, 99
Obstructive sleep apnea (OSA) syndrome, 104
Odds ratio (OR), 161
Optical coherence tomography (OCT), 389
Ovario-Leukodystrophy, 323

P
Parkinson's disease (PD), 6, 7, 350–357,
      360–364, 367–371
   clinical and experimental observations, 359
   diagnosis, 364–365
   different aspects, 350
   dyskinesias, 366
   epidemiology
      incidence and prevalence, 350–351
      risk and protective factors, 351–353
   genetics
      causative genes, 353
      monogenic forms, 354–356
      susceptibility genes, 357
   motor complications, 369
   motor symptoms, 359
   treatment, 367–369
   neuropathological correlates, 357
   non-motor symptoms, 360
      autonomic dysfunction, 361–362
      cognitive impairment, 363–364
      neuropsychiatric disorders, 363
      sensory features, 360–361
      sleep disorder, 362
      treatment, 369–371
   preclinical, 365
   prion propagation, 358
   prodromal, 365
   progression, 365–366
   treatment, 367
   wearing-off, 366
Parkinsonism, 359
Pelizaeus-Merzbacher disease, 324
Physical activity, 100
PICALM, 65
Positron emission tomography (PET) PET, 262
Posterior cortical atrophy (PCA), 25
Power, 161
PPA, *see* Primary progressive aphasia (PAA)
Preclinical disease, 365
Primary progressive aphasia (PPA),
   27, 116, 149
Prion disease, 332
Progranulin, 230
Progranulin (PGRN) models, 200–202
Progressive non-fluent aphasia (PNFA), 124
PSEN1
   critical components, 36
   insArg352, 38
   mutation, 36, 37
   mutations, 38
   phenotypic heterogeneity, 37
   protein, 36
   variations, 37
PSEN2, 39–40
p-value, 161

R
RAB38, 170, 174
Rapid eye movement behavior disorder
   (RBD), 362
Rest tremor, 360
Resting-state electroencephalographic (rsEEG)
   rhythms, 300
Retinal vasculopathy with cerebral
   leukodystrophy (RVCL), 325
RFNG, 176
Rigidity, 360

S
sAPP peptides, 227
Schizophrenia
   clinical perspective, 338
   genetic studies, 338
   inflammatory pathways, 341–342
   natural history, 338
   neurodegeneration, 342–343
   neurodevelopmental hypothesis, 339–341
   prevalence, 338
Semantic dementia (SD), 124, 149
Semantic variant primary progressive aphasia,
   see semantic dementia (SD)

Semantic variant primary progressive aphasia (svPPA), 116
Simple nucleotide variations (SNVs), 155, 156
Small vessel disease
  cerebral microbleeds, 84
  clinical expression, 82
  different expressions, 83
  lacunes, 83
  white matter changes, 82
Smoking, 81, 99, 100, 351
Social network, 101
SOD1, 286
SPECT-DAT scan, 299
Sporadic frontotemporal dementia (FTD), 152
Sporadic FTD
  heterogeneous syndromes, 177
  neurodegenerative condition, 177
  prospective approaches, 178–180
  study design, 177
SQSTM1, 135
Standard neuropsychological test, 118
Stroke, 79
Structural variants (SVs), 155, 157
Subacute dementia, 321–324, 326, 331
  autosomal dominant cerebellar ataxia, 320
  basal ganglia pathology, 326, 331
  dominant disease, 318
  FXTAS, 320
  hereditary spastic paraparesis, 320
  infective, 332
  inflammatory-autoimmune disorders, 332
  leukodystrophy, 321
    adrenoleukodystrophy, 321
    adult polyglucosan body disease, 324
    Alexander disease, 324
    ALSP, 323
    cerebrotendinous xanthomatosis, 323
    Krabbe disease, 322
    metachromatic leukodystrophy, 322
    Ovario-Leukodystrophy, 323
    Pelizaeus-Merzbacher disease, 324
    vanishing white matter disease, 324
  leukoencephalopathy, 321
  lysosomal storage disorders, 326
  microgliopathy, 331
  mitochondrial diseases, 325
  myotonic dystrophies, 318
  neurodegenerative, 318
  Prion disease, 332
  vascular, familial form of, 325
Surrogate biomarker, 235
svPPA, *see* Semantic variant primary progressive aphasia (svPPA)

T
TARDBP, 136–137
Tau, 66
  gene expression, 11
  models, 192–196
  biomarkers, 228
    blood-based measurements, 229
    isoforms, 228
    phospho-tau, 228
    synaptic loss, 229
    total, 228
TBK1, 136
TDP-43, 196–200, 229
Timed walk test (TWT), 386
TMEM106B, 168, 169
Transgene, 188
Transgenic animals, 188
Traumatic brain injury (TBI), 104, 353
TUBA4A, 137
Type I error (α), 161
Type II error (β), 161

U
Ubiquitin, 230
UBQLN2, 137
Uncommon dementias, 314
  See also Dementia

V
Valosin-containing protein (VCP) models, 202–204
Vanishing white matter disease, 324
Variant AD (vAD), 38
VCP-1, 134
Voxel-based morphometry (VBM), 26, 259

W
Wallerian degeneration, 383
Wearing-off, 366
White matter changes, 82
Whole-exome sequencing study (WES), 4
Whole-genome sequencing study (WGS), 4, 289
Wilson disease, 331
World Health Organization (WHO), 94

X
X-linked hypomyelinating leukoencephalopathy, 324

Printed by Printforce, the Netherlands